Sustainable Material Forming and Joining

Sustainable Material Forming and Joining

Edited by
R. Ganesh Narayanan and Jay S. Gunasekera

CRC Press
Taylor & Francis Group
Boca Raton London New York

CRC Press is an imprint of the
Taylor & Francis Group, an **informa** business

CRC Press
Taylor & Francis Group
6000 Broken Sound Parkway NW, Suite 300
Boca Raton, FL 33487-2742

First issued in paperback 2020

ISBN 13: 978-0-367-65661-4 (pbk)
ISBN 13: 978-1-138-06020-3 (hbk)

Visit the Taylor & Francis Web site at
http://www.taylorandfrancis.com

and the CRC Press Web site at
http://www.crcpress.com

Contents

Radha Goyal, Ashish Jain, and Shyamli Singh

Preface

Material forming and joining have been important manufacturing processes for centuries. The main advantages of these processes include material conservation, enhancement of mechanical and metallurgical properties, and their suitability for mass production. However, in the past, little attention has been paid to green and sustainability issues such as energy saving, reduction of pollution, and protecting the environment. Some of the approaches toward sustainable and green manufacturing are as follows: enhanced use of renewable energy, establishment of green buildings, reduction of specific consumption of energy and water and materials, reduction in specific emissions of wastes, effluents, maximize recycling of wastes, water, etc. Considering the importance of the subject, the editors organized a Global Initiative of Academic Networks (GIAN) course on "Green Material Forming and Joining" few years ago. This laid the basic framework for the book. The book is intended to emphasize the importance of sustainability with reference to manufacturing and in particular to material forming and joining processes.

This book has a total of 19 chapters. Chapters 1–8 contribute to the basics of sustainable manufacturing and joining. Chapter 1 provides an introduction to sustainable material forming and joining. The basic material forming and joining processes are also briefly introduced in this chapter, and sustainable manufacturing has been highlighted with the help of few research examples. Chapter 2 covers sustainable material forming with some industry case studies. Chapter 3 covers sustainable joining, which initially highlights the sustainable initiatives followed in conventional joining methods, and later on, emphasizes the importance of novel joining methods like friction stir welding, joining by forming, etc. Chapter 4 deals with improving sustainability in friction and lubrication in metal forming practice. Types of lubricants and additives, selection of lubricants, and lubrication methods are also briefly introduced in this chapter. This chapter also includes a short note on bio-lubricants. Chapter 5 discusses the development of materials for sustainable manufacturing. Chapter 6 discusses steel as a projected sustainable material for future. Chapter 7 deals with strategies to improve the forming quality of sheets. Such forming quality improvement helps in sustainable manufacturing. Chapter 8 discusses the importance of computations and analyses in sustainable manufacturing with some industrial and research case studies.

Chapters 9–15 describe some special topics in material forming and joining that are relevant from a sustainable manufacturing point of view. Chapter 9 introduces hot stamping of sheets, specifically the high-strength steel grades. Chapter 10 covers the hydroforming of tubes and hydropiercing. The importance of computations and analytical modeling is also briefly emphasized in this chapter. Chapter 11 covers some of the recent developments in the field of microforming. Mechanical joining processes and hybrid joining processes are considered crucial in sustainable manufacturing. Chapters 12 and 13 cover a comprehensive discussion on these two topics. Chapter 14 discusses green lubricants and lubrication, including the importance of various biodegradable-based lubricant base stocks. Chapter 15 covers an elaborate discussion on how LASER-based manufacturing helps in green manufacturing.

Chapters 16–19 cover sustainability aspects. Chapter 16 discusses about models for sustainability measurement. Chapter 17 presents the evolution and future of sustainability in a sustainable manufacturing environment. Chapter 18 covers strategies to disseminate the significance of sustainability through education. Finally, Chapter 19 connects sustainability, health, and environment with a case study of the waste management sector.

The authors have given bird's-eye views of these specific topics. Readers are encouraged to refer to the extensive reference database provided in the chapters for in-depth details. The authors and editors have taken extreme care and interest in presenting the concepts and acknowledging the original sources. We believe the manufacturing researchers and academia will welcome the book. Any constructive feedback from the readers is encouraged.

R. Ganesh Narayanan and Jay S. Gunasekera

Acknowledgments

The editors thank MHRD India for introducing GIAN, which paved way for the professional collaboration between the editors and laid the foundation for the book. The support shown by Prof. Gautam Biswas, Director of IIT Guwahati, is acknowledged. We thank all the authors for their chapter contributions. We also appreciate their effort and interest in completing the chapters promptly around their busy schedules. We are grateful to Elsevier, Springer, ASM, and others for permitting to use the data in the journal articles and handbooks. We appreciate and thank those who granted permission to use figures/tables in the book via email.

We also thank the help rendered by Pritam K. Rana, PhD student of IIT Guwahati for helping in formatting the figures in the book. Thanks to Arvind K. Agrawal, Tinu P. Saju, Pritam K. Rana, Saibal K. Barik, who are pursuing PhD at IIT Guwahati, for helping in drawing several figures in the book and during GIAN course preparation. We acknowledge the staff members of CRC Press, Taylor & Francis Group. Finally, we thank our partners/spouses for showing patience during the book preparation and tolerating busy work schedules.

Editors

R. Ganesh Narayanan is an associate professor at the Department of Mechanical Engineering, Indian Institute of Technology (IIT) Guwahati, India. He received his PhD from the IIT Bombay, India. His research interests include metal forming and joining. He has contributed many research articles to reputed journals and international conferences. He has edited a few books, including *Strengthening and Joining by Plastic Deformation* published by Springer Singapore, *Advances in Material Forming and Joining* published by Springer India, and *Metal Forming Technology and Process Modeling* published by McGraw Hill Education, India. He has also edited special issues of journals, including "Advances in Computational Methods in Manufacturing" in the *International Journal of Mechatronics and Manufacturing Systems* and "Numerical Simulations in Manufacturing" in the *Journal of Machining and Forming Technologies*. He has organized two international conferences at IIT Guwahati, namely the International Conference on Computational Methods in Manufacturing (ICCMM) in 2011, and the 5th International and 26th All-India Manufacturing Technology, Design and Research (AIMTDR) Conference in 2014. He also organized a GIAN course on Green Material Forming and Joining at IIT Guwahati in 2016.

Jay S. Gunasekera is an internationally recognized researcher and a specialist in the area of metal forming and manufacturing. He has pioneered discoveries in the design of streamlined extrusion dies, forging and heat treatment problems, and ring rolling.

He was the Chair of the Department of Mechanical Engineering at Ohio University from 1991 to 2006. During his time as the Department Chair, the undergraduate and graduate enrollment increased significantly, a two-semester senior design course was added, two new research laboratories were also developed, two-thirds of the current faculty were hired, and research funding increased exponentially to over two million dollars per year. Dr. Gunasekera has advised over 90 graduate students, many of who are now leaders in their respective fields. Additionally, he helped develop the Center for Advanced Materials Processing (CAMP) and also served as its first director.

Dr. Gunasekera was the American editor of the *Journal of Materials Processing Technology*, published by Elsevier. He has undertaken research and consulting projects for the United States Air Force, General Electric, Pratt & Whitney, General Motors, and a large number of forging companies.

Dr. Gunasekera has published in over 150 technical publications and referenced in many journals and conferences. He was awarded the highest doctorate (DSc) degree by the University of London in 1991 for his contribution in research and publications in the field of manufacturing engineering. He was a Fellow of the City & Guilds of London, the highest honor conferred by that institution, and remains the only Sri Lankan to ever receive this award. He is a Fellow of SME & FRSA, past Fellow of IMechE, IProdE & IEE.

He has been recognized in *Who's Who in Engineering Education*, *Who's Who in the World*, *Who's Who of Emerging Leaders in America*, *Who's Who in American Education*, *The International Who's Who of Intellectuals*, *International Leaders in Achievement*, *The International Directory of Distinguished Leadership*, *Men of Achievement*, and *Who's Who of Engineering Leaders of America*.

Contributors

Nader Asnafi
Mechanical Engineering, School of Science
and Technology
Örebro University
Örebro, Sweden

Shitanshu S. Chakraborty
Materials Processing and Microsystems
Laboratory
CSIR-Central Mechanical Engineering
Research Institute
Durgapur, India

Sumitesh Das
Tata Steel Limited
Jamshedpur, India

Saptarshi Dutta
Department of Mechanical Engineering
IIT Guwahati
Guwahati, India

Tsuyoshi Furushima
Department of Mechanical and
Biofunctional Systems
Institute of Industrial Science
The University of Tokyo
Tokyo, Japan

R. Ganesh Narayanan
Department of Mechanical Engineering
IIT Guwahati
Guwahati, Assam, India

Sayanti Ghosh
Corporate R&D Centre
Bharat Petroleum Corporation Limited
Greater Noida, India

Radha Goyal
Air Quality Management Services (AQMS)
Division
Indian Pollution Control Association
Delhi, India

Jay S. Gunasekera
Department of Mechanical Engineering
Ohio University
Athens, Ohio

Ashish Jain
Indian Pollution Control Association
Delhi, India

Zhengjie Jia
Product Engineering
Litens Automotive Group
Toronto, Ontario, Canada

Debapriya Patra Karmakar
Department of Mechanical Engineering
Indian Institute of Technology Kharagpur
West Bengal, India

Lalit Kumar
Corporate R&D Centre
Bharat Petroleum Corporation Limited
Greater Noida, India

Yuvraj K. Madhukar
Discipline of Mechanical Engineering
Indian Institute of Technology Indore
Madhya Pradesh, India

K. J. Manjunatheshwara
Department of Production Engineering
National Institute of Technology
Tiruchirappalli, India

Prasad Modak
Environmental Management Centre LLP
Ekonnect Knowledge Foundation
Mumbai, India

Ken-ichiro Mori
Department of Mechanical Engineering
Toyohashi University of Technology
Toyohashi, Japan

Suvradip Mullick
School of Mechanical Sciences
Indian Institute of Technology Bhubaneswar
Odisha, India

Gopinath Muvvala
Department of Mechanical Engineering
Indian Institute of Technology Kharagpur
West Bengal, India

N. K. B. M. P. Nanayakkara
Department of Manufacturing and Industrial
    Engineering
University of Peradeniya
Peradeniya, Sri Lanka

Ashish K. Nath
Department of Mechanical Engineering
Indian Institute of Technology Kharagpur
West Bengal, India

Bharat L. Newalkar
Corporate R&D Centre
Bharat Petroleum Corporation Limited
Greater Noida, India

Shivanand M. Pai
Corporate R&D Centre
Bharat Petroleum Corporation Limited
Greater Noida, India

Vivek Rathore
Corporate R&D Centre
Bharat Petroleum Corporation Limited
Greater Noida, India

P. S. Robi
Department of Mechanical Engineering
IIT Guwahati
Guwahati, India

Sagar Sarkar
Department of Mechanical Engineering
Indian Institute of Technology Kharagpur
West Bengal, India

V. Satheeshkumar
Department of Production Engineering
National Institute of Technology
Tiruchirappalli, India

Tetsuhide Shimizu
Graduate School of System Design
Tokyo Metropolitan University
Tokyo, Japan

Tomomi Shiratori
Komatsuseiki Kosakusho Co. Ltd.
R&D Section
Nagano, Japan

Shyamli Singh
Department of Environment and Climate
    Change
Indian Institute of Public Administration
New Delhi, India

S. Vinodh
Department of Production Engineering
National Institute of Technology
Tiruchirappalli, India

Ming Yang
Graduate School of System Design
Tokyo Metropolitan University
Tokyo, Japan

CHAPTER 1

Introduction to Sustainable Manufacturing Processes

R. Ganesh Narayanan
IIT Guwahati

Jay S. Gunasekera
Ohio University

CONTENTS

1.1 CONVENTIONAL MANUFACTURING PROCESSES

Casting, forming, joining, and machining have been the four well-known manufacturing processes to mankind for many decades. Casting has been used to make raw materials since the discovery of structural metals such as copper, iron, etc. In this process, solidified metal blocks, in varied dimensions, are manufactured through melting and solidification. Sand casting, investment casting, die casting, and centrifugal casting are typical examples of casting processes. The solidified metal blocks are then processed to make usable components. In metal (material) forming, plastic deformation has been used to make usable components. A permanent shape is given to the material by using rigid tools. Rolling, forging, extrusion, wire drawing, sheet stamping, shearing, spinning, and tube forming are some of the typical examples of material forming processes. Joining is used to assemble raw materials or fabricated components, either permanently or temporarily, depending on the requirements of application. Welding, adhesive bonding, fastening, and riveting are some of the joining processes used predominantly. Joining involves providing heat and/or pressure to the materials for assembly. Hence, plastic deformation is involved in some of the joining processes. Machining involves removing of material to attain the final shape of the product. Some of the typical machining processes are turning, drilling, milling, shaping, planning, broaching, sawing, etc. Although defined separately, these processes are often combined to manufacture engineering goods. This chapter introduces conventional material forming and joining processes, thus highlighting the importance of sustainability in manufacturing. Finally, some of the sustainable material forming and joining processes have been discussed with some research contributions.

1.1.1 Metal Forming Processes

The metal forming processes are divided into the following two processes: bulk forming and sheet forming. The bulk forming processes include extrusion, forging, rolling, wire drawing, heading, ironing, etc. The deep forming processes include drawing, stretching, bending, shearing, spinning, hydro forming, etc. Both these processes are performed either at room temperature (called cold forming) or at a temperature above the recrystallization temperature (called hot forming) or at intermediate temperature (called warm forming), depending on the formability levels of the materials.

In the **rolling** process, the raw materials in the form of slab, billet, or bloom are rolled into plates, sheets, strips, rods, or tubes by mechanized rollers rotating in opposite directions. The sheet, strip, or foil rolling can be assumed as a plane-strain deformation process in which the deformation in width direction is negligible as compared with thickness and length direction. In the case of rolling thick sheets and slabs, significant deformation along the width, length, and thickness directions may be present. The rolling process is performed in rolling mills and is classified depending on the number of rolls, strands, and shape of the rolled products (Figure 1.1). A two-high roll mill with reversing facility employs rollers rotating in both clockwise and anticlockwise direction (Figure 1.1a). The entry and exit sides can be reversed after each rolling stage, with a reduction in roll gap, making it productive and efficient.

In a four-high roll mill, the actual sheet rolling is accomplished by the primary work roll, while large backup rolls support the primary roll (Figure 1.1b). This helps in large thickness reductions without excessive deflection of the primary work roll. Further heavy thickness reductions are possible in special roll mills like planetary mills without roller deflection (Figure 1.1c). Generally, rolling is accomplished in multiple stages called rolling stands (Figure 1.1d). In each stage, predefined and allowable thickness reduction will be aimed. Thickness reductions more than an allowable limit in each stage creates defects in rolled sheets.

In a **forging** process, a billet is converted into a shaped product with the help of rigid tools. This can be classified into open-die forging and closed-die (or impression die) forging. In open-die forging, the billet has no restriction to flow during deformation and takes the shape until fracture occurs. A simple example for this is simple upsetting operation, through which bulged head of a bolt can be made (Figure 1.2a). On the other hand, in closed-die forging (or impression die forging), a prescribed shape present in the die is given to the billet, making the final product or a semi-finished product (Figure 1.2b). Swaging and radial forging are used to reduce the area of cross-section of the rod or tube either at the end or at any intermediate distance. A metallic container, like a water bottle (Figure 1.3), can be made with the processes. In swaging, a rotating die hammers the workpiece with mechanized up–down movement to deliver a particular shape, while in radial forging, the dies do not rotate, but the rod/tube does (Figure 1.3).

In **extrusion**, a billet undergoes plastic deformation through dies by reducing its area of cross-section. The punch compresses the billet against the die (or die opening) through which the metal gets extruded taking the shape and size of the die opening. The cylindrical rods and tubes, as well as the shaped rods and tubes, are generally made through extrusion process. This is classified into forward extrusion, backward extrusion, and lateral extrusion (Figure 1.4), depending on the direction of metal flow with respect to punch displacement. During forward extrusion, the metal flows in the same direction as that of punch movement. There is a relative movement between billet and the

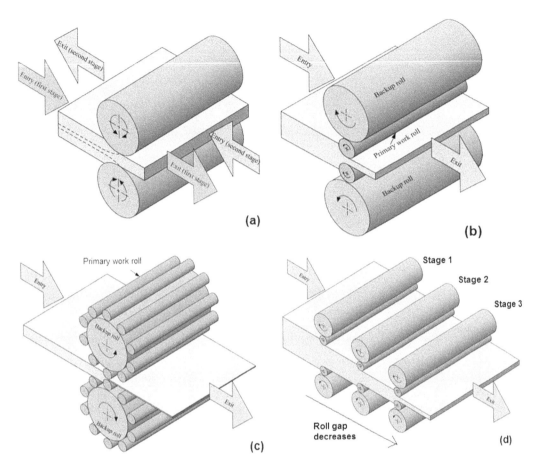

Figure 1.1   Types of roll mill: (a) two-high roll mill, (b) four-high roll mill, (c) planetary roll mill, and (d) roll mill with multiple stands.

die wall, and friction exists between the billet–die wall and the billet–punch interfaces. In backward extrusion, the metal gets displaced in the opposite direction as that of punch displacement, and there is no relative motion between the billet and die walls. Less friction is prevalent during backward extrusion and hence the load requirement is lower in backward extrusion as compared with forward extrusion. In lateral extrusion, the metal flow direction is at some angle to the ram movement that helps the material to get processed at varied strain severity.

In **wire drawing or rod drawing**, force is applied on the exit side of the billet for a product to be made, as opposed to extrusion in which the force is applied on the entry side. In other words, the billet or wire is pulled at the die exit, as opposed to pushing in extrusion of rods. Wires (<5 mm diameter) are made in wire drawing. There is no need to support wire on the entry side during wire drawing, while a support is required for a rod during rod drawing. **Tube drawing** can be conducted either with continuous mandrel or with floating mandrel to support the tube and to control its thickness during deformation (Figure 1.5).

Deep drawing of sheet materials is used to fabricate sheet products by plastically deforming the sheet into the die hole while clamping it with the help of blank holder. Finally, a cup-like component is made without wrinkling and failure. Beverage cans, washbasins, cooking utensils, cylinders, automotive body parts, pressure vessels, etc., are made using deep drawing. The purpose of the blank holder is to minimize wrinkling during this operation. The schematic representation of cup deep drawing

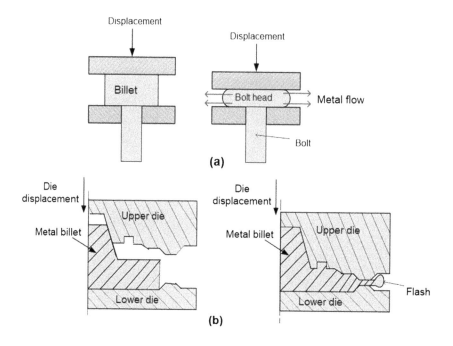

(a)

(b)

Figure 1.2   Forging processes: (a) open-die forging and (b) closed (impression) die forging.

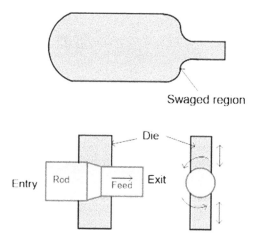

Figure 1.3   Swaging and radial forging.

operation is shown in Figure 1.6a. The deep drawing involves complex deformation with tension and straightening in the cup wall, tension at the bottom, compression, and friction at the flange region (the sheet part lying on the die surface), bending at the punch and die radius. Generally, cup deep drawing cannot be completed in a single stage and the sheet will be redrawn at multiple stages. The operation is called redrawing. The cup can be redrawn by forward redrawing or reverse redrawing (Figure 1.6b,c).

In **stretch forming** (or stretching), a sheet clamped under tension is formed into the die opening with a rigid punch. The sheet is clamped at designated locations with the help of lock beads or drawbeads, and then is wrapped around the punch to give a permanent shape change. Both continuous and discontinuous drawbeads are used for stretching. This operation is primarily used to make

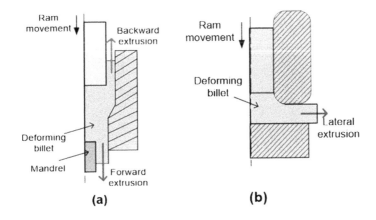

Figure 1.4   Extrusion processes: (a) forward and backward extrusion and (b) lateral extrusion.

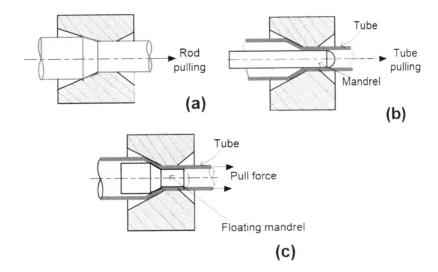

Figure 1.5   Wire, rod, and tube drawing: (a) rod drawing, (b) tube drawing with continuous mandrel, and (c) tube drawing with floating mandrel.

automotive door panels, automotive roof, aircraft skin panels, window frames, etc. The stretching, like deep drawing, is accompanied with bending that helps in reducing the springback of the formed part. In **bending** of sheets, deformation is concentrated in a small curved region, resulting in a nonuniform deformation across the thickness and curvature of the sheet. The bending operations, like V-bending, L-bending, and U-bending, are shown in Figure 1.7. The dies used are different in these operations, resulting in different product shapes, with U-bending resulting in a channel-type component. In the bent region, the sheet outer surface undergoes tension and the inner surface undergoes compression. Springback occurs during bending because of elastic recovery and is compensated by overbending, bottoming the sheet metal, materials processing, etc.

1.1.2 Joining Processes

Most of the traditional joining processes involve fusion welding processes such as arc welding, gas welding, resistance welding, soldering, brazing, and adhesive bonding. A brief overview about

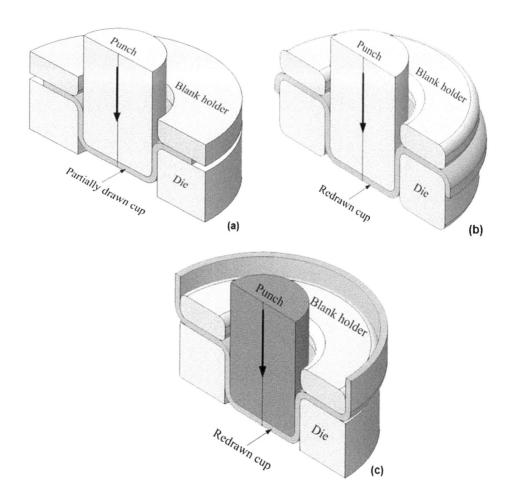

Figure 1.6 Cup deep drawing: (a) single-stage deep drawing, (b) reverse redrawing, and (c) forward redrawing.

these processes is given here. In **arc welding**, an electric arc is generated between the electrodes (consumable or nonconsumable), and the base plate. After melting and solidification, a joint is formed (Figure 1.8a). Temperature of the order of 5,000°C is witnessed in the process. The electrodes used in the process can be either consumable or nonconsumable. If they are consumable, the electrode melts (acts as a filler material) and forms the joint, while if they are nonconsumable, the electrode just generates the arc without melting, but gets depleted in length because of vaporization. The inert gases, such as helium and argon, are used as shielding gases, or fluxes are used for the arc shielding purpose. The flux also provides a protective atmosphere for welding, stabilizes the arc, and reduces spattering. The method of providing flux and shielding gas decides the type of arc welding. For example, with consumable electrodes, shielded metal arc welding uses a consumable electrode consisting of a filler rod coated with chemicals that provide flux and shielding. In gas metal arc welding, the consumable wire electrode is fed from a spool, and shielding gas is provided by a separate hose, and both are combined using a nozzle. In flux-cored arc welding, the electrode wire in the form of tube, with flux core, is fed from a spool, while shielding gas is provided separately. In submerged arc welding, a bare electrode wire is provided separately, while flux is fed in the form of flowing granules, such that the arc and electrode tip are submerged. In tungsten inert gas (TIG) arc welding, a nonconsumable electrode (like tungsten) and shielding gas are provided separately. The sustainability issues in welding come in the form of optimized process parameters, selection of

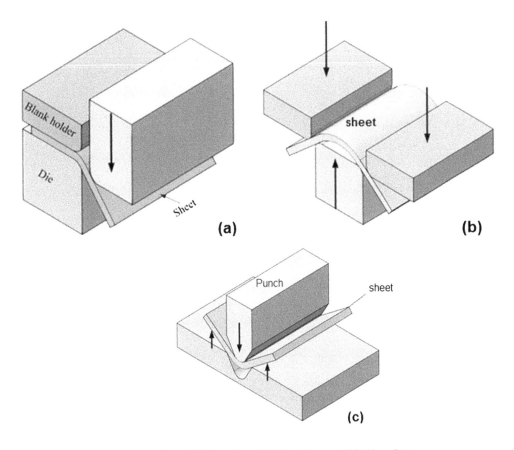

Figure 1.7 Sheet bending operations: (a) L-bending, (b) U-bending, and (c) V-bending.

electrode material, shielding gas usage and recycling, flux usage and removal, and efficient melting or non-melting of base materials.

In gas welding, the oxyacetylene gas welding is used predominantly. Here a flame generated by the combustion of oxygen and acetylene is used to provide heat required for welding (Figure 1.8b). The filler rod is provided separately if needed. The filler rod is coated with a flux that helps to prevent oxidation, thus creating an efficient weld joint. Instead of acetylene, hydrogen, propylene, propane, and natural gas are also used sometimes. In the case of resistance spot welding, two opposing electrodes are used to generate heat at the contacting surfaces creating spot welds (Figure 1.8c). The electrodes used are copper-based and refractory-based, depending on the materials to be welded, requiring better wear resistance to minimize inefficient welding. In the case of resistance seam welding, an electrode wheel is used instead of a rod.

In brazing and soldering, the filler material melts, without melting of base materials, somewhat similar to both fusion welding and solid-state welding. But low-strength joints are created as compared with fusion and solid-state welding. In brazing, the filler has a melting temperature >450°C, but below the melting point of base materials. In soldering, the filler has a melting temperature <450°C. Typical materials used as filler for brazing are 90Al+10Si, 99.9Cu, 60Cu+40Zn, Ni alloys, and silver alloys. Some of the typical solders used are 96Pb+4Ag, 60Sn+40Pb, 96Sn+4Ag. Fluxes are used in both the joining methods. In adhesive bonding, adhesive is used as a filler to bond the metallic surfaces, either with or without the application of pressure (Figure 1.8d). Adhesive is generally a polymer that cures with time, joining the metal surfaces called adherents.

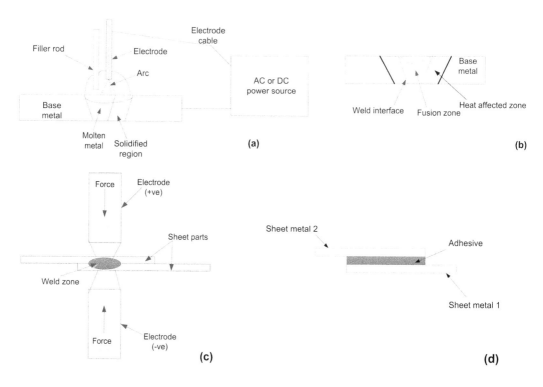

Figure 1.8   Traditional joining processes: (a) arc welding, (b) gas welding, (c) resistance spot welding, and (d) adhesive bonding.

1.2 SUSTAINABLE MANUFACTURING

The processes discussed above deliver products that are used in day-to-day life. In this context, the main query to be answered is: How to design and develop the processes such that they become environmentally friendly? Sustainable and green manufacturing is meant for the purpose. The core of the sustainable and green concept is to develop manufacturing processes, raw materials, computing methods, and policies, including recycling and waste management strategies, pollution control methods, and efficient business development, all aiming at optimum use of resources without creating environmental pollution in any form, while reducing waste. This should include reduced use of raw materials including those that are hazardous, reduced energy utilization, and reduced waste generation. The process developed today should not only yield required products but also be "sustainable" so that it will not affect the environment even after decades. The environmental concern while manufacturing a product has not been present at all since the first industrial revolution. The development of processes did not consider the effects of sustainability and green concepts, though there were innovations and novelty. Nowadays, there is a lot of pressure from government bodies, nongovernment organizations, consumers, and top management of companies to involve sustainability as part of product and process development.

The sustainability elements are to be quantified based on the requirement of the consumer and manufacturer. Jayal et al. (2010) pointed out that, for designing a commercial printer, there are six different sustainability elements including environmental impact, functionality, manufacturability, recyclability, and re-manufacturability, resource utilization and economy, and societal impact. The elements are further divided into numerous sub-elements and these were indexed with high, medium, and low importance to check the severity of such elements on the sustainability of the product through existing data and survey. It is observed from their results that functionality is the most important factor, and societal impact is least important factor from the manufacturer's point of view. On the other hand, societal impact is the most important factor and recyclability/re-manufacturability is the least important factor from the point of view of the customer. The standardization of such sustainability evaluation procedures is crucial and is not unique. Sustainability during the entire product life cycle is required these days. Sustainability during product design, manufacturing, assembly, quality analyses, implementation, and recycling are a must. While implementing this concept, a holistic system-based approach is essential, rather than a localized approach. It is also argued that it is more beneficial to implement the sustainability concept not only at the product level, but also at process and system level by using the following 6R methods: reduce, reuse, recycle, recover, redesign, and remanufacture the products. However, the earlier policies involved only 3R methods: reduce, reuse, and recycle. Figure 1.9 indicates that the stakeholder value evolution with sustainable and green manufacturing concepts is quick and efficient as compared to other concepts.

The following are some examples of products fabricated and processes adopted by considering green and sustainability effects. The following are the two examples of environmentally superior products (Maxwell and van der Vorst, 2003): First, a remote keyless electronic entry unit for cars is made, which is environmentally friendly. This has reduced the hazardous material's use by 40% dematerialization and reduced its energy usage threefold. Second, an upgraded personal computer is manufactured that has extended operational life by about five years, improved reusage and recyclability of components after life, and efficient energy utilization. In both of these examples, as a result of improved environmental friendliness, quick and cheaper products are made, saving costs for the customer.

In the case of manufacturing processes, the sustainability involves elements such as manufacturing cost, power and energy consumption, waste minimization, personal health, work safety, and eco-friendliness, along with the interaction effect. The first three factors are deterministic in nature and analytical models can be used for prediction, while the last three factors are non-deterministic and need soft computing methods. In the case of machining processes, Wanigarathne et al. (2004)

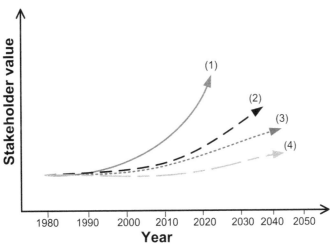

(1): Sustainable manufacturing (6R concept, innovative strategy)
(2): Green manufacturing (3R concept, environmental friendly)
(3): Lean manufacturing (waste reduction strategy)
(4): Conventional manufacturing (substitution based)

Figure 1.9 Evolution of sustainable manufacturing and other manufacturing concepts. (With permission from Jayal et al., 2010.)

proposed a method of assessing the sustainability index as a function of these basic elements. They
have shown, as part of the analyses, that the following equation is sufficiently accurate to predict the
overall sustainability index for a machining process:

$$S = \alpha_{\text{SHE}} S_{\text{SHE}} + \alpha_{\text{OP}} S_{\text{OP}}, \tag{1.1}$$

where S_{SHE} is the sustainability index for safety, health, and environment, assessed using soft computing methods, and S_{OP} is the operational sustainability including cost of machining, power consumption, and waste management; α_{SHE} and α_{OP} are the weighting factors that should be optimized
depending on the machining conditions. Although the model is developed for a machining process,
the same can be used for other manufacturing operations too. The analyses show that the machining
process is best in terms of "eco-friendliness" and worst in terms of "waste reduction" (Figure 1.10).
Some case studies of sustainable manufacturing in machining, casting, die engineering, plastics,
etc., are described here.

Sustainable manufacturing in *machining* includes dry (green) machining, near dry machining,
and cryogenic machining. Dry machining can be achieved either by tool modification or by process
modification. In tool modification, the tool should generate less heat during machining such that
coolant is not required. The tool materials such as high-strength steel (HSS), cemented tungsten carbide, cobalt alloys, coated carbides and coated HSS, cubic boron nitride, and diamond are developed
for the purpose. Different types of coatings also serve the same purpose (Sreejith and Ngoi, 2000).

Another example for dry machining is to optimize the tool dimensions and cutting conditions
during drilling of an aluminum alloy. WC–Co cemented carbide tool has been used by Nouari
et al. (2003). Diamond has been used as a tool-coating material. Without using lubricants, the

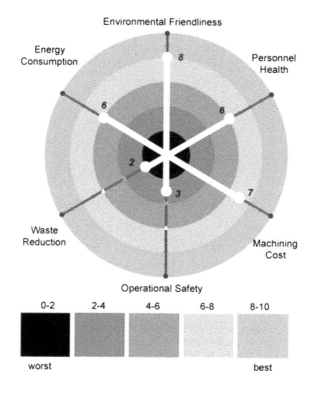

Figure 1.10   A typical sustainability rating system for a machining process. (With permission from Jayal et al.,
2010.)

optimization of tool geometry yielded good surface quality and dimensional accuracy of the drilled material. The drill performed consistently for longer duration. On the other hand, Paul et al. (2001) showed the importance of cryogenic cooling during plain turning of AISI steel over dry and wet machining. Cryogenic cooling by liquid nitrogen jets reduces tool wear and improves tool life and surface finish as compared to dry and wet turning.

Casting is a process that involves heavy production and losses. Tharumarajah (2008) reported that the major contribution of about 87% of intensity of electrical consumption in a die casting process for aluminum is from direct production (e.g., machines, CNC machining centers, and heat treatment plants) and from production services (e.g., cooling system, air compression system lighting). Moreover, yield losses and process losses contribute about 67% and 30%, respectively, in the total cost of losses. The yield losses depend on part and die design and occur mainly due to excess material in runner and flashes. Optimized runner and die design will reduce the yield losses. The process losses are due to issues like misruns and warm-ups.

Misruns are defects that occur because of failure of the molten metal to fill the mold cavity. This occurs due to the low fluidity of the melt, drastic reduction in temperature, mold section is too thin for the melt to fill. Warm-ups are rejected castings before the actual production casting trials become successful. Therefore, yield losses occur because of improper tool design, while process losses are due to inefficient process design and machine maintenance. There are secondary contributing factors like die changeover, lack of knowledge about process and part manufacturing, poor shop floor maintenance, etc., affecting the casting process overall efficiency. These factors finally affect the sustainability of the whole casting process.

The protective coatings on *molds* play a vital role in reducing the corrosion and wear, improving their sustainability. In this context, Peter et al. (2015) evaluated the performance of three thermal spray coatings including two ceramic coatings and a metallic coating. A nickel-based alloy powder and two partially stabilized ZrO_2-based powders are used for making the coatings. The nickel-based coatings performed well showing better wear resistance and lesser friction coefficient as compared with Zr coating. The nickel-based coatings delaminated after 2,000 cycles, while Zr-based coatings delaminated after 1,250 cycles when tested in molten aluminum alloy. The remanufacturing of dies and molds that enhance the life time temporarily and effective use of existing resources should be considered for green and sustainable future. A review on this by Chen et al. (2014) details the repairing strategy for dies and molds, life evaluation strategy, and reliability analyses of the re-made tools.

A simple modification in the pH of the cleaning solution paved way for an economical and environmentally friendly solution to waste management in a zinc *die casting* industry (Park et al., 2002). Hence, the acid use is eradicated, soap is preserved, rinsing is not required, and filtration requirements are minimized during finishing. A practical problem of theft of metallic manhole covers in roads and pavements can be reduced by using nonmetallic covers. In this context, Haggar et al. (2009) showed that the 90% unrecyclable thermoplastics with 10% foundry sand along with steel mesh of diameter 12 mm as reinforcement can be used to make manhole covers that withstand an average load of 112 kN. The cover is made of waste, unrecyclable thermoplastics, and foundry sand. This reduces the unused wastes. These are a few case studies highlighting the importance of sustainability and green issues in the manufacturing sector.

1.3 SUSTAINABLE MATERIAL FORMING AND JOINING

In this section, some case studies on green and sustainable issues in material forming and joining have been introduced. A detailed discussion has been made in the forthcoming chapters. Most of the developments in green and sustainable issues occurred due to the following two reasons: scientific developments and weight reduction of parts. The impact of light weighting of automobile components is enormous. For instance, use of aluminum in the bumper beam of a compact car saved approximately 210 and 190 MJ/kg of aluminum in primary energy and approximately 16 and 15 CO_2 equivalent/kg of aluminum in greenhouse gas, in two different vehicles. Similarly, the aluminum front hood of a large family car saved 170 MJ/kg of aluminum in primary energy and 13 CO_2 equivalent/kg of aluminum in greenhouse gas. It has been analyzed and concluded that the injury rate of the passenger can also be reduced by weight reduction. When the vehicle is made lightweight, with same size, the injury rate is reduced by 15%. By maintaining the same weight, with an increase in the size, the injury rate is further reduced (www.world-aluminium.org/).

On the other hand, some researchers have pointed out the importance of using HSS and advanced high-strength steel (AHSS) for making auto components. One such analysis showed that by the application of multiphase steel (AHSS category) in BMW series car, a body weight of 267 kg can be recognized, which is 17 kg less than the previous model in which multiphase steel was not used. The crash performance is also enhanced in this case. In a new VW Passat, the body weight is reduced by 20 kg when multiphase steel is used (www.worldautosteel.org/). The use of magnesium alloys is also considerable in terms of weight reduction of auto components. A typical analysis shows that about 62% weight reduction is possible when magnesium is substituted for steel, and about 10%–41% when magnesium is substituted for aluminum (Das, 2008), for parts with same stiffness. Although this is the case, producing Mg is costly and Pidgeon process of making Mg is energy intensive and significantly contributes to CO_2 emission.

In the material forming domain, an existing process can be modified or a new process can be developed to achieve sustainability. The modification and newer development can be in the form of tool design and fabrication, lubrication and friction conditioning, materials, process planning, and defect minimization, all leading to energy and waste minimization without spoiling the process efficiency. There are many examples highlighting the significance of lubrication and coating in material forming.

In dry metal forming, lubricants are not used during forming; instead, ceramic, and other coatings on the die surface or self-lubricating systems are used. TiCN, TiC–TiN, and DLC coatings are possible and they are suited for elevated temperatures as well (Reisel et al., 2003). It is known that sheet forming behavior, like limiting draw ratio, depends on the lubricants, and these are sensitive to such coatings as well. There are possibilities of formability reduction when coated tools are used instead of lubricants (Vollertsen and Schmidt, 2014). On the other hand, an efficient coating and proper choice of substrate will improve the cup drawability by delaying fracture, which otherwise would have failed much earlier in dry forming without coating (Murakawa et al., 1995). The galling performance of forming tools improved with these coatings as suggested by Podgornik and Hogmark (2006). Plasma nitriding, DLC coating, and hard TiN coating are tried for the purpose. The performance of such coatings can be improved by metallic interlayers as suggested by Taube et al. (1994). In all these cases, the usage of external lubricants is avoided, thereby reducing their impact on the environment.

In case of processes, hydroforming, incremental forming, microforming, forming with stirring, warm forming, superplastic forming, electromagnetic forming, and press hardening can be classified under sustainable forming. Through these processes, the use of material is minimized, post-forming processes are reduced, deformation of materials at room temperature becomes efficient, and the material properties get enhanced during processing. In hydroforming, a sheet or a tube is plastically deformed with the help of pressure generated internally by liquid medium. In the case of tube hydroforming, an axial force (along the tube length) is provided to avoid thinning and fracture. Some of the typical application includes chassis components such as rear and front axles, engine cradles, exhaust systems, structural body components, etc. By modifying the process from conventional stamping to hydroforming, a weight reduction of 16% and a part cost reduction of 17% are achieved while manufacturing a light truck frame (Hartl, 2005). The dashboard cowl on the Porsche Panamera, made out of austenitic stainless steel, has been hydroformed (Tolazzi, 2010). There were regular complaints about the low productivity of the hydroforming process. This has been improved in the last few years by quick liquid filling technology, lesser transition time from filling to forming, increased forming speed, hydroforming at warm conditions, etc. The internal pressure is drastically reduced at warm forming conditions, reducing the chances of using large tonnage presses for making components.

In incremental forming (ISF), a sheet is formed by a tool incrementally from point to point by following a trajectory on the sheet, instead of deforming the sheet with a large punch in one go in a traditional process. Such incremental deformation allows complex parts to be manufactured, avoiding the use of rigid and dedicated clamping system. This reduces the process time and costs. It is generally seen that the increment forming process allows larger plastic strain to be achieved in the part as compared with conventional forming. Ingarao et al. (2012) suggested, in a comparative energy analysis between single-point incremental forming (SPIF) and traditional stamping in forming a part, that the energy consumption in SPIF is considerably higher than that in the case of conventional stamping. However, SPIF contributes in material savings. The part shape change in SPIF has a significant effect on energy consumption as compared with stamping operations. The exergy analyses suggest that ISF is suitable for prototyping and small production runs up to 300 products from sustainability point of view (Dittrich et al., 2012).

The miniaturized components are made in the fields of biomedical and electronic engineering using microforming. The size effect and friction play a vital role in microforming as compared with macroforming (Engel and Eckstein, 2002; Vollertsen et al., 2006). Microfactories are state-of-the-art, wherein each miniaturized manufacturing system is seen as a factory delivering microcomponents. One of the main objectives of microfactories is to introduce new manufacturing concepts with small-scale facilities, which could significantly minimize energy consumption and pollution, and create an eco-friendly working place (Qin, 2006). A stage will be reached at which a single formed or machined part is replaced by assembling many microformed/machined parts, thus reducing the energy required for each forming. Laser forming, which is classified under die less forming, is one of the predominantly used microforming technologies.

In superplastic forming and warm forming, the sheets are deformed at elevated temperatures at controlled strain-rate ranges. Those materials have lesser ductility at room temperature but show very high ductility/formability during superplastic forming and warm forming. Quick plastic forming is also used in aerospace sector for making sheet parts made of Al alloys (Taub et al., 2007).

Friction stir welding and processing are either used to assemble sheet parts or to process the sheets for property enhancement. The mechanical performance of friction stir processed materials is better than base material in terms of formability, fatigue, and energy savings. There is so much evidence for formability and fatigue performance. Shrivastava et al. (2015) recommended the use of FSW over gas metal arc welding (GMAW) of an Al alloy. They found out that the total energy consumed in joining, and pre- and post-processing for FSW is 40% less than that in GMAW. The life cycle assessment established the fact that FSW resulted in 31% less greenhouse emissions as compared with GMAW. There exist many different processing techniques in which friction stir concept has been used.

Another salient development is the use of electromagnetics for material forming and joining. Kurka and Schieck (2014) highlighted the importance of electromagnetic forming and press hardening for processing and manufacturing lightweight components contributing toward sustainable and green manufacturing.

Arc welding and gas welding have been known to the human community for many decades. These methods are predominantly used in construction and ship building industries. Arc and flame are used as sources of heat generation in these operations. Shielding gas, flux, consumables (coated, uncoated), sample preparation, post-welding preparations, etc., are the main factors that are responsible for an unclean and polluted environment, as well as operator health issues. Welding fumes containing metal oxide particles can damage eye sight and are easily inhaled, thus resulting in a long-term effect on the operator's health. Popović et al. (2014) reported that about 95% of the fumes are generated from the filler metal and flux coating of consumable electrodes, while base material's contribution is minor. They also suggested that that metal-cored wires generate less than half the fumes as compared to flux-cored consumables, and that using shielding gas is better in terms of weld fumes emission as compared to self-shielded wires.

Vimal et al. (2015) proposed sustainability measures for SMAW using a graph theory approach, and suggested the employee skill training on green strategies and involvement program as the most crucial sustainability measure, followed by waste minimization and disposal studies. They also suggested 100% burning for flux and 100% recycling for core wire as the best strategies to take care of sustainability in the case of waste disposal. The life cycle assessment of different welding processes conducted by Sproesser et al. (2015) revealed that apart from welding process efficiency, the joint design and material choice contribute a lot to environmental pollution.

A novel method of joining tube to a plate by friction stir concept has been proposed by Senthil Kumaran et al. (2012). The process consumes lesser power of 0.1874 kWh as compared with TIG welding consuming 0.6755 kWh, reducing the power charge from Rs. 3.377 to Rs. 0.937 (USD 0.047 to USD 0.013). The welding time is drastically reduced from 16 MTS to 0.683 MTS in the case of new method. The joint formed is strong enough as compared with TIG welding. The application of hybrid welding, combination of a high energy CO_2 laser with the metal active gas (MAG) arc welding, is demonstrated in Wieschemann et al. (2003). It combines the advantage of deeper penetration by arc welding and arc stability given by laser welding because of the presence of plasma. The synergistic interaction increases the efficiency of the process while requiring low levels of linear energy. In order to minimize the use of fusion welding processes involving harmful consumables and fluxes, many solid-state welding methods, such as friction processing technologies, including friction stir welding, friction welding, friction spot welding (Fujimoto et al., 2005; Nicholas, 2003; Sahin, 2009), magnetic pulse welding (Kapil and Sharma, 2015; Kimchi et al., 2004), mechanical joining methods like joining by plastic deformation (Alves et al., 2011), self-pierced riveting (Bouchard et al., 2008; He et al., 2008; Porcaro et al., 2006), clinching (Lee et al., 2010; Varis, 2003), etc., are proposed. These processes are described in Figure 1.11. All the process involves plastic deformation as a major component for joining.

1.4 COMPUTER-AIDED ENGINEERING ANALYSES

Computations and computing power in industry environment are inevitable in the present era. Mostly these are conducted for reducing the cost and time involved in conducting experiments at shop floor. Materials savings, power savings during experimentation, reduced machine utilization, minimizing defects, fuel savings, and pollution control are the main factors that aim to cost reduction. Although sustainable and green processes are discussed a lot these days, computations are used for traditional processes as well.

The analytical methods, such as upper-bound elemental technique (UBET), slab method, finite-element methods, etc., are used for metal forming and joining process analyses. UBET was used to design the process and dies during ring rolling and feeder plate design during extrusion (Alfozan and Gunasekera, 2002; Mehta et al., 2001; Ranatunga and Gunasekera, 2006). The upper-bound technique has been used to model the ring rolling process and plastic flow has been analyzed by Alfozan and Gunasekera (2002) to predict the optimum intermediate shapes. In UBET, the deforming workpiece is divided into a number of elements. The total power dissipation is minimized with respect to all the unknown variables to evaluate the optimum velocity field in the rectangular elements. The total power dissipation has contributions from power dissipation due to pure plastic deformation over an element, the velocity discontinuities (internal shear) across the boundaries between elements, and friction at the ring–roll interfaces. Using this method, the near optimum intermediate shapes during ring rolling were predicted with an error of about 8% as compared to experimental data. On the other hand, Mehta et al.'s work concentrated on using finite-element and finite-volume simulation codes such as DEFORM and MSC/SuperForge to analyze the effect of die design (streamlined die

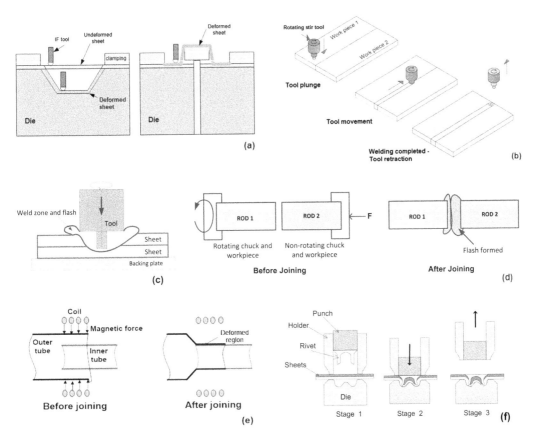

Figure 1.11 Examples of the sustainable forming and joining methods: (a) incremental forming, (b) friction stir welding, (c) friction stir spot welding, (d) friction welding, (e) magnetic pulse welding, and (f) self-pierced riveting.

vs. shear die vs. optimum shear with feeder plate die) on aluminum extrusion with I-section shape. The complexities of die design were brought forward along with the accuracy levels of software used.

The numerical simulation of springback of a vacuum vessel, a double-wall toroidal welded structure, is a practical case study showing the importance of computational methods in process design, including die and punch design for reduced springback (Song et al., 2003). Silva et al. (2015), through an in-house FE model, simulated and designed a process for joining of two tubes by end forming. This process eliminates problems in fusion welding of metal tubes. The tubes are end formed and are mechanically joined to form an integral component. Thus, FE simulations were helpful to optimize the process and tool parameters without much experimental trials. Ingarao et al. (2012), through FE simulations, showed that conventional stamping operations require less deformation energy as compared with SPIF, but SPIF allows materials savings. Satheeshkumar and Narayanan (2015), through FE simulations and necking criterion, demonstrated the importance of modeling adhesive and adhesive bonding during formability evaluation of adhesive-bonded sheets. The forming limit strains were not accurately predicted during tensile and in-plane plane-strain forming tests due to the absence of bonding during FE simulations.

The FE simulation prediction of laser butt welding of magnesium alloys (Belhadj et al., 2010) yielded temperature history, welding width, cooling isotherms that are very much essential for process design. The raw materials, welding parameters, and other welding requirements, if any, can be judged to minimize trials and error costs using an appropriate process design. Thus, a validated model is required. Many researchers have developed theoretical and computational models that are initially validated with limited experiments for their accuracy and further these are used for actual process predictions. The CAE models, numerical models, soft computing models, etc., help in analyzing the processes, either to improve traditional manufacturing processes or to develop new processes that are sustainable.

1.5 SUMMARY

In this chapter, the traditional material forming and joining methods are described briefly. Green and sustainable manufacturing methods have been introduced later with some case studies. Generally, the developments are in the form of manufacturing processes, raw materials, computing methods, policies including recycling and waste management strategies, pollution control methods, and efficient business development. All these factors aim at optimum use of resources without creating environmental pollution in any form, and reduce waste. This includes reduced volume of usage of hazardous raw materials, reduced energy utilization, and reduced waste generation. Sustainability measures are crucial in order to optimize the process for utilizing its success fully. Therefore, strategies should be standardized considering the importance of holistic system-based approach, the life cycle analyses, and not a localized approach. Finally, some examples for green and sustainable material forming and joining in terms of materials and processes are presented. The significance of computing power is also highlighted in bulk forming, sheet forming, and welding.

REFERENCES

Alfozan A., Gunasekera J.S., 2002, Design of profile ring rolling by backward simulation using upper bound element technique (UBET), *Journal of Manufacturing Processes*, 4, pp. 97–108.

Alves L.M., Dias E.J., Martins P.A.F., 2011, Joining sheet panels to thin-walled tubular profiles by tube end forming, *Journal of Cleaner Production*, 19, pp. 712–719.

Belhadj A., Bessrour J., Masse J.E., Bouhafs M., Barrallier L., 2010, Finite element simulation of magnesium alloys laser beam welding, *Journal of Materials Processing Technology*, 210, pp. 1131–1137.

Bouchard P.O., Laurent T., Tollier L., 2008, Numerical modeling of self-pierce riveting—From riveting process modeling down to structural analysis, *Journal of Materials Processing Technology*, 202, pp. 290–300.

Chen C., Wang Y., Ou H., He Y., Tang X., 2014, A review on remanufacture of dies and moulds, *Journal of Cleaner Production*, 64, pp. 13–23.

Das S., 2008, Primary magnesium production costs for automotive applications, *Journal of Metals*, 60, pp. 63–69.

Dittrich M.A., Gutowski T.G., Cao J., Roth J.T., Xia Z.C., Kiridena V., Ren F., Henning H., 2012, Exergy analysis of incremental sheet forming, *Production Engineering Research Development*, 6, pp. 169–177.

Engel U., Eckstein R., 2002, Microforming—From basic research to its realization, *Journal of Materials Processing Technology*, 125–126, pp. 35–44.

Fujimoto M., Inuzuka M., Koga S., Seta Y., 2005, Development of friction spot joining, *Welding in the World*, 49, pp. 18–21.

Haggar S.E., Hatow L.E.I., 2009, Reinforcement of thermoplastic rejects in the production of manhole covers, *Journal of Cleaner Production*, 17, pp. 440–446.

Hartl C., 2005, Research and advances in fundamentals and industrial applications of hydroforming, *Journal of Materials Processing Technology*, 167, pp. 383–392.

He X., Pearson I., Young K., 2008, Self-pierce riveting for sheet materials: State of the art, *Journal of Materials Processing Technology*, 199, pp. 27–36.

Ingarao G., Ambrogio G., Gagliardi F., Lorenzo R.D., 2012, A sustainability point of view on sheet metal forming operations: Material wasting and energy consumption in incremental forming and stamping processes, *Journal of Cleaner Production*, 29–30, 255–268.

Jayal A.D., Badurdeen F., Dillon Jr. O.W., Jawahir I.S., 2010, Sustainable manufacturing: Modeling and opti-
    mization challenges at the product, process and system levels, *CIRP Journal of Manufacturing Science
    and Technology*, 2, pp. 144–152.
Kapil A., Sharma A., 2015, Magnetic pulse welding: An efficient and environmentally friendly multi-material
    joining technique, *Journal of Cleaner Production*, 100, pp. 35–58.
Kimchi M., Shao H., Cheng W., Krishnaswamy P., 2004, Magnetic pulse welding aluminium tubes to steel
    bars, *Welding in the World*, 48, pp. 19–22.
Kurka P., Schieck F., 2014, Innovations in forming technologies, *Achieves of Materials Science and
    Engineering*, 65, pp. 72–76.
Lee C.J., Kim J.Y., Lee S.K., Ko D.C., Kim B.M., 2010, Design of mechanical clinching tools for joining of
    aluminium alloy sheets, *Materials and Design*, 31, pp. 1854–1861.
Maxwell D., van der Vorst R., 2003, Developing sustainable products and services, *Journal of Cleaner
    Production*, 11, pp. 883–895.
Mehta B.V., Al-Zkeri I., Gunasekera J.S., Buijk A., 2001, 3D flow analysis inside shear and streamlined extru-
    sion dies for feeder plate design, *Journal of Materials Processing Technology*, 113, pp. 93–97.
Murakawa M., Koga N., Kumagai T., 1995, Deep-drawing of aluminum sheets without lubricant by use of
    diamond-like carbon coated dies, *Surface and Coatings Technology*, 76–77, pp. 553–558.
Nicholas E.D., 2003, Friction processing technologies, *Welding in the World*, 47, pp. 2–9.
Nouari M., List G., Girot F., Coupard D., 2003, Experimental analysis and optimisation of tool wear in dry
    machining of aluminium alloys, *Wear*, 255, pp. 1359–1368.
Park E., Enander R., Barnett S.M., 2002, Pollution prevention in a zinc die casting company: A 10-year case
    study, *Journal of Cleaner Production*, 10, pp. 93–99.
Paul S., Dhar N.R., Chattopadhyay A.B., 2001, Beneficial effects of cryogenic cooling over dry and wet
    machining on tool wear and surface finish in turning AISI 1060 steel, *Journal of Materials Processing
    Technology*, 116, pp. 44–48.
Peter I., Rosso M., Gobber F.S., 2015, Study of protective coatings for aluminum die casting molds, *Applied
    Surface Science*, 358, pp. 563–571.
Podgornik B., Hogmark S., 2006, Surface modification to improve friction and galling properties of forming
    tools, *Journal of Materials Processing Technology*, 174, pp. 334–341.
Popović O., Prokić-Cvetković R., Burzić M., Lukic U., Beljic B., 2014, Fume and gas emission during arc
    welding: Hazards and recommendation, *Renewable and Sustainable Energy Reviews*, 37, pp. 509–516.
Porcaro R., Hanssen A.G., Langseth M., Aalberg A., 2006, Self-piercing riveting process: An experimental
    and numerical investigation, *Journal of Materials Processing Technology*, 171, pp. 10–20.
Qin Y., 2006, Micro-forming and miniature manufacturing systems-development needs and perspectives,
    Journal of Materials Processing Technology, 177, pp. 8–18.
Ranatunga V., Gunasekera J.S., 2006, UBET-based numerical modeling of bulk deformation processes,
    Journal of Materials Engineering and Performance, 15, pp. 47–52.
Reisel G., Wielage B., Steinhauser S., Hartwig H., 2003, DLC for tools protection in warm massive forming,
    Diamond and Related Materials, 12, pp. 1024–1029.
Sahin M., 2009, Joining of stainless-steel and aluminium materials by friction welding, *International Journal
    of Advanced Manufacturing Technology*, 41, pp. 487–497.
Satheeshkumar V., Narayanan R.G., 2015, Prediction of formability of adhesive bonded steel sheets and
    experimental validation, *Achieves of Civil and Mechanical Engineering*, 15, pp. 30–41.
Senthil Kumaran S., Muthukumaran S., Venkateswarlu D., Balaji G.K., Vinodh S., 2012, Eco-friendly aspects
    associated with friction welding of tube-to-tube plate using an external tool process, *International
    Journal of Sustainable Engineering*, 5, pp. 120–127.
Shrivastava A., Krones M., Pfefferkorn F.E., 2015, Comparison of energy consumption and environmental
    impact of friction stir welding and gas metal arc welding for aluminium, *CIRP Journal of Manufacturing
    Science and Technology*, 9, pp. 159–168.
Silva C.M.A., Nielsen C.V., Alves L.M., Martins P.A.F., 2015, Environmentally friendly joining of tubes by
    their ends, *Journal of Cleaner Production*, 87, pp. 777–786.
Song Y., Yao D., Wu S., Weng P., 2003, Spring-back simulation of sheet metal forming for the HT-7U vacuum
    vessel, *Fusion Engineering and Design*, 69, pp. 361–365.
Sproesser G., Chang Y.J., Pittner A., Finkbeiner M., Rethmeier M., 2015, Life cycle assessment of welding
    technologies for thick metal plate welds, *Journal of Cleaner Production*, 108, pp. 46–53.

Sreejith P.S., Ngoi B.K.A., 2000, Dry machining: Machining of the future, *Journal of Materials Processing Technology*, 101, pp. 287–291.

Taub A.I., Krajewski P.E., Luo A.A., Owens J.N., 2007, The evolution of technology for materials processing over the last 50 years: The automotive example, *Journal of Metals*, 59, pp. 48–57.

Taube K., Grischke M., Bewilogua K., 1994, Improvement of carbon-based coatings for use in the cold forming of non-ferrous metals, *Surface and Coatings Technology*, 68–69, pp. 662–668.

Tharumarajah A., 2008, Benchmarking aluminium die casting operations, *Resources, Conservation and Recycling*, 52, pp. 1185–1189.

Tolazzi M., 2010, Hydroforming applications in automotive: A review, *International Journal of Material Forming*, 3, pp. 307–310.

Varis J.P., 2003, The suitability of clinching as a joining method for high-strength structural steel, *Journal of Materials Processing Technology*, 132, pp. 242–249.

Vimal K.E.K., Vinodh S., Raja A., 2015, Modelling, assessment and deployment of strategies for ensuring sustainable shielded metal arc welding process—A case study, *Journal of Cleaner Production*, 93, pp. 364–377.

Vollertsen F., Niehoff H.S., Hu Z., 2006, State of the art in micro forming, *International Journal of Machine Tools & Manufacture*, 46, pp. 1172–1179.

Vollertsen F., Schmidt F., 2014, Dry metal forming: Definition, chances and challenges, *International Journal of Precision Engineering and Manufacturing-Green Technology*, 1, pp. 59–62.

Wanigarathne P.C., Liew J., Wang X., Dillon Jr. O.W., Jawahir I.S., 2004, Assessment of process sustainability for product manufacture in machining operations, *Proceedings of the global conference on sustainable product development and life cycle engineering*, Berlin, Germany, pp. 305–312.

Wieschemann A., Kelle H., Dilthey D., 2003, Hybrid-welding and the HyDRA MAG + LASER processes in shipbuilding, *Welding International*, 17, pp. 761–766.

CHAPTER 2

Sustainability in Material Forming

Zhengjie Jia
Litens Automotive Group

R. Ganesh Narayanan
IIT Guwahati

N. K. B. M. P. Nanayakkara
University of Peradeniya

Jay S. Gunasekera
Ohio University

CONTENTS

2.1 INTRODUCTION

Environmental sustainability and sustainable manufacturing are widely discussed and investigated due to the requirement of reducing the environmental impact. Sustainable material forming is part of such investigation. The sustainability in material-forming industry generally includes three aspects: (1) environmental sustainability—about planet, (2) economic sustainability—about profit, and (3) social sustainability—about people. Significant research has been done to investigate sustainability in material forming mainly on the environmental sustainability. This includes (1) how to efficiently use materials and resources, (2) how to effectively reduce the energy consumption and wastes during material forming, and (3) how to improve the forming quality with existing resources. The three most critical factors that have excessive impact to environment in material-forming processes are high energy consumption that generates significant CO_2, significant water usage, and generation of nonrecyclable/nondegradable waste materials from production. The material-forming processes should be designed, optimized, and controlled with environmental sustainability as one of the major objectives, and products produced by material forming should not be harmful to people during production and use in terms of health and safety.

In the past decade, material-forming industry has faced new challenges of sustainability requirement to minimize the overall environmental impact under a life cycle perspective, in addition to the traditional requirements for lower cost, better quality, and faster to market. More and more stringent standards and policies have come into effect to accomplish the task. The traditional "trial and error" approach or "try-outs" in material forming has been frequently replaced by sophisticated and efficient analysis methods such as finite-element method/analyses (FEM/FEA) and virtual metal forming. Product quality, production economy, and environmental impact have become fundamental considerations for the material-forming process design in most cases. The growing tie-up between steel makers and vehicle manufacturers through Ultralight Steel Automotive Body (ULSAB) concept aiming to reduce the carbon footmark throughout the life-cycle of the vehicle is one of the best examples. As road vehicles significantly contribute to global energy consumption and greenhouse gas emissions, the European Union (EU) regulators have imposed to reduce average CO_2 emission to about 130 g/km by 2015 and 95 g/km in 2020 across all new vehicles. The impact of vehicle weight on the CO_2 emissions is shown in Figure 2.1 (www.imoa.info). The test data show that a weight reduction of 100 kg saves 0.1–0.5 L per 100 km, which corresponds to 8–12 g/km reduction of CO_2 emission. The data have been fit and it is observed that the CO_2 emission increases following a power law with an increase in vehicle weight. It is also seen that the emission from luxury vehicles is much larger than other class of vehicles, especially the compact cars. A substantial effort has been put into the development of light weight vehicle, including body, components, and structures, in order to achieve the emission target, as a car body using high-strength steels can reduce body weight by more than 100 kg as compared with a car using conventional steel grades (www.imoa.info). Moreover, increased strength is favorable in terms of safety and crash resistance.

In 1994, Audi developed the all-aluminum Audi Space Frame (ASF) system using extruded, die-cast, and flat rolled high-strength aluminum (www.aluminium.org), in order to achieve a significant reduction in vehicle weight. The weight of A8's first-generation ASF aluminum body was reduced from 478 kg (steel body of similar-sized car) to 273 kg, and further it was reduced to 239 kg in the second-generation ASF, and now just 220 kg, which represents a total of 45% reduction in vehicle weight (www.audiusa.com; www.audi-journals.com). In 1999, Audi began to apply the ASF aluminum body-in-white (BIW) technology into the A2, a mass production full-size family passenger car, which converted many iron and steel components into aluminum, including the engine and wheels, and have achieved a diesel fuel consumption performance of 3 L per 100 km (www.audi-world.com). If the aluminum BIW's technology were applied to all passenger vehicles produced in

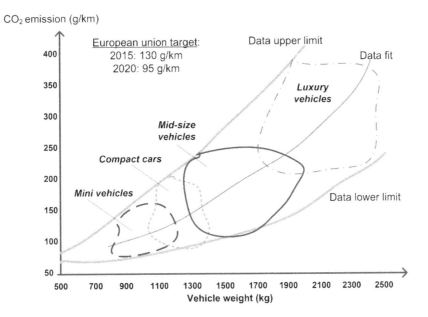

Figure 2.1 Influence of automobile weight on CO_2 emission. (www.imoa.info.)

the USA and Canada, at least an additional 40 million tons of aluminum would be required. On the other hand, savings of about 132–211 million barrels of crude oil can be achieved, with a reduction of about 54–86 million tons of CO_2 equivalent in greenhouse gas (GHG) emission annually (www. aluminium.org).

Sustainability is a concept, which is difficult to measure. Therefore, a comparison of precise sustainability of different material-forming processes can be impractical, but the reduction of CO_2 emission that is directly linked to energy consumption is a measurable target. To achieve the green and sustainable material forming, energy consumption, water usage, and materials' waste in material-forming processes can be managed and reduced by the following factors:

- System-level eco-design and optimization of energy consumption, water usage, and materials' waste through product lifecycle assessment and management (PLA and PLM).
- Virtual forming process development to reduce or eliminate the physical forming tests, and to reduce the total number of forming stages/steps and minimize the real shop floor trials.
- Integrate virtual–physical forming process design and development methodology for lower cost, better quality, faster-to-market, and sustainable material-forming practice.
- Use of new and efficient material-forming technologies and materials.

It is a well-known fact to scientific community that large tonnage of metals is plastically deformed (or formed) every day to fabricate products in automotive, food packaging, aerospace, construction, and agriculture sectors. Because of use of high-strength metals such as steel and cast iron, the energy consumption during forming is substantially high as compared with non-metals such as polymers, ceramics, etc., during part manufacturing. As a result, the industry has been shifting away from steel family in many of the traditional applications, switching to usage of light weight metals such as aluminum and magnesium alloys. However, steel grades account for most of the tonnage of forming in today's industry and certain applications need steel grades as raw material.

2.2 METAL-FORMING PROCESS DESIGN

The various metal-forming processes and their classification are discussed in the first chapter. Such metal-forming processes can be done at room temperature and at elevated temperature; both these temperatures have their own advantages and disadvantage including energy consumption, quality enhancement, and material utilization. Both analytical knowledge and technology information are required to design metal-forming processes. The flow of metal, deformation stresses and strains, microstructure, friction, heat transfer, lubricant types, heat input, and recovery methods, material handling system, tool design, and forming equipment are involved. However, the key is the understanding and control of metal flow, which determines the quality, elimination of defects such as cracks or folds, and the mechanical properties of the final product.

The significant variables in a metal-forming process (Altan et al., 1983) include (1) billet and tool material properties; (2) tool/billet interface conditions such as lubricant/friction and cooling characteristics; (3) deformation mechanics including modeling assumptions, evolution of stress, strain, strain-rate and temperature; (4) equipment conditions including its ram tonnage, rigidity, and accuracy; (5) product complexities such as dimensional accuracy/tolerances, surface texture, and structural properties; (6) ecological conditions including manufacturing facilities and control. Considering all such aspects, metal-forming process design becomes complex and time consuming.

In recent years, some of the above variables have been in severe scrutiny among academicians and industrialists. The final product was the first and foremost expectation in a traditional forming process and the success of the forming process was directly related to the pass/fail criteria of the quality of the product and the cost incurred in manufacturing. However, to meet today's growing concerns, energy consumption, use of resources, and direct and indirect environmental pollution have become more focused topics. In line with this requisite, many of the recent studies on metal forming have been diverted from its traditional path and started focusing on the process-based concerns such as lean adoption in forming, green forming, and sustainability. However, it should be noted that such new paradigm of issues depends on traditional scientific problems such as strain and stress analyses, lubrication effect, temperature effect, etc., and analysis methods.

Minimizing adverse effects of metal forming on the society and the environment has been, therefore, in high demand. In order to minimize the impact on the environment to achieve sustainability in material forming, and to optimize the variables, many modeling methods such as force equilibrium analysis, lower and upper-bound method, FEM/FEA, Taguchi experimental design and statistical methods, axiomatic design approach, geometric methods, and neural networks (NNs), and hybrid methods are used to model the processes. However, each of these methods has its own advantages and disadvantages in terms of accurate prediction and computational simplicity. The slab method has been applied to various types of forming processes for a long time since 1934. Due to simplified assumptions in the slab model, FEM generally provides a much more accurate solution as compared with the slab model.

In recent years, an integrated analysis and design method that combines the advantages of a few of these methods for metal-forming processes has been adopted as a multidisciplinary systematic approach. For instance, Veera Babu et al. (2009, 2010) developed an "expert system" based on FEM-NN hybrid model that can predict the formability of tailor welded blanks. The expert system is developed considering different material grades, and weld conditions. Since the expert system is based on FE simulations, the prediction accuracy depends on the material models and formability criteria. The deliverables from the FE simulations were trained by NN. Until now, the expert system can predict the tensile behavior and deep drawing behavior of welded sheets accurately. Ou et al. (2004) developed an FE-optimization strategy to minimize errors during hot forging of aero engine components. The authors have considered press deflections and die clearance for optimization. They found that the die elasticity governs that aero foil thickness error and press deflections decide the aero foil bow and twist errors. Similarly, Yang et al. (2007) performed FE-NN modeling analyses of manufacturing bevel gears and optimized the effect of a few material properties on maximum forging load and effective stress distribution. With this attempt, one can optimally select the material quality and process conditions for successful gear manufacturing.

There is still lack of knowledge, effective methods, and quantitative studies on modeling the sustainability and its parameters for a formed product in its lifecycle. However, some efforts have been made towards this end as well. Paolo Nava (2009) proposed a method to quantify the environmental impact of a metal-forming process in terms of equivalent CO_2 emissions. The method includes the following four steps: (1) calculation of the mechanical work; (2) conversion of energy into CO_2 emissions; (3) quantification of the non-energy-related CO_2 emissions, such as operation-related energy consumption; and (4) optimization of the total energy consumption on the CO_2 emissions. The author used this method to quantify and minimize the carbon emissions in an open die cold upset forging process to produce a 6061 aluminum alloy disk. In the first step, work done for shearing the raw material was calculated by an approximate analytical model and work done during compression was calculated using FEM. In the second step, the energy-related CO_2 emission was calculated as the summation of individual contributions from shearing and compression, i.e., $CE_S + CE_F$. For this, a parameter known as carbon equivalent signature has been defined. In the non-energy-related CO_2 emissions, global warming potential (GWP) of each lubricant (CE_{LUB}) was considered. The overall CO_2 emissions, $CE_S + CE_F + CE_{LUB}$, from the three steps are optimized in the last step. The results show that to forge a 6061 aluminum alloy disk with the final height of 6 mm and diameter of 30 mm, the worst scenario is that the initial height of 7 mm and using mineral oil will produce a CO_2 emission of 157.4 g, while the optimized case is that the initial height of 10.95 mm and using used cooking oil ester will produce a CO_2 emission of 136.1 g, which is a reduction of about 14% in CO_2 emissions.

Extrusion of Al alloy 7003 with varying Mg content is a crucial task in terms of extrudability and energy consumption. The experimental analyses of Jo et al. (2002) reveal that Mg content (in wt%) have an adverse effect on the extrusion pressure and total energy consumption. The total energy consumption is increased from 16,639 to 16,652 MJ/t with increasing the Mg content from 0.5% to 1% in AA7003. Moreover, for a particular Mg content, the total energy consumption (1) decreased in the case of indirect extrusion as compared with direct extrusion and (2) decreased with a decrease in extrusion rate from 10 to 5 mm/s. It is also clear from the analyses that the CO_2 emission can be minimized from 225 to 194 kg when induction heating system is used instead of a heater with light oil. Thus, lower Mg content, indirect extrusion, lower extrusion rate, and induction heating system are the requirements of sustainable extrusion of AA7003 material. Modification in manufacturing process will also reduce energy consumption and CO_2 emissions. A helical pinion gear was manufactured by cold extrusion process, instead of traditional machining route. Yun et al. (2014) followed the LCA technique as per ISO14000 and equation 2.1 to quantify sustainability in the gear manufacturing operation:

$$G_j = \frac{\sum_i (m_{ij} \times e_i)}{p_j}, \tag{2.1}$$

where G_j is the quantity of carbon emissions generated by fuel usage during part manufacturing, m_{ij} is the fuel and energy used to manufacture products in a year, e_i is the fuel emission factor, and p_j is the yearly production of products. About 25% and 49% of energy consumption for the single- and double-type extrusion processes, respectively, has been attained in comparison with the traditional machining process. The billet preparation for making gear through machining consumed about 9 MJ of energy, while for extrusion it is about 6–7.3 MJ. The CO_2 emission was also minimized by 40% by adopting cold extrusion route. Because of strain hardening, the hardness of the final part has enhanced by 35% in the case of cold extrusion. Reduction in energy consumption and emission, and improvement in strength of the final part are examples of sustainable metal forming.

In the past decade, one of the popular approaches is to use the strategy of "virtual metal form-ing" to design, analyze, and optimize the metal-forming processes in a virtual world, instead of physical experiments or real shop floor testing. This not only reduces the energy consumed in physical tests but also reduces the costs and curtails the time to market, thereby making the metal-forming process environmentally and economically sustainable. The virtual metal forming using computer-aided engineering simulations has been widely employed to investigate the complicated metal-forming process, which takes into account all the input variables listed earlier. FEA, as a tool, plays an important role during such activity. This will help us in lowering the environmental impact from resource extraction and energy consumption. Several real industrial examples that demonstrate the impressive achievements are given here. Hadley Group, UK uses FEA to reduce the number of the traditional physical testing to achieve the sustainable cold rolling forming process for the improvement of the rolling product quality (English, 2013).

Hadley Group, UK developed the UltraSTEEL process, a patented cold roll forming process (dimples are introduced on the sheet before cold roll forming, Figure 2.2) to produce the strip steel for building and construction industries. In the past years, in order to understand and improve the mechanical and structural properties of the strip steel produced by the UltraSTEEL process, Hadley has to invest in a complete set of tooling, costing at $30,000–$150,000, to produce proto-types and to perform physical testing. This is expensive in terms of dollars and is eco-unfriendly as more physical prototype testing means more steel materials wasted and more energy consumed in prototype testing. Due to the complicated geometry formed by the UltraSTEEL process, theoreti-cal analysis and calculations cannot be used to accurately predict the performance and behavior. Finally, Hadley successfully addressed this challenging task by using nonlinear FEA to analyze the UltraSTEEL process. Such a green and sustainable metal-forming practice has resulted in the increase of $4 million in sales (English, 2013).

Another example is that Pilsen Steel, Czech uses FEA to eliminate the multiple heating–cooling–reheating cycles in physical testing to achieve the sustainable forging forming process for the ingots cracking issue in a forging operation (Tikal, 2013). Pilsen Steel, Czech Republic, a leading producer of castings, ingots, and forgings, had an issue with ingots cracking in a forging process. Pilsen Steel was to cool the ingots in water to between 500°C and 600°C after casting. The ingots were then placed in the forging furnace at 1,100°C–1,200°C (Figure 2.3). During forg-ing, multiple cracks were found in the ingots. In order to investigate the origin of the formation of longitudinal cracks in 34CrNiMo6 steel ingots in the forging process, both the ingots heating process and forging operation were needed to be investigated, which would require multiple times of heating–cooling–reheating cycles. In addition to the expensive nature of the trials, the energy consumption could be significantly high. Pilsen successfully used nonlinear FEA to analyze the thermal stress in the ingots during heating and to predict the parameters for the soaking process

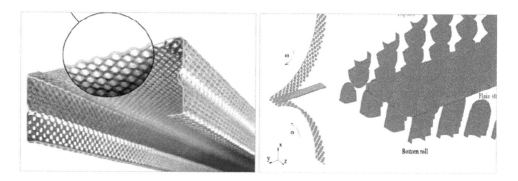

Figure 2.2   Strip steel produced by the UltraSTEEL process. (With permission from English, 2013.)

Figure 2.3 34CrNiMo6 steel ingots. (With permission from Tikal, 2013.)

used to cool the ingots after casting before placed in the furnace. The ingot's cracking issue was successfully solved by increasing the temperature of the ingot before the forging process and a significant heating/reheating energy was saved during the analyses, in which FEA plays an important role in the sustainable forging process (Tikal, 2013).

The importance of virtual metal-forming analyses can also be understood from the FEA of 3D transfer die simulation, and stamping tool design. Transfer die problems and relevant solutions cause disruption on shop floor production and delays that affect profits in long run. By using FEA of transfer die problems, tool fabricators and stamping analysts can troubleshoot and successfully run the transfer die stamping trials before the real production. In the 3D transfer die simulations, kinematic analyses can be conducted, crashes can be avoided, and strokes per minute and transfer motion can be optimized. Using such CAE facility, a supplier of metal stampings has improved the productivity rate, and another supplier has increased 11–15 parts per min by optimizing the motion curves and timing (Hansen, 2015). FEA software used for stamping analyses is capable of predicting the shape and size of initial sheet, sheet failures such as thinning, wrinkling, necking, etc., die/tool design like drawbead location and addition, blank holder and addendum modifications, friction, and lubricant requirements, etc. (Hedrick, 2002). It can even simulate the actual crash event and predict the final strength of the part. In many case studies, one step FE simulation gives quick and acceptable answer and solves problems well before real manufacturing (Goodmonson, 2010; Hansen, 2015).

2.3 ASPECTS OF GREEN FORMING

A number of aspects in material forming can be taken into account in minimizing the impact to the environment. The sustainable and green material forming can be achieved by using the following factors:

- Reduce energy consumption in each forming stage by reducing the preheating temperature, using warm forming instead of hot forming whenever possible, and well controlling the thermal energy released to the atmosphere.
- Reduce energy consumption by reducing the total number of forming stages.

- Reduce energy consumption by decreasing the forming load through uniform straining using optimum product/die geometries, lubrication to reduce friction, etc.
- Reduce energy consumption by reducing/eliminating trimming and finishing processes.
- Aiming for near net shape forming so that energy consumed in the post-forming operations can be minimized.
- Reduce material wastes by reducing redundant work, such as for flash formation.
- Reduce water usage.
- Reduce nonrecyclable/nondegradable materials in the forming processes, such as using biodegradable vegetable oils instead of petroleum-based lubricants when possible.
- Reduce waste and nonconforming products through quality enhancement.
- Reduce or eliminate rework on defects.
- Reduce handling and excessive movements and transportation inside and outside the facilities.
- Reduce the use of inappropriate processes and high energy-consuming technologies

The latter approaches are from the classical lean technology through which the metal-forming industry can be heavily enriched. It is obvious that on top of direct approaches on transforming the metal-forming system into green, the lean adaptation is capable of enhancing the outcome further through a "system" based approach.

2.3.1 Energy Consumption in Hot, Warm, and Cold Forming vs. Green Forming

The energy consumption throughout the material-forming operations has a significant effect on the green and sustainable material forming. Many different forms of energy, such as thermal energy, mechanical energy, and electric energy, are used in material-forming processes. The energy consumed in the forming operation essentially includes (1) the thermal energy in preheating the stock/billet, (2) the work done against yield stresses to achieve plastic deformation, (3) the work done against friction, and (4) indirect energy consumption in tool preparation and handling. The total energy consumed in a forming process is the summation of all these different forms of energies. In addition, the actual energy consumed would also be greatly affected by how the rolls, dies, and presses are set, types of presses, etc.

In general, the energy consumed is significantly dissimilar in different forming conditions. In hot forming, such as hot forging, hot rolling, and hot extrusion operations, the thermal energy consumed in preheating the stock/billet to a temperature above the recrystallization temperature of the material accounts for a major portion of the total energy consumption, but the work done during deformation is significantly reduced due to softness of the stock. On other hand, in cold forming, extra thermal energy in preheating is not a prerequisite, but relatively a large amount of mechanical/electrical energy is consumed to achieve plastic deformation. The energy consumed in warm forming for both heating the stock and the plastic deformation is at intermediate level, and therefore optimization of the total energy intake should be critically attempted rather than focusing on energy consumed on one of the aspects.

The form of energy used in the metal-forming process acts as a constraint for the journey toward green. In hot forming processes, almost all the thermal energy used during preheating process is released back to the environment ultimately. Therefore, use of warm forming or cold forming instead of hot forming whenever possible will greatly eliminate the thermal energy in billet preheating and the thermal energy loss to the atmosphere. However, in cold forming, due to the increased use of mechanical energy, a direct energy dump to the environment is not a concern, but a similar level of indirect environmental effect would have created during the energy generation. In addition, the recrystallization does not occur in warm or cold forming, and thus subsequent heat treatment which is needed to restore material properties is eliminated as the original material properties are not greatly distorted. Therefore, the total energy input to the operation is reduced and a greener forming process is achieved. Moreover, in hot forming, the metals become strain-rate sensitive, and high forming speed will result in surface defects. Optimizing the forming speed for a given reduction ratio in hot forming becomes a crucial activity. The objective of sustainable forming is achieved by carefully controlling the stock temperature as to optimize the material flow. For example, a research project undertaken by one of the authors for American Axle & Manufacturing Co in Detroit, USA revealed that the temperature of the steel billet can be lowered by carefully re-designing the dies. The research indicates that it was possible to reduce the number of forging stages, and still to fill the dies while reducing the temperature of the billet, which results in the lower energy consumption and the better product quality. In another example, during cold extrusion, the die design was optimized using FEA to reduce the load requirement, finally reducing the energy consumed for extrusion (Joun and Hwang, 1993). Two different die profiles, (1) die profile represented by small line segments, and (2) die profile described by cubic spline, are used for optimization. With the optimum die profile, strain distribution was made uniform and energy consumed during cold extrusion was minimized.

However, forming load in cold and warm forming will be much higher than that in hot forming, meaning that more mechanical/electric energy will be used in the cold and warm forming operations to achieve plastic deformation. In addition, some annealing may be needed between cold forming steps, meaning that some additional thermal energy will be consumed in cold forming, which is not a requirement in a hot forming. Therefore, there is a trade-off, and the increased heating energy intake in hot forming process and increased mechanical energy in cold forming process should be carefully balanced to achieve total energy optimization. Often the trade-off is decided by material selection. Some metals like Ti are formable at elevated temperature to make useful products. In such situations, optimum design of cooling ducts becomes prominent. Hot stamping of quenchable boron manganese alloyed steel 22MnB5 of 1.75 mm thickness was attempted by Hoffmann et al. (2007). FE-based thermal analyses of two optimized duct geometries (diameters), small and large cooling ducts, reveal that cooling rates of about $40°C/s$ with smaller duct and $33°C/s$ with larger duct, which are greater than the desired minimum cooling rate of $27°C/s$, are possible. The smaller duct provided better cooling rate. The true benefit to the environment can be estimated by the virtual metal-forming analysis, like the examples quoted above, in terms of equivalent CO_2 emissions.

2.3.2 Number of Forming Stages vs. Green Forming

Reducing the number of forming stages will significantly reduce the total energy consumption because the energy consumed in the extra forming stages and the cumulative thermal energy released to the atmosphere (energy waste) between consecutive forming steps will be eliminated. Furthermore, the need of heating between forming states will also diminish. In addition, reduced number of forming stages also means the reduced handling effort, the possible savings through the overheads of the forming process, and the increased tool life in terms of number of products manufactured. Additional benefits associated with social sustainability such as less exposure to forming presses enhances health and occupational safety concerns. A typical example of reducing the number of stages of forging operations by converting warm forging to cold forging is done by Ku and Kang (2014) during the manufacture of outer race of constant velocity joints. The nine-stage multistage warm forging process is converted to a six-stage forging process that combined the backward extrusions as one operation, and the necking, ironing, and sizing are combined into another operation. Thus, the proposed cold forging consists of six operations only. The final process design and tool life analyses using FEA proved that the new process is capable of manufacturing the outer race within acceptable tolerances.

2.3.3 Amount of Plastic Deformation

Energy required for plastic deformation and the process temperature are interrelated: the higher the forming process temperature, the more thermal energy consumed in preheating, the more heat energy lost to the atmosphere. But lesser the forming load, lesser the mechanical/electric energy required for plastic deformation and longer life of forming equipment. The objective is to optimize the total energy consumption, including thermal, mechanical, and electric energies. This will be challenging to all the classic analytic modeling and analysis techniques. FEA and virtual metal forming can be used to investigate the interactions and the relationships between these variables (temperature and deformation) and the objective function (the minimum overall total energy consumption). In general, energy consumption in rolling, forging, extrusion and drawing can be reduced by reducing the forming load, and optimizing the product/die geometries. FEA and soft computing methods have been widely used to improve die geometries and product geometries to achieve the green material-forming goal. Just to quote an example for improving the efficiency of product manufacturing, a two-stage tube drawing process for the manufacture of straight-type cowl cross bar of automobile was analyzed by FEA (Kim et al., 2014). It has been found that the reduction ratio that decides the amount of plastic deformation exceeded the 30% allowable limit in each stage. Increasing the number of stages was not recommended because of economic concerns. Hence, a process modification that involves an external compressive force was recommended and later examined using FEA. For the same reduction ratio, a compressive force of 10% of forming load produced a defect free cross bar, but with 4% thinning at a particular location of the tube. By increasing the compressive force, the thinning is reduced to about 0.6% yielding a defect free part in two stages, without modifications in reduction ratio.

2.3.4 Parasitic Energy Loss due to Friction

Friction is a critical factor that greatly affects the material-forming operations and has a significant effect to green forming practice due to parasitic energy loss and the use of lubricants. The energy consumed to overcome the friction between billet and forming tool contact surfaces is largely a redundant work and parasitic energy loss. Friction must be carefully controlled because too high or too low friction results in defects in the finished products. Lubrication is typically used in reducing the friction in forming processes. Liquid lubricants, such as mineral oils, can effectively reduce friction, but not suitable in hot and cold forming because most of the lubricants chemically breakdown and lose lubrication properties at high temperatures, and also tend to breakdown in high pressures. In these cases, solid film lubrication is highly appropriate. In certain forming processes, such as warm forming, it may be possible to substitute harmful petroleum-based lubricants with more environmentally friendly counterparts, such as biodegradable vegetable oils, coconut or palm seed oils, animal fat, and used cooking oil. It is worth of undemanding the environmental effects due to the excessive use of lubricants.

GWPs of some lubricants are described by Dettmer (2006) and Nava (2009). The GWP is the quantity of equivalent CO_2 emissions connected to the life cycle of the lubricants including resource extraction, manufacturing, and discarding. It is defined as grams of CO_2 equivalent per gram of lubricant. By knowing the GWP, the CO_2 emissions due to lubricant use (CE_{LUB}) can be assessed as follows:

$$CE_{LUB} = m_{LUB} \times GWP_{LUB}, \qquad (2.2)$$

where m_{LUB} is the mass of the lubricant used and GWP_{LUB} is for the lubricant.

It is evident from the data given by Nava (2009) that out of all the lubricants, mineral oil (Castrol) has the highest GWP of 3.08 g CO_2/g_{LUB} and used cooking oil ester has the lowest of about 0.6 g CO_2/g_{LUB}, while other lubricants such as rapeseed oil ester, palm oil ester, or animal fat ester, fall in between. On the other hand, rapeseed oil ester and palm oil ester are costlier, valuing about 3.3 euro/kg, as compared with mineral oil (Castrol) costing about 1.5 euro/kg.

Nava (2009) also compared the environmental impact and frictional properties of two eco-friendly lubricants namely palm oil ester and used cooking oil ester, and Castrol mineral oil in the ring compression test of Al6061-T6. It is found that used cooking oil ester and Castrol mineral oil deliver a friction coefficient of 0.14 and palm oil ester of about 0.16. A small difference in the tribological performance between them has been revealed.

Baosteel Stainless Steel Co., Ltd., Shanghai, China achieved stable batch production of ferritic stainless steels for automotive exhaust systems using the tandem cold rolling mill (TCM), instead of the Sendzimir mill (www.worldstainless.org/). In order to replace traditional rolling oil, Baosteel optimized the ingredients in the emulsion to reduce the potential surface scratches on the strip surface and also optimized the cold rolling process to control the profile during processing. Baosteel finally achieved the increased production efficiency, the reduced energy consumption, the better surface quality, and the properties of the product. Baosteel has successfully transferred this technology into the production of other super-pure ferritic stainless steels (www.worldstainless.org/). In the recent past years, use of nanoparticles in lubricants is of interest to many research groups. Optimum nanoparticle concentration will yield better friction and lubrication characteristics. Zareh-Desari and Davoodi (2016) studied the lubrication performance of vegetable oil based nanolubricants for metal-forming applications. Rapeseed oil and soybean oil with SiO_2 and CuO nanoparticles at varying concentrations are compared during ring compression tests of pure copper and Al6061 samples. It is observed that 0.5–0.7 wt% concentration of nanoparticles is the optimum level for achieving minimum friction coefficient in all the cases. Moreover, the deformation load is reduced significantly when nanoparticles are added as compared with traditional vegetable oils. Reduction in deformation load is an indication of reduction in energy consumption, and sustainable manufacturing. Kim and Altan (2008) evaluated the performance of five different lubricants based on their maximum load requirement during strip drawing tests. The sheet used is a dual-phase steel (DP 590/600), with two different Zinc coatings—galvannealed and galvanized. Their performances are tabulated in Table 2.1. The polymer-based lubricant with EP additives and water-soluble dry film lubricants are found to be most effective in all the situations. Therefore, selection of lubricant and additives (if any) plays a vital role in deciding the forming performance.

Nanayakkara (2007) carried out extensive tests with water-soluble lubricants manufactured by FUCHS lubricants and perceived significant improvements in friction properties while ensuring lesser environmental impacts.

2.3.5 Minimizing Material Waste toward Green Forming

Elimination of material waste is another milestone along the journey toward green forming. The material scraped at the end of the forming process is a significant waste and has a significant effect

Table 2.1 Performance of Lubricants from Strip Drawing Tests on DP Steel (Kim and Altan, 2008)

Lubricant	Performance from Strip Drawing Tests			
	Uncoated Die		PVD Coated Die	
	Galvannealed Sheet	Galvanized Sheet	Galvannealed Sheet	Galvanized Sheet
Polymer-based with extreme-pressure additives	Most effective	Most effective	Most effective	Most effective
Water-soluble dry film lubricant	Most effective	Most effective	Most effective	Most effective
Water-free dry film lubricant	Not acceptable	Most effective	Reasonable	Acceptable
Synthetic	Not acceptable	Not acceptable	Reasonable	Acceptable
Straight oil	More friction	Reduced friction	More friction	Reduced friction

on environmental sustainability, economic sustainability, and social sustainability, as all of these wasted materials, such as flash in forging, have gone through all the forming processes and have consumed a substantial amount of energy in preheating and plastic deformation, in addition to the increased costs to the business, increased waste to human living environment and health. In other words, the elimination of material waste will realize more products to be manufactured using the same volume of raw materials with the same amount of energy. Eliminating or minimizing material waste can be achieved by the use of optimum billet size. Such analyses using experiments are time consuming and resource intensive. In these situations, FEA will be helpful. Khaleed et al. (2011) analyzed the manufacturing of front and back hubs of an autonomous underwater vehicle by flash-less cold forging and experimental validation has been done. The volume of the initial billet has been optimized by modifying the diameter of the billets using volume constancy principle between final forged product and initial billet. Underfilling and flash are the two defects that are monitored by FEA. Such FEA yielded an optimum billet diameter of 27 mm and length of 26 mm in the case of front hub, and 28.2 mm diameter with a blind hole diameter of 5 mm in the case of back hub. The thermal analyses also reveal that the optimum billet size is acceptable in terms of maximum temperature achieved. The products are made successfully by experiments as well. Similar attempts have been made to optimize the sheet raw material for reducing material wastes and enhancing die life during sheet stamping. Many computer-aided methods and algorithms are developed for raw material optimization. Nye (2000) developed an exact algorithm based on Minkowski Sum concept to optimize the strip orientation and strip width. Such algorithm would be helpful in optimum nesting, finally minimizing the sheet wastes.

2.3.6 Indirect Energy Consumption in Material Forming

The energy consumed in material forming also includes the different forms of energies spent on material handling, moving, transportation, etc., in material-forming production operation. Those indirect energy consumptions do not directly contribute to the material plastic deformation, but may represent a significant amount of energy consumed in the production operation, which has an impact to the green forming goal. This portion of energy consumption can be minimized by proper process design, effective layouts, and enhanced scheduling in the design stages and the press shop to achieve the green objectives in material-forming processes.

2.3.7 Continuous Improvement of Quality toward Green Forming

Quality is a growing concern in manufacturing today not only to satisfy consumers and end users but also due to other issues leading toward sustainability and environment. After the world war, Japan gave a different dimension to quality. With the help of Edward Deming, Genichi Taguchi, and other experts, the field has been enriched. Furthermore, one of the leading manufacturers of automobiles, Toyota, has focused on their customized manufacturing concepts, for example, Toyota Production System (TPS) [X2]. Through the approaches like Lean, Six Sigma, etc., elimination of overall wastes of the process was engrossed and has gained huge advantages on cost of operation to become competitive. The outcome has been tremendous looking at these approaches through the view of energy conservation. Quality is primarily elimination of wastes; however, indirectly it is the saving of resources which would have been consumed by the defects. In metal forming, one of most contributive operations in automotive manufacture is following green manufacturing. Continuous improvement, or in technological terms *kaizen*, covers all the aspects of improvements, including defects and resource utilization leading to sustainability. Therefore, it is apparent that by adopting novel concepts such as Lean, Six Sigma, and Continuous Improvement, sustainability in the metal-forming industry can be enhanced.

2.4 GREEN ROLLING

Rolling consumes significant energy. Green rolling is one of the most important aspects in the sustainable material forming. Any energy efficiency improvement in rolling can produce a significant benefit to the environment and business. Outokumpu Oyj, Tornio, Finland (www.worldstainless.org/) sets a long-term target in its energy efficiency and low carbon program in 2010, to reduce the Group's emission profile by 20% by 2020. Outokumpu's facility in Tornio is well known for its modern production facility and unique energy efficiency. Since late 2011, Outokumpu's Tornio Works converted 50 separate refrigeration compressors in the cold rolling plant to a centralized cooling system, and installed new cooling towers and absorption coolers to utilize process waste heat as primary energy. Such a modified, novel system resulted in 11 gigawatt hours' (GWh) reduction of the annual electricity consumption from 15 to 3.9 GWh and the reduction of 6,700 tons of CO_2 emissions. Another example is that Yieh United Steel Corporation (YUSCO) Kaohsiung, Taiwan, uses natural gas in place of fuel oil to reduce both energy consumption and CO_2 emissions from its steam system of the cold rolling processes (www.worldstainless.org/). YUSCO achieved the reduction of the proportion of fuel oil burned for the total heat demand from 89% in 2009 to 67% in 2011, and a 22% increase in the proportion of natural gas (clean energy) in 2009–2011, which results in the reduction of 3.3 million kg of CO_2 emissions. In addition to energy consumption, parasitic energy loss, material waste, etc., are crucial in rolling as well. Some critical facts related to the green rolling are specifically discussed below.

2.4.1 Thermal Energy Consumption in Rolling Processes

It is a practice that the initial steps of the rolling process are carried out as hot rolling and the final/finishing stages are usually done in cold condition. If the desired deformation is predetermined, some of the intermediate stages may be carried out as warm rolling. In hot rolling, preheating and heat loss to the atmosphere has a significant effect on the green rolling practice. Heat loss to the atmosphere should be minimized through proper process sequencing, layouts, and delays in handling, etc. However, in warm rolling and cold rolling, these thermal energy consumptions can be optimized or eliminated. In addition, several other facts also should be taken into account in the optimization of the total energy consumption, such as cooling rates, the need of heat treatment, and the heat absorption to the rolls and presses. The case study analyses done by Worrell et al. (2001) on the energy use and CO_2 emissions in US steelmaking industry till 1994 reveal that hot rolling takes about 157 PJ of fuel energy, 34 PJ of electrical energy, 191 PJ of final energy, 263 PJ of primary energy, and 3.7 MtC of CO_2 emissions, which falls next to iron making and boilers in integrated steelmaking. On the other hand, cold rolling and finishing takes only 43 PJ of fuel energy, 15 PJ of electrical energy, 58 PJ of final energy, 89 PJ of primary energy, and 1.3 MtC of CO_2 emissions. During integrated steel production, many state-of-the-art technologies were introduced to reduce CO_2 emissions. In the integrated hot rolling, CO_2 emissions reduction in kgC/t are 7.18 in hot charging, 3.59 in hot rolling mill, 8.38 in recuperative burners, 1.91 in furnace insulation, 3.95 in controlling oxygen levels and VSDs on combustion air fans, 0.39 in rolling mill drives, and 0.46 in heat retrieval. Similarly, a considerable reduction of about 10 kgC/t of CO_2 emission is observed in integrated cold rolling and finishing (Worrell et al., 2001).

2.4.2 Amount of Plastic Deformation and Roll Forces

As discussed earlier, energy required for plastic deformation and the process temperature are interrelated: the higher the rolling process temperature, the more thermal energy consumed in preheating and lost to the atmosphere, but the lesser rolling forces on rolls and the lesser mechanical energy for plastic deformation. The objective is to optimize the overall energy consumption. In general, energy consumption can be reduced by reducing the roll forces and optimizing the product/roll geometries. In addition, reducing roll forces in hot working will reduce roll deflection and increase the roll life. Under high roll forces, the roll stand may deflect to such an extent that the roll gap can open up significantly. As a result, the roll gap should be reduced to compensate for the deflection. Some of strategies to reduce roll forces (Kalpakjian and Schmid, 2009), which will mainly lead to process efficiency and less energy consumption, are given as follows:

- Effective lubrication to reduce friction at the deforming zone.
- Using small-sized rolls to reduce the contact area for narrow billets. However, roller deflection could have an adverse effect. In addition, guide rollers could be employed.
- Considering lesser reductions/roll pass to reduce the contact area. However, it will increase the number of passes to attain the required reduction.
- High-temperature rolling to soften the material. But the additional thermal energy input will trade off some mechanical energy savings. The optimal condition should be determined by minimizing the total energy in both the thermal and mechanical energies.
- Applying tensions on the entry and exit of the strip, but this adds additional work to the process. The optimal condition should be determined by minimizing the total energy in both the additional work due to added front/back tensions and the work savings due to reduced roll forces.

Roll life also depends on selection of roll material and roll manufacturing methods. In hot rolling mills, untimely failure of rolls is a major concern as it critically affects the mill operation and production schedule. Maintenance of roll and rolling mills and analyses of such failures are time consuming and are inevitable for extending the service life of rolls. Palit et al. (2015) investigated two roll failure case studies: (1) roller in the fifth stand of a hot strip mill made of Indefinite Chilled Double Poured (ICDP) cast iron failed in the neck region, and (2) roller in the sixth stand made of the same material failed because of subsurface defect. In the first case, the microstructural analyses revealed segregation of ferrite around graphite flakes forming large clusters in pearlite matrix. This would have originated through improper roll manufacturing process. Such clusters would have reduced the strength of the roller. In the second case, subsurface defect got revealed after the roll diameter was reduced to 566.5 mm from original 620 mm. The defect was a lumpy entrapment in the shell and its chemical composition is very different than the neighboring regions. This is again claimed to have originated from the inappropriate roll manufacturing process. By proper selection of roll material, appropriate roll manufacturing methods, reducing forming load, etc., the energy consumed to fabricate new rolls can be conserved due to less need of replacement of rolls.

2.4.3 Lubrication to Reduce Friction in Rolling

Lubrication is typically used in reducing the friction in rolling processes. Kalpakjian and Schmid (2009) have stated that the 20% of the forces applied in rolling process is expended to overcome friction. Therefore, the use of lubricants is evident in rolling process. As discussed earlier, liquid lubrication, such as mineral oils, can be used for effective reduction of friction. However, high temperature in hot rolling and high pressure in cold rolling makes such lubricants not suitable in hot and cold rolling. In order to secure lubrication to reduce friction, and in turn to reduce rolling work, solid film lubrication is highly appropriate. In general, the working temperature affects the interface friction and evolution of strength of the rolled strips.

Making aluminum strips involves three rolling stages. In the first stage, an ingot of several hundred millimeters thick is reduced to about 30 mm in a hot reversing mill. The ingot is rolled 15–20 times in the reversing mill. The lubricants used in hot reversing mill are formulated to generate high friction to reduce the refusal cases, at the entry of the roll. These lubricants are very stable emulsions containing petroleum sulfonate, and nonionic emulsifiers with low levels of film strength additives, additives like oxidation inhibitors, corrosion inhibitors, and antifoaming agents. In the second stage, using a hot continuous tandem mill, the sheet is rolled to 2–5 mm thickness, typically through two to six stands. For this stage, lubricants containing petroleum oils and either anionic or nonionic emulsifiers with high levels of triglyceride film additives are used. The oil concentration for both the stages in the lubricant is about 5%. These lubricants have shorter lives because of hot rolling conditions and careful monitoring is required. The coil is then annealed and allowed to cool. After this, in the third stage, cold rolling is done to reduce thickness to 0.3–0.9 mm. Cold rolling-annealing cycle is possible to reduce the hardening effect. In cold rolling stage, straight petroleum oils with small amounts of fatty acids, esters, alcohols are used as lubricants. In the hot rolling stages, the lubricants have 100 mm^2/s viscosity, while it is 3–7 mm^2/s in cold rolling.

2.4.4 Arrangement of Rolling Mills vs. Green Rolling

The arrangement of the rolling mill directly affects the energy intake to achieve the desired product. Three main arrangements of the rolling mills are reversing mills, high rolling mills, and continuous mills. Each has its own advantages and disadvantages in terms of production rates, costs, and energy consumption. Energy consumption in each of the processes are dependent on the setup conditions, amount of plastic deformation, and the condition (hot, cold, or warm) of the stock being rolled.

In a reversing mill, at each pass, the rolls have to come to a momentary stop and then accelerate backward to their regular rolling speed, which causes a significant kinetic energy loss due to stopping, significant addition of mechanical energy due to acceleration, thermal energy loss due to time delay in reversing direction operation, and an indirect energy loss due to additional handling required. In addition, reversing mills have limitations associated with the length of the billet, but have great advantages in the use of resources, i.e., rolls, space, etc. Thus, it is ideal for the product with low production rates and shorter lengths of work pieces.

High rolling mills have fewer number of rolls as compared to continuous mills. However, in continuous mills, the handling effort is reduced and the rolling process happens in a seamless manner, so it is ideal for mass or a flow manufacture of long rolled products. On the other hand, cluster mills can well balance roll deflection to obtain uniform rolled cross sections and therefore, to reduce scraps and rework, and thus to indirectly reduce net energy intake. Since the arrangements are dissimilar and the energy is consumed in a different way in each of the above four arrangements, the process design, layouts, and systems planning for the optimization of the energy consumption should be done on a system level, rather than a local operation or stage level, in order to achieve the best arrangement for the rolling and to achieve the green rolling goal.

2.5 GREEN FORGING AND EXTRUSION

In addition to energy consumption, parasitic energy loss, material waste, etc., discussed in earlier, the flash formation is of special concern in green forging practice. In a closed die forging, the excess billet materials (flash) is squeezed out through the parting off surface of die halves. This flash will be subsequently trimmed and followed up with a finishing process. The flash results in the material waste and the extra energy consumed in forming and trimming the flash as well as in finishing the product. The total energy in deformation is a function of both stress and the strain. Managing the strain effectively is critical to eliminate redundant work in the process. Straining can be managed through proper die design. Many techniques have been used to optimize the forging process and die design, such as smoother die corner radii and smoother die walls to allow easy flow of materials through the cavities, and to minimize the flash and energy. Ranatunga et al. (2001) used upper-bound elemental technique to optimize the billet geometry during disk forging of a rotor like turbine component. The die filling and extrusion of flash are optimized for successful component making with optimum extrusion height. A scaling factor for die fill and extrusion height has been presented. A typical case study has been demonstrated. Similar attempt made by Shahriari et al. (2008) using FEA for optimizing the finisher die forging of a super alloy reveals that the flash height and outer draft angle influence the load requirement significantly, while other parameters like web thickness and inner draft show moderate effect, and inner fillet and outer fillet radius show insignificant effect. A larger flash height and outer draft yield lesser load for forging the part. Lubrication is a typical way of optimizing forging loads. Fellers and Hunt (2001) have emphasized the use of soaps and silicones and polytetrafluoroethene-based low friction lubrication in reducing energy loss. Therefore, minimizing the flash by the process and die optimization will not only achieve a sustainable forging process but also the cost reduction and quality improvement.

Similar efforts are made by researchers to optimize extrusion outputs as a function of material, temperature, lubricants, and die design. Extrusion is typically carried out in hot condition, but few products such as automotive axles can be cold extruded using vegetable-based oils. A refined palm oil can be a potential lubricant for cold extrusion of pure aluminum as compared with paraffinic mineral oils. Reduced load, good part surface finish, and uniform strain distribution are the characteristics of extrusion when the palm oil is used for extrusion (Syahrullail et al., 2011). Furthermore, the use of the warm forming approach—warm extrusion will reduce the net energy intake significantly. Because of temperature effect, the lubricant used for cold extrusion may not be appropriate for warm and hot extrusion. A separate optimization analyses should be followed for the purpose. In addition, the use of dry film lubrication as illustrated earlier is also applied in extrusion process.

2.6 GREEN WIRE DRAWING

In drawing operation, the total energy consumed is the aggregate of thermal energy required for heating the stock, mechanical energy transferred for deformation and overcoming friction, and indirect energy inputs accounted for handling and feeding the wire and tool (die) preparation, etc. The similar considerations and techniques discussed earlier can be used to manage the thermal energy required for heating the stock in a warm drawing and the lubrication (e.g., dry films) to reduce friction for the green drawing operation. In addition,

- Drawing life of the dies has a critical importance. When excessive reduction ratios are achieved, it negatively affects the tool life due to excessive rubbing.
- It has been found that if the draw ratio is more than 45%, the lubricants may chemically breakdown (Kalpakjian and Schmid, 2009).
- Another way of optimizing the drawing force is the bundle drawing (Kalpakjian and Schmid, 2009) in which the wires are drawn in a bundle reducing the tool cost as well. Through this, the effective drawing force may be a multiple of the number of wires, but friction and the use of equipment could be greatly saved, which reduces the parasitic loss due to friction.

2.7 GREEN STAMPING

In sheet stamping operations, sustainability is attained by adopting new deformation process, process conditions like working temperature, strain-rate, lubrication, and friction, etc., tool design, and material characteristics. Sheet materials like steel grades are generally cold stamped, while some aluminum alloys, magnesium alloys, and titanium are hot stamped as they show enhanced formability at elevated temperatures. Computations help in such optimization activities. Takuda et al.'s (2003) results indicate the capability of FE simulations to predict the desired draw ratios of SS304 during warm deep drawing conducted at 120°C of about 2.7. The martensite content prediction at various locations of cup is also consistent. El-Morsy and Manabe (2006) conducted FE simulations of warm deep drawing of AZ31 sheet with the blank holder and the die at 300°C, keeping the punch and sheet at room temperature, and in another case with all the parts at uniform temperature. The case with the assumption of uniform temperature shows early failure. The other case with different temperatures shows a successful cup forming.

An appropriate tool design minimizes stamping defects such as wrinkling, earing, springback, etc. Springback is of concern in stamping industries. This can be compensated by suitable die design methods and computations. If not minimized, springback disrupts the sheet assembly operations and creates mismatch between the mating sheets. Joinability becomes a problem then, questioning the existence of sustainable manufacturing. "Displacement adjustment method (DA)" is used in many stamping situations for springback prediction (Gan and Wagoner, 2004). In this case, the displacement of surface nodes is followed defining a new die surface in a direction opposite to springback error. The sheet is deformed to the new target surface and checked for error in dimensions. This is performed iteratively until springback error is minimized to a desired level. Springforward method, in which the sheet is displaced iteratively depending on the force equilibrium after forming, is compared. Forming operations such as arc bending, U-channel bending, compound curvature bending, and arbitrary 2D shape are done to ensure the efficiency and accuracy of the proposed DA method. The quality of results from DA method is better than springforward method.

The lubricants suggested for cold stamping and hot stamping of variety of sheet materials are available in ASM handbook (Semiatin et al., 2006). These are meant for sustainable forming. For example, for Al alloys, graphite suspension can be used during hot forming, while synthetic solutions, emulsions, lanolin suspensions, water suspensions, soap solutions, mineral oil, fatty oils can be used in cold forming. Similarly, for carbon and low-alloy steels, graphite suspension can be used during hot forming, while emulsions, soap pastes, water, fatty oils plus mineral oils, polymers, conversion coating, molybdenum disulfide, graphite in grease, synthetic solutions can be used in cold forming. The lubricant applying methods, its recovery and disposing strategies contribute a lot to sustainability.

2.8 SUMMARY

In this chapter, an overview of sustainable material forming has been presented with numerous practical examples from the industry and research case studies. The importance of part-making at room temperature and elevated temperatures, defect minimization, new process design and material selection, load reduction, etc., in bulk forming and sheet forming for the purpose of energy and waste savings are elaborated. Some of the developments in material forming are discussed separately in the forthcoming chapters.

Sustainability not only includes CO_2 emission reduction, but also waste reduction, which means cost reduction. Thus, the environmental sustainability in material-forming industry is proportional to economic sustainability in material forming. The more sustainable material-forming process, the less costs to the business. Green material-forming benefits not only the environment but also the business. The virtual material forming can be used to design, analyze, and optimize material-forming processes without much involvement of physical experiments and tests, and therefore to help reducing both thermal energy and mechanical energy to yield a sustainable manufacturing environment. As the core tool of the virtual material forming, CAD/CAM/CAE provides the integration of design, analysis, and manufacturing functions into a system. CAD/CAM/CAE not only reduces costs in the process development but also leads time from concept to design to manufacture. In addition, other routine and monotonous (but important) tasks such as the preparation of bills of materials, costing, production scheduling, etc., can be also performed automatically in the same systems.

Environmental sustainability, economic sustainability, and social sustainability should be considered simultaneously. The Design-to-Environment and Design-to-Cost should be combined to form up the Integrated Management System for environmental factors, quality, costs, and occupational health and safety to achieve a sustainable material forming.

REFERENCES

Altan, T., Oh, S., Gegel, H.L. 1983. *Metal Forming, Fundamentals and Applications.* American Society for Metals, Metals Park, OH.

Dettmer, T. 2006. Nichtwassermischbare Kühlschmierstoffe auf Basis nachwachsender Rohstoffe (Non-water-miscible cooling lubricants based on renewable raw materials). Dr.-Ing. Dissertation, Technische Universität Braunschweig, Vulkan Verlag, Essen, Germany.

El-Morsy, A.W., Manabe, K.I. 2006. Finite element analysis of magnesium AZ31 alloy sheet in warm deep-drawing process considering heat transfer effect. *Materials Letters* 60: 1866–1870.

English, M. 2013. Simulation helps increase sales by $4M - Nonlinear FEA validates new cold roll forming process. *Simulating Reality Magazine* III: 2–3.

Fellers, W.O., Hunt, W.W. 2001. *Manufacturing Processes for Technology*, Second Edition, Prentice Hall, Upper Saddle River, New Jersey.

Gan, W., Wagoner, R.H. 2004. Die design method for sheet springback. *International Journal of Mechanical Sciences* 46: 1097–1113.

Goodmonson, G. 2010. Simulation empowers toolmaking creativity. *Stamping Journal*, May/June issue, 16–19.

Hansen, M. 2015. 3-D transfer die simulation eliminates transfer die headaches. *Stamping Journal*, November/December issue, 28–29.

Hedrick, A. 2002. Taking advantage of simulation technology. *Stamping Journal*, September/October issue, 60–62.

Hoffmann, H., So, H., Steinbeiss, H. 2007. Design of hot stamping tools with cooling system. *Annals of the CIRP* 56: 269–272.

Jo, H., Cho, H., Lee, K., Kim, Y. 2002. Extrudability improvement and energy consumption estimation in Al extrusion process of a 7003 alloy. *Journal of Materials Processing Technology* 130–131: 407–410.

Joun, M.S., Hwang, S.M. 1993. Optimal process design in steady-state metal forming by finite element method-II. Application to die profile design in extrusion. *International Journal of Machine Tools and Manufacture* 33: 63–70.

Kalpakjian, S., Schmid, S.R. 2009. *Manufacturing Engineering and Technology*, Sixth Edition, Pearson, India.

Khaleed, H.M.T., Samad, Z., Othman, A.R., Mujeebu, M.A., Abdullah, A.B., Zihad, M.M. 2011. Work-piece optimization and thermal analysis for flash-less cold forging of AUV propeller hubs - FEM simulation and experiment. *Journal of Manufacturing Processes* 13: 41–49.

Kim, H., Altan, T. 2008. R&D Update: Lubrication and galling in stamping of galvanized AHSS - Part III - B-pillar simulations and the Strip Drawing Test (SDT). *Stamping Journal*, January/February issue, 14–15.

Kim, H., Youn, J., Rhee, H. 2014. Development of combined tube drawing process for straight-type cowl cross bar of automobile. *International Journal of Precision Engineering and Manufacturing* 15: 2093–2099.

Ku, T., Kang, B. 2014. Tool design and experimental verification for multi-stage cold forging process of the outer race. *International Journal of Precision Engineering and Manufacturing* 15: 1995–2004.

Nanayakkara, N. K. B. M. P. 2007. Experimental analysis of friction in forming automotive coated steels. PhD thesis, Deakin University, Australia.

Nava, P. 2009. Minimizing carbon emissions in metal forming. Master degree thesis, Department of Mechanical and Materials Engineering, Queen's University, Ontario, Canada.

Nye, T.J. 2000. Stamping strip layout for optimal raw material utilization. *Journal of Manufacturing Systems* 9: 239–248.

Ou, H., Lan, J., Armstrong, C.G., Price, M.A. 2004. An FE simulation and optimization approach for the forging of aeroengine components. *Journal of Materials Processing Technology* 151: 208–216.

Palit, P., Jugade, H.R., Jha, A.K., Das, S., Mukhopadhyay, G. 2015. Failure analysis of work rolls of a thin hot strip mill. *Case Studies in Engineering Failure Analysis* 3: 39–45.

Ranatunga, V., Gunasekera, J.S., Frazier, W.G., Hur, K. 2001. Use of UBET for design of flash gap in closed-die forging. *Journal of Materials Processing Technology* 111: 107–112.

Semiatin, S.L., Marquard, E., Lampman, H. 2006. *ASM Handbook, 14B: Metalworking: Sheet Forming*, ASM International, OH, USA.

Shahriari, D., Amiri, A., Sadeghi, M.H., Cheraghzadeh, M. 2008. Optimal closed die finish forgings for Nimonic 80-A alloy using FEM method. *International Journal of Material Forming* 1: 29–32.

Syahrullail, S., Zubil, B.M., Azwadi, C.S.N., Ridzuan, M.J.M. 2011. Experimental evaluation of palmoil as lubricant in cold forward extrusion process. *International Journal of Mechanical Sciences* 53: 549–555.

Takuda, H., Mori, K., Masachika, T., Yamazaki, E., Watanabe, Y. 2003. Finite element analysis of the formability of an austenitic stainless steel sheet in warm deep drawing. *Journal of Materials Processing Technology* 143–144: 242–248.

Tikal, F. 2013. Getting to the root cause of cracks - Heat transfer analysis helps solve tough forging problem. *Simulating Reality Magazine* III: 4–5.

Veera Babu, K., Ganesh Narayanan, R., Saravana Kumar, G. 2009. An expert system based on artificial neural network for predicting the tensile behavior of tailor welded blanks. *Expert Systems with Applications* 36: 10683–10695.

Veera Babu, K., Ganesh Narayanan, R., Saravana Kumar, G. 2010. An expert system for predicting the deep drawing behavior of tailor welded blanks. *Expert Systems with Applications* 37: 7802–7812.

Worrell, E., Price, L., Martin, N. 2001. Energy efficiency and carbon dioxide emissions reduction opportunities in the US iron and steel sector. *Energy* 26: 513–536.

www.aluminum.org/sites/default/files/Aluminum_The_Element_of_Sustainability.pdf.

www.audi-journals.com/press/en/main.html.

www.audiusa.com/us/brand/en/exp/commitment/environmental/environmental_protection_at_audi.html#source=http://www.audiusa.com/us/brand/en/exp/commitment/environmental/environmental_protection_at_audi.html&container=page.

www.audiworld.com/news/02/aluminum/content1.shtml.

www.imoa.info/download_files/sustainability/IMOA_CarBody_Chassis_Construction.pdf.

www.worldstainless.org/Files/issf/non-image-files/PDF/ISSF_Sustainability_ Case_Studies.pdf.

Yang, T.S., Hwang, N.C., Chang, S.Y. 2007. The prediction of maximum forging load and effective stress for different material of bevel gear forging. *Journal of Mechanical Science and Technology* 21: 1566–1572.

Yun, J., Jeong, M., Lee, S., Jeon, J., Park, J., Kim, G.M. 2014. Sustainable production of helical pinion gears: Environmental effects and product quality. *International Journal of Precision Engineering and Manufacturing-Green Technology* 1: 37–41.

Zareh-Desari, B., Davoodi, B. 2016. Assessing the lubrication performance of vegetable oil-based nano-lubricants for environmentally conscious metal forming processes. *Journal of Cleaner Production* 135: 1198–1209.

CHAPTER 3

Sustainability in Joining

R. Ganesh Narayanan
IIT Guwahati

CONTENTS

3.1 INTRODUCTION

Joining is a process of assembling individual parts to form a single system. From the beginning of human civilization, several joining techniques have been developed as per the requirements and suitability of mankind. Such joining methods include fusion welding processes (or traditional joining), adhesive bonding, soldering and brazing, and mechanical fastening. In the last few decades, the material consumption in various forms has greatly increased to meet several of our demands. Naturally modifications in the joining techniques were introduced and newer joining methods were developed for precise, efficient, light weight, and faster production. These predominantly include joining involving plastic deformation (otherwise solid-state welding) such as friction stir processing/welding, friction welding, self-pierced riveting, clinching, joining by end forming, electromagnetic joining, ultrasonic welding, and hybrid joining methods. Automation is also incorporated into such developments. Nowadays, more efficient use of materials and energy has become the main concern for environmental sustainability. Although newer joining methods have been developed for many years, the concern about sustainability has taken over only in the last decade. In this context, the traditional joining methods, and their advantages and disadvantages in terms of sustainability requirements, are presented initially. Later some of the recently developed joining methods are described with their contribution toward sustainable manufacturing. Several joining case studies made by researchers to make traditional and recent developments sustainable and eco-friendly are also discussed.

3.2 TRADITIONAL JOINING METHODS AND SUSTAINABILITY

Arc welding is a fusion welding process in which the materials are joined by melting with the help of an electric arc generated between an electrode and workpieces using electric power supply. The joint is made by solidification at the interface. Sometimes it is required to provide pressure during arc welding and filler materials are used to provide strength to the welded joint. Various types of joints, such as butt joints, corner joints, tee joints, lap joints, and edge joints are possible by arc welding. While most of these joints are groove welds, fillet welds are also fabricated. The quality of electrodes and filler rods constitutes a major portion in making arc welding process sustainable and green.

The electrodes can be of consumable type or nonconsumable type. In the first case, consumable electrodes act as filler metal during welding, while in the second case, tungsten electrodes are used, which resists melting by arc. The consumable electrodes are of two forms: (1) rods with length of about 250–400 mm with a diameter of 8–10 mm and (2) wires with a diameter of 1–6 mm. Using wire electrodes is sustainable as there is no need to change it regularly during welding. Wires are fed continuously into the welding zone from spools. Regular changeover is mandatory in the case of rods, as these are available in finite lengths. The consumable electrodes melts (or consumed) and added to the weld joint as filler metal. While using nonconsumable electrodes, filler metal in the form of wire is supplied separately during welding. Since high temperatures are involved in arc welding, shielding of arc, and weld pool is important in order to avoid chemical reaction with oxygen, nitrogen, and hydrogen in air. Inert gases such as helium and argon are used as shielding gases for the purpose. Arc welding of ferrous metals involves using oxygen and carbon dioxide, usually in combination with Ar or He. Fluxes are also used for the same purpose. The flux stabilizes the arc and reduces spattering other than providing protective atmosphere for weld pool. Flux will be seen in the form of scale after solidification above the weld region. Like filler material, shielding gas and flux are potential candidates for making the process greener and more sustainable.

Shielded Metal Arc Welding (SMAW), Gas Metal Arc Welding (GMAW or Metal Inert Gas arc welding-MIG), Flux Cored Arc Welding (FCAW), and Submerged Arc Welding (SAW) are the arc welding processes that use "consumable electrodes." The arc welding process that uses "nonconsumable electrodes" includes Gas Tungsten Arc Welding (GTAW) and Plasma Arc Welding (PAW). The salient features of the arc welding processes and potential candidates for improving sustainability are presented in Table 3.1. Beyond certain well-accepted advantages, due to complex vaporization–condensation–oxidation processes, various fumes and gases are generated during arc welding. The fumes contain toxic gaseous pollutants such as carbon dioxide (CO_2), carbon monoxide (CO), nitrogen oxides (NO_x), sulfur dioxide (SO_2), and ozone (O_3) that are hazardous to the workers (Popovic et al., 2014).

There is much evidence showing modifications made in the arc welding process to take care of environment concerns and sustainability. Dennis et al. (1996) presented a method of reducing the concentration of hexavalent chromium in the welding fume during metal core arc welding by the addition of active metals such as zinc, magnesium, and aluminum to welding wires containing 10% Cr. The wires containing 1% Zn show a significant reduction in fume formation rate and the concentration of hexavalent chromium that are well below the case without Zn. This performance is better than the cases with Mg and Al addition. There is an optimum voltage of 18 V at which the effect is achieved, beyond which concentration of hexavalent chromium has increased. The optimum voltage will be different for other active metals. The 1% Mg wire produced the highest hexavalent Cr formation rate over the voltage range tested. In another example, Topham et al. (2010) compared the concentration of hexavalent chromium in the fumes generated during GTAW of ER308L stainless steel using a silica precursor, tetraethyloxysilane (TEOS), and another case of using argon shielding gas. About 45% of hexavalent chromium reduction has been achieved when 3% TEOS is added to the shielding gas. The mechanism behind such effect is found to be the formation of inert

Table 3.1 Arc Welding Methods and Candidates for Sustainability

Arc Welding Process	Salient Features	Candidates for Sustainable Welding
(i) Using Consumable Electrodes		
SMAW	• Uses consumable electrode consisting of an outer filler metal coating that acts as flux and shielding • Filler material should be chemically compatible with metals welded • Most widely used arc welding method as it is less expensive and portable • Steels, stainless steels, cast irons, and some nonferrous alloys are arc welded	• Filler material • Flux as coating on filler material • Removal of scale formed by solidification of coating (flux) • Choice of suitable material for reduced power consumption
GMAW/MIG welding	• Consumable bare metal wire used as electrode • Shielding is accomplished by shielding gas • No flux is used • Argon and helium are used for nonferrous metals, while CO_2 is used for ferrous materials • Better deposition rates than SMAW • Easy to automate	• No scale formation as flux is not used • Consumable wire material • Reduction of electrode wastes as wires are used instead of rods
FCAW	• Electrode is a continuous consumable tube that contains flux and alloy elements in its core • Continuous supply of consumable tube is possible • Use of shielding gas is optional (called as self-shielded FCAW) • Hybrid of SMAW and GMAW • Primarily used for welding steels and stainless steels • High-quality welds that are smooth and uniform are formed	• Scale removal because of flux usage • Reduction of electrode wastes as wires are used instead of rods • Shielding gas use is optional
SAW	• Uses continuous, consumable bare wire electrode • A shield of granular flux provides arc shielding • Granular flux completely submerges the welding operation • Flux is solidified as a slag layer above the joint • High-quality deeper welds are possible because of thermal insulation and slow cooling • Applications in welding large diameter pipes, boilers, I-rails • Low carbon steel and stainless steels can be welded	• Extra machine features such as hopper, vacuum system for recovering granular flux may hinder sustainability development • Since the welding is submerged, preventing sparks, spatter, and radiation that are hazardous becomes viable—acts as window for sustainability and green manufacturing • Thick steel plates (of 25 mm) can be welded with good quality welds improving the process efficiency • Efficient recovery of granular flux
(ii) Using Nonconsumable Electrode		
GTAW	• Uses nonconsumable tungsten electrode and inert gases (e.g., helium, argon) for arc shielding • Filler is not mandatory • No flux usage • Predominantly used for aluminum and stainless-steel welding	• No scale formation and removal, as flux is not used • No weld spatter because no filler metal is transferred across the arc • Electrode utilization is efficient and change is not frequent
PAW	• Uses a high-velocity stream of inert gas, such as argon or argon–hydrogen mixtures, into the arc region to form a high velocity, intensely hot plasma arc stream • Any material can be welded because of constriction of arc by plasma • Good arc stability and excellent weld quality	• Handling high temperature and plasma formation makes the process costly • No scale formation and removal, as flux is not used • Good arc stability improves process efficiency and becomes sustainable

silica layer around aerosol preventing further oxidation of chromium during welding. Moreover, nitrate proportion decreased by about 53%, i.e., from 83 ppm when argon is used as shielding gas to 39 ppm when TEOS is added to the shielding gas.

Mohan et al. (2015) reported that through TiO_2 coating on the electrodes, the concentration of Fe, Mn, Ni, Si, Cr, and Ti has significantly reduced in the fumes generated as compared with the case without coating. The contribution of sintering temperature during electrode fabrication is the highest in reducing the concentration of elements as compared with the effect of molar concentration, dipping time, and quantity of poly ethylene glycol. It is also observed that to have a sustainable effect on fumes, the coating morphology, and the crystallite size of the deposit should be controlled. A harder weld zone is obtained in the case of using coated electrodes as compared with uncoated ones, which is another achievement in the direction of sustainable joining. Praveen et al. (2005) reviewed the advancements in pulse GMAW and suggested to control internal disturbances mainly the dynamic response of power source and external disturbances by implementing nonlinear relationships between welding parameters considering power source dynamics to make process efficient and stabilize arc, both aiming for sustainable and green process. A sustainable welding process can also be achieved with the implementation of quality systems using intelligent microprocessor control in combination with automatic feedback control. Pires et al. (2010) shown that GMAW variants like shielding gases and filler wire composition can be controlled to make the GMAW process sustainable and green. Because of arc stability, decrease in arc temperature, decrease in heat input, the GMAW-P, and cold metal transfer have shown a reduction in CO emissions as compared with GMAW. The effect of filler wires on fume generation rate reveals that (1) the solid wire with and without Cu coating performs almost similarly—about 2,100 to 3,000 mg/kg at 150 A current, and about 5,500 to 6,100 mg/kg at 300 A current, and (2) both the filler wires reduce the fume generation rate as compared with two other filler wires, namely Green 207 and Green 201. The shielding gas mixtures, such as $Ar+5\%CO_2+4\%O_2$ and $Ar+5\%O_2$, show reduced fume formation rate as compared with $Ar+18\%CO_2$ and $Ar+8\%CO_2$ for the filler wire Green 207.

In *gas welding* (or *flame welding*), various fuels are combined with oxygen to generate heat required for welding. The fuels used are acetylene, methylacetylene-propadiene, hydrogen, propylene, and propane. If acetylene is used, which is predominantly the case; it is called oxyacetylene gas welding. Filler rod is not mandatory, but if used, it is coated with flux responsible for protecting weld pool from atmosphere. In oxyacetylene gas welding, combustion occurs in two stages. In the first stage, the products are CO and H_2, which further undergoes reaction in the second stage, resulting in CO_2 and heat generation. The combined heat generated is about $50 \times 10^6 J/m^3$ of acetylene. In gas welding, the important sustainability and green contribution to the process comes in the form of fuel (e.g., acetylene or propane) and the type of flame (e.g., neutral or reducing or oxidizing) used for combustion. Methods of storing acetylene and materials for hoses are also crucial. The combination of acetylene and oxygen is highly flammable, and hence oxyacetylene welding is not eco-friendly and unsafe for workers.

A few initiatives to improve the sustainability of the process are presented here. Amza et al. (2013) studied the effect of type of flame on the concentration of CO, NO, NO_2, and SO in the fumes generated during oxyacetylene flame welding of steel (C%: maximum 0.17). A filler wire with composition same as that of steel is used. Table 3.2 gives the summary of results generated highlighting the influence of flame type on the sustainable production and green environment by controlling the pollutant gases generated. In the table, pollution coefficient is defined by the ratio of the total mass of materials used in the welding process to the mass of the material deposited in the weld seam. The total mass is the combination of the mass of the filler wire, mass of acetylene remaining outside the reaction in the welding bath, mass of oxygen released into the atmosphere, mass of losses in the air and in the soil, and the mass of unnoticeable losses. Thus, the selection of flame type during gas welding depends on which gas concentration to be reduced and which one is acceptable. This should be done with utmost care for maintaining sustainability.

Table 3.2   Influence of Flame Type on the Pollution Level and Gas Concentration (Amza et al., 2013)

Flame Type	Concentration (Evaluated in ppm)		
Gases Generated	Normal Flame	Reducing Flame	Oxidizing Flame
CO	Lowest	Highest	Moderate
NO	Moderate	Lowest	Highest
NO$_2$	Highest	Highest	Moderate
SO	Lowest	Highest	Moderate
Pollution coefficient	Highest	Lowest	Moderate

Muñoz-Escalona et al. (2006) reported that the best surface is produced when oxy-propane is used for cutting of a 25.4-mm-thick AISI 1045 carbon steel plate and it has more influence on hardness within a particular welding speed as compared with oxyacetylene cutting. On the other hand, oxygen consumption has increased by 45% when oxy-propane gas welding is used as compared with oxyacetylene welding. Though this is the case, oxy-fuel welding is expensive (in cost per hour) when acetylene is used as a fuel as compared with propane. To be specific, the operational cost of using oxyacetylene is 80% (app.) larger as compared with oxy-propane. Hence, the selection of fuel during gas welding not only depends on quality of welds and properties but also it depends on the operating cost and oxygen consumption. Such strategies matter a lot for maintaining sustainability.

Electron beam (EB) welding is a fusion welding process in which the heat required for welding is generated by a highly focused, high-intensity stream of electrons impinging against the work metal. The electron gun operates at high voltage to accelerate the electrons, which is of the order of 10–150 kV. The power density is very high for the process as the weld area is about 2 mm^2. EB welding can be performed at high-vacuum, medium-vacuum, and non-vacuum conditions. Except in high-vacuum welding, in which both electron generation and welding happens in vacuum, the other two cases either has partial vacuum or no vacuum for welding. Of course, weld quality has to be compromised depending on the case. Certain demerits such as high initial machine cost, precise joint preparation requirement, and restrictions associated with performing welding in a vacuum, safety norms followed because of generation of X-rays make it unsuitable for shop floor production. On the other hand, good quality welds with deep penetration and less distortion can be made with EB welding.

Laser beam welding is a noncontact-type fusion welding process in which two or more metal pieces are joined with the help of a laser beam. Filler material is not used, and shielding gases such as helium, argon, nitrogen, and carbon dioxide are used to prevent oxidation. Some of the advantages of the process include fabrication of narrow welds with high welding speeds, high energy density, and suitable for high production rate. Some of the disadvantages like difficulty in welding high reflective and high thermal conductive materials such as aluminum, copper, and their alloys, maintaining accurate weldment position, etc., should be taken care during actual applications, and the machine is expensive.

Investigations on improving the quality of EBW process and welds fabricated can be related to sustainable manufacturing and green initiation. A comparison between EBW-high vacuum, CO$_2$ laser welding, and TIG welding shows that EBW-high vacuum is more suitable for welding Ti (CP) sheets without formation of defects (Yunlian et al., 2000). Wanjara et al. (2006) revealed that by double-pass welding sequence during EB welding of 1-mm-thick Ti–6%Al–4%V, (1) the undercut of 15 μm is removed and (2) the underfill depth at the weld surface is reduced from 37 to 11 μm. Similarly, for a 3.2-mm-thick Ti alloy, the second pass has reduced the underfill depth at the surface from 73 to 38 μm. The radius of curvature at the region of undercuts is observed to increase because of the second welding pass. In the case of mechanical properties, the second pass in EBW has recovered the ductility loss after first welding pass as compared with base Ti–6%Al–4%V sheet material. The parent material has a total elongation of 14.8% which has decreased to 11%

after single-pass EBW, and later increased to about 14.5% after the second pass. But TIG welding has detrimental effect on the ductility of the base material from 11% to 3.5% after first pass, and later to 3.2% after the second pass. The reduction of undercuts and underfill, and improvement in ductility of weldments, demonstrate the weld integrity improvement indicating sustainable joining practice. A practical attempt of EBW of Ti alloy gas bottle of 17.5 mm thickness has been performed by Saresh et al. (2007) through two-pass double-side procedure to obtain a full penetration weld having consistent mechanical properties and microstructure as that of normal EB welding. This route has been demonstrated and tested for its acceptable performance. The single-pass EB welding of 17.5 mm Ti alloy did not produce appropriate weld for the said application. The mechanical properties of the double-pass EB welded joints are greater and also consistent in the weld cross-section. A uniform hardness distribution across the parent metal, HAZ, and the fusion zone has been obtained. The manufacturing route demonstrated is efficient and can be implemented in actual production schedule with the available EBW machine meant for single-pass thinner (thickness of 12 mm) materials.

In the case of laser welding, Liu et al. (2006) demonstrated the efficiency of a hybrid welding technique that combines a low-power (approximately 400 W) laser source with arc generation during welding of Mg alloy. The results indicate that (1) the joint produced by TIG welding has the least tensile strength of about 85% of the base metal, and the laser alone was slightly better than TIG (about 90% of the base material), (2) the joint made by hybrid laser-arc welding exhibited tensile strength of 95% of the base metal and more than 100% ductility as that of base metal, much larger than TIG welding, and (3) penetration of the hybrid welding joint is four times deeper than that of laser welding, while it is two times than that of TIG welding. The hybrid laser double GMA welding proposed by Wei et al. (2015) should also be noted for its contribution toward improvement in homogeneous alloying elements distribution and weld metal microstructure. Such improvements in the joint fabrication have an significant impact on the mechanical properties in a positive manner. This has helped in the betterment of sustainable manufacturing reducing the laborious work and unsafe work environment.

In *resistance spot welding (RSW)*, the heat generated by resistance to current flow at the contact between electrode and workpiece is used to fabricate joint. It is generally classified under fusion welding process. The electrodes apply pressure at the interacting surfaces to have a compatible joint. In RSW, no shielding gases, flux, filler metal are used and the electrodes are nonconsumable. In this way, RSW is better than other fusion welding processes that involve shielding gases, flux, and filler metal, in terms of sustainability and eco-friendly production. The current used is very high (5,000–20,000 A), while voltage is relatively low, say less than 10 V. The duration of the current is about 0.1–0.5 s. The high-speed adaptability for automation and no necessity for edge preparation make the process advantageous. Some critical parts of smaller size pose difficulties because of inaccessibility for electrodes on both the sides. Thicker parts cannot be welded efficiently. The quality of weld depends on the resistance at the faying surfaces, which in turn depends on surface finish, cleanliness, contact area, and pressure applied. Therefore, there is a wide scope for improving the process capability in terms of sustainability and green manufacturing.

In this context, the effect of electrode force, welding current and welding time on the tensile-shear strength of Ti sheets has been examined by Kaya and Kahraman (2012). Their results show that shear strength increased with all the parameters resulting in stronger joints at higher levels. The formation of twins increased with increasing electrode force, welding current, and welding time has been revealed through microstructural examinations. The evaluation of electrode wear rate during RSW online will be useful in estimating the electrode life so that repeated fabrication and cost involved in such fabrication can be minimized resulting in sustainable manufacturing. Electrode pitting (EP) and electrode tip enlargement (ETE) are two important indexes that quantify electrode wear. As a result of this, electrode tip morphology will change resulting in defective and undesired weld. In this situation, Zhang et al. (2008) evaluated electrode life in terms of number of welds—1,220, 3,160, 5,980, 9,560, for four different steel grades. The results also reveal that a smaller radius of electrode face at the start of welding results in a higher electrode wear rate indicating an optimum electrode radius a requirement for efficient manufacturing. In a similar line of thought, Wang et al. (2016) presented the influence of EP and ETE morphology on the lap shear fracture load. It is found that EP beyond a critical level reduced the lap shear fracture load, and the incomplete fusion zone size increases with an increase in the EP size.

An important contribution from Tang et al. (2003) explores the effect of RSW machine characteristics such as machine stiffness, friction, and moving mass on the weld quality. Through successful experimentation and FE analyses, the following guidelines are suggested to design an RSW machine for improving the weld quality.

- Higher machine stiffness is essential and should be ensured during structural design of RSW machines. Higher stiffness delivers good weld quality by proper contact interface, good electrode alignment, and provides large forging force. In fact, a criterion relating to the machine stiffness and electrode misalignment is required instead of standardizing the machine stiffness.
- Optimum contact area is required to guarantee good weld quality and limited axial electrode misalignment. Equation 3.1 relates the actual contact area (A_r), in terms of percentage of electrode face area, and the axial electrode misalignment (Tang et al., 2003). The RSW machine should be designed and developed to have proper overlap area:

$$A_r = \frac{2R^2 \arccos\left(\dfrac{d}{2R}\right) - 2d\sqrt{R^2 - \left(\dfrac{d}{2}\right)^2}}{\pi R^2},$$ (3.1)

where R is the radius of electrode face and d is the axial misalignment.
- Like axial misalignment, angular misalignment should be restricted, especially for some types of welding guns like long-arm scissors guns, in order to avoid strong pressure and current concentrations. Even a slight axial and angular misalignment will generate huge asymmetrical pressure distribution. Such misalignments can be minimized by using dome shaped electrodes.
- Friction should be kept as low as possible as it is unfavorable to weld quality. To do so, the sliding mechanism can be replaced by roller guide to support the moving parts of RSW machine.
- Moving mass of RSW machine has no effect on the weld quality. Still it is better to keep the mass as less as possible to enhance the electrode life and machine portability for ergonomic considerations.

Mechanical fasteners are mechanical structures that are used to create macroscopic interlocking between mating joint elements, with the help of fasteners. The fasteners can be bolted joints, rivets, press fitting, shrink and expansion fits, and simple methods such as stapling, stitching, etc. Some of the advantages these fasteners are as follows: fastening is independent of the material of the parts to be joined, disassembly possible without damaging the parts, no changes to the chemical composition or microstructures of parts joined, no generation of fumes and emissions, and relatively low capital and operating costs. The disadvantages of these fasteners include requirement of additional machining operation for making holes resulting in increased energy consumption, generation of stress concentration due to presence of holes and sudden transition of geometries, leaking and corrosion acceleration because of open nature of the joint, and weight addition.

3.3 SUSTAINABLE AND GREEN JOINING METHODS

The joining technologies having very less negative impact on environment can be grouped as sustainable and green joining technologies. Although there is much evidence in the previous section in favor of modifying traditional welding methods such as arc welding, gas welding, and RSW for sustainable manufacturing and eco-friendliness, many new/novel joining methods are also developed along with materials' development aiming at sustainability and green engineering. Most of these methods either minimize energy consumption or material consumption or both, in a way reducing environmental hazards. The solid-state welding methods are directly related to sustainable and green manufacturing, as generation of fumes, emissions, usage of shielding gas, flux, and electrode, consumable wastes are either very minimal or totally avoided. In the section, solid state and a few hybrid joining methods are described with their merits and demerits. Some research case studies are presented, supporting their contribution toward sustainable and green manufacturing. Most of the solid-state welding methods involve plastic deformation during welding, and welding happens in solid state without melting and solidification of parent materials.

3.3.1 Friction-Based Joining Methods

Friction stir welding/processing (FSW/FSP) is a solid-state welding process in which a specially designed rotating pin is first inserted into the contacting edges of the sheets to be welded with some axial force with a proper tilt angle and then traversed along the interface to form a joint (Figure 3.1). Some important advantages include, low distortion of workpiece and dimensional stability, relatively good surface finish, better mechanical properties in the weld zone, and fine microstructure in the weld zone. Elimination of grinding wastes, elimination of consumables such as rods, wires, fillers, and shielding gas, no formation of fumes, emissions, and radiation, improved material usage, and diverse material combination are some of the benefits that improve the environmental friendliness of the process. This results in suitable weight reduction, decreased fuel consumption, and minimized energy requirement as compared with laser welding (Mishra and Ma, 2005). Some of the disadvantages are as follows: welding speeds are moderately slow, rigid clamping of work pieces required, backing plate required, and formation of keyhole at the end of each weld. In friction stir spot welding (FSSW), the translational movement of the tool is absent (Figure 3.2). Similar concept is used in friction welding of rods. In this process, a rotating rod is made to contact a stationary rod with considerable axial force. The interface friction, heat generation, and axial force create a metallurgical joint through plastic deformation (e.g., forging) at the end of the rods (Figure 3.3). Linear friction welding (LFW) is another joining process in which a linearly reciprocating metal block in contact

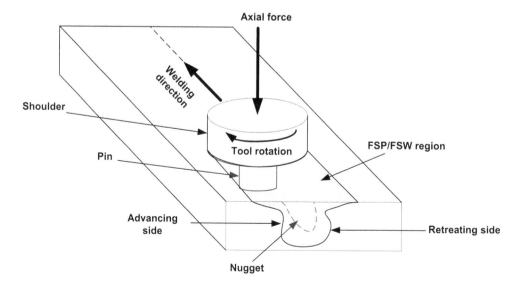

Figure 3.1 Friction stir welding of sheets and plates.

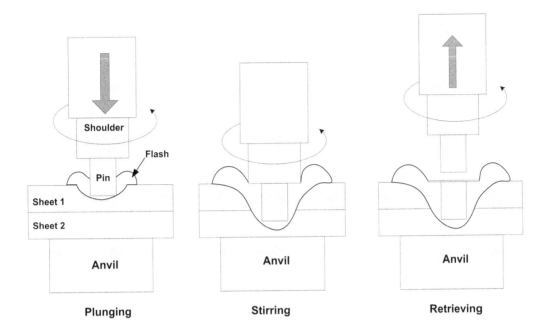

Figure 3.2 Friction stir spot welding of sheets and plates.

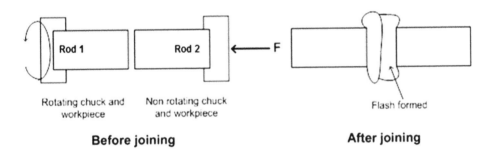

Figure 3.3 Friction welding of rods.

with a stationary metal block under the action of axial pressure undergoes plastic deformation at the interface, with friction heat generation, to fabricate a joint (Figure 3.4). At the end of the process, the reciprocating motion is stopped, but axial pressure is applied for joint making. Some of the existing facilities of LFW can operate with amplitude of 0.05–3 mm, with a frequency range of 10–1,000 Hz. The non-axisymmetric components can be welded by using LFW.

Considerable research work has been performed in FSW and FSSW aiming at understanding the importance of various process conditions for sustainable and efficient manufacturing in terms of reduced defect formation and improved mechanical properties. In the case of FSSW of Al 6061 sheets, Su et al. (2006) demonstrated that the clamping mechanism and tool design contribute to energy utilization. It is observed that, while using a steel tool, the clamp and anvil support only 12.6% of the total energy generated during FSSW is transferred for welding the sheets. When using a mica clamp and anvil support, about 50% of the energy generated is transferred. The remaining energy generated during tool rotation dissipates in the tools, anvil, support plate, and environment. Moreover, about 4% of the total energy generated is contributed toward formation of

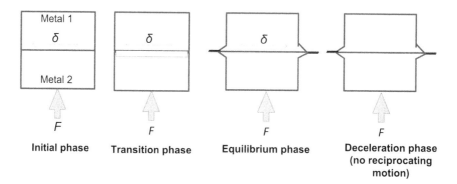

Figure 3.4 Linear friction welding of metals.

the stir zone, while the remaining gets dissipated through the tools, anvil, clamp, and surrounding atmosphere. By improving the energy utilization and reducing the energy dissipation, the softer zone width and hence the distortion are reduced. It is also observed that for a tool geometry, the rotating pin has taken about 70% of the energy generated during spot welding.

Similar attempts made by Awang and Mucino (2010) through FEA reveal that the friction at the tool and interface dissipates almost 97 kJ of energy (Figure 3.5), while other factors show an insignificant effect. With a decrease in rotational speed from 3,000 to 2,500 to 2,000, the dissipated energy by friction is reduced from 1.471 to 1.347 to 1.188 kJ, respectively. This shows a reduction of about 20% when rotation speed is reduced to 2,000 rpm from 3,000 rpm. Similarly, when plunge rate is increased from 1 to 10 mm/s, the friction dissipation energy is reduced to 14%, while plastic dissipation energy to 40%, both are huge reductions. Such results point out a fact that parameter optimization is very much required to minimize energy dissipation during FSSW process, though defect-free welds can be produced in a wide range of parameters.

Minimizing the defect formation reduces the material wastes, energy utilization and has cost benefits. In this context, Chen et al. (2006) presented the influence of tool tilt angle on the defect formation. It has been suggested through experiments that lower tilt angle because of insufficient plastic flow of material and higher tilt angle because of formation of weld flash, defect formation is enhanced. The lazy "S" defect is due to the disordered mixing of the oxide layer on the initial butt

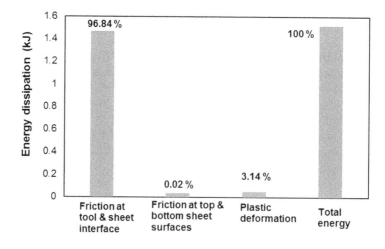

Figure 3.5 Energy dissipation during FSSW of Al6061-T6 sheet. (Awang and Mucino, 2010.)

surface on the retreated or advanced side of FSW of Al alloy 5456. Kim et al. (2006) have shown that by controlling the downward plunge force, at different rotational speed and welding speed, three types of defects are formed: (1) flash formation due to the excessive heat input, (2) cavity or groove caused by insufficient heat input, and (3) cavity as a result of abnormal stirring. Ramulu et al. (2013) through their experimental analyses reveal that a higher welding speed (80–120 mm/min), higher rotation speed (1,300–1,500 rpm), and higher plunge depth (1.85–2 mm) are favored for producing a weld without internal defects during FSW of 6061T6 sheets. A simple criterion, equation 3.2, has been developed relating torque (τ) and axial force (F) to welding parameters (p) such as welding speed, traverse speed, and plunge depth, by which parameter range for defect-free weld formation can be identified online during welding. Hence, energy dissipation and material wastes during defective weld formation can be minimized leading to sustainable manufacturing:

$$\left(\frac{\partial \tau}{\partial p}\right)_{\text{defective}} > \left(\frac{\partial \tau}{\partial p}\right)_{\text{defect-free}} \quad \text{and}$$

$$\left(\frac{\partial F}{\partial p}\right)_{\text{defective}} > \left(\frac{\partial F}{\partial p}\right)_{\text{defect-free}} \tag{3.2}$$

where ∂p is change in welding speed or tool rotation speed or plunge depth

FSW of harder materials like steel is difficult mainly from tool wear point of view. Obtaining optimized process and tool parameters for successful weld formation without much tool wear is cumbersome. Thomas et al. (1999) showed the feasibility of welding 12-mm thick 12% chromium alloy low carbon steel. Under restricted process conditions, successful defect-free welds were fabricated and tensile and bend tests exhibited their quality equivalent to that of parent base material. Friction welding of hard metal rods like stainless steel to stainless steel, stainless steel to aluminum alloy, and stainless steel to titanium are also attempted by many groups (Paventhan et al., 2011; Satyanarayana et al., 2005).

3.3.2 Mechanical Joining Methods

Self-piercing riveting (SPR) and clinching (Figure 3.6) are high-speed mechanical fastening process for spot joining of sheet materials using a semi-tubular rivet to form a mechanical joint. As the name suggests, predrilled holes are not required, allowing the joint to be made rapidly in one operation. SPR has some advantages including possibility to join dissimilar materials, low energy process, no fumes or emissions, good tool life, and little or no part distortion. However, its disadvantages include the introduction of additional consumable items to form joint and therefore weight addition, requiring access to both sides of the joint for completion, etc.

Clinching, on the other hand, is a method of joining different metal parts (mainly sheets) by a process of local deformation without use of any additional joining elements with the application of a punch and a die. Both SPR and clinching involve plastic deformation and locking of sheets for joint formation. Like SPR, clinching produces clean joint (no burrs, sparks, fumes) and can fasten pre-painted and coated materials without reworking, and eco-friendly. However, there are some restrictions for its application. It cannot be used for joining plastics, hard materials, and low ductile materials. The materials to be joined need to have high ductility (Martinsen et al., 2015). In steel grades, the joint strength (maximum force from tensile tests) is about 4–5 times lesser for clinched joints as compared with SPR joints (Kascak et al., 2012). Moreover, clinched joints fail early as compared with SPR joints showcasing a lower deformation capacity.

Mori et al. (2012) showed the superiority of SPR joint over clinching and RSW joint in terms of static and fatigue tests. The fatigue strengths for the clinching and SPR are improved by slight slip at the interface between the sheets. The stress concentration at the edge of the weld because of

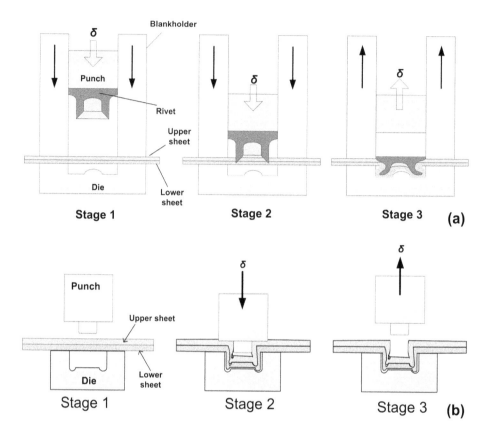

Figure 3.6   (a) Self-pierced riveting and (b) clinching of sheets.

complete bonding has reduced the fatigue strength of RSW joint. In general, SPR and clinching can be preferred over RSW of Al sheets. Varis (2006) suggested a cost analysis method to understand the effectiveness of SPR and proposed a clinching processes. In the clinching process, the unit costs decrease as the tool service life increases, and the total costs decrease. On the other hand, the unit costs in SPR are constant and the total costs increase directly in proportion to the number of joints.

3.3.3 Surface-to-Surface Joining Methods

Adhesive bonding is a surface-to-surface joining method. In this process, similar or dissimilar grade materials are joined using an adhesive, which adheres to the surfaces of the two metals to be joined, transferring the forces from one metal to the other. Dissimilar metals can be joined, good noise and vibration damping properties, uniform stress distribution over a larger area, seals the joint against moisture and debris admission are some advantages of adhesive bonded joints. However, there are some disadvantages, mainly against eco-friendliness and sustainability, including usage of epoxy or solvent-based systems giving rise to considerable environmental concerns, health and safety hazards involved in the use of adhesive, thus implying significant costs in providing adequate fume extraction, protective clothing and adequate provision for fire protection storage during fabrication, inadequate shelf-life of the adhesive, and poor peel strength as compared with other joining methods.

Roll bonding (or accumulated roll bonding, ARB) (Figure 3.7) is a solid-state welding process, in which bonding is made by plastic deformation of the sheets to be bonded via rolling. During bonding, surface expansion results in exposing the surfaces of virgin metal. Once the rolling

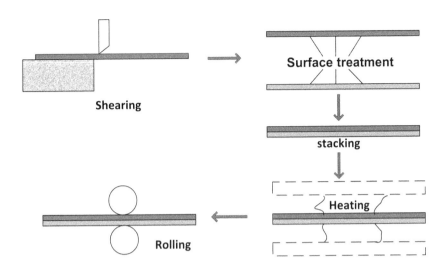

Figure 3.7   Accumulated roll bonding of sheets.

pressure reaches a large, critical value, the virgin material extrudes the fractured layer, resulting in joint formation. Roll bonding is done at room temperature, called cold roll bonding, and at elevated temperature, called hot roll bonding. In accumulative roll bonding, bonding is accomplished by several passes of rolling, each pass accounting for a prescribed thickness reduction. Sandwich and multicomponent materials can be made by ARB leading to tailored made properties. Ultra-refined grains are produced during ARB as demonstrated by many researchers.

Hausol et al. (2010a) demonstrated the production of two different Al sandwich structures: one between AA1050A and AA5754 made at room temperature with 50% thickness reduction, and the other AA6014-AA5754 sandwich made at 230°C via two ARB cycles with 50% thickness reduction/cycle. By doing so, specifically in the second case, the strength of AA5754 is fully preserved and serrated yielding vanished. Hence, the application of ARB processed materials is extended showcasing the importance of intelligent materials design in sustainable and green manufacturing. Hausol et al. (2010b) established the fabrication of multicomponent structures with fibers, particles, and foils as reinforcements through ARB, which is another example for ARB applications in sustainable and green manufacturing. The parameters, such as amount of thickness reduction, initial sheet thickness, rolling speed, rolling direction, and heat treatments (before and after rolling), significantly affect the roll bonding characteristics. Jamaati et al. (2010) found during cold roll bonding (CRB) that larger thickness reductions, lower initial thickness, rolling speed, annealing treatment before, and/or after the CRB process are the deciding factors that improved bond strength evaluated through peel tests.

3.3.4 Joining of Sheets and Tubes by Plastic Deformation

Joining by forming is a separate category of joining of metallic structures in which plastic deformation is used for joining. Mechanical interlocking happens during such joining methods. Since fumes and emissions are not generated, protective gases and fluxes are not used, the process is a better option for sustainable manufacturing. However, the strength of the joint should be checked as there is no metallurgical bonding between the joining structures. Rotary swaging, joining by end forming, SPR, and clinching belong to the category. SPR and clinching are discussed in the previous section. Zhang et al. (2014) performed rotary swaging (Figure 3.8) method to join the tubes of different diameters. Rotary swaging is a type of incremental forming process, where dies around the

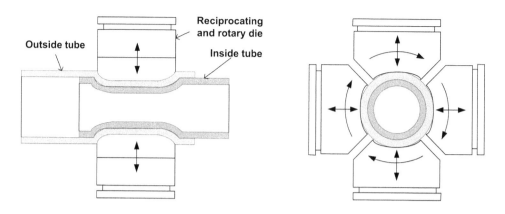

Figure 3.8 Joining tube to tube by rotary swage forming. (With permission Zhang et al., 2014.)

workpiece move simultaneously in both radial and axial directions relative to the workpiece. The swaging dies perform high-frequency radial movement with short strokes. The process is capable of joining different materials, with different interface characteristics. It is a clean joining process because of no fume generation and no external consumables used. But the process is restricted to join tubes of different diameters.

Alves et al. (2014) proposed another method of joining tube to tube by plastic instability through tube bending at the end of the tube (Figure 3.9). Here joining is accomplished in one stroke by a

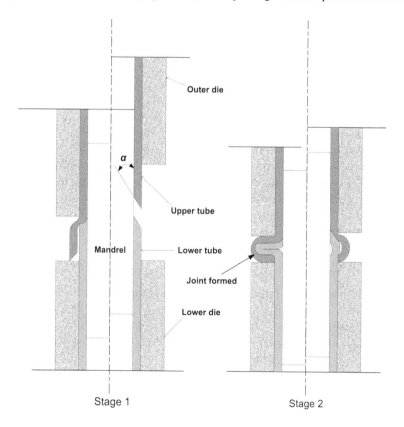

Figure 3.9 Joining tube to tube by tube bend formation. (With permission Alves et al., 2014.)

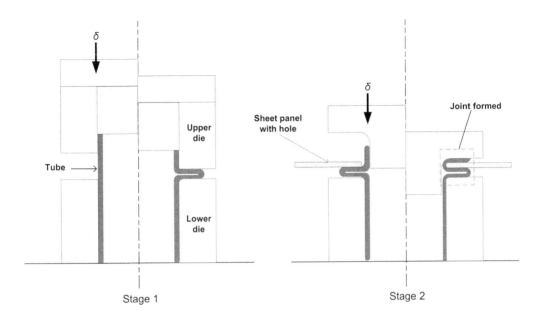

Figure 3.10 Tube to sheet joining by tube bend formation. (With permission Alves et al., 2011.)

sequence of two different tube forming operations such as expansion and compression beading at the end of the tube. Alves et al. (2011) suggested a novel method of joining tube to sheet by tube end forming where a mechanical interlock is produced by compression beading and external inversion (Figure 3.10). The proposed joining process offers significant advantages as compared with conventional methods such as mechanical fastening, welding, or structural adhesive bonding. These include less-consumable consumption (eliminates filler materials and shielding gases), high levels of repeatability in production line, energy saving process, no thermal residual stress at joint region, capable of joining dissimilar materials, and lesser capital investment. Similarly, Agrawal and Narayanan (2017) suggested end curling of tube and sheet for joining (Figure 3.11). In this process, a tube is made to plastically deform in the die groove by providing displacement to the tube in a universal testing machine. The tube, after end curling, enters into the bent region of the sheet. At the process completion, a neck is formed in the undeformed region of the tube, just above the sheet, to obtain a compact joint. Silva et al. (2015) proposed a similar process for joining of tubes by their ends. This has several advantages, such as easy accommodation of different thickness tubes, self-alignment of the tubes to be joined, requirement of chamfering in one of the two tubes, and good structural rigidity for the transport of gases and fluids, etc.

3.4 COMPUTATIONAL ANALYSES AND RELEVANCE TO SUSTAINABLE JOINING

Computational assistance is required to evaluate the mechanical behavior, temperature evolution, joinability, load requirement, deformation level, etc., so that the product can be joined efficiently without much time delay and resource loss. The state-of-the-art of computations is such that the process can be analyzed at the design stage itself for selecting material, optimized process conditions, evaluating power consumption, tool/die design, evaluation of mechanical properties, minimization of material waste, and production schedule. In many situations, specifically in real product analyses, computations replace experiments. Hosseinpour and Hajihosseini (2009) highlighted many issues, as listed below, that are generally addressed by computer simulation in manufacturing, more from industrial perspective.

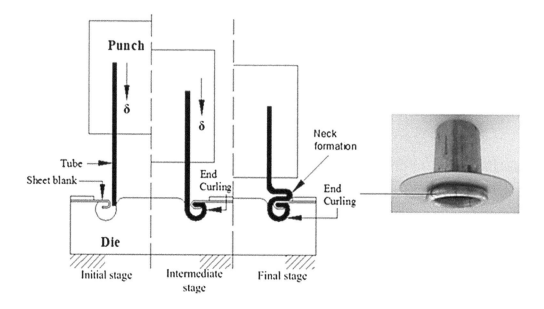

Figure 3.11   Joining a tube to a sheet by end curling. (With permission from Agrawal and Narayanan, 2017.)

- Number and type of machines for making a product
- Number, type, and physical arrangement of electronics and mechanical equipment, such as transporters, conveyors, fixtures, etc.
- Location and size of inventory buffers
- Evaluation of a change in product volume
- Evaluation of fitness of new equipment on an existing manufacturing system
- Evaluation of capital investments and labor requirements
- Time-in-system and bottleneck analysis
- Production scheduling and inventory policies
- Control strategies (e.g., for an automated guided vehicle system, robot applications, etc.)
- Reliability analysis and quality control policies, etc.

This section highlights a few scientific works demonstrating the importance of computations in joining processes. The validated theoretical or computational models can be used instead of rigorous experiments, paving way for sustainable manufacturing and preserving green and eco-friendliness.

Fan and Shi (1996) developed a 2D axisymmetric model to predict the heat transfer and fluid flow in GTAW by considering the electrode tip geometry. The mass continuity equations, momentum conservation, and energy conservation equations are solved for analyses. The governing equations are discretized using a finite-volume approach. Finally, the temperature contours, distribution of electromagnetic force, arc current, electro tip angle, and arc length are predicted and the agreement is quite good with measured data. The numerical prediction of laser butt welding characteristics of Mg alloys was done by developing a transient nonlinear 3D model (Belhadj et al., 2010). The model is capable of predicting the temperature history from the beginning of welding till thermal equilibrium is reached. Similarly, temperatures at different locations can also be evaluated. The predictions such as temperature history, weld width, and cooling isotherms are in good agreement with experiments.

The soft computing methods, such as neural networks and genetic algorithms, along with theoretical methods, contribute a lot to process optimization and analyses. A neural network model coupled with finite-element analysis was developed by Fratini et al. (2009) to predict the average

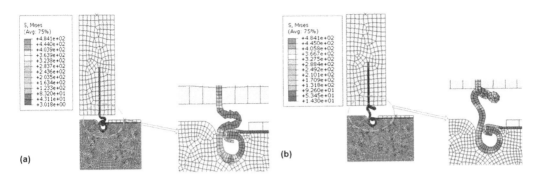

Figure 3.12 FE simulations of sheet to tube joining process shown in Figure 3.11: (a) successful case and (b) unsuccessful case. (With permission from Agrawal and Narayanan, 2017.)

grain size during FSW of AA2139-T8 sheets. The weld configurations such as butt joint, lap joint, and T-joint are analyzed. The strain, strain-rate, temperature evolved during FSW, and Zener–Hollomon parameter are given as input for grain size prediction. The experimental comparison is found to be satisfactory with numerical results. Porcaro et al. (2006) performed FE simulations using a 2D axisymmetric model in LS-DYNA for joining two sheets using rivet and tools. During modeling, an implicit solution technique together with an r-adaptive method has been used. The effect of mesh size, adaptive time interval, friction conditions, and material failure on the joint prediction are studied. The results and validation support the efficiency of modeling through accurate prediction. Here sheet failure is modeled in terms of evolution of thinning using critical sheet thickness parameter.

The sheet to tube joining method by end curling proposed in Figure 3.11 is actually designed and analyzed by 2D FE simulations to arrive at the optimum set of parameters (e.g., die groove radius, friction coefficient, etc.) for making successful joints. Such FE simulations are shown in Figure 3.12. One of the simulations predicts the joint formation to be successful (Figure 3.12a), while the other one is unsuccessful as neck forms well above the sheet (Figure 3.12b). About 60 such FE simulations are conducted for process design. Later a few cases are performed at laboratory scale to prove the concept. Such novel joining methods that are not governed by equations can be modeled realistically by FE simulations and other modeling methods before lab scale trials. Thus, the process becomes sustainable and eco-friendly, as physical wastage of consumables, tube and sheet materials, lubricants, etc., are minimized.

3.5 SUMMARY

Sustainable joining is possible in traditional welding and recently developed welding processes. Shielding gas and flux, consumables, electrode, filler, fuel for combustion, stability of the machine, and cleanliness of the materials are the candidates for maintaining sustainability and green manufacturing in traditional welding processes such as arc welding, gas welding, EB welding, laser welding, etc. Most of the novel welding processes are solid-state in principle, in which use of shielding gas, flux, consumables, fuel, etc., is absent, making these sustainable and eco-friendly. The friction and plastic deformation play a vital role for joint fabrication. Friction-based joining methods such as friction stir welding, friction welding, and mechanical joining methods such as self-pierce riveting, clinching, joining of structures by end forming, and adhesive bonding belong to the category. The heat energy, mechanical energy, and deformation energy requirements are different for these processes as compared with traditional processes. In most cases, these are comparable, and in some cases, the solid-state welding processes are more sustainable. In this chapter, many case studies are provided to bring out the candidates for sustainable joining. It is evident from the case studies that independent of the joining process, parameter optimization for sustainable joining is very much important. Computations using analytical models, FEA, and soft computing methods help in optimizing such processes for sustainable joining.

REFERENCES

Agrawal, A.K., Narayanan, R.G., 2017. Joining of a tube to a sheet through end curling. *Journal of Materials Processing Technology* 246: 291–304.

Alves, L.M., Dias, E.J., Martins, P.A.F., 2011. Joining sheet panels to thin-walled tubular profiles by tube end forming. *Journal of Cleaner Production* 19: 712–719.

Alves, L.M., Silva, C.M.A., Martins, P.A.F., 2014. End-to-end joining of tubes by plastic instability. *Journal of Materials Processing Technology* 214: 1954–1961.

Amza, G., Dobrotă, D., Groza, D. M., Paise, S., Apostolescu, Z., 2013. Research on environmental impact assessment of flame oxyacetylene welding processes. *Metalurgija* 52: 457–460.

Awang, M., Mucino, V.H., 2010. Energy generation during friction stir spot welding (FSSW) of Al 6061-T6 plates. *Materials and Manufacturing Processes* 25: 167–174.

Belhadj, A., Bessrour, J., Masse, J.E., Bouhafs, M., Barrallier, L., 2010. Finite element simulation of magnesium alloys laser beam welding. *Journal of Materials Processing Technology* 210: 1131–1137.

Chen, H.B., Yan, K., Lin, T., Chen, S.B., Jiang, C.Y., Zhao, Y., 2006. The investigation of typical welding defects for 5456 aluminum alloy friction stir welds. *Materials Science and Engineering A* 433: 64–69.

Dennis, J.H., French, M.J., Hewitt, P.J., Mortazavi, S.B., Redding, C.A.J., 1996. Reduction of hexavalent chromium concentration in fumes from metal cored arc welding by addition of reactive metals. *The Annals of Occupational Hygiene* 40: 339–344.

Fan, H.G., Shi, Y.W., 1996. Numerical simulation of the arc pressure in gas tungsten arc welding. *Journal of Materials Processing Technology* 61: 302–308.

Fratini, L., Buffa, G., Palmeri, D., 2009. Using a neural network for predicting the average grain size in friction stir welding processes. *Computers & Structures* 87: 1166–1174.

Hausol, T., Hoppel, H.W., Goken, M., 2010a. Tailoring materials properties of UFG aluminium alloys by accumulative roll bonded sandwich-like sheets. *Journal of Materials Science* 45: 4733–4738.

Hausol, T., Maier, V., Schmidt, C.W., Winkler, M., Hoppel, H.W., Goken, M., 2010b. Tailoring materials properties by accumulative roll bonding. *Advanced Engineering Materials* 12: 740–746.

Hosseinpour, F., Hajihosseini, H., 2009. Importance of simulation in manufacturing. *World Academy of Science, Engineering and Technology* 51: 285–288.

Jamaati, R., Toroghinejad, M.R., 2010. Investigation of the parameters of the cold roll bonding (CRB) process. *Materials Science and Engineering A* 527: 2320–2326.

Kascak, L., Spisak, E., Mucha, J., 2012. Evaluation of properties of joints made by clinching and self-piercing riveting methods. *Acta Metallurgica Slovaca* 18: 172–180.

Kaya, Y., Kahraman, N., 2012. The effects of electrode force, welding current and welding time on the resistance spot weldability of pure titanium. *International Journal of Advanced Manufacturing Technology* 60: 127–134.

Kim, Y.G., Fujii, H., Tsumura, T., Komazaki, T., Nakata, K., 2006. Three defect types in friction stir welding of aluminum die casting alloy. *Materials Science and Engineering A* 415: 250–254.

Liu, L., Hao, X., Song, G., 2006. A new laser-arc hybrid welding technique based on energy conservation. *Materials Transactions* 47: 1611–1614.

Martinsen, K., Hu, S.J., Carlson, B.E., 2015. Joining of dissimilar materials. *CIRP Annals - Manufacturing Technology* 64: 679–699.

Mishra, R.S., Ma, Z.Y., 2005. Friction stir welding and processing. *Materials Science and Engineering: R: Reports* 50: 1–78.

Mohan, S., Sivapirakasam, S.P., Santhosh Kumar, M.C., Surianarayanan M., 2015. Welding fumes reduction by coating of nano-TiO_2 on electrodes. *Journal of Materials Processing Technology* 219: 237–247.

Mori, K., Abe, Y., Kato, T., 2012. Mechanism of superiority of fatigue strength for aluminium alloy sheets joined by mechanical clinching and self-pierce riveting. *Journal of Materials Processing Technology* 212: 1900–1905.

Muñoz-Escalona, P., Payares, M.C., Dorta, M., Diaz, R., 2006. Analysis and influence of acetylene and propane gas during oxyfuel gas cutting of 1045 carbon steel. *Journal of Materials Engineering and Performance* 15: 684–692.

Paventhan, R., Lakshminarayanan, P.R., Balasubramanian, V., 2011. Prediction and optimization of friction welding parameters for joining aluminium alloy and stainless steel. *Transactions of Nonferrous Metals Society of China* 21: 1480–1485.

Pires, I., Quintino, L., Amaral, V., Rosado, T., 2010. Reduction of fume and gas emissions using innovative gas metal arc welding variants. *International Journal of Advanced Manufacturing Technology* 50: 557–567.

Popovic, O., Cvetkovic, R.P., Burzic, M., Lukic, U., Beljic, B., 2014. Fume and gas emission during arc welding: Hazards and recommendation. *Renewable and Sustainable Energy Reviews* 37: 509–516.

Porcaro, R., Hanssen, A.G., Langseth, M., Aalberg, A., 2006. Self-piercing riveting process: An experimental and numerical investigation. *Journal of Materials Processing Technology* 171: 10–20.

Praveen, P., Yarlagadda, P.K.D.V., Kang, M.J., 2005. Advancements in pulse gas metal arc welding. *Journal of Materials Processing Technology* 164–165: 1113–1119.

Ramulu, P.J., Ganesh Narayanan, R., Kailas, S.V., Reddy, J., 2013. Internal defect and process parameter analysis during friction stir welding of Al 6061 sheets. *International Journal of Advanced Manufacturing Technology* 65: 1515–1528.

Saresh, N., Gopalakrishna Pillai, M., Mathew, J., 2007. Investigations into the effects of electron beam welding on thick Ti-6Al-4V titanium alloy. *Journal of Materials Processing Technology* 192–193: 83–88.

Satyanarayana, V.V., Reddy, G.M., Mohandas, T., 2005. Dissimilar metal friction welding of austenitic–ferritic stainless steels. *Journal of Materials Processing Technology* 160: 128–137.

Silva, C.M.A., Nielsen, C.V., Alves, L.M., Martins, P.A.F., 2015. Environmentally friendly joining of tubes by their ends. *Journal of Cleaner Production* 87: 777–786.

Su, P., Gerlich, A., North, T.H., Bendzsak, G.J., 2006. Energy utilisation and generation during friction stir spot welding. *Science and Technology of Welding and Joining* 11: 163–169.

Tang, H., Hou, W., Hu, S. J., Zhang, H.Y., Feng, Z., Kimchi, M., 2003. Influence of welding machine mechanical characteristics on the resistance spot welding process and weld quality. *Welding Research* 116-S–124-S.

Thomas, W. M., Threadgill, P. L., Nicholas, E. D., 1999. Feasibility of friction stir welding steel. *Science and Technology of Welding and Joining* 4: 365–372.

Topham, N., Mark, K., Yu-Mei, H., Chang-Yu, W., Sewon, O., Kuk, C., 2010. Reducing Cr^{6+} emissions from gas tungsten arc welding using a silica precursor. *Journal of Aerosol Science* 41: 326–330.

Varis, J., 2006. Economics of clinched joint compared to riveted joint and example of applying calculations to a volume product. *Journal of Materials Processing Technology* 172: 130–138.

Wang, B., Hua, L., Wang, X., Song, Y., Liu, Y., 2016. Effects of electrode tip morphology on resistance spot welding quality of DP590 dual-phase steel. *International Journal of Advanced Manufacturing Technology* 83: 1917–1926.

Wanjara, P., Brochu, M., Jahazi, M., 2006. Thin gauge titanium manufacturing using multiple-pass electron beam welding. *Materials and Manufacturing Processes* 21: 439–451.

Wei, H.L., Li, H., Yang, J.L., Gao, Y., Ding, X.P., 2015. Arc characteristics and metal transfer process of hybrid laser double GMA welding. *International Journal of Advanced Manufacturing Technology* 77: 1019–1028.

Yunlian, Q., Ju, D., Quan, H., Liying, Z., 2000. Electron beam welding, laser beam welding and gas tungsten arc welding of titanium sheet. *Materials Science and Engineering A* 280: 177–181.

Zhang, Q., Kaiqiang, J., Dong, M., 2014. Tube/tube joining technology by using rotary swaging forming method. *Journal of Materials Processing Technology* 214: 2085–2094.

Zhang, X.Q., Chen, G.L., Zhang, Y.S., 2008. On-line evaluation of electrode wear by servo gun in resistance spot welding. *International Journal of Advanced Manufacturing Technology* 36: 681–688.

CHAPTER 4

Friction and Lubrication in Sustainable Metal Forming

N. K. B. M. P. Nanayakkara
University of Peradeniya

R. Ganesh Narayanan
IIT Guwahati

CONTENTS

4.1 INTRODUCTION

Metal forming, which is a widely used process of shaping of metals, has undergone remarkable innovations and technological uplifts since its inception. The process economy, quality of the formed part, and the concerns of its effects on the environment have been the main targets behind the centuries-long existence of this process. As an integral component in a forming system, the lubrication too has undergone continuous developments and innovations of technology to satisfy the above requirements.

Traditional metal-working lubricants are sourced directly from mineral oils. These lubricants have been effectively used in metal-forming applications over the past decades. However, with the evolution of the technology together with the rising requirements of the product, the lubricants used in the metal forming have also undergone drastic improvements. These improvements have been further elevated by the social and technological concerns. These uplifts in the lubrication technology are both product focused and the process focused. The product-focused needs of lubrication and related technologies are mainly on neutralizing the adverse effects of lubrication on the product, for example, degradation of the quality due to reactions with the lubricant such as corrosion. In addition, targeting the issues such as wear, galling, and material transfer through the selection and development of lubrication systems also focuses on the formed product. Advancements such as self-lubricating materials have been developed to enhance the quality of the product. Furthermore, in certain products, such as automotive skin panels, deterioration of the surface coatings during the forming operation is a quality-related concern. In the recent past, dry films and coating friendly lubrication have been introduced to minimize such damages to the coatings.

The process-based advancements are of two main types: process economy-based innovations and development and environment-focused novelties. The process economy is a function of the cost of lubricant, the amount of lubrication needed, and the technology of application. These factors are on spotlight in order to optimize the economy of the process, thus to maintain the cost of the product, which is a demanding trend in today's manufacturing. This trend has further enriched due to the cost of tooling and the tool life. Specifically, solid lubricants have come into the picture in recent years. More and more innovative concepts of lubrication have started appearing throughout the industry with promising lubricity and other key expectations. However, recently, the emphasis on the environment has been somewhat overweighed compared to the other concerns, and therefore, sustainability has become a focal point in selecting lubricants for forming processes. Therefore, minimizing the adverse effects on the environment is considered as a major goal in metal-forming lubrication. The main expectations through the sustainable lubrication therefore include the minimizing or elimination of damages to the environment and optimizing the overall energy consumption in the lubrication system. Furthermore, the reusability or recyclability of the lubricants is also encouraged without scarifying the focus on quality of the product. Health and safety requirements are also considered as an integral component of sustainability, and therefore they are too unified into sustainable lubrication.

This chapter is focused on the effects of friction and lubrication in metal forming, the mechanisms of lubrication, and an account on lubricants used in metal forming leading toward sustainability of the lubrication in metal-working processes.

4.2 FRICTION IN METAL FORMING

In metal-forming processes, friction and lubrication are mutually related. Friction is an inherent characteristic in all forms of metal-forming processes and lubrication is often used to control friction in forming applications. During metal-forming processes, a rigid metal body (die) is in contact with the workpiece, which is the deformable body. The friction between these two bodies is important to transfer of forces from the die to the workpiece, which controls the materials flow. Thus, the friction between the workpiece and the tool surfaces has a significant effect on manufacturing of defect-free formed parts. Controlled friction is necessary to achieve desired shapes and geometries while assuring mechanical properties in formed parts. Localized stress concentrations may develop within the workpiece if the friction is not controlled sufficiently. These localized stresses will contribute to degradation of mechanical properties of the product and sometimes product defects. Draw die features and carefully selected process considerations, such as surface properties and lubrication, are needed to be designed and installed in industrial forming operations where quality is definite requirement. Therefore, a proper understanding of friction characteristics between the work and tool interfaces will allow the selection or design of the most suitable process conditions, especially the lubrication requirements for a given forming operation. Furthermore, these selections will finally ensure extended requirements such as process economy, environmental, and sustainability concerns.

Behavior of friction in metal forming should be carefully understood before taking process-related decisions. In certain cases, friction is a relative measure, for example, in a given die geometry, friction profiles may be created to illustrate the regions of "high" and "low" friction. However, in many aspects quantified friction is desired in selecting process conditions. A number of analytical models are available to represent friction in metal-forming operations. These models have been developed with certain assumptions and approximations and their validity is a subjective concern based on the type and the nature of the forming operation.

4.2.1 Friction Theories

In mechanics, the friction coefficient, μ, is expressed with the aid of frictional force (F) that is required to slide the two surfaces apart and the normal reaction force (N) between contacting surfaces. This mathematical representation of the coefficient friction is known as the Coulomb's law of friction and is illustrated in Figure 4.1.

Coulomb's law of friction is one of the fundamental mathematical models in assessing friction; however, it will have certain limitations when applying for metal-forming operations.

The Coulomb's coefficient of friction can be defined as follows:

$$\mu = F/N \tag{4.1}$$

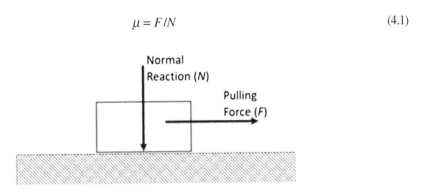

Figure 4.1   Basic illustration for friction couple for Coulomb's friction.

The Coulomb's law of friction has been extensively used as an initial predictive model in the analysis of interface friction in many engineering applications. In most of the practical circumstances, interfacial stresses are used to represent the friction coefficient rather than relating it with normal and sliding forces, provided that the effective contact area is the same in each case. Initially, the coefficient of friction is presented considering the shear force at the boundary and the normal compressive force at the interface as follows:

$$\mu = \frac{F_S}{F_N} \tag{4.2}$$

where F_S is the interfacial shear force and F_N is the compressive force between the workpiece and tool surface. Once the interfacial stresses are introduced, the coefficient of friction is presented as follows:

$$\mu = \frac{\tau_F}{P}, \tag{4.3}$$

where τ_F is the interfacial shear stress and P is the compressive stress between the two surfaces. The above relationship leads to a standard form of representation:

$$\tau_F = \mu P, \tag{4.4}$$

In addition to the classical Coulomb's law, constant friction theory is also a fundamental model is assessing friction and the friction is defined as follows:

$$\tau_F = \frac{m\sigma_0}{\sqrt{3}}, \tag{4.5}$$

where m is the friction factor and σ_0 is the flow stress of the material. Equation 4.5 is obtained by assuming von-Mises yield function (Dixit and Narayanan, 2013).

In contrast to these classical theories, the coefficient of friction appears to be a function of the kinetics of the sliding system. The coefficient of friction slightly deceases after the interfacial sliding commences and this is known as the static to dynamic transition of friction.

There are number of deficiencies arise in applying direct interface friction relationships in forming applications particularly when lubrication presents. In lubricated sliding, which is predominant in metal-forming applications, the frictional behavior is complicated due to the presence of the stick-slip phenomenon and the lubricant shear at the interfaces. The lubricated state (τ) is represented by the film-lubricated model:

$$\tau = \eta\dot{\gamma} \tag{4.6}$$

where η is the lubricant viscosity and $\dot{\gamma}$ is the shear strain-rate within lubricant.

4.2.2 Validity of Friction Theories in Metal Forming

The above relationships are defined for dry contact conditions and they are predominantly true for unlubricated conditions at relatively low surface pressures. In lubricated friction, however, this behavior changes dramatically. The Coulomb's law of friction breaks down in applied metalworking processes especially when there is a lubricant present at the interface. The nature of the lubricant and the lubricated surfaces will drastically change the metal flow. The pressure–viscosity characteristics of the lubricant play a key role in determining the friction and alter the sliding or pulling force. Other than the viscosity of the lubricant, the variable pressure distribution, which is likely to be present at contact interfaces of different part geometries, results in complications in the friction.

4.3 LUBRICATION IN METAL FORMING

4.3.1 Lubrication Regimes

The coefficient of friction in lubricated processes is believed to depend on speed, pressure, and lubricant viscosity. These parameters can be described by the Stribeck curve. Under lubricated conditions, the variation of the coefficient of friction, μ, with a combined parameter Z (which is a function of contact pressure, speed, and the lubricant viscosity) has been defined (Sanchez, 1999; Emmens, 1988) and given by

$$Z = \frac{\eta V}{P} \tag{4.7}$$

where η is the lubricant's viscosity and P and V are the surface pressure and the sliding speed, respectively. Therefore, it is clear that the Stribeck curve provides essential knowledge of how friction varies with process parameters. In particular, based on the variation of friction, it is evident that the viscosity of the lubricant is dominant in Stribeck model. According to the behavior of the coefficient of friction against Z, three diverse lubrication regimes are identified in lubricated friction (Figure 4.2).

These three regimes are known as thick-film lubrication, thin-film lubrication, and boundary lubrication. Specific properties and characteristics of these lubrication regimes are explained below.

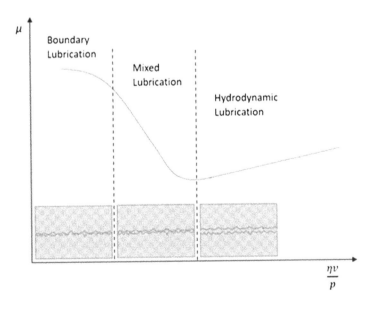

Figure 4.2 Stribeck curve showing the lubrication regimes and a schematic of lubricant entrapment.

However, one, or more, of these regimes may operate depending upon the geometry, contract conditions and the size of the die-workpiece interface (Emmens, 1988; Wilson, 1979).

4.3.1.1 Thick-Film Lubrication

This is also known as the hydrodynamic lubrication and this type of lubrication exists when there is a thick film of lubricant trapped between the tool and the workpiece, which completely averts tool-workpiece contact. Therefore, the viscosity of the lubricant and the degree of deformation govern the necessary conditions to overcome friction. In hydrodynamic lubrication, the trapped lubricant between the asperities is sufficient enough to support the tooling load, instead of the pressure between the two metallic interfaces being carried by the contacting asperities (Anderson and Russo, 1994; Wilson and Sheu, 1988; ASM Handbook, 1988).

4.3.1.2 Thin-Film Lubrication

In thin-film lubrication, the trapped lubricant between tool and the workpiece is comparatively thin. Some metal asperity contacts are present in this regime and therefore the loads are partially carried by the surface asperities between the tool and the workpiece. The lubricant itself supports the tooling loads partially and therefore the frictional sliding is controlled by two aspects: the lubricant and the contacting surfaces. Because of this combined phenomenon, this lubrication regime is also called mixed lubrication regime and frequently observed in sheet metal-forming applications.

4.3.1.3 Boundary Lubrication

In boundary lubrication regime, physical adherence of the lubricant to the metal surfaces occurs, and relatively a thin lubricant film is available between the two metal interfaces. In boundary lubrication regime, almost the total of the load transferred from tooling is held by the surface asperities. However, direct contacts over metallic surfaces are prevented by the presence of tightly adhering boundary films formed on the surfaces as a result of a chemical reaction or physical absorption (Wilson and Sheu, 1988). The adherence and shear strength of the lubricant film governs the effectiveness of the lubrication. In addition, boundary films can reduce friction by preventing strong adhesion between the metal surfaces.

4.3.2 The Need of a Lubricant

It is well known that a lubricant is primarily used to overcome or to control friction between the tool and the workpiece. Some of the lubricants are specifically developed for metal forming instead of common applications in processes such as machining. The lubricants are available for the use in either at room temperature or at higher temperatures. However, in general, the lubricants have some common functions, and may vary considerably from one metal-working operation to another. The main function of lubricants used in metal forming is to control friction and tool wear (ASM Handbook, 2006). There are many supplementary requirements expected from forming lubricants: cooling material and tool (like billet and die in forging, roll and sheet in rolling, etc.) are among them. Furthermore, the lubricants support to produce good surface quality in formed parts without smudge, defects, etc. The residue film layers on metallic surfaces helps to minimize bacterial attack and oxidation before the succeeding operations, for example, the manufacturing deep drawn cans for food packaging and prevents the formed parts from corrosion and rust for extended storage. In certain situations, lubricants provide compatibility with lubricants in further operations, such as joining and surface coatings.

4.4 TYPES OF LUBRICANTS

Lubricants used in metal-forming applications have been categorized in several ways: classifying according to the physical state of the lubricant is the most common method. Traditional metal-forming lubricants are fluids, mainly mineral oils. For special applications, dry films are also used and some of the dry films are self-lubricated solutions.

4.4.1 Fluid Lubricants

Fluid lubricants contain primary ingredients such as oils, water, and synthetic compounds along with additives to improve the effectiveness of lubrication. The fluid lubricants are generally present as solutions, emulsions, and pastes. Viscosity, density, and compressibility of the fluid lubricants are the major properties of concern in fluid lubrication if used for cold working. Generally, pressure and temperature alter the viscosity of a fluid lubricant. Fluid lubricants are relatively economical and are mainly of two types: synthetic oils and mineral oils. Mineral oils are traditional extracts from fossil oil and the enhanced composition of the same is referred to synthetic oils. Synthetic oils are generally of higher purity and quality compared with mineral oils because of advanced processing and filtering. In contrast, water-based and water-soluble lubricants are preferred in modern environment focused forming industry. Oil-based lubricants are generally blends of constituents of paraffin, naphtha, and aromatics, while animal based oils, vegetable oils, and most of the synthetic oils too fall under the same in category.

4.4.1.1 Pastes

In certain situations, pastes may be categorized under solid lubricants since they display certain solid properties. Pastes are stabilized blends of mineral or synthetic oils, greases, waxes, and sometimes, soaps. Powdered or needle shaped hard waxes and hard soaps also belong to this category.

4.4.1.2 Solutions

A fluid in which several components are mutually soluble is a solution. Many types of solutions are used for sheet-forming applications and they are mainly differentiable as per the base, which is the main constituent in the solution. Aqueous solutions are characterized by higher cooling capacity and support cleanliness during operations. Compounded oils are used in situations where corrosion protection is important, biological contamination should be contained, and thick-film lubrication is required. However, cleanliness and the welding of parts will be difficult. Synthetic compounded oils are relatively costly, but are suitable for high-temperature applications.

4.4.1.3 Emulsions

Emulsion is a mixture in which one immiscible fluid is suspended in another in the form of droplets (Kajdas et al., 1990). In sheet-forming applications, the continuous phase is water, while the suspended phase is oil or a synthetic fluid. This also contains solid lubricants such as graphite, mica, or sodium carbonate, and additives. In all cases, surfactants are present to take care of emulsion stabilization. Some advantages include easy removal from parts and equipment, and provide superior cooling effect. The disadvantages include vulnerable to biological contamination, have less effective lubrication and corrosion deterrent properties than other lubricants. More details on fluid lubricants can be obtained from ASM handbook (2006).

4.4.2 Lubricant Additives

The properties of neat oils may not satisfy a wide range of expectations in metal working. Therefore, several additives, which elevate critical properties, are used in commercial applications. Additives generally serve to improve properties: the load-carrying capacities and bonding shear strength, and viscosity–temperature–pressure characteristics. Another function of lubricant additives is to reduce the likeliness of corrosion of work surfaces. The additives create two types of layers: physical adsorption layers and chemical reaction layers. Friction modifiers in lubricants bond to the metal surface without causing chemical reactions. These types of physically acting additives have the inherent property of temperature dependence. The effect of these additives ranges from a total absence of physical adsorption up to a stable chemical bond (formation of metallic soap). With a combination of carefully selected additives, lubricants can be optimized for the use in a particular application.

Additives like film-strength additives, extreme pressure (EP) additives, suspended solids, corrosion inhibitors, oxidation inhibitors, defoamers, and antimicrobial agents also elevate the applicability of lubricants in metal-working operation. Viscosity of a lubricant is the measure of resistance to flow at a given temperature and the kinematic viscosity is often used to describe the lubricant's viscosity. In metal-working operations, viscosity of the lubricants determines the separation of tool from the workpiece and that controls friction and wear. Optimum lubrication in terms of viscosity is very important in deciding the applicability of lubricant and it depends on the ability of lubricant to penetrate and remain in the contact zone until the forming process is completed.

The main function of additives is to control lubricity, in a way controlling friction and wear, protect workpiece surfaces during contact, prevent corrosion of tool, workpiece, and lubricant handling system. These additives include fatty acids, esters, alcohols, amines, amides, and alkyl acid phosphates. They get adsorbed in the tool-workpiece interface and prevent metal-to-metal contact, welding of asperities, and degradation of surfaces.

EP additives are generally used in severe metalworking processes involving ferrous materials. The EP additives are compounds containing sulfur, chlorine, phosphorous, or combinations of any of these. Some of the examples of this type are sulfurized triglycerides, chlorinated hydrocarbons, chlorinated esters, phosphate esters, and alkyl acid phosphates (ASM Handbook, 1992).

Suspended solids are in the form of very fine powders suspended by either mechanical agitation or emulsifiers. Solids include graphite, MoS_2, metal powders, metal halides, metal oxides, and mica, suspended in water or oil carriers, for forging and extrusion operations. The solids act as lubricant, while the carriers act as coatings on die (ASM Handbook, 1992).

Emulsifiers contain portions that are hydrophilic (compatible with water), and lipophilic (compatible with oil) within the same molecule. The emulsifier's tendency to be more compatible with water or oil is determined by hydrophile–lipophile balance (HLB). At higher HLB values, in a scale of 0–30, water compatibility is more, while oil compatibility is less. Different lubricants like petroleum oils, animal or vegetable oils, waxes, synthetic oils, etc., have different HLB values at which stable emulsion occurs, and this decides the applicability of the lubricant. For instance, it can be seen that the required HLB to emulsify paraffinic petroleum oil is 10, whereas it is 14 for castor oil. Emulsifiers are categorized as cationic, anionic, amphoteric, and nonionic. In this case, the anionic and nonionic are used in metal-working operations (ASM Handbook, 1992).

Corrosion inhibitors like amine-borates, amine carboxylates, amine alkyl acid phosphates, and sulfonates are used to prevent ferrous metal corrosion, while benzotriazole and tolyltriazole are used in the case of nonferrous metals (ASM Handbook, 1992).

Antioxidants are formulated into metal-working lubricants to minimize oxidative degradation of the lubricant into acidic products that tend to corrode metal surfaces including workpiece and tool. They are included not more than 0.5% in the lubricants. Commonly used antioxidants are butylated hydroxytoluene, butylated hydroxy anisole, and phenyl naphthyl amine. Antioxidants are helpful at elevated temperature forming environments like hot forging and rolling as oxidation is prevalent at these temperatures.

Defoamers avoid formation of foam during metal-working operations. Silicones, long-chain alcohols, triglycerides, and water-insoluble polyglycols are some of the examples of defoamers (ASM Handbook, 1992). These have appropriate dispersibility and surface tension. The defoamer acts at the gas–fluid interface to break down the elastic film of the fluid, and thus air is released.

Antimicrobial agents are considered to destroy attacks from living organisms, including bacteria, fungi, and yeast. Some of the antimicrobial agents used regularly are biocides like phenolic materials and formaldehyde-release agents, and fungicides (ASM Handbook, 1992). The parameters such as concentration, influence on emulsion stability, and eco-friendliness decide the applicability of such agents. These agents should be handled with caution because of the toxicity toward human.

4.4.3 Solid Lubricants

Solid lubricants are favored when difficult to form situations in which film strength more than what is possible with solutions and emulsions is required. Solid-film lubricants are generally preferred so that the film thickness can be controlled by end application rather than by process conditions alone. Dry soaps, and metallic and polymer coatings, are industrially used dry films in metal forming. However, buildup and segregation of lubricant on the tooling and regular cleaning of tools are some disadvantages. The mechanism of solid-film lubrication is essentially thick-film lubrication. Solid-film lubricants such as polyethylene, polyvinylchloride, and polytetrafluoroethylene are used in practice. These are applied to the workpiece as a powder, coating, or as thin sheet for giving adequate protection against wear and friction. Other lubricants include MoS_2, graphite, sodium carbonate, teflon, nylon, etc. (ASM Handbook, 2006). Some of the important solid lubricants are described below.

4.4.3.1 Dry Soaps

Soaps are found to be used predominantly on steels. On stainless steels, they are suitable only with coated tools. They are not used for nonferrous metal lubrication. Soap coating does not protect the workpiece and tool against corrosion or rough handling. However, it has the advantage of being water soluble and does not interfere much with welding. For good adhesion, the surface must be cleaned, and freshly pickled. Sometimes waxes are used for lubrication, but it can best be removed by organic solvents.

4.4.3.2 Polymer Coatings

Thermoplastic polymer coatings such as polyethylene, polypropylene, polytetrafluoroethylene (PTFE), polyvinylchloride (PVC), polystyrene, polyimide, and polymethyl methacrylate (PMMA) are used as lubricants for metal-working purposes. Polyethylene and polypropylene are used in the form of thin sheets of few μm thickness along with mineral oil for lubrication. PTFE and PVC are also available as sheets and coatings. PMMA can be deposited from either solvent solution or emulsions and can be modified to suit the boundary lubrication condition. Polyimides have excellent temperature stability. Polymers are used mostly in the form of coatings for lubrication. Since polymer coatings show increased stretchability, the formability of sheets is also improved equivalent to hydraulic bulging (ASM Handbook, 2006). Forming forces are considerably reduced during deep drawing. Under low-pressure conditions, coated sheet can be formed dry (no external lubrication), while under high-pressure conditions, external lubrication is essential. Then compatibility becomes an issue. For instance, PVC is attacked by external lubricants containing chlorine. Water-based lubricants and waxes are used for the purpose.

4.4.3.3 Metal Coatings

Metal coatings, such as tin-coated sheets and zinc-coated sheets, are used predominantly. Cracking or flaking of coating may occur, particularly in severely deforming zone. Die pickup should be taken care with caution. Hard coatings, like TiN, TiC, Al_2O_3, and multilayer coatings, like TiC/Al_2O_3/TiN, are also used as lubricants reducing the wear and friction (ASM Handbook, 2006). These are deposited through physical vapor deposition and chemical vapor deposition techniques. Diamond-Like Carbon (DLC) coatings show low-friction property suitable for dry metal forming and dry machining. A case study has revealed the use of titanium and chromium atoms in the substrate diffuse to coating surface and leads to the formation of a mesh-like heave structure comprised of Cr_2O_3 and $Cr_2Ti_7O1_7$ phases. It was found that this annealed coating exhibits the excellent self-lubricating behavior with friction coefficients below 0.3 from 25°C to 1,000°C (He et al., 2017).

4.5 SELECTION OF LUBRICANTS FOR METAL WORKING

Too-low or too-high friction is not preferred in metal forming. The friction should be maintained within the predefined boundaries as suits for a given product and the process. This varying requirement leads to the need of proper mechanism of controlling friction because controlled friction contributes to the economy of the forming process and the quality of the finished products. In metal-forming processes, the optimum friction is achieved through the careful selection of process parameters, and therefore lubrication is a highly deterministic factor since the lubricant is considered as one of the main approaches of controlling friction. In developing the lubrication system for a metal-forming process, the selection of a lubricant is of utmost importance. However, the selection process of a lubricant is a complicated task and many constraints are to be taken into account. The following are the main determinants in selecting a forming lubricant.

4.5.1 Material Pair in Contact

The material pair is a highly influential factor in selecting lubricants, which includes the workpiece and the die or tooling. Lubricants used in the presence of different material pairs are not similar in numerous aspects. In many circumstances, the chemistry of the material pair is a deterministic factor to secure product requirements and extended die life. Corrosion inhibitors in lubricants play an important role in order to satisfy forming of corrosion-free products, especially for materials sensitive to corrosion, for example, ferrous-based materials used in products and dies. However, the use of water-based lubricants, which is a demanding trend in satisfying environmental concerns, has resulted in severe limitations on the use with material pairs such as steel due to the formation of surface rust, both on products and on dies. In some other applications, the form of the lubricant too depends on the material pair. Dry lubricants such as clay minerals, graphite, iron oxide, and other films such as molybdenum disulfide are popular in forging applications due to their superior properties. The use of these dry films too is a concern with some workpiece materials due to the facts such as their relative hardness and also similar effects on chemical reactions. Table 4.1 gives a brief account on lubricants used in forming common sheet materials.

4.5.2 Nature of the Forming Process

The selection of lubricants largely depends on the nature of the process since the lubricant entrapment and the transfer of forming loads are directly related to the geometry of the stock and the tool. The geometry alters the magnitude and the direction of the normal loads on the sliding surfaces, thus controlling the load requirement for frictional sliding. The magnitude of press loads,

Table 4.1   Lubricants for Forming of Variety of Sheet Materials (ASM handbook, 1992)

Materials	Lubricants
Ferrous metals	Straight oils or emulsions containing fatty materials and extreme pressure (EP) additives
Aluminum alloys, copper alloys	Straight oils or emulsions containing fatty acids, esters, or alcohols
Titanium alloys	Halogenated hydrocarbons like fluorocarbons

degree of deformation, development of interfacial shear stresses and strain rates, and the working temperatures significantly differentiate the two processes: bulk forming operation, such as forging rolling or a sheet-forming operation, such as stamping or deep drawing. In bulk forming operations, since the volumetric deformation is severe, the role of the lubricant is different from the counterpart, the sheet-forming operations. In sheet-forming operations, the contact surface area is relatively larger than many of the bulk forming processes and this results in more frictional contact.

Majority of the industrial sheet deformation processes are carried out in cold conditions, and therefore oil lubricants are widely selected. In addition, several solid films and self-lubricating interfaces are also solutions for lubrication in the recent past. However, the fluid films are associated with traditional deficiencies such as cleaning efforts, health and safety concerns, and negative effects on the environment, especially in the case of the synthetic oils.

Special care should be taken in introducing lubricants in bulk forming, especially in hot forming to retain the effectiveness of the lubricant. Two important case studies with aluminum rolling lubricants and steel warm forging lubricants are presented with some details. Making aluminum strips involves three rolling stages. In the first stage, an ingot of several hundred millimeters thick is reduced to about 30 mm in a hot reversing mill. The ingot is then rolled 15–20 times in the reversing mill. The lubricants used in hot reversing mill are formulated to generate high friction to reduce the denial cases, at the entry of the roll. These lubricants are very stable emulsions containing petroleum sulfonate, and nonionic emulsifiers with low levels of film-strength additives, oxidation inhibitors, corrosion inhibitors, and antifoaming agents. In the second stage, using hot continuous tandem mill, the sheet is rolled to 2–5 mm thickness, typically through two to six stands. For this stage, lubricants containing petroleum oils, and either anionic or nonionic emulsifiers with high levels of triglyceride film additives are used. The oil concentration for both the stages in the lubricant is about 5%. These lubricants have shorter lives because of hot rolling conditions. In the third stage, cold rolling is done to reduce thickness to 0.3–0.9 mm. In cold rolling stage, straight petroleum oils with small amounts of fatty acids, esters, alcohols are used as lubricants (ASM Handbook, 1992).

In forging of billet at room temperature is commonly done by using phosphating, in which a Zn phosphate film that helps in retaining soap lubricants during forming is used. In the case of stainless steels, oxalate films are used for lubricant retention. At warm forging temperatures, graphite dispersed in either water or oil carrier is used as lubricant. Phosphate coatings are ineffective in these situations. The lubricant is sprayed on the dies and billet to ensure complete coating. More details can be seen in ASM Handbook (1992, 2006).

4.5.3 Actual Normal Contact Stresses in Forming

The degree of achievable deformation and thus the need of press loads which determines the tonnage of the press gives technical requirements of the lubrication. The resultant area of the load-bearing surface alters the normal contract stress. As per the classical laws of friction, interfacial shear stress needed for the material flow over the contact surfaces is a resultant of these normal contact stresses. Therefore, the magnitude of the interfacial shear stress, and thus the normal contact stress, is accountable in application of controlling methodologies of friction mainly in introducing a lubricant in controlling friction. Furthermore, the load-bearing capacity of the lubricant is a deterministic factor in the selection of a lubricant. The interfacial stresses transfer into the lubricant system as fluid pressure. The capacity of withstanding the pressure is therefore important in matching a lubricant for a given application. Viscosity alters the capacity of load bearing in a liquid lubricant. Furthermore, EP additives modify the performance of a lubricant at high-pressure forming, especially in bulk forming applications.

4.5.4 Surface Characteristics of the Workpiece Material

Although the surface contact and the surface properties are believed to have no effect on conventional friction theories, the effect of such parameters cannot be ignored in metal-forming operations. Surface roughness and the topography of the surfaces have a significant effect on the friction between sliding interfaces. Thus, the surface characteristics of the workpiece material as well as the changes taken place on these parameters during the process directly influence the lubricant selection. In association to surface properties, the wear and transfer of materials, in terms of friction welding or galling, are of utmost importance since they alter friction in great extent. The roughness of the surface contributes to the lubricant entrapment and results in different lubrication regimes. Thus, it is a well-established fact that the frictional behavior in metal forming is a function of roughness of the two surfaces moving against each other. It is generally accepted that the friction coefficient increases with an increase in surface roughness. Furthermore, when a metal is being deformed, the surface roughness will generally increase with the degree of deformation and, therefore, increased roughness enhances the friction when sliding progresses.

The effect of texture of the surface is explained with the differentiation of the frictional stresses. Frictional stresses at the tooling–workpiece interface may be broken into adhesion and ploughing components. In metal-forming processes, the friction is mainly generated through the adhesion at contacting asperity peaks (Bowden and Taylor, 1953). However, both components are important in assessing coefficient of friction and the selection of lubrication (Wilson, 1991).

The texture of the surface facilitates the lubricant entrapment and balancing friction in different ways. The surface topography together with pressure and temperature conditions will lead to complicated frictional responses. Increased roughness results in more asperity contacts, and thus shifts the lubrication regime. As an example, with the special emphasis to deep drawing where the surface pressure on the material which is being deformed is much lower than the most of other forming processes can lead to different conditions of friction. Under these circumstances, the lubricant balances the friction by two actions. The pressure in the lubricant generated by the flow decreases the friction by relieving the load on the roughness peaks, while the friction is partially managed by the flow of lubricants through roughness created micro channels in the surface. However, this flow is only a minor source of friction in high-pressure applications. Furthermore, the tribology and friction drastically changes due to the flattening of asperities as the sheet passes through number of die and punch surfaces in multipass or multidraw of sheet-forming processes. Surface asperities are generally deformed during the forming process due to contact pressure, and therefore the frictional conditions may vary during the process. Some high spots of asperities can project from the surfaces, despite the tool pressure during forming, and these tend to alter the lubrication regime. The form and the height of the asperities of the surface can have a significant effect on the properties of a sliding contact running under mixed lubrication (Anderson and Russo, 1994) which is more common in applied sheet-forming processes. In contrast, in the same research (Anderson and Russo, 1994) it is found that in the thick-film lubrication regime, the fluid film separation of the surfaces is such that there are no asperity contacts and consequently the wear is low and the friction is mainly determined by the shear of the lubricant film. Therefore, the smoother the contact surfaces the higher the load that can be supported, but on smooth surfaces, local scratches may have a strong negative effect on the load level when the transition occurs from mild-to-severe sliding (Anderson and Russo, 1994). Even in the boundary lubrication regime (where the asperity contacts are very likely), the high interface pressures and plastic deformation of the substrate may result in changes in the effective hardness of the asperities: therefore, the Coulomb's law will break down (Wilson and Sheu, 1988). Therefore, in all lubrication regimes, the surface conditions are very much significant in controlling friction and the selection of lubricants and their behavior.

Therefore, it is clear that the surface roughness, friction, and lubrication are interrelated in sheet-forming operations. The surface topography and texture is another parameter that influences the friction, and thus the selection process of lubricants.

4.5.5 Pressure, Temperature, and Speed Characteristics of the Process

Temperature and the average or the maximum pressure in the forming system are key determinants in the selection process of lubricants. Pressure-absorbing capacity of the lubricant is essential due to possible severe compressive stresses in the forming process. The capacity to withstand against the pressure is a function of lubricant properties, mainly the density and the chemical composition. However, denser lubricants sometimes are not preferred for thin-film applications due to its limitations of the application methodologies. Certain lubricants chemically breakdown once exposed to high pressures, which could be obvious in bulk forming operations that undergo severe deformation pressures. Table 4.2 summarizes the suggested lubricants in cold forming and hot forming conditions for varied sheet materials.

Flash point of the lubricant determines the safe operation temperatures of the lubricant. Specially in the processes carried out at temperatures such as in hot working, and sometimes, warm working, the excessive heat can cause certain fluid lubricants to ignite. In addition, the fluid lubricants may be subjected to chemical breakdown even at moderate temperatures. Therefore, process temperature is one of the key determinants in selection of forming lubricants. The constraint with the process

Table 4.2   Lubricants Used for Hot Working and Cold Working Processes

Material	Cold Forming Lubricant	Hot Forming Lubricant
Aluminum alloys	Synthetic solutions, emulsions, lanolin suspensions, water suspensions, soap solutions, mineral oil, fatty oils	Graphite suspension
Copper alloys	Emulsion, fatty oils, mineral oils, soap suspensions, water suspensions, tallow suspensions, synthetic solutions	Pigmented pastes, graphite suspensions
Magnesium alloys	Solvent plus fatty compounds, mineral oils plus fatty compounds	Graphite plus molybdenum disulfide, soap plus water, tallow plus graphite
Nickel and nickel alloys	Emulsions, mineral oils plus EP additives, water plus chlorine additives, conversion coatings plus soap	Graphite suspension, molybdenum disulfide suspension, resin coating plus salts
Refractory metals and alloys	Copper plating	Molybdenum disulfide suspension
Carbon steels and low alloy steels	Emulsions, soap pastes, water, fatty oils plus, mineral oils, polymers, conversion coating, molybdenum disulfide, graphite in grease, synthetic solutions	Graphite suspension
Stainless steels	Fatty oils, mineral oil, water, polymers, conversion coating plus soap, mineral oil plus additives, pigmented soaps	Graphite suspension
Titanium and titanium alloys	Water, pigmented soaps, polymers, conversion coating plus soap	Graphite suspension, MoS_2 suspension

temperature has paved the extensive use of dry films in high-temperature applications. In addition, self-lubricating materials are also proposed. The temperature dependent behavior of viscosity of fluid lubricants affect the performance of lubrication in complex manner. The temperature results in the reduction of the viscosity and it enhances the fluid to flow along microchannels on the surface and enhance the lubrication. The lubricant flow along the surface micro channels is an accountable factor in thin-film lubrication. However, with the drop in the viscosity, lubricant tends to spread thinner and the lubricant loses its load-bearing capacity and reduces the effectiveness of thick film and sometimes, mixed lubrication regimes with possible asperity contacts.

Speed of draw is another influential process variable on lubrication. In most of the conventional metal-forming applications, wet-lubricated surface contacts are maintained unless otherwise dry film lubricated or the workpiece is self-lubricating. In wet-lubricated contacts, as depicted by the dependencies in Stribeck curve, the speed of relative sliding is judgmental for the coefficient of friction. Furthermore, the viscosity of the lubricant controls the velocity of sliding along the frictional path. This can be explained by the phenomena that the interfacial shear at the wet contact controls the relative velocity of the sliding pair altering frictional characteristics. Therefore, the velocity of sliding is deterministic in the selection of forming lubricants.

4.5.6 Application Method of Lubricants

In metal-forming operations, the lubricants are introduced into the forming system in different ways. Most of these are direct on surface applications or depositions. Method of application is a practical concern in selection of lubricants.

Drip method. Drip method is the simplest and the cheapest method to apply a lubricant on a workpiece surface, which is simply by dripping the lubricant on the work surface. However, dripping is not the most suitable method of application due to its practical limitations in achieving controlled film thicknesses in friction sensitive operations.

Roll coating. Roll coating is another technique of application where a lubricant applied on the surfaces of the blank while it is moving it between two rotating rollers under controlled pressure, allowing a precise control of the amount of lubricant. This method is generally limited for sheet-forming applications and the applications where thin stocks are formed.

Electro deposition. Electro deposition is a relatively expensive application process where a lubricant is deposited on the work surface using an electric charge. Electro deposition is a fast method of application and minimizes waste. This process is only applicable for large-scale operations since it requires high capital investment.

Airless spraying. Airless spraying is a precise technique of application of forming lubricants. Precise amount of lubricant can be applied on local areas, with a minimal waste of lubricant. However, the method is limited for low viscosity lubricants.

Mops and sponges. Another low-cost method is the sue of mops and sponges which is still used in small-scale operations. The control and the consistency depend on the skill of the worker and very often lead to waste of lubricants and sloppy work environments.

4.5.7 Environmental and Health and Safety Concerns

Environmental effects and health and safety concerns are among the highly demanding trends in selecting metal-forming lubricants in today's world. Many of these are formulation-related properties of lubricants and not process-related characteristics. Few decades ago, synthetic lubricants were the most commonly used friction inhibitors in metal-working industry. These usually display reasonable friction control. However, today, they are not among the best options because many of them do not display certain operational and post operational compatibilities and standards. Some of these standards include the environmental effects such as the ability to dispose and the pollution and reusability which gives all overall sustainability of the lubrication. On top of these environmental concerns, the occupational safety and satisfied health-related issues during usage, especially the operator and user related issues also have become significant. Biological contamination and worker skin infection are among the main health and occupational safety related problems with lubricants.

In addition to the above parameters, there are many complicated aspects making the selection process highly versatile. The cost of the lubricant and the cost of the lubrication process, which includes the cost of disposal as well, play crucial role in lubricant selection.

4.6 LUBRICANTS FOR SUSTAINABLE METAL FORMING

Lubricant selection shall not be restricted to technical feasibility. However, the traditional selection of metal-working lubricants have limited to a few process focused or the product focused concerns. Sustainability of lubricants in metal forming has been a growing demand in the manufacturing industry. The use of fossil oil-based lubricants may not be the right choice since fossil-based products are diminishing sources. In addition, fossil oils usually poor in properties related to cleaning and disposal, thereby not a satisfactory solution. Furthermore, the instability in oil prices has made alternate lubricants a better choice in automotive and other heavy industries. Environmental friendliness, consistency, durability, and energy efficiency issues are being sensitized by policy makers regularly in various forums across the globe. Developing light-weight materials, efficient processes, alternative fuels, effective waste minimization, and many more strategies are followed to address environmental concern. Maintaining sustainability and green manufacturing are considered seriously by many leading industries.

The environmental and ecological problems associated with oil-based hot forging lubricants have led over the past several years to the development or attempted development of more desirable water-based lubricating compositions for use in forging. As per the patent (Kratzer, 1983), it is evident that past attempts directed to water-based compositions have involved graphite, clay minerals, iron oxide, and other materials such as molybdenum disulfide. However, the attempts have in most instances not been fully satisfactory for numerous reasons, such as a failure to properly lubricate the forging die under actual operating conditions or because the water present in the composition did not adequately wet the metal surfaces involved. One important move in the direction is to develop 'biodegradable lubricants' from vegetables, crops, edible items, etc., that are applicable for various industries (Nagendramma and Kaul, 2012).

4.6.1 Vegetable Oils

Vegetable oils are a renewable source of oils, and they are biodegradable and human friendly. However, out of more than 350 oil-bearing crops, only a few are being used for industrial and domestic applications. Palm, soya bean, sunflower, coconut, safflower, rapeseed, cotton seed, and peanut oils are considered as potential bio lubricants. Please refer to Table 4.3 for the oil content in percentage of volume of some edible and nonedible oil seeds (Mobarak et al., 2014). Coconut has about 65% oil content and tops in the list. Most of the required properties as discussed earlier which are crucial properties of these natural oils are presented in Table 4.4 (Mobarak et al., 2014).

The vegetable oils have the following advantages over mineral oils resulting in excellent lubricity. They do possess a higher viscosity index ensuring effectiveness at elevated temperatures and the increased load bearing. Vegetable oils in general have low pour point providing good lubrication for cold starts. One another remarkable advantage is the higher flash point that ensures safety from prevent catching fire during leakage and the ability to maintain lubricant properties at elevated temperatures. One of the significant points is that these vegetable oils have a lesser tendency for reaction with most of the metals inhibiting corrosion and oxidation.

Table 4.3   Oil Contents in Natural Seeds Used for Extracting Oil Lubricants

Edible Oils		Nonedible Oils	
Species	Oil Content (% of Volume)	Species	Oil Content (% of Volume)
Rapeseed	38–46	Jatropha	40–60
Palm	30–60	Neem	30–50
Peanut	45–55	Karanja	30–50
Olive	45–70	Castor	45–60
Corn	48	Mahua	35–50
Coconut	63–65	Linseed	35–45
		Moringa	20–36

Source: With permission from Mobarak et al. (2014), Elsevier.

Table 4.4   Comparison of Properties of Vegetable Oils with Mineral Oils

Properties	Vegetable Oils	Mineral Oils
Density in kg/m^3 at 20°C	940	880
Viscosity index	100–200	100
Shear stability	Good	Good
Pour point, °C	−20 to +10	−15
Cloud flow	Poor	Poor
Miscibility with mineral oils	Good	NA
Solubility in water	No	No
Oxidation stability	Mediocre	Good
Hydrolytic stability	Poor	Good
Sludge formation	Poor	Good
Seal swelling tendency	Slight	Slight

Source: With permission from Mobarak et al. (2014), Elsevier.

The most important consideration with the vegetable oils is related to the disposal and cleaning. Most of the common vegetable oils are biodegradable and nontoxic while satisfying high level of cleanliness. These lubricants, in particular, coconut oil, carry a reasonable anti-wear capability, which is a demanding need for metal forming. Finally, these are, in general, edible and carry no odors, which satisfies a good occupational safety requirement. Beyond these advantages and disadvantages, vegetable oils possess good physio-chemical properties, as shown in Table 4.5.

4.6.2 Self-Lubricating Films

Self-lubricating tool coatings are identified as a sustainable approach in lubrication since it minimizes major environmental concerns as cleaning efforts. The modification of tooling through tool design and fabrication with lubrication and friction conditioning is an approach to energy and waste minimization without spoiling the process efficiency. There are many examples that highlight the significance of self-lubrication tool coatings in metal forming. Under such dry metal-forming applications, lubricants are not used during forming, instead, ceramic and other coatings on the die surface or self-lubricating systems are used. TiCN, TiC–TiN, and DLC coatings are possible to be used at elevated temperatures (Reisel et al., 2003). Furthermore, these efficient coatings and proper choice of substrate will improve drawability in deep drawing by delaying fracture, which otherwise would have failed much earlier in dry forming without coating (Murakawa et al., 1995). Another advantage is that the galling performance of forming tools is improved with these coatings as suggested by Podgornik and

Table 4.5   Physio-Chemical Properties of Edible and Nonedible Vegetable Oils

Vegetable Oils	Kinematic Viscosity at 40°C (mm²/s)	Oxidation Stability 110°C (h)	Cloud Point (°C)	Flash Point (°C)
(i) Edible Type				
Palm	5.72	4	13	165
Sunflower	4.45	0.9	3.42	185
Coconut	2.75	35.4	0	112
Soybean	4.05	2.1	1	176
Linseed	3.74	0.2	−3.8	178
Olive	4.52	3.4	—	179
Peanut	4.92	2.1	5	177
Rape seed	4.45	7.5	−3.3	62
Rice bran	4.95	0.5	0.3	—
(ii) Nonedible Type				
Jatropha	4.82	2.3	2.75	136
Karanja	4.8	6	9	150
Mahua	3.4	—	—	210
Neem	5.2	7.2	14.5	44
Castor	15.25	1.2	−13.5	260
Tobacco	4.25	0.8	—	166

Source: With permission from Mobarak et al. (2014), Elsevier.

Hogmark (2006). The performance of such coatings can be further improved by metallic interlayers as suggested by Taube et al. (1994). In all these cases, the usage of external lubricants is avoided, thereby reducing their impact on the environment. Moreover, usage of extra lubricants is also minimized.

The total energy consumption is a significant factor in finding sustainable solutions in lubrication in metal forming. The energy inputs start from the manufacturing, application and the disposal processes, directly. However, indirect energy consumption also to be taken into account. Manufacturing of mineral or fossil-based lubricants devour great deal of energy from the extraction process. Refining and blending are too consuming energy and later the method of application. Cleaning and disposal processes directly or indirectly consumes additional energy to worsen the environmental concerns. Thus, the total energy consumption from the manufacture of the lubricant until the disposal is a deterministic factor. In general, biological solutions do consume lesser energy in manufacture relative to the counterparts. In addition, they are renewable and satisfy other environmental and safety concerns. Therefore, lubricant selection and the use of renewable sources of lubrication in metal working lead to sustainability in metal forming.

4.7 SUMMARY

The friction in metal-forming applications was identified as a variable that needs to be controlled in assuring quality of the formed parts as well as to elevate the process economy. Different models for the analysis of friction were presented and evaluated their suitability in assessing lubricated friction in metal-forming applications. Lubrication was understood as the main strategy of controlling friction. Thus, the effect of the lubricant on coefficient of friction at the sliding interfaces during metal forming was discussed in the early sections of the chapter.

Later, the types of lubricants were presented with special emphasis toward improving lubricants to suit different process conditions. This led to the selection criteria of lubricants for metal-forming processes. Number of different factors important in selection of lubricants were discussed including the material-based determinants such as the materials in the contact pair, the surface texture and topography together with nature of the process, for example, the sheet forming or bulk forming, process conditions such as tool pressure, forming temperature, etc. Finally, the need of including environmental concerns in lubricant selection was highlighted giving way to sustainable lubricants and lubrication.

REFERENCES

Anderson, S., Russo, E.S. 1994. The influence of surface roughness and oil viscosity on the transition in mixed lubricated sliding steel contacts, *Wear*, 174: 71–79.

ASM Handbook, 1988. Metal Working: Bulk Forming, Vol. 14, 1st Edition.

ASM Handbook, 1992. Friction, Lubrication, and Wear Technology, Vol. 18, Metal working lubricants by Joseph T. Laemmle, pp. 255–277.

ASM Handbook, 2006. Metal Working: Bulk Forming, Vol. 14B, 9th Edition, Ed. Semiatin, S.L.

Bowden, F.P., Taylor, D.T. 1953. *The Friction and Lubrication in Solids*, 1: 98–100. Oxford University Press, USA.

Dixit, U.S., Narayanan, R.G. 2013. *Metal Forming Technology and Process Modelling*, McGraw Hill Education, New Delhi, pp. 1–568.

Emmens, W.C. 1988. The influence of surface roughness on friction. Controlling sheet metal forming process. 15th Biennial Congress International Deep Drawing Research Group, Dearborn, USA, 16–18 May 1988, ASM International, NADDRG, 1988, pp. 63–70.

He, N., Li, H., Ji, L., Liu, X., Zhou, H., Chen, J. 2017. Reusable chromium oxide coating with lubricating behavior from 25 to 1000°C due to a selfassembled mesh-like surface structure, *Surface and Coatings Technology*, 321: 300–308.

Kajdas, C., Harvey, S.S.K., Wilusz, E. 1990. *Encyclopedia of Tribology*, 15. Elsevier, North Holland.

Kratzer, T.L., 1983. Water-based metal forming lubricant composition and process, Patent, US 4401579 A.

Mobarak, H.M., Mohamad, E.N., Masjuki, H.H., Kalam, M.A., Al Mahmud, K.A.H., Habibullah, M., Ashraful, A.M. 2014. The prospects of bio-lubricants as alternatives in automotive applications, *Renewable and Sustainable Energy Reviews*, 33: 34–43.

Murakawa, M., Koga, N., Kumagai, T. 1995. Deep drawing of aluminium sheets without lubricant by use of diamond-like carbon coated dies, *Surface and Coatings Technology*, 76–77: 553–558.

Nagendramma, P., Kaul, S. 2012. Development of ecofriendly/biodegradable lubricants: An overview, *Renewable and Sustainable Energy Reviews*, 16: 764–774.

Podgornik, B., Hogmark, S. 2006. Surface modification to improve friction and galling properties of forming tools, *Journal of Materials Processing Technology*, 174: 334–341.

Reisel, G., Wielage, B., Steinhauser, S., Hartwig H. 2003. DLC for tool protection in warm massive forming, *Diamond and Related Materials*, 12: 1024–1029.

Sanchez, R. 1999. Experimental investigation of friction effects enhanced by tool geometry and forming method on plain strain sheet metal forming, *Tribology Transactions*, 42: 343–352.

Taube, K., Grischke, M., Bewilogua, K. 1994. Improvement of carbon-based coatings for use in the cold forming of non-ferrous metals, *Surface and Coating Technology*, 68–69: 662–668.

Wilson, W.R.D., 1979. Friction and lubrication in bulk metal forming processes, *Journal of Applied Metalworking*, 1: 1–19.

Wilson, W.R.D., 1991. Friction Models for Metal Forming in the Boundary Lubrication Regime. *Journal of Engineering Materials and Technology*, ASME, 113: 60–68.

Wilson, W.R.D., Sheu, S. 1988. Real area of contact and boundary friction in metal forming, *International Journal of Mechanical Science*, 30: 475–489.

CHAPTER 5

Development in Materials for Sustainable Manufacturing

Saptarshi Dutta and P. S. Robi
IIT Guwahati

CONTENTS

5.1 INTRODUCTION

Continuous pursuit of higher quality of life, population explosion, scarcity of natural resources, high levels of business competitions, environmental degradation, greenhouse emission, and unequal social equities have drawn the attention of industries to develop strategies that are sustainable for the success of the industries. The United Nation's 1987 report of the World Commission on "Environment and Development: Our Common Future" defined sustainable development as "development that meets the needs of the present without compromising the ability of future generations to meet their own needs."

Manufacturing firms across the world are facing the challenge of producing more products at lower cost using minimum resources and generating less waste and environmental pollution. Sustainable manufacturing is defined as the creation of manufactured products that use processes that minimize negative environmental impacts, conserve energy, and natural resources, are economically sound, and are safe for employees, communities, and consumers (US Department of Commerce, 2011). A good sustainable manufacturing strategy will focus on: (1) promote social well-being, (2) ensure economic stability, (3) minimize resource depletion, and (4) minimize environmental degradation. The sustainable manufacturing strategy has resulted in a number of sustainable manufacturing practices, which can be viewed at three stages: (1) product level in which the concept of recover, redesign, and remanufacture has been added to the traditional concept of reduce, reuse, and recycle, thereby changing the strategy from single life cycle to multiple life cycle; (2) process level focused on the technological improvement leading to reduction in resource utilization and waste generation, and (3) system development based on the organization and supply chain management (Jayal et al., 2010; Jawahir and Dillon, 2007).

Sustainable manufacturing has become an integral part of global economy and plays a critical role in the modern life. Although several brain storming efforts and intense deliberations over sustainable manufacturing have taken place over recent few decades, researchers are yet to converge at a common definition of sustainable manufacturing (Dornfeld, 2009; Haapala et al., 2013; Wang et al., 2016; Millar and Russell, 2011; Despeisse et al., 2012; Jawahir and Bradley, 2016). The fact is that there is no simple way of explaining how to develop sustainable products. The simplest way of defining it is that it is a product, which will give as little impact as possible on the environment during its life cycle (Ljungberg, 2007). The period of life cycle begins from the extraction of raw material and goes through various stages of modification, production, use, and final recycling. The key areas identified for sustainable development include environment, equity, and futurity. During the life cycle, the ecosystem must be balanced, where the perennial issue of poverty and equal usage of natural sources should be solved, in order to maintain stable societies with equity. Finally, the development should be carried out keeping in view the future generations.

The recent method of understanding the sustainability of a product is by the Triple Bottom Line (TBL) approach in which the corporate performance is assessed based on economic, social, and environmental parameters (Elkington, 1997). Therefore, the business firm correlates the traditional economic bottom line with the impact of the business on the society and environment, as shown schematically in Figure 5.1.

Any business will be sustainable only if it is economically viable. The ecological side gives emphasis on factors such as clean landscapes, pure air and water, minimum environmental pollution, reduced greenhouse effect, etc. The societal or equity side can explain if a certain product can lead to social welfare regardless of people's background and whether it promotes proper health, safety, etc.

As per the United Nations statistics, the world human population is expected to reach 9.3 billion in 2050 from the current 7 billion. This will have a strong impact mainly by increasing urbanization

Figure 5.1 The integral elements of sustainable manufacturing.

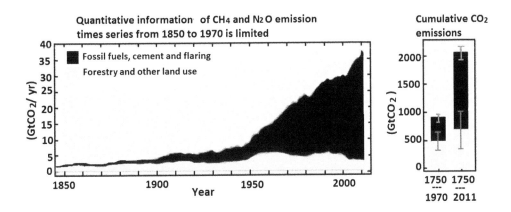

Figure 5.2 Global CO_2 emissions. (From Pachauri et al., 2014.)

and affluence in the developed and developing nations. The interdependence of life and the ecosystem in which it resides are the driving force for the sustainable growth and development. Sustainable manufacturing has emerged as a globally recognized mandate and it includes green or environment-friendly manufacturing practices that facilitate the self-healing and self-replenishing capability of natural ecosystems. Figure 5.2 clearly indicates how carbon dioxide emissions have increased significantly over the past decades and burning of fossil fuels as one of the major contributors to this effect. The concept of sustainability manifests itself more in the development of new materials and manufacturing technology. In the modern era of technological development, the main emphasis for sustainable manufacturing is focused on selection of materials, manufacturing processes, and recycling or reuse of materials that ultimately leads to environmental sustainability. This highlights the need to assess various environmental impacts of the business.

5.2 ENVIRONMENTAL IMPACT ASSESSMENT

With limited resources and serious environmental impacts, it is obvious that a more sustainable life style will become more and more important. It is generally difficult to assess the environmental impacts of metal production processes because of the many inputs and outputs involved. Life cycle assessment (LCA) has been a useful tool for identifying the part of a metal that has a significant effect on the environment (Finnveden, 2000; Heijungs, 1992; Hundal, 2001). Therefore, a product is evaluated step by step from the extraction stage till the expiry of its intended life. During the life cycle of a product, different stages are passed, such as material extraction, manufacturing, packaging, transportation, product use, recycling, and/or disposal. All these stages will have a certain environmental impact, which is mainly caused by the flow of the materials through different stages (Hui, 2002).

Eco-design is an approach aimed at reducing the environmental impact for any manufacturing industry. This is basically aimed at (1) selection of materials having very low environmental impact, (2) design of clean production processes, (3) minimizing or avoiding hazardous and toxic materials, (4) energy audit for maximizing the efficiency during production and use, and (5) implementation of a proper waste management system and recycling (Ljungberg, 2007). Since the material determines the use of our natural resources as well as the amount of energy used for the production and use of the product (Ljungberg, 2003; Ashby, 1999; Ashby and Johnson, 2002), the selection of materials for a certain product is of vital importance.

5.3 MATERIAL SELECTION

Selection of materials plays a crucial role in ensuring the sustainability of manufacturing a product. Introduction of a new material can even lead to elimination of age-old processing techniques and/or materials. Typical example is the development of new nickel base super-alloys for high temperature gas turbine blade applications. Earlier, the super-alloy materials were manufactured by forging process. The emergence of new alloys resulted in the development of Investment casting technique. This was followed by the advancement of solidification technology which lead to the development of directional solidification and subsequently single-crystal casting technique. The recent technological development in high temperature gas turbine applications are the introduction of blade cooling and thermal barrier coatings. Concomitantly, new materials that can withstand very high temperatures were also developed. Each of these stages resulted in improved efficiency of the gas turbine engine. Another example is the development of catalytic converters for petrol engines resulting in less usage of carburetors in petrol engines (Ljungberg, 2007).

Some of the major guidelines for selection of materials for sustainable manufacturing are as follows: (1) the material properties namely strength, Young's modulus, density, ductility, toughness, formability, corrosion resistance, etc., (2) efficiency of the manufacturing processes, (3) power consumption during manufacturing and usage, (4) material cost, (5) minimum environmental impact, and (6) recyclability and/or reuse.

For structural applications, materials exhibiting a combination of high strength, light weight, reasonable ductility and high toughness along with good corrosion resistance are the best choice especially for products with long life span. The last six decades have seen the development of metallic as well as nonmetallic materials having low density. Weight reduction in transportation sector has been identified as one of the key issues for meeting stringent requirements on greenhouse gas emissions. Weight reduction in automobiles can be achieved either by design optimization or by replacing heavier steel components with lighter metals. The development of materials such as aluminum, magnesium, and titanium, having high specific strength and stiffness, is given high thrust for use in sectors such as automobile and aircraft, as well as space application. Manufacturing of metal foams and honeycomb structures results in reduced material usage, very high stiffness, increased payload capacity, reduced fuel consumption, and reduced waste emissions. The last several decades of research were focused on development of the following materials.

5.3.1 High-Performance Steels

Automotive industry has vigorously pursued weight reduction as a major technical target compared to any other industrial sectors. The need to improve crash safety and to reduce consumption of fuel and emissions has been the motivating factor for this weight reduction. The intensive usage of lighter materials has the potential to reduce weight significantly. The state-of-the-art car bodies consist of up to 80% by weight high-strength steel (HSS). By this approach, the weight of the car body is reduced by around 40–100 kg compared to that of a traditional car body which is primarily made of mild steel (Lüdke and Pfestorf, 2006).

Lighter vehicles reduce fuel consumption, provide quick acceleration, smaller braking distance, and exhibit better agility. However, in addition to high strength, the light-weight selection of steel is also determined by the ease with which it can be welded, formed, and worked upon by other manufacturing techniques. Service conditions necessitate that the particular steel should possess enough durability during loading conditions, impact resistance at the lowest temperatures which is expected during service, high stiffness, and better resistance to corrosion. In addition to technical benefits, the related cost is also a key factor during the decision-making stage. This explains why HSS remains the favored material for vehicle construction, because not only is the vehicle weight reduced and the technical performance augmented, but the total production cost is also lowered.

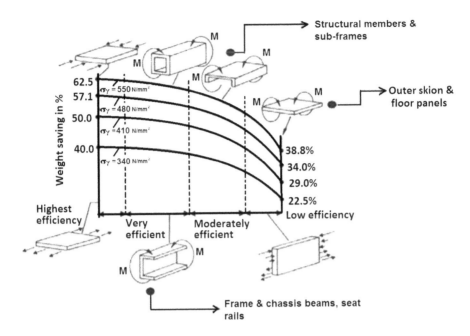

Figure 5.3 Weight-saving potential by replacing a 200-MPa yield strength steel with high-strength steels for varied loading conditions. (Reprinted with permission from Hardy Mohrbacher, 2013, Springer Nature.)

Figure 5.3 shows the weight reduction potential by increasing the strength of steel for the different load condition. It is evident from Figure 5.3 that increasing the yield strength of steel from 200 to 550 MPa makes possible to have a component weight reduction up to 62.5% for structural parts.

One of the major advantages of using HSS for manufacturing a vehicle is that the present manufacturing technologies can be used to a great extent and there is no need for new equipment. However, in alternative materials, usually a completely new manufacturing approach is required and involves a substantial amount of capital for developing novel manufacturing equipment. In the case of HSS, some precise upgrades have to be made, and upgradation of knowledge is essential. Some of these upgrades include higher capacity press due to high strength, upgradation of tool materials, and adaptation of new welding procedures. More recently, the technology of hot-stamping has been extensively introduced for making car body panels. The classic conflict of high strength and good formability can be solved by this technology. With the help of hot-stamping technology, steel components having high strength (i.e., ranging from 1,500 to 2,000 MPa) can be produced even after retaining their complex component shapes (Mohrbacher, 2013). In the case of lower strength level, a good variety of steels having improved cold formability has been developed which can be used in conventional press stamping lines. In order to achieve metallurgical optimization, microstructural fine tuning can be done which is achieved by modifying the chemical composition and/or by thermo-mechanical processing. In this respect, two specific alloying elements, niobium, and molybdenum, have proven to be particularly beneficial when added to steel at very low levels. Ferritic or bainitic single-phase microstructure is preferred for these alloys to minimize edge damage and other related failure during peripheral stretching. Strengthening of ferritic microstructure is achieved by grain refinement and precipitation hardening. Niobium is a very strong grain refiner and is added to steel in the range of 0.03–0.1 wt% (Cuddy and Raley, 1989). Niobium also enhances the strength by precipitation hardening which is achieved due to the precipitation of nano-sized NbC particles uniformly in the ferrite matrix (Gray and Yeo, 1968). Molybdenum up to 0.3

wt% promotes transformation of pearlite to bainite, thereby increasing the toughness of the steel (Wang and Kao, 1993), in addition to impeding the coarsening of the precipitates of NbC and TiC (Funakawa and Seto, 2007).

A typical automobile chassis and frame components are exposed to high static and dynamic loads. These parts generally involve rigorous welding operations. Hence, for good weld ability, steels having low carbon equivalent is vital. High-strength low alloy (HSLA) steel is usually the ideal material. HSLA specifications characteristically limit the maximum carbon content to 0.1%. In the case of automotive applications, HSLA steel with yield strength ranging up to 700 MPa is currently used. Frame and chassis components possess shape complexity. In order to manufacture a frame and chassis component, forming methods like profiling, press-brake bending, stretch flanging, as well as hole cutting and expansion are involved. The presence of hard phases like martensite or pearlite in the microstructure intensifies the possibility of edge damage.

If all materials are considered which are available for vehicle construction, steel has undoubtedly the lowest CO_2 emissions on an equal part weight basis. Alternative materials have higher emission levels during their primary production. However, their inherent lower density enables weight reductions to the level of 40%–60%, which reduces the total emissions during the life of the vehicle. A weight reduction of around 25% is achieved if intensive use of advanced HSS is carried out. Metallurgical optimization of HSS reveals a greater processing window and more robustness during its production and manufacturing. This in turn results in a more efficient use of the material having lower scrap rates, which in turn reduces CO_2 emissions during the manufacturing phase. Niobium and molybdenum, which act as an alloying element, can be supplemented either singly or in various combinations to make a noteworthy contribution to attain this optimization.

5.3.2 Aluminum Alloys

Aluminum alloys are widely used to replace heavy structures in the automobile industry. The main advantage of aluminum alloys is the high strength and light weight as compared to steels. They are very ductile and malleable and hence can be easily processed by casting followed by secondary bulk processing techniques such as rolling, drawing, extrusion, etc. The cast aluminum–silicon alloys are extensively used as engine block and piston material in internal combustion engines. The wrought aluminum alloys can be hot/cold deformed easily to any shape by sheet metal forming techniques making it the ideal choice for automotive bodies. The properties of wrought high-strength aluminum alloys such as 2XXX, 6XXX, and 7XXX series alloys can be tailored by a sequence of thermo-mechanical treatments. Aluminum–copper alloys, developed around one century back, still remain unchallenged. The influence of addition of other alloying elements on the microstructure and mechanical properties of aluminum alloys has been well established. Information regarding the maximum mechanical properties that can be achieved in commercial aluminum alloys has almost reached a saturation. The current trends in increasing the strength of commercial aluminum alloys are by micro-alloying, i.e., by the addition of trace amounts (<0.1 wt%) of elements such as Sn, Ag, Sc, etc. The recent investigations are centered around development of these materials, characterization of microstructure, and mechanical properties under various heat treatment conditions and development of deformation mechanism maps of micro-alloyed aluminum alloys (Mandal and Robi, 2018; Banerjee et al., 2010).

5.3.3 Magnesium Alloys

Magnesium is a fairly strong, silvery-white metal that is around 30% lighter than aluminum. Magnesium in the pure form has a density of 1.81 g/cm^3 as compared with 2.68 g/cm^3 and 7.87 g/cm^3 for aluminum and steel, respectively. Due to its extreme light weight, magnesium alloys are used as a replacement of steel/cast-iron components for applications in engines, steering wheel, trimmings, train driving, etc., in automobiles (Schmidt et al., 2004). A 3.0 L V6 engine made of magnesium alloy weighs 30 kg, whereas the same block weighs 84.6 and 39 kg, respectively, when made using cast iron and aluminum. This replacement in few cases results in almost 64% of the weight of a car by aluminum alloys. Aluminum alloy also finds extensive use in computers and consumer electronic parts. In addition to these, the benefits of using magnesium are its high shock resistance, dent resistance, and noise and vibration damping characteristics. The main challenge in replacing heavier components like cast iron, steel, or aluminum by magnesium is the poor corrosion resistance of magnesium especially in moist environment. The main manufacturing areas that one has to focus attention while replacing components of heavier metal by magnesium in automobile applications include the development of special magnesium alloys, use of environmentally friendly cover gas during it melting, and/or holding and recycling of magnesium waste during casting process (Ma et al., 2003).

5.3.4 Biocomposites

Industrial ecology, eco-efficiency, and green chemistry are guiding the development of the next-generation materials, products, and processes. Considerable growth has been seen in the use of biocomposites (natural fiber composites) derived from local and renewable resources in the domestic sector, building materials, aerospace industry, circuit boards, and automotive applications over the past few decades. These sectors necessitate high durability and environmental friendliness instead of high strength or stiffness. The research attempts aimed at the development and characterization of polymer matrix composites using a variety of natural fibers as reinforcement materials indicate high potential use of biocomposites for various applications. Several natural fibers have been investigated with polymer as matrix material to produce composite materials that are competitive with synthetic fiber composites (Fiore et al., 2011). The agricultural wastes having high marketing appeal can be used to prepare fiber-reinforced polymer matrix composites for commercial use. The growing global environmental and social concern, high percentage of exhaustion of petroleum resources, and new environmental regulations have forced the search for new composites that are compatible with environment. The main advantages of biocomposites include (1) environmental friendliness, (2) cost-effectiveness compared to synthetic fiber composites, (3) fewer greenhouse emissions, and (4) ability to be recycled/reused.

The growing environmental consciousness coupled with high performance at reasonable costs has seen a rapid growth in the development of novel materials based on natural fibers. Natural fibers are an alternative resource to artificial fibers as reinforcement for polymeric materials for the production of low-cost, renewable and eco-friendly composites due to their specific properties, advantages in health issues, and recyclability. Most of the natural fibers are single-cell materials and contain cellulose and lignin as the major constituents (Mohanty et al., 2002). The natural fibers are classified as follows: (1) seed fibers obtained from coir, kapok, cotton, banana, etc., (2) bast fibers obtained from plants like flax, hemp, kenaf, etc., (3) core fibers derived from plants such as kenaf, hemp, and jute, (4) leaf fibers obtained from the leaves of abaca, sisal, pineapples, and (5) other types of fibers obtained from wood and roots. Several other fibers that are obtained from natural resources namely oil palm, banana, sisal, jute, wheat, flax straw, sugarcane, cotton, silk, bamboo, and coconut have been proved to be good and effective reinforcement in thermosetting and thermoplastic resin matrices (Siva et al., 2012). Natural fiber composites have been used for a wide variety of structural applications due to their high specific strength and modulus compared to bulk metals (Netravali and Chabba, 2003). These applications range from household to more sensitive and specialized areas such as spacecraft and aircraft (Hassan et al., 2010). Kenaf fiber-reinforced composite is an alternative biocomposite material that is used particularly in applications such as building and construction due to lightweight and low cost (Ozturk, 2010).

Composites made up of soy oil-based resin and recycled paper in the form of paper sheets from cardboard boxes are used in manufacturing composite structures. Structural unit beams made from these recycled paper composites exhibit stiffness and strength required for the construction of roof (Dweib et al., 2006). Fire retardant-treated waste paper board reinforced composites reveal very good incombustible properties and can be used as interior finishing material or as insulation board. Silk fiber-reinforced composites have become a promising biomaterial for engineering and biomedical application (Cheung et al., 2009). The crashworthiness characteristics of woven natural silk-reinforced epoxy composites tubes were explored. Investigation of the energy absorption response of triggered and non-triggered woven natural silk-reinforced epoxy composite rectangular tubes by carrying out the axial quasi-static crushing test revealed progressive failure instead of catastrophic failure (Eshkoor et al., 2013).

Emergence of nanotechnology has revolutionized modern manufacturing by offering a wide range of functionalities that are possible by nanomaterials. Novel functionalities can be achieved via atomic- and molecular-level designs and methods. To achieve these functionalities, manufacturing is poised to become ever more complex and sophisticated. This will require sustainable practices to be built in production and the use or reuse of nanomaterials. Because of the growing push to increase the use of nanomaterials and nanoprocessing, it would be vital to carefully assess their effect on and interaction with the environment.

5.3.5 Eco-Materials

Materials are regarded as one of the basic elements for the development of a society and economy. Research and development of materials are generally carried out keeping in mind its performance during service. But often research is not directed toward the large amount of resource and energy that it consumes, the pollution that it creates, the emission caused by its production, and other-related stages like mining, transportation, etc. Thus, eco-materials are defined as those materials that enhance the environmental improvement throughout the whole life cycle while maintaining good performance (Halada and Yamamoto, 2001). In order for any material to become an eco-material it should satisfy the following prerequisites: (1) energy-saving ability, i.e., it should be capable of reducing total lifecycle energy consumption of a system; (2) resource-saving ability, i.e., total material consumption should be minimal, (3) reusability, (4) recyclability, (5) structural reliability, (6) chemical stability, (7) biological safety, i.e., it must not cause negative effects to the ecological system, and (8) substitutability, i.e., it should possess the ability to be used as an alternative materials. Few materials that have been projected as potential eco-materials are discussed in the subsequent paragraphs.

a. **Eco-cement**: A considerable amount of energy is consumed and a large amount of greenhouse gases are emitted while producing Portland cement. One ton of CO_2 is produced by 1 ton of Portland cement (Wallah and Rangan, 2006). Eco-cement is manufactured by substituting a part of the raw materials for cement production with municipal wastes. In the eco-cement, approximately 50% of municipal wastes replace the cement raw materials.

b. **Geopolymers**: These were developed in 1978 and are a class of inorganic polymers with three-dimensional network of alumina–silicate materials similar to zeolites (Palomo et al., 1999). They play the role as binders having lower CO_2 emission as compared with Portland cement. Geopolymers are environment friendly. They pose as a promising material and can be used as an alternative to ordinary Portland cement (OPC) (Deventer et al., 2012). Five times less CO_2 is produced while manufacturing geopolymers as compared with Portland cement. No Portland cement is used while producing geopolymer concrete. It can be used as an alternative to Portland cement concrete, since it achieves a significant reduction in the energy consumption and CO_2 emission. Several attempts have been made to produce geopolymeric green cement using ground granulated blast furnace slag, coal fly ash, and clay––bricks as industrial raw materials (El-Gamal and Selim, 2017).

c. **Wood-ceramics**: Wood or woody materials are used for making wood-ceramics. They are porous carbon materials which are obtained by carbonizing wood or woody materials in a vacuum furnace at high temperature (400°C–3000°C). The wood or woody materials are impregnated with phenol resin using ultrasonic vibration before carbonizing. Soft amorphous carbon is formed from the wood or woody materials during carbonizing and hard glassy carbon is formed from the impregnated phenol resin. Wood-ceramics retain the pores which were there originally in the wood or woody material. Therefore, wood-ceramics are porous amorphous carbon/glassy carbon composite materials (Okayama et al., 1997). Wood-ceramics can be used as electromagnetic shield materials, heater for floor heating, and humidity sensors (Halada, 2003). Wood-ceramics and Si powder sintered at 1600°C in vacuum can produce SiC of 1–10 μm of grain size and hence used for manufacturing biomorphic silicon carbide-based ceramics (Guanjun et al., 1999). Since few wood-ceramics exhibit good self-lubricating properties they can be used for bearing applications (Hokkirigawa et al, 1996).

d. **Bioplastics**: Plastic bags have become a common element in our daily life. Majority of the thermoplastics available in the market today are derived from petroleum-based resources which are non-biodegradable and harmful to the environment and marine life. Bioplastics have attracted major attention as potential replacement of conventional fossil-based plastic products. Bioplastics can be prepared from natural polymers, such as proteins, polysaccharides, and lipids, or a combination of these compounds. Attempts have also been made to produce bioplastics from fish proteins. Fish proteins possess certain characteristics such as plasticity, elasticity, ability to form networks, and the ability to serve as a good oxygen barrier, which have been proven to be advantageous for bioplastic preparation (Zavareze et al., 2013). Research has been carried out to synthesize successfully bioplastic using keratin from chicken feathers (Ramakrishnan et al., 2018).

5.3.6 Fly Ash

The use of industrial wastes is a healthy practice to dispose the waste and conserve the available resources for future generations (Kisku et al., 2017). Fly ash is an industrial waste, which is generally dumped as land filling causing ecological imbalance. They are produced from coal-fired power plants. The pulverized coal is blown with air into the boiler's combustion chamber where it gets ignited resulting in fly ash. This is subsequently carried away along with the flue gas. The disposal of fly ash presents a challenge because of the staggering amount produced by coal-fired power plants. If not properly disposed of, fly ash can cause water and soil contamination, thereby affecting the ecological system. The USA, China, and India together account for nearly 70% of the total coal consumption throughout the world (Yao et al., 2015). According to CEA Report (2016), nearly 166 million tons of fly ash is generated from 132 thermal plants annually in India. About 56% of fly ash is utilized effectively through different methods, and the remaining fly ash is still a concern to the society. Maximum amount of fly ash produced is used for replacement of cement thereby consuming the generated industrial waste as well as reducing the requirement for cement clinker. A high-quality fly ash with low carbon content is used as mineral additives in cement and concrete production (Nadesan and Dinakar, 2017).

Fly ash can also find its application in structural fills and embankments, small fills for road shoulders, and large fills for highway embankments. Lower quality fly ash which has higher and variable carbon content is generally used in land filling (Chandra and Berntsson, 2002). The ease of handling and compaction of fly ash reduces both the construction time and equipment costs, thus making it more economical than soil and rock. Soil improvement techniques, such as soil stabilization and drying, can be done with the help of fly ash. It can also be used as a mineral filler in hot mix asphalt (HMA). Mineral filers increase the stiffness of the asphalt mortar matrix, thus improving the rutting resistance of pavements and the durability of the mix. Fly ash, if properly managed, can be recycled and put to beneficial reuse, avoiding the hazards that result from storage of the material. The fly ash can be recycled and reused as one of the most abundant industrial by-products on earth in a variety of applications, including cement production, paving material, as well as in structural fills and embankments. The number of ways in which fly ash can be beneficially reused makes it a valuable material, allowing companies in the coal sector to save both money and energy.

5.3.7 Waste Utilization from Paper and Pulp Industry

Industries generate huge amount of solid wastes. Utilization of these waste products from the industrial processes is of great importance both from the environmental point of view and from financial aspect. Adequate solutions should be identified for its disposal so that it does not create any negative environmental impacts. These solid wastes can be reused as raw materials which in turn will put less focus on the extraction of natural resources. The two viable options for reducing the environmental impact are as follows: (1) reutilize these wastes in the same processes which created them and (2) convert the industrial wastes to value added products.

Solid wastes from industries are generally disposed of in landfills. The construction sector is aiming at ways to implement the principles of sustainable development. Solid wastes from recycled paper company generally include films made from polymers, adhesive ribbons, narrow ribbons, elastics, metals, glass, textile, and others. Plates used in the construction industry, similar to that of plywood, were manufactured using these solid wastes (Pelegrini et al., 2010). Disposal of these materials result in wastage of resources in addition to the fact that they cause environmental pollution. The results of study on the waste plate presented values close to those of the commercial agglomerated wood plates. It exhibited low water absorption and excellent flexibility in comparison to the wood plates. Incorporation of paper pulp waste to manufacture non-structural elements in the form of a plaster–pulp composite material was also studied (Agulló et al., 2006). Ahmadi and Al-Khaja (2001) carried out an analysis on the use of paper waste sludge obtained from paper industry, which can be used as a replacement to the mineral filler material, in various concrete mixes for non-structural masonry construction.

Research attempts were carried out to find alternative methods for treating industrial waste because of its large amount of generation and the environmental problems that it creates. Martínez et al. (2012) studied the possibility of recovering two forms of waste from the paper industry, i.e., sludge from the purification of waste water and residue which is obtained from cleaning the pulp. The sludge and residue in turn were used in ceramics. This helped in saving both raw materials and energy, since these waste products had high calorific content. They showed that increased content of waste increased linear expansion and water absorption which in turn reduced resistance to compression. In developing countries, there is a large accumulation of unprocessed wastes which in turn raise environmental concerns. Converting unmanaged wastes into building materials not only helps in reducing the problems related to pollution but it also leads to reduced cost of buildings not having to compromise on structural strength. Attempts are going on to reuse the byproducts and residues from pulp and paper industry in cement plant and brickworks, agricultural use and composting, anaerobic treatment, recycling, etc. Raut et al. (2012) used recycled paper mill waste to make energy absorbing light-weight bricks. They investigated both the physical and mechanical properties of brick samples which were made with paper pulp and binder. The test results showed that these bricks are economical and weigh half as much as conventional clay bricks.

The sludge residue is produced during the manufacturing process of paper. The treatment of waste waters at sewage treatment stations produces sludge as its final product. This waste sludge can be used as a raw material, in the production of clay bricks (Cusidó et al., 2015). They found that by increasing the content of paper sludge in the clay mixture the thermal and acoustic insulation properties of the material are improved.

5.4 CONCLUSION

Global demand for materials and energy is expected to rise sharply over the next several decades. Over utilization of resources, overconsumption, pollution, and overpopulation will continue to pose challenges in the near future. For sustainable growth, it is imperative to protect the environment, conserve resources, and develop technology that minimizes or eliminates environmental load. The appropriate material selection and suitable design play a vital role here. In this context, an important goal of sustainability is to train and educate a new generation of workforce who can think and act holistically.

This chapter attempts to provide some ideas on the development of certain materials that are influenced by sustainable practices. The evolving applications of new steels grades in frame, chassis, and other auto parts were discussed. A brief overview of the development in aluminum and magnesium alloys is also discussed in the chapter. For a sustainable development of materials industry, development of eco-materials is essential. Greater eco-efficiency, zero emission and higher recycling are the foundation for eco-products processing. Reuse of fly ash and sludge from paper and pulp industry in the construction sector seems to be a fruitful answer to the challenging problem of reutilizing environmental pollutants and waste products.

Educating people to live in a sustainable way, bringing into place new legislations which will curb waste and emission control of hazardous products, buying products of those companies which operate with an ecological aim, and increased research and development of sustainable products will go a long way towards providing a sustainable future for us and generations to come.

REFERENCES

Agulló L., Aguado A., Garcia T., 2006, Study of the use of paper manufacturing waste in plaster composite mixtures. *Building and Environment*, 41, 821–827.

Ahmadi B., Al-Khaja W., 2001, Utilization of paper waste sludge in the building construction industry, *Resources, Conservation and Recycling*, 32, 105–113.

Ashby M.F., 1999, *Materials Selection in Mechanical Design*, 2nd ed. Oxford: Butterworth -Heinemann.

Ashby M.F., Johnson K., 2002, *Materials and Design: The Art and Science of Materials Selection in Product Design*, Oxford: Butterworth-Heinemann.

Banerjee S., Robi P. S., Srinivasan A., Kumar L.K.,2010, High temperature deformation behaviour of Al–Cu–Mg alloys micro-alloyed with Sn, *Materials Science and Engineering: A*, 527(10–11), 2498–2503.

Central Electricity Authority, 2016, Report on Flyash Generation at Coal/Lignite Based Thermal Power Stations and Its Utilization in the Country for the Year 2015–2016, New Delhi.

Chandra S., Berntsson L., 2002, *Lightweight Aggregate Concrete: Science, Technology, and Applications*, Norwich and New York: Noyes Publications.

Cheung H.Y., Lau K.T., Ho M.P., Mosallam A., 2009, Study on the mechanical properties of different silkworm silk fibers, *Journal of Composite Materials*, 43, 2521–2531.

Cuddy L., Raley J., Austenite grain coarsening in microalloyed steels, *Metallurgical Transactions A*, 14(10), 1983–1989.

Cusidó J.A., Cremades L.V.B., Sorianoc C., Devant M., 2015, Incorporation of paper sludge in clay brick formulation: Ten years of industrial experience investigated the physical and chemical properties of mixtures of clay and paper sludge, *Applied Clay Science*, 108, 191–198.

Despeisse M., Mbaye F., Ball P.D., Levers A., 2012, The emergence of sustainable manufacturing practices, *Production Planning & Control*, 23, 354–376.

Deventer J.S.J., Provis J.L., Duxon P., 2012, Technical and commercial progress in the adoption of geopolymer cement, *Minerals Engineering*, 29, 89–104.

Dornfeld, D., 2009, Opportunities and challenges to sustainable manufacturing and CMP, *Proceedings of MRS Spring Meeting*, San Francisco.

Dweib M.A., Hu B., Shenton III H.W., Wool R.P., 2006, Bio-based composite roof structure: Manufacturing and processing issues, *Composite Structures*, 74, 379–388.

El-Gamal S.M.A., Selim F.A., 2017, Utilization of some industrial wastes for eco-friendly cement production, *Sustainable Materials and Technologies*, 12, 9–17.

Elkington J., 1997, *Cannibals with Forks: The Triple Bottom Line of 21st Century Business*, Gabriola, Island: New Society Publishers.

Eshkoor R.A., Oshkovr S.A., Sulong A.B., Zulkifli R., Ariffin A.K., Azhari C.H., 2013, Comparative research on the crashworthiness characteristics of woven natural silk/epoxy composite tubes, *Materials & Design*, 47, 248–257.

Finnveden G., 2000, On the limitations of life cycle assessment and environmental systems analysis tools in general, *International Journal of Life Cycle Assessment*, 5(4), 229–238.

Fiore V., Valenza A., Di Bella G., 2011, Artichoke (*Cynara cardunculus* L.) fibres as potential reinforcement of composite structures, *Composites Science and Technology*, 71, 1138–1144.

Funakawa Y., Seto K., 2007, Stabilization in strength of hot-rolled sheet steel, *Tetsu to Hagane*, 93, 49–56.

Gray J.M., Yeo R.B.G., 1968, Columbium carbonitride precipitation in low-alloy steels with particular emphasis on 'precipitation-row' formation, *Transactions of the ASME*, 61, pp. 255–269.

Guanjun O., Zhang M.R., Zhihao J., 1999, Mechanical properties and microstrucure of SiC made from natural woods, *Proceedings of 4th International Conference on Ecomaterials*, Gifu 1999, Society of Non-traditional Technology, Tokyo, pp. 301–304.

Haapala K.R., Zhao F., Camelio J., Sutherland J.W., Skerlos S.J., Dornfeld D.A., Jawahir I.S., Clarens A.F., Rickli, J.L., 2013, A review of engineering research in sustainable manufacturing, *Journal of Manufacturing Science and Engineering*, 135, 041013–041013.

Halada K., 2003, Progress of ecomaterials toward a sustainable society, *Current Opinion in Solid State and Materials Science*, 7, 209–216.

Halada K., Yamamoto R., 2001, The current status of research and development on eco-materials around the world, *MRS Bulletin*, 26(11), 871–878.

Hassan A., Salema A.A., Ani F.N., Bakar A.A., 2010, A review on oil palm empty fruit bunch fiber-reinforced polymer composite materials, *Polymer Composites*, 31, 2079–2101.

Heijungs R., (editor), 1992, *Environmental Life-Cycle Assessment of Products; Guide and Backgrounds* (two volumes), Leiden: CML.

Hokkirigawa K., Okabe T., Saito K.J., 1996, Friction properties of new porous carbon materials: Woodceramics, *Journal of Porous Materials*, 2, pp. 237–243.

Hundal M., (editor), 2001, *Mechanical Life Cycle Handbook: Good Environmental Design and Manufacturing*. New York: Marcel Dekker.

Hui I.K., Lau H.C.W., Chan H.S., Lee K.T., 2002, An environmental impact scoring system for manufactured products, *The International Journal of Advanced Manufacturing Technology*, 19, 302–312.

Jawahir I.S., Bradley R., 2016, Technological elements of circular economy and the principles of 6R-based closed-loop material flow in sustainable manufacturing, *Procedia CIRP*, 40, 103–108.

Jawahir IS., Dillon Jr. O., 2007, Sustainable manufacturing processes: New challenges for developing predictive models and optimization techniques. In: 1st International Conference on Sustainable Manufacturing. Montreal, Canada.

Jayal A.D., Badurdeen Jr. F., Dillon O.W., Jawahir I.S., 2010, Sustainable manufacturing: Modeling and optimization challenges at the product, process and system levels, *CIRP Journal of Manufacturing Science and Technology*, 2(3), 144–152.

Kisku, J.H., Ansari M., Panda S.K., Nayak S., Dutta S.C., 2017, A critical review and assessment for usage of recycled aggregate as sustainable construction material, *Construction and Building Materials*, 131, 721–740.

Ljungberg L.Y., 2003, Materials selection and design for structural polymer, *Materials & Design*, 24, 383–390.

Ljungberg L.Y., 2007, Materials selection and design for development of sustainable products, *Materials & Design*, 28, 466–479.

Lüdke B., Pfestorf M., 2006, Functional design of a lightweight body-in-white. In: Hashimoto S., Jansto S., Mohrbacher H., Siciliano F. (Eds.), *International Symposium on Niobium Microalloyed Sheet Steel for Automotive Application*, Warrendale, PA: TMS, p. 27.

Ma Q., Graham D., Zheng L., St John D.H., Frost M.T., 2003, Alloying of pure magnesium with Mg-33.3Zr master alloy, *Materials Science and Technology*, 19, 156–162.

Mandal P.K., Robi P.S., 2018, Influence of micro-alloying with silver on microstructure and mechanical properties of Al-Cu alloy, *Materials Science and Engineering: A*, 722, 99–111.

Martínez C., Cotes T., Corpas F.A., 2012, Recovering wastes from the paper industry: Development of ceramic materials, *Fuel Processing Technology*, 103, 117–124.

Millar H.H., Russell S.N., 2011, The adoption of sustainable manufacturing practices in the Caribbean, *Business Strategy and the Environment*, 20, 512–526.

Mohanty A.K., Misra M., Drzal L.T., 2002, Sustainable bio-composites from renewable resources: Opportunities and challenges in the green materials world, *Journal of Polymers and the Environment*, 10, 19–26.

Mohrbacher, H., 2013, Reverse metallurgical engineering towards sustainable manufacturing of vehicles using Nb and Mo alloyed high performance steels, *Advances in Manufacturing*, 1(1), 28–41.

Nadesan M.S., Dinakar P., 2017, Structural concrete using sintered flyash lightweight aggregate: A review, *Construction and Building Materials*, 154, 928–944.

Netravali A.N., Chabba S., 2003, Composites get greener, *Materials Today*, 6, 22–29.

Okayama T., Okabe T., Saito K., Endo H., Yamamura A., 1997, Manufacturing methods of wood ceramics from waste paper, Proceedings of 3rd International Conference on Eco materials, Tsukuba Japan, 1997, Society of Non-traditional Technology, Tokyo, pp. 186–189.

Ozturk S.J., 2010, Effect of fiber loading on the mechanical properties of kenaf and fiberfrax fiber-reinforced phenol-formaldehyde composites, *Journal of Composite Materials*, 44, 2265–2288.

Pachauri R.K., Meyer L.A., Core Writing Team (Eds.), 2014, Climate Change 2014: Synthesis Report. Contribution of Working Groups I, II and III to the Fifth Assessment Report of the Intergovernmental Panel on Climate Change, IPCC, Geneva, Switzerland, 151, p. 3.

Palomo A., Grutzeck M.W., Blanco M.T., 1999, Alkali-activated fly ashes, a cement for the future, *Cement and Concrete Research*, 29, 1323–1329.

Pelegrini M., Gohr Pinheiro I., Valle J.A.B., 2010, Plates made with solid waste from the recycled paper industry, *Waste Management*, 30, 268–273.

Ramakrishnan N., Sharma S., Gupta A., Alashwal B.Y., 2018, Keratin based bioplastic film from chicken feathers and its characterization, *International Journal of Biological Macromolecules*, 111, 352–358.

Raut S.P., Sedmake R., Dhunde S., Ralegaonkar R.V., Mandavgane S.A., 2012, Reuse of recycle paper mill waste in energy absorbing light weight bricks, *Construction and Building Materials*, 27, 247–251.

Schmidt W., Dahlqvist E., Finkbeiner M., Krinke S., Lazzari S., Oschmann D., 2004, Life cycle assessment of lightweight and end-of life scenarios for generic compact class passenger vehicles, *International Journal of Life Cycle Assessment*, 9, 405–416.

Siva I., Winowlin Jappes J.T., Suresha B., 2012, *Polymer Composites*, 33, 723–732.

US Department of Commerce, 2011, Sustainable Manufacturing Initiative website, http://trade.gov/competitiveness/sustainablemanufacturing/index.asp.

Wallah S.E., Rangan B.V., 2006, Low-Calcium Fly Ash-Based Geopolymer Concrete: Long-Term Properties. Research Report GC 2, Faculty of Engineering, Curtin University of Technology, Perth, Australia.

Wang E.J., Lin C.Y., Su, T.S., 2016. Electricity monitoring system with fuzzy multi objective linear programming integrated in carbon footprint labeling system for manufacturing decision making, *Journal of Cleaner Production*, 112(Part 5), 3935–3951.

Wang S., Kao P., 1993, The effect of alloying elements on the structure and mechanical properties of ultra-low carbon bainitic steels, *Journal of Materials Science*, 28, 5196.

Yao Z.T., Ji X.S., Sarker P.K., Tang J.H., Ge L.Q., Xia M.S., Xi Y.Q., 2015, A comprehensive review on the application of coal flyash, *Earth-Science Reviews*, 141, 105–121.

Zavareze E.R., Halal S.L.M., Silva R.M., Dias A.R.G., Prentice-Hernández C., 2013, Mechanical, barrier and morphological properties of biodegradable films based on muscle and waste proteins from the Whitemouth croaker (*Micropogonias furnieri*), *Journal of Food Processing and Preservation*, 38, 1973–1981.

CHAPTER 6

Steel
A Sustainable Material of the Future

Sumitesh Das
Tata Steel Limited

CONTENTS

6.1 INTRODUCTION

The term "sustainability" can be defined as "the avoidance of the depletion of natural resources in order to maintain an ecological balance." Tata Steel, one of the major steel companies, defines sustainability as "an enduring and balanced approach to economic activity, environmental responsibility, and societal benefit" [1]. Sustainability is about meeting the challenges of ensuring that future generations can enjoy the same kind of life styles people enjoy today. This naturally involves taking a long-term perspective on balancing economic, environmental, and social impacts of business.

The economic growth of a nation is dependent on the consumption of steel. Since the industrial revolution in the 18th century, steel has been a material of choice across various industrial segments, e.g., automotive, construction, white goods, packaging, energy production, tools, and healthcare. It is commonly said that "everything around is either made of steel or manufactured by an equipment made of steel." The demand of steel was met with the invention of the Bessemers' process—this was scalable and was the basis for the sustained development in the economies of all developed countries. An interesting statistic is the stocks of steel per person, based on their current wealth (GDP per person). It suggests that as a person's income increases, they build up their stock of steel. As per the National Steel Policy, the overall per capita steel consumption in India is only 30 kg against a world average of 121 kg. In rural India, the consumption is 0.6 kg.

In 2016, the total production of crude steel was 1630 million tons. The bulk of the production came from China, followed by Europe. Figure 6.1 shows the crude steel production in the various geographies.

Steel is consumed in different segments—predominately in the construction segment. The next major consumer is the automotive segment followed by the white goods segment. Figure 6.2 shows the broad areas of consumption of steel.

It is interesting to note that the physical properties of steel demanded in the construction industry vary from those of the automotive industry. The automotive industry has a high demand for superior surface quality. Similarly, malleable and ductile properties of steel are needed by the automotive industry. Steel becomes a material of choice because of the infinite opportunities that it offers by a simple combination of its constituent elements such as C, Mn, Si, P, As, B, etc.

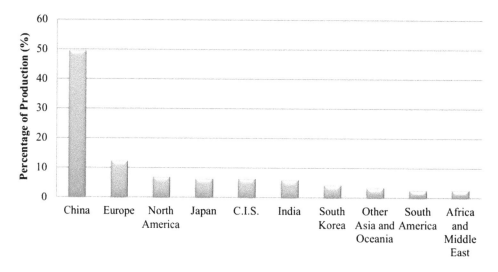

Figure 6.1   Country- or region-wise distribution of crude steel production [1].

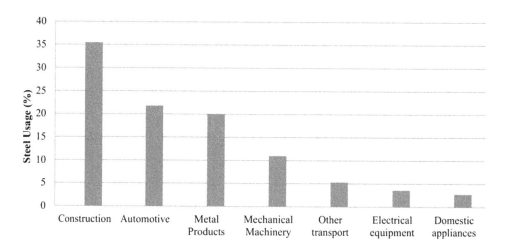

Figure 6.2   Steel usage in various applications for a developed country [2].

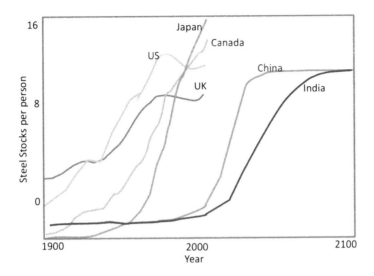

Figure 6.3 Projected growth in steel consumption in the next 100 years [3].

Figure 6.3 shows the projected demands in steel consumption for various economies globally. The consumption has already peaked in the developed countries, e.g., UK, Japan, US, Canada. Steel consumption is already high because of its usage in construction, ship building, engineering, and automotive. Most of the steel has already been used for construction sector and very few new mega projects are on the anvil. The only major consumable is the automobile. However, with the advancing population and advanced technologies such as autonomous cars and lighter stronger materials, it is expected that the demand for steel will at best stabilize.

However, the scenario is very different in developing countries: steel is used to build new roads, railway lines, multi-story buildings, and bridges. There is also a major requirement of steel to put up basic infrastructure for gas pipelines, water, and sanitation. It is expected that the demand for steel will grow substantially in China and India with the increasing living standards. Most of the growth will come from new industrializing nations: Brazil, China, India, Iran, and Mexico.

Once steel is produced, it becomes an everlasting product as it is 100% recyclable. However, there are three major challenges that emerge from the steel consumption projections, especially for the developing countries. First, rapid urbanization occurred due to major population increase requires a sustainable steel policy. Second, the steel industry is one of the largest energy-intensive sectors worldwide. Finally, the industry is responsible for the emission of 4%–5% of the total anthropogenic CO_2. It is therefore important to understand the growth of the steel industry and the impact of CO_2 emissions if alternate technologies that have sustainability at its core are not developed.

6.2 ROUTES OF STEEL PRODUCTION

There are two main routes for steel production: the first route is the Blast Furnace–Basic Oxygen Furnace (BF–BOF) and the other route is the Electric Arc Furnace (EAF). In the BF–BOF route, raw materials such as iron ore, coal, limestone, and sometimes recycled steel are used. The iron ore is first reduced to iron, also called hot metal or pig iron, followed by conversion to steel in the BOF. The steel is cast in a caster into slabs, blooms, or billets. The slabs are rolled into coil and plates. The blooms and billets are rolled into sections or bars. The BF–BOF route accounts for nearly 70% of steel production worldwide.

The EAF uses electricity to melt recycled steel. Sometimes other sources of metallic iron such as Direct-reduced iron (DRI) are used. Alloy addition is done to adjust the steel to the final chemical composition. Beyond steelmaking, the downstream processes of casting and rolling are similar to the BF–BOF route. The EAF route accounts for nearly 29% of steel production globally. The key raw materials needed in steelmaking include iron ore, coal, limestone, and recycled steel. Inputs for the two main steel production routes are described below. The integrated (BF–BOF) route typically uses 1,400 kg of iron ore, 800 kg of coal, 300 kg of limestone, and 120 kg of recycled steel to produce 1,000 kg of crude steel.

The EAF route typically uses 880 kg of recycled steel, 16 kg of coal, and 64 kg of limestone to produce 1,000 kg of crude steel.

6.3 NEED FOR SUSTAINABLE MEASURES

During the conversion of iron ore to steel, the major emissions are those of carbon dioxide (CO_2), sulfur oxides (SO_x), nitrogen oxides (NO_x), and dust to air. In addition, emissions to water streams also take place. Figure 6.4 shows the major sources of CO_2 generation in an integrated steel production unit.

Due to consistently increasing demand of iron and steel products for human needs, fossil-fuel energy use and CO_2 emission will continue to grow in this industry. Therefore, there is a strong motivation to develop and implement energy-efficient and low-carbon technologies as well as carbon reduction programs for this industry. Enhancing the development and deployment of high-temperature waste heat energy recovering technologies along a desired combination of carbon capture and storage (CCS) technologies will be the effective solution to reducing CO_2 emissions from iron and steel production. The main aim of this chapter is to provide a comprehensive overview of the worldwide carbon reduction programs and new CO_2 breakthrough technologies for energy saving and CCS in iron and steel-making processes by collating updated information from a wide range of sources. Also, a discussion on the selection of the appropriate technology and their barriers and stages of development and deployment is presented. There are a number of factors that limit the role of using biomass in CO_2 abatement. Thus, implementation of CCS technology in coal-based integrated steel plant would be an efficient means for sustainable green iron and steel manufacturing.

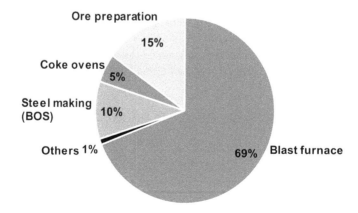

Figure 6.4 Source of carbon dioxide and other emissions in an integrated steel plant. (Reprinted with permission from *International Journal of Greenhouse Gas Control*, Elsevier 2013, 19: 145–159).

6.4 LIFE CYCLE ANALYSIS AS A TOOL FOR SUSTAINABILITY

All major steel companies have institutionalized measures to minimize and reduce the emission foot print. The steel industry adopts a life cycle assessment (LCA) approach to assess the footprint of the emissions taking place during the steel-making process (Figure 6.5) [4]. LCA is a technique to assess environmental impacts associated with all the stages of a product's life from raw material extraction through materials processing, manufacture, distribution, use, repair and maintenance, and disposal or recycling. One of the institutions actively involved in the LCA is WorldSteel. Its data base contains the "cradle-to-gate" input and output data pertaining to environment. The typical inputs include the following:

- Use of resources such as raw materials, energy, and water
- Land, air, and water emissions.

LCA studies enable informed material selection decisions and more eco-efficient products by identifying potential areas to reduce the product's environmental footprint.

6.5 ENVIRONMENTAL MANAGEMENT SYSTEMS

Steel plants usually employ a holistic environmental management system to monitor and improve its environmental performance and increase its operational efficiency.

6.5.1 Water and Air Management

The average consumption and discharge for integrated steel plants are 28.6 m³/ton of steel and 25.3 m³/ton of steel, respectively. However, the average consumption and discharge for the EAF route are 28.1 m³/ton of steel and 26.5 m³/ton of steel, respectively. Water consumption and

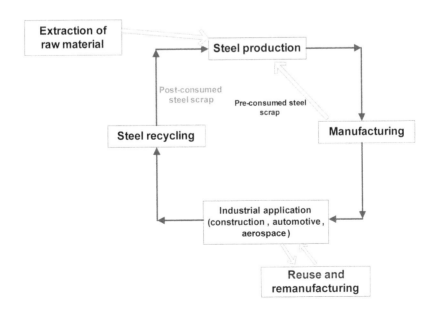

Figure 6.5   Life cycle analysis of steel.

discharge are close to each other and few losses occur in the process, indicating an overall efficient use of water. In most cases, water loss is caused by evaporation. The steel plants in water-scarcity areas are able to recycle and reuse around 98% of their water by using advanced technologies.

A key aspect of steel industry environmental protection is to minimize emissions to the air. Emission sources are mapped and monitored. Process improvements can then be identified and implemented with the goal of reducing emissions. The control mechanisms to reduce emissions can include the following:

- baghouse filtration systems
- chemical treatment
- thermal oxidization
- scrubber systems
- dust suppression.

6.5.2 Byproduct Gases

The steel industry continuously strives to use the by-product gases from coke ovens, blast furnaces, and basic oxygen furnaces. Coke oven gas contains around 55% of hydrogen. These gases can be used as complementary fuel resources and fulfill around 40%–60% of the total energy requirements. Byproduct gases can also be leveraged for power generation. It is reported that the byproduct gases from the BOF saves the equivalent of around 300 million cubic meters of natural gas.

6.5.3 Slag

Around 400 million tons of iron and steel slags are generated each year. Slags are a mixture of calcium oxide, silica, magnesium oxide, alumina, and iron oxides. Although the slags are lighter than the liquid metal, they float and can be easily removed. The Blast Furnace slags are mainly found in the following three forms: air-cooled, granulated, and pelletized or expanded. These forms of blast furnace slags are classified according to their mode of cooling. Air-cooled slag is hard and dense and is especially suitable for use as construction aggregate. It is also used in ready-mix concrete, road bases and surfaces, roofing, and mineral wool as insulation. Granulated slag forms sand-sized particles of glass and is primarily used to make cementitious material. Slag also helps in bringing down the cost of cement. The recovery rate for slag varies from around 80% for steelmaking slag to nearly 100% for ironmaking slag. However, one of the technical challenges is to reduce or eliminate the free lime content. Once the lime is removed, slag has a potential to be used as fertilizers, in cement and concrete production, and for waste water treatment.

6.6 ENERGY MANAGEMENT

Steel production is energy intensive, right from the reduction of iron ore to iron and finally to steel. For this, reducing agents such as coal, coke, and sometimes natural gas is used. Coke is prepared by carburizing coal in the absence of oxygen at high temperatures. This is the primary reducing agent of iron ore.

In economic terms, energy purchases account for around 40% of the basic steel production. In an integrated steel plant, around 50% of the energy input is derived from coal, 35% from electricity usually from thermal power plants, and 10% from natural gases and other recycled gases. The efficient use of energy has always been one of the steel industry's key priorities. Steel companies have implemented various technologies that have cut the energy consumption by 50% per ton of steel over the past few decades (Figure 6.6).

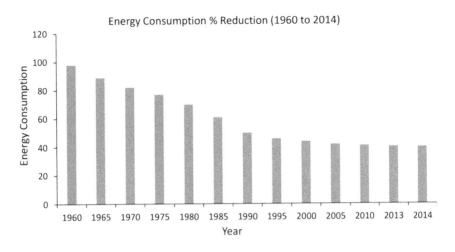

Figure 6.6   Indexed global energy consumption/ton of crude steel production [5].

6.7 ALTERNATE TECHNOLOGIES: A COMPARISON CHART

The ironmaking process in an integrated steel plant has remained relatively the same over the past few decades. The blast furnace is a proven technology with unlimited scalability. However, its dependency on high-quality coking coal is a challenge in its effort to be a sustainable process. The pressure to utilize lean ores has driven the steel research fraternity to work on alternate technologies. The technologies seek to address the issue of fine coal and high ash. Table 6.1 shows a comparison of the alternative technologies of steel production.

6.8 STEEL AS A SUSTAINABLE SOLUTION BEYOND THE PRODUCTION UNIT

There are four possible ways in which steel reduces the energy use during its entire life cycle and also the associated emissions.

Table 6.1   Comparison of Alternative Technologies

Process in Steel Industry	Advantages	Disadvantages
Corex	This is again a proven technology	Very high operational costs due to high coal and oxygen requirements. The maximum capacity of working unit size is 0.8 Mtpa. Coal of a specific ash content can be used
Hismelt	Agglomeration and coke-making operations are not required	Steel majors are still in the process of proving the efficiencies of these processes at scale and so high risk is involved
Finex	Agglomeration and coke-making operations are not required	
Fastmet	Option of using iron ore fines	This approach work for coal with very low ash and volatile matter. In case coal is used that contains high ash, the end product would contain higher amounts of sulfur and phosphorus. Hence the end product cannot be used directly in steelmaking
Corex–Midrex combinations	Lower capital and operational costs	Coal with low-ash content is required. Coupled with this, there is high risk and complexity

a. *Light weighting.* New steel grades, typically the high-strength steels (HSS) and advanced high-strength steels (AHSS), help car makers to reduce the vehicle weight by 25%–40% as compared with the conventional mild steel. Under these circumstances, the vehicle weight reduces by 170–270 kg (for a typical family car). This corresponds to an overall savings of 3–4.5 tons of greenhouse gases over the total life cycle of the vehicle.

b. *Increased product life cycle.* Structures made out of steel, e.g., buildings and bridges, last for 40–100 years, thereby reducing the need for replacement.

c. *Recycling.* This is steel's unique advantage. Steel is almost 100% recyclable and can be theoretically recycled infinitely without any loss of quality. More than 650 million tons is recycled annually. For every 1 kg of steel scrap recycled into new steel, this helps to save around 1.4 kg of iron ore, 0.7 kg of coal, and 0.1 kg of limestone.

d. *Energy generation.* Over 20-year period, a 3-MW wind turbine has the potential to deliver around 80 times more energy than that used in the production of steel that goes into making the turbine.

6.9 CASE STUDIES: SUSTAINABILITY IN STEEL INDUSTRIES

Most papers refer to technologies such as CO_2 sequestration, use of Coke Dry Quenching (CDQ), use of AHSS, etc., [6] as methods of sustainability in a steel industry. One of the sustainability measures is to adopt measures to reduce the amount of ash in coal used in the steel industry. Table 6.2 shows the ash content in Indian coals as compared with coals in other countries. This high content of ash leads to higher power consumption and higher CO_2 release in a steel plant.

The conventional method is to beneficiate the incoming high-ash coal Beneficiation technologies such as separation using magnets, gravity, and fluids. However, the challenge is to remove the ash from the fine parts of the coal circuit and yet maintain the yield. This challenge was recognized as a common problem across the various integrated steel industries and a research program was initiated by the Steel Development Fund under the Steel Ministry [7]. Under this program, advanced flotation technologies are being worked on.

One of the promising researches is to use chemicals that react with the inter-locked minerals of the coal matrix so that they can be easily removed [8]. This technology can also be applied to non-coking coals. As of 2017, the work is at a pilot scale and are able to achieve an ash percent of around 10%–12% (i.e., reductions of 60%) with yields of 70%. It is estimated that use of such low-ash coal can improve power plant efficiencies to 53%. At the same time, the CO_2 emissions reduce by 24%.

The second example is from the products wherein steel provides the ultimate recyclability. As sand and water becomes scarce, the use of concrete and cement in buildings comes under threat. However, the growing population requires that housing must be provided to all, especially in developing countries such as India and Brazil.

The World Steel Association has recognized this need for sustainable housing. It has promoted steel's usage in buildings through various standards such as the National Green Building Standard (ICC-700) for residential buildings, ASHRAE Standard 189.1 for commercial construction, and the

Table 6.2 Coal Types: Carbon and Ash Contents [6]

Coal Type	Content (% Weight)	
	Carbon	Ash
Antracite (best quality coal)	72–87	6.9–11
USA	73–74	7.2–13
China	48–61	28–33
India	30–50	30–50
Lignite (lean coal)	35–45	6.6–16

US Green Building Council's Leadership in Energy and Environmental Design (LEED) program for all types of buildings [9].

The most common form of steel structures used for building the frames for the buildings is light gauge cold-formed sections. These steel sections have been commonly used because of its strength, formability, and functionality. The formed sections also reduce the need to carry out additional activities at the site, thereby reducing the water and energy consumption. Steel buildings are also energy neutral. Also, steel's ability to be recycled and reclaimed enables it to score highly on the sustainability scale.

6.10 CONCLUSION

With growing awareness on greenhouse emissions and stringent policy measures being adopted by many governments, the steel industry has been continuously striving to cut down on its emissions. Only then, it will continue to remain the first choice among materials across segments. Development of a new technology that is sustainable and at the same time is able to meet the throughput of a steel plant takes decades. Keeping this in mind, several research groups across the world have pooled their expertise together to develop technologies that will enable the steel industry to continue to be sustainable in the future.

Some of the major research and investment is happening in the following programs:

- the EU (ultra-low CO_2 steelmaking, or ULCOS, supported by 10 EU companies and miners)
- Japan (Course 50, Japanese Iron and Steel Federation)
- the US (the American Iron and Steel Institute)
- Canada (the Canadian Steel Producers Association)
- South America (ArcelorMittal Brazil)
- South Korea (POSCO)
- China (Baosteel) and Taiwan, China (China Steel)
- Australia (BlueScope Steel/One Steel and CSIRO coordination).

So far, nearly one billion USD have been invested in these programs. Identification of the most promising technologies that can potentially reduce CO_2 emissions by 50% or more have been completed. As on 2017, the work is focused on examining the feasibility of the technologies at various production levels, from laboratory work to piloting, building demonstration units, and eventually commercial implementation.

REFERENCES

1. www.worldsteel.org/media-centre/press-releases/2017/world-crude-steel-output-increases-by-0.8--in-2016. html.
2. www.steelonthenet.com/consumption.html.
3. https://theconversation.com/iron-ore-still-has-an-important-role-to-play-in-australias-economy-54476.
4. www.worldsteel.org/steel-by-topic/life-cycle-thinking.html.
5. www.worldsteel.org/en/dam/jcr:f07b864c-908e-4229-9f92-669f1c3abf4c/fact_energy_2016.pdf.
6. www.counterview.net/2017/07/reliance-thinktank-report-low-calorific.html.
7. http://steel.gov.in/sites/default/files/research_main.pdf.
8. www.tata.com/article/inside/EgfM18htxAk%3D/TLYVr3YPkMU%3D.
9. www.buildings.com/news/industry-news/articleid/12364/title/building-with-sustainable-green-materials.

CHAPTER 7

Strategies to Improve the Forming Quality of Sheets

R. Ganesh Narayanan
IIT Guwahati

CONTENTS

7.1 INTRODUCTION

Improving the forming quality of sheet is part of sustainable manufacturing. Such quality improvement will result in use of lightweight sheet grades, efficient production practice, reduced usage of raw materials, quality products, enhanced product life time, improved crash performance, reduced post-forming operations, improved surface characteristics, and reduced power and fuel consumption. All these factors will contribute in maintaining sustainability not only in part making but also in part management. Generally, a sheet grade is replaced by another sheet grade of better forming quality to improve the part performance. For example, Figure 7.1 which is based on 2006 data shows that by using multiphase steel in place of advanced high-strength steel (HSS) and deep drawing quality steel, a weight reduction of about 17 kgs has been realized in a BMW series car as compared with an older model. In addition, it has been reported that the torsion and bending stiffness have increased and crash performance has improved. Such sheet grade replacement is not possible always for all the automotive applications. Hence, there is a definite requirement of sheet formability modifications on the existing sheet grades. In this context, materials processing, forming process, changes in mechanical properties, tooling, and lubrication are some of the approaches by which the forming quality of sheets are improved. In this chapter, such approaches have been discussed with examples from existing research case studies.

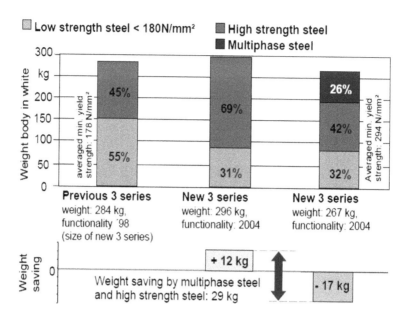

Figure 7.1   Weight reduction of BMW series cars with and without use of multiphase steel. (With permission from www.autosteel.org/sustainability2/past-gdis-presentations/gdis-2005.aspx; The Application of Multiphase Steel in the Body-in-White by Markus Pfestorf.)

7.2 ROLLING METHOD AND CONDITIONS

The sheet grades determined for industrial stamping operations are rolled prior to stamping. The rolling strategies and conditions affect the microstructure and mechanical properties of such sheets improving their formability. The differential speed rolling, structure-controlled rolling, change in rolling temperature, and rolling direction are a few strategies that will be discussed.

Differential speed rolling (DSR) is a variety of asymmetric rolling process in which the sheet is rolled by upper and lower rolls with different rotational speeds (Figure 7.2). Using rolls with different diameters and different materials (will change the friction conditions at the contact interface) are the other ways to follow asymmetric rolling. Rolling speed differentiation coefficient or roll speed ratio, which is the ratio of upper roll to lower roll speed, is an important factor to be controlled. The deformation geometry is changed during DSR. The neutral points where the roll velocity equals sheet velocity shift during DSR (Figure 7.2). For slow roll, the neutral points shift toward the entrance of the roll gap, while for fast roll, the neutral points shift toward the exit. This particular condition has the advantage of lowering the roll force and imparting high through thickness shear strain to the sheet. In material processing, the two important purposes of DSR are (1) to strengthen the sheet materials by grain refinement through high plastic strain build-up and (2) to control the deformation texture that affects the material anisotropy and other material properties (Polkowski, 2016).

Huang et al. carried out DSR on hot extruded AZ31B Mg alloy sheet of thickness 4 mm that was heated at 723 K for 24 h under argon atmosphere (Huang et al., 2009). The DSR was performed at a rotational speed ratio of 1.167 without lubrication. The plate of 4-mm thickness was reduced to 2 mm by six passes at 703 K, and further rolled to 1 mm by eight passes at 573 K. The sheet was rotated after each rolling pass. The conventional rolling was also conducted by following the same rolling procedure. It is observed from their data that the DSR processed sheet exhibited a lower yield strength (0.2% proof stress), a larger elongation at fracture, a smaller r value, and a larger n value specifically in the rolling direction. The enhancement in the total elongation is mainly due to the improvement in the n value leading to a diffused strain localization in the form of necking.

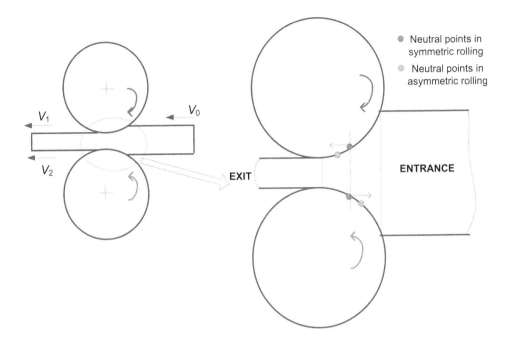

Figure 7.2 Schematic of DSR with neutral point shift.

As a result, the Erichsen values, evaluated through Erichsen cup tests, of the DSR processed sheet significantly increased as compared to conventionally rolled sheets, from 2.6 to 4 at room temperature. It has increased further from 4.1 to 7.6 at 423 K. The improvement is about 1.5 and 1.9 times, respectively. Similar attempt of DSR on commercial pure Ti at rolling temperatures from room temperature to 773 K revealed that with increase in rolling temperature, the improvement in Erichsen values for DSR processed sheet is considerable as compared with the conventionally rolled sheets. The Erichsen values increased slowly from 12.2 to 12.7 for the conventionally rolled sheets, while these values significantly increased from 12.9 to 14.7 in the case of DSRed sheets (Huang et al., 2010). Kim et al. (2007) also revealed the strengthening and improvement in ductility of AZ31 plates through DSR using a roll speed ratio of 3:1 and appropriate annealing.

Single roller drive rolling is another asymmetric rolling method to establish larger shear deformation in through thickness direction of the sheet. In this process, in a conventional rolling mill, one roll is driven by a motor and another roll (idle roll) is allowed to rotate freely as it is disconnected from the driving shaft. In the first pass, a large amount of shear deformation is introduced on the sheet in the driven side. In the second pass, the driven roll is made idle, and hence large shear deformation is introduced on the other side of the sheet. After the second stage, a unidirectional equal shear deformation is introduced by rolling (Sakai et al., 2001). The work done by Sakai et al. (2001) on AA5052 sheet of 3-mm thickness showed that at any annealing temperatures between 310°C and 460°C, the sheet is almost isotropic with average plastic strain ratio equal to 1, while the planar anisotropy (ΔR) is −0.14 and is much lower than the conventionally rolled sheet. This indicates that the press formability and drawability of sheets rolled by single roller drive rolling (unidirectional rolling, in the case of Sakai et al., 2001) is better than that of conventional rolling. Similar investigation done by Chino et al. (2002) on AZ31 Mg alloy processed by single roller drive rolling at 673 K showed better press formability (evaluated by conical cup test) at annealed condition as compared with the annealed sheet rolled by conventional rolling at 673 K. The larger average plastic strain ratio and lower planar anisotropy yielded such improvement in the press formability of the annealed sheet processed by single roller drive rolling.

Cross-rolling of sheets is a new shear rolling method to improve the press formability of sheets and plates (Figure 7.3). In this method, the rolls will be positioned at some angle "θ" to the TD in the sheet plane (TD-RD). Because of the thrust force along the axial direction of the roll, the intense shear deformation will be introduced in the through thickness or normal direction (ND) of the sheet. Chino et al. (2006a, 2007) studied the cross-rolling effect on the formability of the AZ31 Mg alloy sheet of 6 mm and 5 mm thickness. The following four cases are compared: (1) as-cross-rolled specimen, (2) as-normal-rolled specimen, (3) cross-rolled specimen, annealed at 673 K for 30 min, and (4) normal-rolled specimen, annealed at 673 K for 30 min. During experiments, θ is kept as 7.5°. The Erichsen cup tests show that the formability of cross-rolled sheet is larger than conventionally rolled sheets at all the testing temperatures between 433 and 493 K. The formability is almost doubled for cross-rolled sheets. In the meantime, the differences in average plastic strain ratio and planaranisotropy between them are insignificant.

Rolling direction plays a vital role in controlling the micro-texture of sheets and hence the formability and other mechanical properties. Generally, unidirectional rolling is performed. However, Zhang et al. (2013) and Chino et al. (2006b) proposed changing rolling directions after every rolling pass as shown in Figure 7.4. In route A, the rolling sequence is not changed after each pass. In

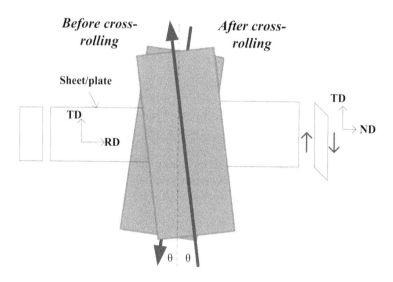

Figure 7.3 Cross-rolling of sheets. (With permission from Chino et al. 2007, Elsevier.)

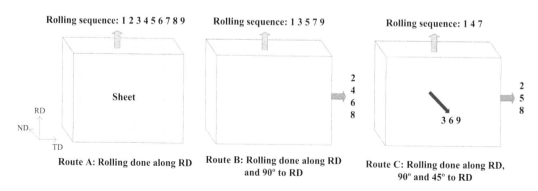

Figure 7.4 Schematic of rolling sequence to improve the formability of sheets. (With permission from Zhang et al., 2013, Elsevier.)

routes B and C, the rolling direction is changed after each pass by 45° or 90° to the RD. Zhang et al. showed that for an Mg alloy of 3 mm thickness, routes B and C show fine-grained microstructure, higher strength, and larger fracture elongation. Consequently, the press formability of such sheets has enhanced by about 28% and 31% in routes B and C, respectively, as compared with route A. The Erichsen cup value of AZ31 Mg alloy fabricated by route B is about 1.6 times higher than that of sheet fabricated by route A at 493 K and 513 K (see Chino et al., 2006b).

Rolling temperature affects the stretchability of sheets. There is so much evidence that the formability can be enhanced by varying the rolling temperature and heat treatment done prior to rolling. For example, Chino and Mabuchi (2009) showed that an Mg alloy heated at 723 K showed better stretchability as compared with the sheet heated at 663 K for a particular time. Both the sheets are unidirectionally rolled at 353 K to attain 20% reduction. Thus, a 5-mm thick sheet is converted to 1-mm thick sheet after seven rolling stages. An Erichsen value of 4.7 mm has been achieved when heated at 723 K as compared with 3.4 mm at 663 K. It is proposed that the superior formability of the specimen heated at 723 K is because of the enhancement of thickness-direction strain resulting from the weak basal texture intensity. Huang et al. (2011) showed that by increasing the rolling temperature from 723 to 828 K at the last rolling pass improves the stretch forming of an AZ31 Mg alloy considerably, after annealing at a particular temperature. There is about 50% improvement in Erichsen value (from 4.5 to 8.6) by increasing the rolling temperature from 723 K to 798 K. Increasing the last rolling temperature to 828 K does not improve the formability due to the unchanged textures and mechanical properties.

Equal channel angular rolling (ECAR) is another rolling process that induces considerable shear deformation on the sheets. In this process, the sheet is rolled in a normal twin roller and at the end of the rolling operation it is passed through a channel having a change in deformation angle (called oblique angle). The process parameters, such as rolling route, rolling pass, clearance between sheet and channel, mold structure, sheet pre-heated temperature and time, and preheating temperature of mold influence the microstructure, and hence the mechanical properties. AZ31 Mg alloy has been processed through this method by Chen et al. (2007). For a 1.6-mm thick sheet, a twin-roller hot rolling mill with roll diameter of 360 mm each with a gap of 1.5 mm is used. The channel has a gap of 18 mm at the entrance and exit of the mold with an oblique angle of 115° and with an oblique radius of 2 mm. After ECAR, the yield strength decreases with clear strain-hardening behavior, and elongation increases gradually with increase in number of rolling passes. The Erichsen value has improved from 4.18 for as-received sheet to 6.26 for ECAR sheet after four passes, and the limiting drawing ratio has increased from 1.2 to 1.6. Such improvement in formability is because of nonbasal crystal orientation. For the same material, Cheng et al. (2007) showed the importance of channel height to sheet thickness in changing the crystal orientation from basal plane to nonbasal plane. The limiting draw ratio has improved from 1.2 to 1.6 mm for an ECARed sheet.

7.3 FRICTION STIR WELDING AND PROCESSING

In FSW/FSP, a rigid cylindrical tool with a pin at the end is used to stir the sheets at the interface edge to form a permanent joint. The mechanical properties like strength and ductility of the weld region improve depending on the material. Such improvement in mechanical properties enhances the overall formability of sheets. The post-FSW formability of AA5052-O sheet studied by Sato et al. (2004) reveals that the plane-strain forming limit of the stir zone increased up to a grain size of 10 μm, after which it decreases. The contributing microstructural features are larger grain size and lower dislocation density and subboundaries. Ramulu et al. (2012a, 2013) showed, through formability studies using limiting dome height (LDH) tests of FSWed Al 6061-T6 sheet grades, that the forming limit curves can be improved by controlling rotation speed, welding speed, plunge depth, and shoulder diameter. It can be observed that the forming limit of FSWed sheet increases with an increase in the tool rotation speed from 1,300 to 1,400 rpm for a constant welding speed, while it decreases slightly with an increase in welding speed from 90 to 100 mm/min for a constant tool rotation speed. In addition, the forming limit increases significantly with an increase in shoulder diameter from 12 to 18 mm and plunge depth from 1.85 to 1.9 mm. Figure 7.5 shows the effect of rotational speed and shoulder diameter on the FLD evaluated in Ramulu et al.'s work.

Lee et al. (2009) studied the effect of FSW on the weld zone ductility of five different sheet grades. FSWed sheets made of AA5083-H18, AA5083-O, and AZ31 materials show better formability as compared with the base sheets, while those made of DP590 and AA6111-T4 show reduced formability. They also pointed out that for proper design of TWBs, the weld zone strength and ductility are important factors in deciding the strain localization. Superplasticity has been observed in some of the Al alloys after FSP/FSW. Ma et al. (2003) showed elongations of 1,280% at 525°C and 1×10^{-1}/s, and 1,210% at 550°C at the same strain rate for Al–4Mg–1Zr extruded bar after FSP. For AA2095 sheet material, elongations of 267% and 282% are observed at the weld nugget made with welding speeds of 3.2 and 4.2 mm/s, respectively (Salem et al., 2002). These are some examples showing the improvement in formability of materials after FS processing.

7.4 FORMING AT ELEVATED TEMPERATURE

Warm forming and hot forming of metals play a vital role in deciding the ductility and flow stress during deformation. Warm working is generally done at about $0.3T_{\mathrm{m}}$, while hot working is done within the range of $0.5T_{\mathrm{m}}$–$0.75T_{\mathrm{m}}$, where T_{m} is the melting point (absolute temperature) for a specific metal. Sheet forming defects like springback is minimized during forming at elevated temperatures. It is known that Mg alloys show restricted use in automotive sector because of lower formability at room temperature. Its formability improves considerably at elevated temperatures, but practicing that needs process understanding. Hariharasudhan et al. (2004) performed experiments and finite-element simulations of round cup forming and rectangular pan forming of AZ31B sheet at elevated temperatures. Tensile tests reveal that the ductility has improved from 16% to 35% when testing temperature increased from 25°C to 235°C with almost halved flow stress value at 235°C. Out of four different temperatures, 150°C, 200°C, 250°C, and 300°C, the limiting draw ratio is maximum at 200°C. At this temperature, lesser thinning is observed as compared with other temperatures. The finite-element simulation results match well with the experimental data in terms of thinning and load evolution behavior. However, the predicted thinning values are larger than the experimental values. Li et al. (2013) developed FLCs for AZ31B Mg alloy at 150°C, 200°C, and 300°C by hemispherical dome height tests. It is observed that the FLC improves considerably with increase in temperature. A significant increase of about 20% major strain is seen in balanced biaxial stretching strain-path for each temperature increment. Hence, there is about 35% improvement in

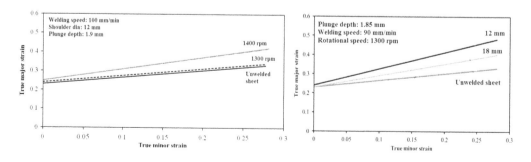

Figure 7.5   FLD showing the effect of rotational speed and shoulder diameter on the FLCs of FSWed sheets (Ramulu, 2012b).

major strain when compared between 150°C and 300°C. In plane-strain strain-path, about 12% improvement in major strain is witnessed for each temperature increment. The failure mode is brittle in nature at room temperatures, while it is ductile above 200°C. The same material, AZ31, is also tested for its formability (or FLC) by hemispherical dome height tests at 200°C, 250°C, and 300°C and at strain rates of 0.001, 0.01, and 0.1/s by Chan and Lu (2014). The FLCs improve considerably with increase in temperature and decrease in strain rate. Moreover, the influence of strain rate on FLCs is significant at elevated temperatures and this happens because of improving strain-rate sensitivity index (m) at elevated temperatures.

Some of the aluminum alloys show low formability at room temperature. These are formed at elevated temperatures, mainly using warm forming as it is below recrystallization temperature, for better formability. Three Al alloys, Al 5754, Al 5182+1% Mn, and Al 6111-T4, are tested for warm formability by Li and Ghosh (2004) at temperatures from 200°C to 350°C. Al 5754 and Al 5182+1% Mn show considerable formability improvement during biaxial testing with increase in temperature, when compared to Al 6111-T4. The formability improvement is good enough at 200°C that it is comparable to A–K steels formed at room temperature. It is also suggested to maintain the die temperature 50°C larger than the punch temperature to obtain an optimum formability of Al alloys. Toros et al. (2008) reviewed the importance of warm forming of Al–Mg alloys done at 200°C to 300°C in order to improve the formability and surface quality.

Hot and warm forming of HSS and ultra HSS are attempted by many authors; for example, warm forming of a cold rolled high-strength low alloy steel (HC300LA) by Sen and Kurgan (2016), warm and hot stamping of HSS and ultra HSS grades by resistance heating by Mori et al. (2005), warm stretching of TRIP aided dual-phase steel (TDP steel) by Sugimoto et al. (1995), hot forming of ultra HSS steel-22MnB5 by Liu et al. (2011), hot and warm forming, and springback reduction of HSS by Yanagimoto and Oyamada (2006). A considerable reduction in springback of 22MnB5 steel at elevated temperatures (700°C–950°C) (Yanagimoto and Oyamada, 2006) and significant improvement of draw ratio by 22% for 1.2 mm thick HC300LA sheet and by 21% for 1.5 mm thick sheet compared to forming at room temperature (Sen and Kurgan, 2016) are some data highlighting the importance of warm and hot forming for formability improvement.

7.5 INCREMENTAL SHEET FORMING

In incremental sheet forming (ISF), a rigid tool with hemispherical head is used to impart localized deformation on the sheet by following a predefined tool path. The tool moves horizontally and then vertically to form the desired component. Since only a small portion of sheet just below the tool head deforms, the load requirement is much less. ISF can be performed either with die or without die. The case with die is used for giving particular complex shape to the sheet (Figure 7.6a,b). Two rigid tools can also be used for ISF as shown in Figure 7.6c. The parameters such as tool type, tool size, feed rate, friction at the tool-sheet interface, tool rotational speed, incremental depth, and mechanical properties of the sheet influence the formability of sheet synergistically. Hence, by controlling the parameters, the forming limit of the sheet metal can be improved.

In this context, Kim and Park (2002) revealed that for 1050 Al sheet of 0.3mm thickness, the forming limit is higher (1) for a tool with rotating head without friction as compared to hemispherical head with friction—showing 5% improvement, (2) for smaller tools of 10mm and 5mm diameter as compared to 15 mm—showing about 5%–20% improvement depending on tool movement with respect to the rolling direction, and (3) for lower feed rates of 0.1mm as compared to 0.3 and 0.5 mm—showing improvement of 12%–20% depending on the tool movement with respect to the rolling direction. Also there is a considerable difference in forming limit when the tool path is along the rolling direction or perpendicular to the rolling direction. Forming of sheet at elevated temperatures improves its formability. This is true for laser-assisted

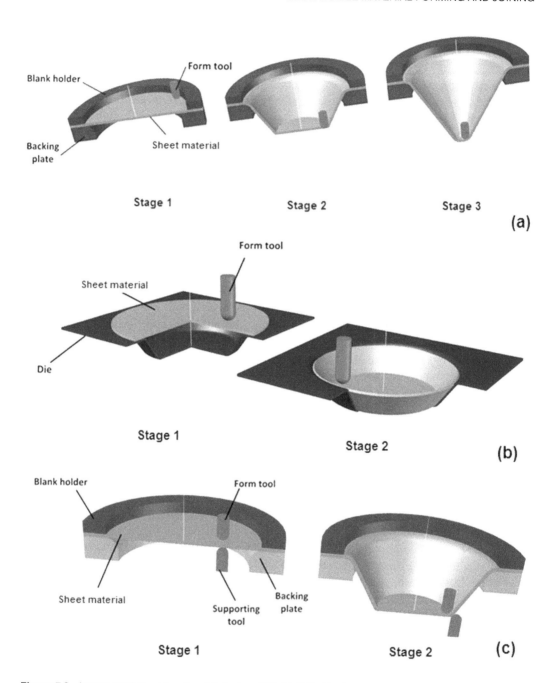

Figure 7.6 Incremental sheet forming: (a) die-less ISF, (b) ISF with die, and (c) ISF using double tools.

ISF, as observed from the work of Duflou et al. (2007). In their work, Duflou et al. used a 500-W Nd–YAG laser system with glass fiber beam delivery as a heating source. A localized heat spot is created on the region of deformation, without heating the outside region. Thus, a softer region is generated before the onset of plastic deformation. Load reduction for Al 5182 of 1.25 mm thickness, and formability improvement for TiAl6V4 sheets of 0.6 mm thickness have been shown. At 20°C, it is possible to form a cone-shaped sheet part with slope of 32°, and cracks appeared before reaching 30 mm depth. On the other hand, by varying the spot diameter and the laser energy input level, cones with slope angles up to 56° are generated improving the formability of TiAl6V4 sheets.

Jeswiet and Young (2005) developed a new method of developing FLD for ISF for any material grade. The forming limit strains of five different shapes like cone, hyperbola, dome, pyramid, and lobe, are obtained and plotted in the same graph as a master FLC of that sheet grade. These five different shapes generated five different strain paths slightly different from each other. But, on the whole, the maximum forming limit strain of about 300% has been achieved for a 3xxx Al alloy sheet using ISF. This is generally unachievable in conventional sheet forming methods. Davarpanah et al. (2016) investigated the effect of ISF (single point) on the mechanical properties of the deformed polymer, for an amorphous PVC and a semicrystalline polyamide. The deformed polymer (both PVC and polyamide) showed improvement of about 50%–60% in ductility depending on the rotation speeds and incremental depths as compared with undeformed polymer. Lasunon and Knight (2007) demonstrated that by using double point incremental forming (DPIF), the formability of sheet can be improved as compared to single point incremental forming (SPIF). When comparing products with the same wall angle, (1) the effective strain in SPIF parts is larger than from DPIF, (2) sharp corners are made properly by DPIF, (3) wall thickness attained is larger in DPIF yielding deeper parts, all showing that DPIF is better than SPIF in terms of formability.

7.6 HYDROFORMING

In hydroforming, a material in the form of sheet or tube is plastically deformed by internal fluid pressure. In tube hydroforming, during bulging, the wall thickness of the tube decreases, thus resulting in tube failure or bursting. The decrease in wall thickness is minimized by providing tube compression in the axial direction in the presence of the internal fluid pressure. However, the axial compression should be controlled and compatible with increase in fluid pressure to avoid tube wrinkling, even at lower pressure. Many research groups have initiated strategies to improve the formability of materials in hydroforming.

Pulsating hydroforming is a method to improve the forming limit of tubes. In pulsating hydroforming, the internal pressure is oscillated within a pressure range (Figure 7.7). Mori et al. (2004, 2007) showed that by pulsating hydroforming, the formability of tubes improves because of uniform expansion of tubes as opposed to what is observed in constant internal pressure hydroforming. Mori et al. (2007) varied the internal pressure at 17.7 ± 3 MPa ($P_0 \pm \Delta P$) with 0.67 cycles per unit punch displacement for a tube of 38.1 mm outer diameter, 1.1 mm wall thickness, and 200 mm tube

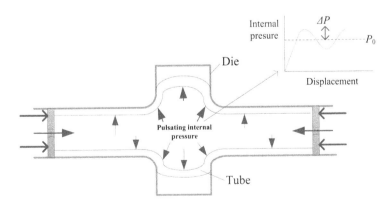

Figure 7.7 Schematic of pulsating hydroforming and internal pressure variation. (With permission from Mori et al., 2004.)

length. Under this condition, a uniform expansion of tube is seen, while a bulged tube is obtained at a higher constant pressure of 20.5 MPa, and a wrinkled tube at a lower constant pressure of 17.7 MPa. The wall thickness measurement revealed a uniform distribution in the case of pulsating hydroforming with minimized thinning as compared to constant pressure hydroforming. In this method, appearance of small wrinkling at low pressure and disappearance of wrinkling at high pressure is observed, which finally improved the deformation of tubes with lesser thinning. Defects like bursting are also avoided by pulsating hydroforming with control over amplitude and cycle number (Mori et al., 2004).

Xu et al. (2015) investigated the same method on austenitic stainless steel and demonstrated that formability improves because of increase in fraction of strain-induced martensite, which is nothing but transformation-induced plasticity effect, delaying the neck formation. Beyond a critical pulsating load of 20 MPa, axial feeding process can be incorporated during the loading and unloading process to enhance the formability.

There is another effective method to improve the formability of tubes in hydroforming—by using *useful wrinkles* in the expanding area of tube. Generally, wrinkles are formed when the axial force exceeds a critical limit, which is not compatible with the internal pressure. But such wrinkles can be useful to improve the tube's formability. Yuan et al. (2007) showed that the loading paths (axial distance vs. internal pressure) and number of wrinkles decide the final tube forming quality. The loading paths are varied between 3 and 9 MPa internal pressures, while the number of wrinkles is varied from 0 to 3. The loading path with 6 MPa and 3 number of wrinkles provided a good formed tube with a lesser thinning rate of 8.5%. Moreover, a process window (Figure 7.8) suggested by them revealed that the window can be improved, by useful wrinkles requiring lower internal pressure, from the existing window created by the case without wrinkles. Beyond a critical internal pressure, bursting of tubes will happen, while folding will occur below a minimum pressure.

Use of *magnetorheological fluid* (MRF) as pressurizing medium with magnetic field instead of conventional mineral oil is found to enhance the deep drawability of a deep drawing quality steel because of enhancement of sealing limit. Rosel and Merklein (2014) demonstrated this by experimental and finite-element simulations. For the same flange contact pressures during sheet hydroforming, each of 6 N/mm² and 22 N/mm², it is observed that the dome height reaches to about

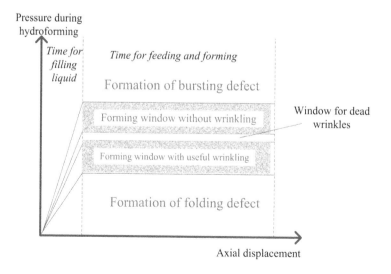

Figure 7.8 Process window for the optimum use of wrinkles during hydroforming. (With permission from Yuan et al., 2007.)

(1) 15 mm for conventional mineral oil, as opposed to 20 mm when MRF (with 0.4 T magnetic flux density) is used, in the case of 6 N/mm², and (2) 30 mm full dome height in case of 22 N/mm² for both the fluids, but with smaller corner radius in case of MRF. Slightly larger internal pressure has been used for the experiments in the case of MRF. Moreover, minimum sheet thickness in the critical location of the sheet is 0.67 mm when MRF is used, which is smaller than 0.75 mm achieved with mineral oil.

Hydroforming at elevated temperatures (warm and hot hydroforming) improves the formability of a material. In this method, hydroforming is carried out at elevated temperatures to combine the advantages of both hydroforming and hot/warm forming conditions. A few research groups including Keigler et al. (2005) on AA5182 sheet material, Liewald and Pop (2008) on AZ31 tubes, Chen et al. (2015) on thermomechanical treatment with sheet hydroforming on AA2219, and Seyedkashi et al. (2014) on warm hydroforming of AA6061 tubes demonstrated the method. In general, the hydroforming done at elevated temperatures minimizes the nonuniform thickness distribution and delays the fracture initiation in order to improve the formability. The process can be conducted in two modes: isothermal hydroforming and nonisothermal hydroforming. In isothermal hydroforming, all the tools including punch, blank holder, die, and high-pressure oil are maintained at the same temperature during forming. There is no heat exchange in the whole process. In the nonisothermal hydroforming, the punch is kept at lower temperature by cooling water circulation and other tools like blank holder, die, and liquid chamber are at higher temperature. Lang et al. (2012) exploited the difference in the two modes and found that the nonisothermal hydroforming is better in terms of part formability. With nonisothermal hydroforming, 132 mm drawn part height can be attained when punch is kept at 50°C and die at 250°C, while only 31 mm is achievable in isothermal hydroforming. A sheet of 260-mm initial diameter is used for experiments. With increase in punch temperature to 150°C and 250°C, the cup height is found to decrease considerably in nonisothermal hydroforming.

Similarly, *electromagnetic assisted hydroforming* (EMAH) is another way to produce sheet parts with reduced thinning and better die corner filling. Chu et al. (2014) revealed that in EMAH, in order to form a part with rectangular cross-section, compounding action of internal pressure and electromagnetic force at desired levels are required. Internal pressure plastically deforms the material, while electromagnetic force applied on a straight segment generates a petal-like preform that helps in better cornet filling by reducing the contact friction. Electromagnetic force is induced by a coil with a discharge voltage of 2.2 kV and capacitance of 777 µF for an AA5052-O tube. It evident from Table 7.1 that the pressure required in EMAH is much less than that of conventional hydroforming to form the same radius. This means that sharp corner radius can be reached at a lower pressure in EMAH without failure.

Table 7.1   Relation between Corner Radius and Internal Pressure: Conventional Hydroforming vs. EMAH (Chu et al., 2014)

Internal Pressure (MPa)	Corner Radius (mm)	
	Conventional Hydroforming	EMAH
4	13.5	13.5
10	11.5	6
20	8	3.5
30	6.5	3
40	5.8	2.5
50	5	2
60	4.5	2

7.7 SANDWICH SHEET FORMABILITY

Lightweight materials and structures are used in many applications due to the requirement of fuel savings and environmental concerns. Sandwich sheets made of metal–polymer combination are a potential candidate for lightweight material applications. A metal–polymer sandwich sheet consists of two layers of metallic sheets (steel or aluminum) as skin materials and a polymer core are all joined by adhesive. The sandwich sheet offers lower density, higher specific flexural stiffness, better dent resistance, and good damping behavior. However, their mechanical performance and joinability gets affected.

In this context, formability of sandwich sheets in comparison to monolithic base sheets is a topic to be analyzed. The base material properties, adhesive properties, and adhesion quality all affect the formability synergistically. Satheeshkumar and Narayanan (2014, 2015a) showed that the formability of adhesive-bonded sheets can be improved by changing the adhesive properties. The adhesive properties are modified by varying the hardener to resin ratios in the epoxy and acrylic adhesive system. The epoxy adhesive with 1:1 hardener to resin ratio exhibits larger ductility as compared to the one with 0.6:1. The other ratios fall in between. Similar results are found in the case of acrylic adhesives when the hardener to resin ratio is varied between 0.8:1 and 1.2:1. When adhesive-bonded sheets are made of such adhesives, the overall formability or ductility also improves with increase in hardener to resin ratio (Table 7.2). Similar results are observed in strain-hardening exponent and forming limit strain. The improvement in ductility/elongation and limit strains is mainly due to the conversion of resin-rich formulation to hardener-rich formulation, making the sample more ductile.

Forming limit diagrams are developed by Kim et al. (2003) for AA5182/polypropylene/AA5182 sandwich sheets, Liu et al. (2013) for AA5052/polyethylene/AA5052 sandwich sheet, and Aghchai et al. (2008) for two layers AA1100-st12 sheets bonded by polyurethane adhesive. The results from their experimental and modeling work reveal the improvement of forming limit of sandwich sheets as compared with monolithic sheets. According to Kim et al., strain-hardening exponent, strain rate sensitivity index, and thickness of the sheet are the factors governing the change in forming limit of sandwich sheet. They showed that the higher strain rate sensitivity of the sandwich sheet is compensated by the lower strain-hardening exponent, and hence greater thickness of the sandwich sheet is the main reason for its improved formability as compared with that of the base sheet. Liu et al.'s results indicate a considerable increase in forming limit of sandwich sheet as compared to AA5052 parent sheet and the forming limit increases with increase in core layer thickness. There is

Table 7.2 Improvement in Ductility of Base Sheets Constituting Adhesive-Bonded Sheets (Satheeshkumar, 2015b)

Adhesive-Bonded Sheet	Hardener to Resin Ratio	Improvement in Ductility as Compared to Base Sheet (%)[a]	
		Deep Drawing Quality Steel	SS316L Sheet
Epoxy based	1:1	7.5	24.9
	0.9:1	6.5	7.9
	0.8:1	5.5	7.5
	0.7:1	5.5	0.6
	0.6:1	3.5	0.2
Acrylic based	1.2:1	6.5	7.9
	1.1:1	5.5	4.9
	1:1	3.5	2.9
	0.9:1	2	0
	0.8:1	0.5	0

[a] for 13 mm gauge length.

about 6% increase in limit strain between 1.5-mm thick and 2-mm thick sandwich sheet, and about 3% improvement between 2-mm and 3-mm thick sandwich sheets. On the other hand, Aghchai et al.'s work show that the forming limit of AA1100-st12 two-layered sheet is better than the lower formable sheet, AA1100, and hence using the two-layersheet improves the formability of a low formable sheet.

7.8 TOOLING: FLEXIBLE FORMING AND BLANK HOLDING

In *multipoint forming*, a sheet is plastically deformed by several rigid punches or pins. Each punch concentrates deformation over a small region, instead of using a single traditional die in conventional forming operations. Several complex sheet surfaces that are corrugated can be generated using multipoint forming press and each punch can be controlled by computer interface. Moreover, the deformed sheet can be changed in its shape locally by using some of the punch elements later on, rather than fabricating another die for the purpose. There are two important defects, namely dimpling and buckling, that are commonly seen during multipoint forming. Both these defects are generated by smaller contact area between the punch and sheet which is again decided by the shape of the punch tip. Dimples are formed when the round-edged punch delivers sufficient pressure on the sheet such that localized thinning occurs during forming. Although such dimples are smaller than the overall sheet area, these will act as failure source during forming. Buckling arises because of too small contact area between the punches and sheet such that there are noncontact regions between the punches. The regions are deprived of clamping pressure and destabilization occurs in the form of buckling.

The main advantage of multipoint forming is that in this there is nothing called die and related design and fabrication process. The die equivalent is the shape taken by arrangement of set of punches that are controlled by computer interface. The cost involved in design and fabrication of stamping dies is eliminated. In this process, the punches have rounded ends that are simple and easy to design and fabricate. But in this method, relating the initial sheet shape to final shape requires standard design procedures and real-time control of movement of punches is crucial to decide the specific forming path.

There are four different multipoint forming methods (Figure 7.9) as described in Li et al. (1999). In multipoint die forming, as shown in Figure 7.9a, the relative motion between the punches is avoided. The initial shape delivered to the sheet is decided by the arrangement of punches before the forming starts. This is similar to conventional die forming with single die, except that in this case, small punches decide the shape locally. In the multipoint half die forming (Figure 7.9b), either the upper set of punches or the lower set is fixed, while the other one is passive. The desired shape is delivered by the fixed punches and the passive punches move down to take the shape of the formed sheet. The multipoint press forming is flexible among the types (Figure 7.9c) and thus the desired shape has been imparted by relative movement of punches (all active punches) in the lower and upper sides. The initial shape is not decided before forming like in the first two cases; rather, it is decided during forming such that the local deformation path of individual punches can be controlled by computer interface. Here each punch acts like a small press. In the fourth case (Figure 7.9d), the active punches decide the shape of the component, while the passive punches are forced to move down as per the movement of active punches.

Furthermore, Li et al. (2002) introduced two more variants of multipoint forming: namely multipoint forming with variable blank holding force (BHF) and sectional multipoint forming. In the case of multipoint forming with variable BHF, the BHF application region is divided into small segments. Each segment is controlled separately by servo valves. Use of this method on forming an aluminum sheet of 1 mm thickness eliminates wrinkles fully as compared to the case of multipoint forming without BHF. In the sectional multipoint forming (Figure 7.10), the sheet is deformed

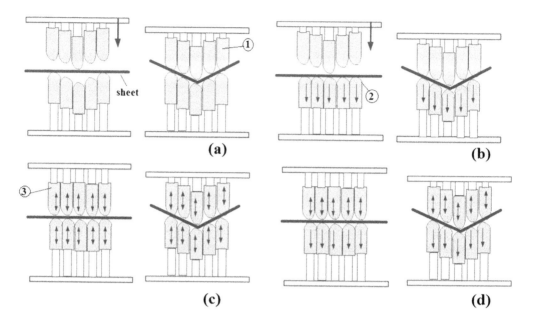

Figure 7.9 Multiple point forming: (a) multipoint die forming: having rigid punches, (b) multipoint half die forming: having rigid and passive punches, (c) multipoint press forming: having active punches, and (d) multipoint half press forming: having both active and passive punches [1: fixed punch, 2: passive punch, 3: active punch]. (With permission from Li et al., 1999.)

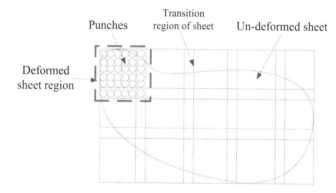

Figure 7.10 Sectional multipoint forming. (With permission from Li et al., 2002.)

section by section so that a multipoint forming press with smaller bed size can be used for stamping large sized sheet parts. In an example quoted by Li et al. (2002), a sheet of size 3 m has been deformed using section multipoint forming with the forming area of about 140 mm × 140 mm. There are three regions: deformed sheet region, transition region of sheet, and undeformed sheet region, in a sheet during sectional multipoint forming and the transition region, which is nothing but a sheet region which is deformed in the previous stage and will be deformed again in the next stage, play a crucial role in formability analyses.

Tan et al. (2007) introduced the following three forming modes to fabricate titanium alloy cranial prosthesis: (1) sheet in contact with rigid punches directly during forming, (2) sheet sandwiched between polyurethane elastic cushion and not contacting the rigid punches directly, and (3) using steel padding with BHF, along with the sandwiched sheet and not contacting the rigid punches directly. When the retiary sheet is formed with direct contact with punches, defects like dimples and wrinkles are formed because of pointed contact area. Moreover, the sheet surface is rough. In the case with sandwiched sheet, since the direct contact is avoided, such defects are minimized in the formed sheet, though they exist. Large elastic deformation occurs in the elastic cushion, and the elastic cushion fills the gaps between nearby punches. This distributes the concentrated loads over a larger region, averaging out the deformation. In the third case, the steel padding is sandwiched between polyurethane elastic cushions and Ti retiary sheet and it is held by BH pressure. The steel padding deforms along with the retiary sheet. It can be observed that the third case delivers a successful product without wrinkles and dimples.

The multipoint forming can also be done in a closed look forming set up (Liu et al., 2008). In this process, there is feedback between the controls and the sheet shape after each forming cycle. The desired shape and feedback shape are compared to minimize the error. Depending on the error achieved, the forming step size is decided. Thus, the shape error is minimized to reduce springback after each forming cycle. A laser scanner implemented in the feedback system will be helpful for precision manufacturing. The scanned data can also be used for reverse engineering. Davoodi and Zareh-Desari (2014) also showed that the springback of sheet in multipoint forming can be controlled by sheet property and thickness, anisotropy ratio, elastic layer thickness, elastic layer hardness, and number of punch elements, through experiments and numerical simulations.

In *rubber pad forming*, there is a rigid die and a flexible rubber pad, which are involved in plastically deforming a sheet. The shape provided in the rigid die is imparted to the sheet with the help of rubber pad as shown in Figure 7.11. The presence of rubber pad distributes the deformation uniformly and improves the formability of sheets, as compared to the conventional dies. The design and manufacture of only one of the dies is necessary. This minimizes the total fabrication cost. The precise assembly of rigid die is not required in this case. The time and cost involved in forming press is greatly minimized.

Bipolar plate for a proton exchange membrane fuel cell is fabricated by rubber pad forming. Liu and Hua (2010), Lim et al. (2013), and Jeong et al. (2014) analyzed the effect of some crucial tool and process conditions on the quality of bipolar plate after forming. The process parameters such as rubber hardness, rubber thickness, punch speed and pressure, internal and external radii of die, and draft angle of the die all contribute to the forming quality of the plate. The forming depth improved in proportional to rubber pad thickness, punch velocity and pressure, outer radius, and draft angle. When the internal radius is too small, it is difficult to fill the die cavity during forming. The rubber pad hardness is not an important parameter during forming of the plate. During such fabrication, Liu and Hua have reported to observe uneven thickness distribution and chances of failure in the die corner. A scaled down version of a support rib for an aircraft wing or tail flap has been fabricated by rubber pad forming using an aluminum alloy with less thinning and good surface finish (Browne and Battikha, 1995).

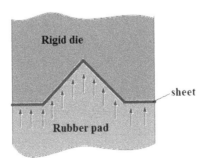

Figure 7.11 Rubber pad forming with rigid die.

Variable BHF also improves the quality of a sheet product by controlling the localized deformation. The BHF is varied with respect to punch displacement during forming. Such variation will follow a strategy. For instance, when the wrinkling height exceeds a critical limit, the BHF will be increased to eliminate the deviation. On the other hand, if it is less than lower limit, then an optimum BHF will be identified to suppress the wrinkle formation. A closed-loop PID control strategy can be followed for the purpose (Zhong-qin et al., 2007). The strain paths can be modified from drawing to stretching using variable BHF strategy. This improves the forming limit and drawability of the sheet, reduces the springback and wrinkling, and improves the dent resistance. The success of variable BHF is prominently seen in forming tailor-welded sheets. Here two sheets of different thickness and quality are welded to form a component. The weld line gets displaced to a critical region for a constant BHF during the forming process. Such movement can be avoided if larger BHF is given to the thinner/weaker sheet and smaller BHF to the thicker/stronger sheet. Another method is providing clamping pins along the weld line and localized holding force can be controlled by computer interface. Chen et al. (2008) demonstrated four different schemes: flat BH with and without clamping pins, and stepped BH with and without clamping pins, for minimizing weld line movement during box drawing experiments. Their results show the superiority of stepped BH with clamping pins over other cases. Similar attempt made by Kinsey et al. (2000) for a door inner showed that the localized clamping has improved the stamping ability of the welded sheets without failure initiation and reduced weld line movement.

7.9 LUBRICATION

Lubrication and friction between the sheet and tools are interrelated. Good lubrication reduces friction and improves the sheet forming quality. The deformation path, forming limit, surface finish, friction coefficient, strain distribution, change during sheet forming, while adhesion, material transfer, cold welding, scratching, and galling reduce with lubricants and friction conditions. Moreover, the lubricants used pose economic and environmental challenges including generation of pollutants into waste stream, difficulty in disposal of used lubricants, development of health hazards for the users, specifically in case of synthetic lubricants, etc. This paved way for the development of new lubricants that are eco-friendly and forming friendly.

Lovell et al. (2006) developed a new lubricant, which is a combination of boric acid crystals (5 wt%) and canola oil, which are natural and eco-friendly for sheet forming applications. The performance of the lubricant is compared to that of canola oil, transmission oil, and an unlubricated case. The new lubricant performed better than the other cases with friction coefficient as low as 0.3 evaluated with a strip friction test simulator. The average roughness (R_a) value attained is about 0.59 µm, after forming, the lowest value among the four cases. The wheat flour particles in varying concentrations were dispersed in water and their effect, as lubricant, on drawability is evaluated by Yoshimura et al. (2002). It is observed that the wheat flour lubricant performed better than grease and press oil in terms of Erichsen number and drawn cup height. For 30% wheat flour concentration, the largest Erichsen number has been attained, i.e., 14 mm, 15 mm, and 11 mm for a Ti alloy, stainless steel and commercial steel, respectively. Similar results are seen during deep drawing operation.

Rao and Wei (2001) demonstrated the use of boric-acid dry films for drawing and stretching applications of Al alloys. The performance of dry film of saturated ethanol solution with boric acid is compared with that of chemically pure oleic acid, commercial solid lubricants such as MoS_2 and Teflon, and graphite+oil lubricants. The boric-acid-based lubricant performs at par with commonly used lubricants like Teflon and graphite+oil in terms of maximum drawing force and dome height for all the Al alloys tested. The procedure of applying the boric acid lubricant is simple and is suitable for wide range of forming speeds. In another attempt, Rao and Xie (2006) presented the importance of boric acid over some commonly used lubricants through ring forging, cold strip rolling, deep drawing, and stretching experiments. The performance of boric acid is better in sheet forming applications (in terms of dome height and maximum drawing force) as compared with rolling and forging.

Sheets and dies are coated with permanent coatings for press formability applications. Dies coated with titanium nitride (TiN) or vanadium carbide (VC) for forming HSS sheets (Abe et al., 2014) and diamond-like carbon (DLC) coating on Al 5052 sheets (Horiuchi et al., 2012) are typical examples. In Abe et al.'s work, dies with TiN coating has been performed with chemical vapor deposition and physical vapor deposition. Dies with VC coating performed better than TiN coating in preventing material seizure and fracture during drawing of HSS grades. The VC-coated die is suitable for drawing cup having small surface roughness in the low drawing speed (~8 mm/s) and low ironing ratio. Ironing ratio is defined by the ratio of difference between the sheet thickness and clearance between punch and die to sheet thickness. Horiuchi et al. reported that the DLC coating reduced the friction coefficient by about 40% as compared to a commonly used oil lubricant and an unlubricated case. The maximum drawing force is also reduced considerably in the case of DLC-coated sample as compared with oil-lubricated case during Al deep drawing. Liu et al. (2009) revealed the effect of angle between rolling direction and sliding direction during strip draw testing and the effect of lubricant additives on the friction coefficient. They found that friction coefficient is at its minimum when the angle is 30°, and it is at its maximum when the sliding direction is parallel to rolling direction. Hence, sheet deforming direction with respect to rolling direction is also crucial in deciding the final forming behavior. Moreover, independent of the angle, the friction coefficient tested with lubricant with additives is lower as compared to lubricant without additives.

7.10 SUMMARY

In this chapter, strategies to improve the formability of sheets are presented. Rolling methods and conditions, and friction stir processing, change the raw sheet characteristics through alterations in microstructure and mechanical properties. Enhancement in sheet formability in terms of limiting draw ratio, forming limit, and ductility are observed from available case studies. Other methods, such as incremental forming, forming at elevated temperature, hydroforming, multipoint forming, blank holding, lubricants, etc., change the forming quality through alterations in deformation pattern. Forming limit improvement by incremental forming and forming at elevated temperature, defect minimization in hydroforming and multipoint forming, drawability improvement and defect minimization using blank holding, and surface characteristics improvement with lubricants are also witnessed. Following such strategies with appropriate process design will enhance the forming quality of sheets, thus improving the sustainability and eco-friendliness in industry practice.

REFERENCES

Abe, Y., Ohmi, T., Mori, K., Masuda, T. 2014. Improvement of formability in deep drawing of ultra-high strength steel sheets by coating of die. *Journal of Materials Processing Technology* 214: 1838–1843.

Aghchai, A.J., Shakeri, M., Mollaei-Dariani, B. 2008. Theoretical and experimental formability study of two-layer metallic sheet (Al1100/St12). *Proceedings of the Institution of Mechanical Engineers, Part B: Journal of Engineering Manufacture* 222: 1131–1138.

Browne, D.J., Battikha, E. 1995. Optimization of aluminium sheet forming using a flexible die. *Journal of Materials Processing Technology* 55: 218–223.

Chan, L.C., Lu, X.Z. 2014. Material sensitivity and formability prediction of warm-forming magnesium alloy sheets with experimental verification. *International Journal of Advanced Manufacturing Technology* 71: 253–262.

Chen, W., Lin, G.S., Hu, S.J. 2008. A comparison study on the effectiveness of stepped binder and weld line clamping pins on formability improvement for tailor-welded blanks. *Journal of Materials Processing Technology* 207: 204–210.

Chen, Y.Z., Liu, W., Yuan, S.J. 2015. Strength and formability improvement of Al-Cu-Mn aluminum alloy complex parts by thermomechanical treatment with sheet hydroforming. *Journal of Metals* 67: 938–947.

Chen, Z.H., Cheng, Y.Q., Xia, W.J. 2007. Effect of equal-channel angular rolling pass on microstructure and properties of magnesium alloy sheets. *Materials and Manufacturing Processes*, 22: 51–56.

Cheng, Y.Q., Chen, Z.H., Xia, W.J., Zhou, T. 2007. Effect of channel clearance on crystal orientation development in AZ31 magnesium alloy sheet produced by equal channel angular rolling. *Journal of Materials Processing Technology* 184: 97–101.

Chino, Y., Mabuchi, M., Kishihara, R., Hosokawa, H., Yamada, Y., Wen, C., Shimojima, K., Iwasaki, H. 2002. Mechanical properties and press formability at room temperature of AZ31 Mg alloy processed by single roller drive rolling. *Materials Transactions* 43: 2554–2560.

Chino, Y., Sassa, K., Kamiya, A., Mabuchi, M. 2006a. Enhanced formability at elevated temperature of a cross-rolled magnesium alloy sheet. *Materials Science and Engineering A* 441: 349–356.

Chino, Y., Lee, J.S., Sassa, K., Kamiya, A., Mabuchi, M. 2006b. Press formability of a rolled AZ31 Mg alloy sheet with controlled texture. *Materials Letters* 60: 173–176.

Chino, Y., Sassa, K., Kamiya, A., Mabuchi, M. 2007. Microstructure and press formability of a cross-rolled magnesium alloy sheet. *Materials Letters* 61: 1504–1506.

Chino, Y., Mabuchi, M. 2009. Enhanced stretch formability of Mg–Al–Zn alloy sheets rolled at high temperature (723 K). *Scripta Materialia* 60: 447–450.

Chu, G.N., Chen, G., Chen, B.G., Yang S. 2014. A technology to improve formability for rectangular cross section component hydroforming. *International Journal of Advanced Manufacturing Technology* 72: 801–808.

Davarpanah, M., Bansal, S., Malhotra, R. 2016. Influence of single point incremental forming on mechanical properties and chain orientation in thermoplastic polymers. *ASME Journal of Manufacturing Science and Engineering* 139: 021012-1–021012-9.

Davoodi, B., Zareh-Desari, B. 2014. Assessment of forming parameters influencing spring-back in multi-point forming process: A comprehensive experimental and numerical study. *Materials and Design* 59: 103–114.

Duflou, J.R., Callebaut, B., Verbert, J., De Baerdemaeker, H. 2007. Laser assisted incremental forming: Formability and accuracy improvement. *Annals of the CIRP* 56: 272–276.

Hariharasudhan, P., Ngaile, G., Altan, T. 2004. Finite element simulation of magnesium alloy sheet forming at elevated temperatures. *Journal of Material Processing Technology* 146: 52–60.

Horiuchi, T., Yoshihara, S., Iriyama, Y. 2012. Dry deep drawability of A5052 aluminum alloy sheet with DLC-coating. *Wear* 286–287: 79–83.

Huang, X., Suzuki, K., Chino, Y. 2010. Improvement of stretch formability of pure titanium sheet by differential speed rolling. *Scripta Materialia* 63: 473–476.

Huang, X., Suzuki, K., Chino, Y., Mabuchi, M. 2011. Improvement of stretch formability of Mg–3Al–1Zn alloy sheet by high temperature rolling at finishing pass. *Journal of Alloys and Compounds* 509: 7579–7584.

Huang, X., Suzuki, K., Watazu, A., Shigematsu, I., Saito, N. 2009. Improvement of formability of Mg–Al–Zn alloy sheet at low temperatures using differential speed rolling. *Journal of Alloys and Compounds* 470: 263–268.

Jeong, M.G., Jin, C.K., Hwang, G.W., Kang, C.G. 2014. Formability evaluation of stainless steel bipolar plate considering draft angle of die and process parameters by rubber forming. *International Journal of Precision Engineering and Manufacturing* 15: 913–919.

Jeswiet, J., Young, D. 2005. Forming limit diagrams for single-point incremental forming of aluminium sheet. *Proceedings of the Institution of Mechanical Engineers, Part B: Journal of Engineering Manufacture* 219: 359–364.

Keigler, M., Bauer, H., Harrison, D., De Silva, A.K.M. 2005. Enhancing the formability of aluminium components via temperature controlled hydroforming. *Journal of Materials Processing Technology* 167: 363–370.

Kim, K.J., Kim, D., Choi, S.H., Chung, K., Shin, K.S., Barlat, F., Oh, K.H., Youn, J.R. 2003. Formability of AA5182/polypropylene/AA5182 sandwich sheets. *Journal of Materials Processing Technology* 139: 1–7.

Kim, W.J., Lee, J.B., Kim, W.Y., Jeong, H.T., Jeong, H.G. 2007. Microstructure and mechanical properties of Mg–Al–Zn alloy sheets severely deformed by asymmetrical rolling. *Scripta Materialia* 56: 309–312.

Kim, Y.H., Park, J.J. 2002. Effect of process parameters on formability in incremental forming of sheet metal. *Journal of Materials Processing Technology* 130–131: 42–46.

Kinsey, B., Liu, Z., Cao, J. 2000. A novel forming technology for tailor-welded blanks. *Journal of Materials Processing Technology* 99: 145–153.

Lang, L., Liu, B., Li, T., Zhao, X., Zeng, Y. 2012. Experimental investigation on hydromechanical deep drawing of aluminum alloy with heated media. *Steel Research International* 83: 230–237.

Lasunon, O., Knight, W.A. 2007. Comparative investigation of single-point and double-point incremental sheet metal forming processes. *Proceedings of the Institution of Mechanical Engineers, Part B: Journal of Engineering Manufacture* 221: 1725–1732.

Lee, W., Chung, K.H., Kim, D., Kim, J., Kim, C., Okamoto, K. Wagoner, R.H., Chung, K. 2009. Experimental and numerical study on formability of friction stir welded TWB sheets based on hemispherical dome stretch tests. *International Journal of Plasticity* 25: 1626–1654.

Li, D., Ghosh, A.K. 2004. Biaxial warm forming behavior of aluminum sheet alloys. *Journal of Materials Processing Technology* 145: 281–293.

Li, M.Z., Cai, Z.Y., Sui, Z., Yan, Q.G. 2002. Multi-point forming technology of sheet metal. *Journal of Materials Processing Technology* 129: 333–338.

Li, M., Liu, Y., Su, S., Li, G. 1999. Multi-point forming: A flexible manufacturing method for a 3-d surface sheet. *Journal of Materials Processing Technology* 87: 277–280.

Li, W., Zhao, G., Ma, X., Gao, J. 2013. Study on forming limit diagrams of AZ31B alloy sheet at different temperatures. *Materials and Manufacturing Processes* 28: 306–311.

Liewald, M., Pop, R. 2008. Magnesium tube hydroforming. *Materialwissenschaft und Werkstofftechnik* 39: 343–348.

Lim, S.S., Kim, Y.T., Kang, C.G. 2013. Fabrication of aluminum 1050 micro-channel proton exchange membrane fuel cell bipolar plate using rubber-pad-forming process. *International Journal of Advanced Manufacturing Technology* 65: 231–238.

Liu, C., Li, M., Fu, W. 2008. Principles and apparatus of multi-point forming for sheet metal. *International Journal of Advanced Manufacturing Technology* 35: 1227–1233.

Liu, H., Liu, W., Bao, J., Xing, Z., Song, B., Lei, C. 2011. Numerical and experimental investigation into hot forming of ultra-high strength steel sheet. *Journal of Materials Engineering and Performance* 20: 1–10.

Liu, J., Liu, W., Xue, W. 2013. Forming limit diagram prediction of AA5052/polyethylene/AA5052 sandwich sheets. *Materials and Design* 46: 112–120.

Liu, X., Liewald, M., Becker, D. 2009. Effects of rolling direction and lubricant on friction in sheet metal forming. *Journal of Tribology* 131: 042101: 1–042101: 8.

Liu, Y., Hua, L. 2010. Fabrication of metallic bipolar plate for proton exchange membrane fuel cells by rubber pad forming. *Journal of Power Sources* 195: 3529–3535.

Lovell, M., Higgs, C.F., Deshmukh, P., Mobley, A. 2006. Increasing formability in sheet metal stamping operations using environmentally friendly lubricants. *Journal of Materials Processing Technology* 177: 87–90.

Ma, Z.Y., Mishra, R.S., Mahoney, M.W., Grimes, R. 2003. High strain rate super plasticity in friction stir processed Al-Mg-Zr alloy. *Materials Science and Engineering A* 351: 148–153.

Mori, K., Maeno, T., Maki, S. 2007. Mechanism of improvement of formability in pulsating hydroforming of tubes. *International Journal of Machine Tools and Manufacture* 47: 978–984.

Mori, K., Maki, S., Tanaka, Y. 2005. Warm and hot stamping of ultra high tensile strength steel sheets using resistance heating. *CIRP Annals - Manufacturing Technology* 54: 209–212.

Mori, K., Patwari, A.U., Maki, S. 2004. Improvement of formability by oscillation of internal pressure in pulsating hydroforming of tube. *CIRP Annals - Manufacturing Technology* 53: 215–218.

Polkowski, W. 2016. Differential Speed Rolling: A New Method for a Fabrication of Metallic Sheets with Enhanced Mechanical Properties, Chapter 5. *Progress in Metallic Alloys*, Dr. Vadim Glebovsky (Ed.), InTech. doi:10.5772/64418.

Ramulu, P.J., Kailas, S.V., Narayanan, R.G. 2012a. Influence of tool rotation speed and feed rate on the forming limit of friction stir welded AA6061-T6 sheets. *Proceedings of the Institution of Mechanical Engineers, Part C: Journal of Mechanical Engineering Science* 227: 520–541.

Ramulu, P.J., 2012b. Forming behavior of friction stir welded sheets, PhD thesis, IIT Guwahati, India.

Ramulu, P.J., Narayanan, R.G., Kailas, S.V. 2013. Forming limit investigation of friction stir welded sheets: Influence of shoulder diameter and plunge depth. *International Journal of Advanced Manufacturing Technology* 69: 2757–2772.

Rao, K.P., Wei, J.J. 2001. Performance of a new dry lubricant in the forming of aluminum alloy sheets. *Wear* 249: 86–93.

Rao, K.P., Xie, C.L. 2006. A comparative study on the performance of boric acid with several conventional lubricants in metal forming processes. *Tribology International* 39: 663–668.

Rosel, S., Merklein, M. 2014. Improving formability due to an enhancement of sealing limits caused by using a smart fluid as active fluid medium for hydroforming. *Production Engineering Research and Development* 8: 7–15.

Sakai, T., Hamada, S., Saito, Y. 2001. Improvement of the r-value in 5052 aluminum alloy sheets having through-thickness shear texture by 2-pass single-roll drive unidirectional shear rolling. *Scripta Materialia* 44: 2569–2573.

Salem, H.G., Reynolds, A.P., Lyons, J.S. 2002. Microstructure and retention of superplasticity of friction stir welded superplastic 2095 sheet. *Scripta Materialia* 46: 337–342.

Satheeshkumar, V., Narayanan, R. G. 2014. Investigation on the influence of adhesive properties on the formability of adhesive-bonded steel sheets. *Proceedings of the Institution of Mechanical Engineers, Part C: Journal of Mechanical Engineering Science* 228: 405–425.

Satheeshkumar, V., Narayanan, R.G. 2015a. In-plane plane strain formability of adhesive-bonded steel sheets: influence of adhesive properties. *International Journal of Advance Manufacturing Technology* 76: 993–1009.

Satheeshkumar, V., 2015b, Forming of Adhesive Bonded Steel Sheets, PhD thesis, IIT Guwahati, India.

Sato, Y.S., Sugiura, Y., Shoji, Y., Park, S.H.C., Kokawa, H., Ikeda, K. 2004. Post-weld formability of friction stir welded Al alloy 5052. *Materials Science and Engineering A* 369: 138–143.

Sen, N., Kurgan, N. 2016. Improving deep drawability of HC300LA sheet metal by warm forming. *International Journal of Advanced Manufacturing Technology* 82: 985–995.

Seyedkashi, S.M.H., Moslemi Naeini, H., Moon, Y.H. 2014. Feasibility study on optimized process conditions in warm tube hydroforming. *Journal of Mechanical Science and Technology* 28: 2845–2852.

Sugimoto, K., Kobayashi, M., Nagasaka, A., Hashimoto, S. 1995. Warm stretch-formability of TRIP-aided dual-phase sheet steels. *ISIJ International* 35: 1407–1414.

Tan, F.X., Li, M.Z., Cai, Z.Y. 2007. Research on the process of multi-point forming for the customized titanium alloy cranial prosthesis. *Journal of Materials Processing Technology* 187–188: 453–457.

Toros, S., Ozturk, F., Kacar, I. 2008. Review of warm forming of aluminum–magnesium alloys. *Journal of Materials Processing Technology* 207: 1–12.

Xu, Y., Zhang, S.H., Cheng, M., Song, H.W. 2015. Formability improvement of austenitic stainless steel by pulsating hydroforming. *Proceedings of the Institution of Mechanical Engineers, Part B: Journal of Engineering Manufacture* 229: 609–615.

Yanagimoto, J., Oyamada, K. 2006. Springback-free isothermal forming of high-strength steel sheets and aluminum alloy sheets under warm and hot forming conditions. *ISIJ International* 46: 1324–1328.

Yoshimura, H., Torikai, S., Nishihara, T., Toshiji, N., Inouchi, N. 2002. Application of wheat flour lubricants to the press forming process. *Journal of Material Processing Technology* 125–126: 375–378.

Yuan, S., Wang, X., Liu, G., Wang, Z.R. 2007. Control and use of wrinkles in tube hydroforming. *Journal of Materials Processing Technology* 182: 6–11.

Zhang, H., Huang, G., Roven, H.J., Wang, L., Pan, F. 2013. Influence of different rolling routes on the microstructure evolution and properties of AZ31 magnesium alloy sheets. *Materials and Design* 50: 667–673.

Zhong-qin, L., Wu-rong, W., Guan-long, C. 2007. A new strategy to optimize variable blank holder force towards improving the forming limits of aluminum sheet metal forming. *Journal of Materials Processing Technology* 183: 339–346.

CHAPTER **8**

Computations and Sustainability in Material Forming

Zhengjie Jia
Litens Automotive Group

R. Ganesh Narayanan
IIT Guwahati

CONTENTS

8.1 INTRODUCTION

In recent years, the sustainability issues in material-forming industry have become a fundamental requirement, in addition to the traditional requirements for lower cost, better quality, and faster to market. The material-forming industry is now required to minimize the overall environmental impact under a life cycle perspective. Extensive research has been conducted in this area to study the material-forming sustainability including how to efficiently use materials and resources, and how to effectively reduce the energy consumption and wastes in the material-forming process. The material-forming process should be designed, optimized, and controlled with environmental sustainability in mind. The traditional "trial and error" approach and "try-outs" in material forming have been frequently replaced by sophisticated and energy-efficient material-forming techniques, where computations play a significant role. As the demand for efficiently comprehensive analysis, robust design, and optimum control of material-forming processes consistently increases, the computations have become inevitable for the sustainable material forming and joining. This chapter mainly focuses on these computations and how they significantly affect the sustainability issues in material-forming industry.

A sustainable material-forming process should demonstrate high efficiency in energy consumption, minimum material waste, low cost, high quality, and minimum impact to society. Sustainability assessment in material-forming processes is a multiobjective and interdisciplinary task, and is a great challenge due to its inherent complexity and uncertainty. Although the effective methods and quantitative studies on how to properly model the sustainability and its parameters for a formed product in its lifecycle are lacking, the computational methods including analytical methods, numerical methods, and virtual material-forming approaches have been widely used by industries. Methods such as static equilibrium method (or slab method), slip line field solution, upper-bound method and upper-bound element technique (UBET), finite-element method (FEM) and finite-difference method (FDM), computational fluid dynamics (CFD), multibody dynamics (MBD), Taguchi experimental design technique, axiomatic design approach, geometric methods, statistical methods, and their synthesis are applicable. The computations have been used as the basic means for a multidisciplinary systematic approach, to design and optimize material-forming processes for environmental sustainability, economic sustainability, and social sustainability. Each method has its own advantages and disadvantages. The selection of a method depends on the forming process being modeled. The slip line field solution uses the assumption of plane-strain conditions and is usually limited to 2D analysis of rigid ideally plastic materials. A system approach is understood as a better way of analysis of material-forming processes. It allows study of the input/output relationships and the effects of process variables on product quality and process economics. A multidisciplinary systematic approach has been adopted in attempts to understand the fundamental aspects of the problems in the analysis, design, and control of material-forming processes, and has provided a more fundamental understanding of material-forming processes. However, in the past decade, the computer-based virtual material-forming approach has become more and more popular. It takes the advantages of the power of numerical methods and the benefits of the rapid growing of computer technologies in the past decades. The virtual material-forming approach can take into account all the input variables such as the billet geometry and material, the tooling geometry and material, the conditions at the tool/billet interface, the mechanics of plastic deformation, the equipment used, the characteristics of the final product, and the plant environment where the process is being conducted, to investigate the complicated material-forming process in one simulation process. It is a virtual experiment with the ability to control all the testing parameters for a truly parametric investigation virtually, which is usually very difficult for a physical experiment, analytical method, and theoretical analysis. Use of virtual material forming not only reduces the energy consumed in physical tests but also reduces the costs and shortens the time to market, thereby making the material-forming process sustainable in terms of both environmental and economic sustainability. In addition to the role of the computations in the analysis, the computer system is also widely used in many routine and monotonous, but important tasks such as the preparation of bills of materials, costing, production scheduling, etc., which greatly enhanced the efficiency in the material-forming processes management and business management toward a sustainable material-forming practice.

Although it is a challenging task, many computations have been proposed to handle the sustainability issues in the past years. Liu (2011) proposed a computational tool, Industrial Sustainability Evaluation and Enhancement (ISEE), to facilitate industrial practice on engineering sustainability. Liu embedded the basic sustainability principles into a system approach for sustainability assessment and decision-making. The tool is capable of identifying the most desirable design for sustainability improvement, by analyzing uncertain data and sustainability status quo to predict the future performance and to evaluate design alternatives using various sustainability metrics. Falsafi et al. (2016) proposed a new computational scheme to address a cold recyclability issue of sheet-metal products based on the assessment of their post-manufacture residual formability. Conventional recycling of sheet-metal wastes requires melting metals and additional processes to make final products. Both melting process and additional forming process consumes significant energy. Falsafi's scheme is to recycle sheet-metal wastes without melting them by facilitating design for sustainability. To further reveal the importance of the computations in material-forming sustainability, several major analytical methods, numerical methods, and virtual material-forming approaches and their applications are presented in the following sections.

8.2 COMPUTATIONS AND SUSTAINABLE EXTRUSION PROCESS

Extrusion is a net shape metal-forming process, in which the product of desired cross-section is obtained by forcing a billet through a die. Extrusion has high material utilization and produces parts with superior metallurgical and material properties. Extrusion process allows excellent surface finish and closely controlled dimensions in long products that have constant cross-sections. However, extrusion process is carried out with significant energy consumption. The green extrusion requires a system-level eco-design and optimization of energy consumption and materials waste. Extrusion process outputs such as extrusion load, equivalent strain rate, and temperature are first to be offline optimized using computation tools and then is well controlled in order to achieve the best quality, minimum material waste, and minimum energy consumption to achieve economic sustainability and environmental sustainability.

The upper-bound theorem, as an energy method, has been used in metal forming, especially extrusion, for many years (Hosford, 2007, Hwang et al., 2003, Jia et al., 1996, Gunasekera et al., 1989, Gunasekera and Jia, 1998, Kudo, 1960, 1961, Kobayashi, 1964, Avitzur, 1968, Cramphorn et al., 1976, Kiuchi et al., 1981, Kiuchi, 1986, Yang et al., 1985, Hoshino et al., 1985, Phogat et al., 2012, Jia, 1995a). More details will be discussed in the following sections. One of the successful examples is the streamlined die design method developed by Gunasekera and Hoshino (1982) based on upper-bound theory. Typical extrusion processes are carried out using flat-faced dies or so-called square dies under hot conditions with or without lubricants. But for certain harder alloys like steel and titanium, converging dies are used with streamlined surfaces. In the case of extrusion where the cross-section is different in shape at entry and exit, the material flow does not remain on the same radial plane, which contains the longitudinal axis, so that a 3D approach is necessary. The streamlined die design method can be used to construct smooth curved streamlines with zero gradient along the extruding direction at the entry and the exit of the die as against straightly converging dies for the extrusion of regular polygonal sections from cylindrical billets through straight converging dies. These dies would produce changes of flow-direction at inlet and outlet of the dies. These were taken into account as velocity discontinuities. The profile of the dead metal zone is, in general, a shaped envelope which minimizes the redundant shear. The streamlined die method produces the die configuration which requires no energy dissipation at the entry and the exit of die, which makes that the streamlined die method a sustainable extrusion die design method.

8.2.1 Upper-Bound Method and UBET

The upper-bound method widely used for extrusion analyses for many years is a modeling technique that considers the total energy rate of the system and gives the velocity distribution by minimizing the total energy rate with respect to the unknown variables that are the velocities on the surfaces, which divide the billet into several rigid blocks. Since the velocity fields for 3D metal-forming problems are very complex, the upper-bound method is usually used to solve 2D problems. The upper-bound method is a valuable technique in solving metal-forming problems, since an upper-bound ensures a conservative effect. The upper-bound method involves construction of a kinematically admissible velocity field for a deformation process, and a simultaneous minimization of total energy rate provides the "upper bound solution" for the process.

The UBET is an energy rate minimization method based on the upper-bound theorem. In the UBET, a kinematically admissible velocity field is determined for each elemental region. The total velocity field is obtained by assembling and connecting the kinematically admissible velocity field of each element, and the best velocity field which describes the actual metal flow is then obtained by minimizing the total rate of energy consumption (or total power dissipation) in each element. However, for the general 3D metal-forming problems (e.g., 3D shear-die aluminum extrusion), the velocity field becomes very complex and the general 3D UBET-elements such as brick and prismatic elements must be used. If taking account of the effects of time-dependent heat generation and heat transfer during metal-forming processes, the analysis even becomes much more complex. In 1998, an UBET proposed by Jia (1998) based on the upper-bound theorem is a generic concept for the numerical analysis of metal-forming processes. It combined the 3D UBET model with the temperature module T-FORM and the material modeling techniques to take into account, at each time step, the effects of temperature changes on the flow stress of material, the heat generation due to plastic deformation and friction on the interface between billet and tools, and the heat transfer within billet and between billet and tools during metal-forming processes. The UBET model considers the elemental strain-hardening effects in an incremental manner and can be applied to rigid-plastic materials to simulate the 3D plastic flow characteristics. It is, therefore, potentially applicable to the analysis of simple and complex metal-forming processes. It employs the "flexible polyhedron" optimization technique, one of the direct search methods, to perform efficient minimization of the total rate of energy dissipation for the best velocity field, to obtain knowledge needed for design of metal-forming processes. This UBET model is capable of predicting the velocity fields, strain-rate distributions, and temperature increase in the billet during metal-forming processes. This in turn predicts important process parameters, such as forming loads, maximum forming speed subject to temperature rise, safe process regions, and so on.

In the UBET analysis, the plastically deforming region is subdivided into simple rectangular, triangular, brick, and/or prismatic elements linked together with shear surfaces. On the shear surface, the velocity is discontinuing, $u_1 \neq u_2$. The discrete quantity is $\Delta u = u_{1s} - u_{2s}$. But the normal component of the velocity in both sides of the shear surface must be equal, $u_{1n} = u_{2n}$, to ensure material continuity as shown in Figure 8.1. The power dissipation is calculated in each element for discrete shear between elements and for friction sliding on the interface between element and tool. The total velocity field is obtained by assembling and connecting the kinematically admissible velocity field of each element, which ensures a smooth transition of material flow and satisfies volume constancy and boundary conditions. The best velocity field which describes the actual metal flow is then obtained by minimizing the total power dissipation $\dot{W}_t$ with respect to the unknowns K_i (Jia et al., 1996, Gunasekera et al., 1998):

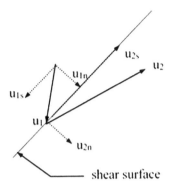

Figure 8.1 Velocity on shear surface.

$$\dot{W}_t = \dot{W}_p(K_i) + \dot{W}_k(K_i) + \dot{W}_f(K_i) \tag{8.1}$$

where $\dot{W}_p(K_i)$ is the power dissipation due to pure plastic deformation over a deforming volume (UBET-element), $\dot{W}_k(K_i)$ is the power dissipation due to velocity discontinuities (internal discrete shear) along boundaries between UBET-elements, $\dot{W}_f(K_i)$ is the power dissipation due to friction at interfaces between billet and tool and K_i are the parameters (unknowns) for the optimization of velocity fields.

In order to solve $\dot{W}_p(K_i)$, $\dot{W}_k(K_i)$, and $\dot{W}_f(K_i)$, 3D velocity fields and strain-rate fields in the rectangular and triangular ring elements in a cylindrical coordinate system (r, θ, z), and in the brick (hexahedron) and prismatic elements in Cartesian coordinate system (x, y, z) are determined first, based on the assumptions, (1) The velocity is uniformly distributed on UBET-element boundary surfaces; (2) u, v, and w are linearly varied within UBET-elements along X, Y, and Z, respectively; (3) the elastic deformation is small as compared to plastic deformation and is therefore neglected; (4) the material is incompressible (volume constancy): $\dot{\varepsilon}_x + \dot{\varepsilon}_y + \dot{\varepsilon}_z = 0$; (5) the material is isotropic, homogeneous, rigid-plastic, and obeys von Mises yield criterion; (6) the flow stress is the function of effective strain, effective strain rate and temperature; (7) the friction shear stress τ is expressed by a constant friction factor m, i.e., $\tau = mk = m\dfrac{\bar{\sigma}}{\sqrt{3}}$; and (7) tools (dies) are rigid.

For example, for a brick (hexahedron) element in Cartesian Coordinate System (x, y, z) as shown in Figure 8.2, 3D general kinematical admissible velocity fields can be determined as follows (Jia, 1998):

$$
\begin{aligned}
u &= \frac{u_{i+1} - u_i}{x_{i+1} - x_i} x + \frac{u_i x_{i+1} - u_{i+1} x_i}{x_{i+1} - x_i} \\[2mm]
v &= v_j + \left(\frac{u_{i+1} - u_i}{x_{i+1} - x_i} + \frac{w_{k+1} - w_k}{z_{k+1} - z_k} \right)(y_j - y) \\[2mm]
w &= \frac{w_{k+1} - w_k}{z_{k+1} - z_k} z + \frac{w_k z_{k+1} - w_{k+1} z_k}{z_{k+1} - z_k}
\end{aligned} \tag{8.2}
$$

3D effective strain-rate fields in a brick element can be defined by (Jia, 1998)

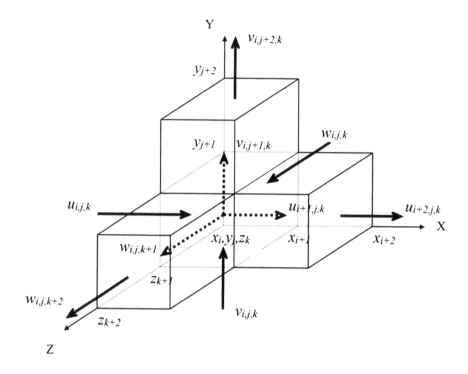

Figure 8.2　Brick element (Jia, 1998).

$$\dot{\varepsilon}_x = \frac{\partial u}{\partial x} = \frac{u_{i+1} - u_i}{x_{i+1} - x_i}$$

$$\dot{\varepsilon}_y = \frac{\partial v}{\partial y} = -\left(\frac{u_{i+1} - u_i}{x_{i+1} - x_i} + \frac{w_{k+1} - w_k}{z_{k+1} - z_k} \right)$$

$$\dot{\varepsilon}_z = \frac{\partial w}{\partial z} = \frac{w_{k+1} - w_k}{z_{k+1} - z_k}$$

$$\dot{\gamma}_{xy} = \dot{\gamma}_{yz} = \dot{\gamma}_{zx} = 0$$

$$
\begin{aligned}
\bar{\dot{\varepsilon}} &= \sqrt{\frac{2}{3}\left(\dot{\varepsilon}_x^2 + \dot{\varepsilon}_y^2 + \dot{\varepsilon}_z^2 + \frac{1}{2}\dot{\gamma}_{xy}^2 + \frac{1}{2}\dot{\gamma}_{yz}^2 + \frac{1}{2}\dot{\gamma}_{xz}^2 \right)} \\
&= \frac{2}{\sqrt{3}}\left[\left(\frac{u_{i+1} - u_i}{x_{i+1} - x_i} \right)^2 + \left(\frac{(u_{i+1} - u_i)(w_{k+1} - w_k)}{(x_{i+1} - x_i)(z_{k+1} - z_k)} \right)^2 + \left(\frac{w_{k+1} - w_k}{z_{k+1} - z_k} \right)^2 \right]^{\frac{1}{2}}
\end{aligned}
\tag{8.3}
$$

The power dissipation $\dot{W}_\mathrm{p}$ due to pure plastic deformation can be expressed as the function of effective stress $\bar{\sigma}$ and strain rate $\bar{\dot{\varepsilon}}$ for the given UBET-element volume v (Jia, 1998):

$$\dot{W}_p = \int_v \bar{\sigma} \cdot \dot{\bar{\varepsilon}} \cdot dv = 2k \int_{z_k}^{z_{k+1}} \int_{y_j}^{y_{j+1}} \int_{x_i}^{x_{i+1}} \frac{2}{\sqrt{3}} \left[\left(\frac{u_{i+1} - u_i}{x_{i+1} - x_i} \right)^2 + \left(\frac{(u_{i+1} - u_i)(w_{k+1} - w_k)}{(x_{i+1} - x_i)(z_{k+1} - z_k)} \right)^2 \right.$$

$$\left. + \left(\frac{w_{k+1} - w_k}{z_{k+1} - z_k} \right)^2 \right]^{\frac{1}{2}} dx \cdot dy \cdot dz$$

$$= \frac{4}{\sqrt{3}} k \left(\left(u_{i+1} - u_i \right)^2 \left(y_{j+1} - y_j \right)^2 \left(z_{k+1} - z_k \right)^2 \right. \tag{8.4}$$

$$+ \left(u_{i+1} - u_i \right)^2 \left(w_{k+1} - w_k \right)^2 \left(x_{i+1} - x_i \right) \left(y_{j+1} - y_j \right)^2 \left(z_{k+1} - z_k \right)$$

$$\left. + \left(w_{k+1} - w_k \right)^2 \left(x_{i+1} - x_i \right)^2 \left(y_{j+1} - y_j \right)^2 \right)^{\frac{1}{2}}$$

The power dissipation $\dot{W}_k$ due to internal shear along boundary surfaces between brick elements with velocity discontinuities as shown in Figures 8.3 and 8.4 is defined as follows (Jia, 1998):

$$\text{on } xy \text{ plane:} \quad \dot{W}_k = \int_s \tau \cdot |\Delta U| \cdot ds = k \iint_s \sqrt{\left(\Delta V_{xy} \right)^2 + \left(\Delta U_{xy} \right)^2} \, dx \cdot dy \tag{8.5}$$

$$\text{on } yz \text{ plane:} \quad \dot{W}_k = \int_s \tau \cdot |\Delta U| \cdot ds = k \iint_s \sqrt{\left(\Delta V_{yz} \right)^2 + \left(\Delta W_{yz} \right)^2} \, dy \cdot dz \tag{8.6}$$

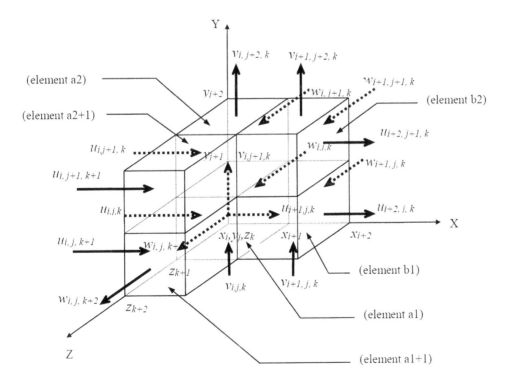

Figure 8.3 The discontinuity of velocities between brick elements (Jia, 1998).

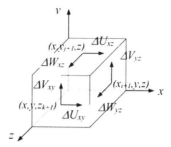

Figure 8.4 The discontinuity of velocities on the boundary surfaces of a brick element (Jia, 1998).

on xz plane: $\dot{W}_k = \int_s \tau \cdot |\Delta U| \cdot ds = k \iint_s \sqrt{(\Delta U_{xz})^2 + (\Delta W_{xz})^2}\, dx \cdot dz$ (8.7)

where ds is the area of shear-surface. For xy plane, $ds = dx \cdot dy$; for yz plane, $ds = dy \cdot dz$; and for xz plane, $ds = dx \cdot dz$.

The power dissipation $\dot{W}_f$ due to friction at interfaces between element (billet) and tools (dies) is defined as follows (Jia, 1998):

$$\dot{W}_f = \int_s \tau \cdot |\Delta U| \cdot ds = k \iint_s \sqrt{(\Delta W)^2 + (\Delta U \cos\theta - \Delta V \sin\theta)^2}\, dy \cdot dl$$

$$= k \iint_s \sqrt{(\Delta W)^2 + \left(\frac{\Delta V}{\sin\theta}\right)^2}\, dy \cdot dl$$

(8.8)

where dl is the length of the incline edge, and

$$\sin\theta = \frac{z_{k+1} - z_k}{\sqrt{(z_{k+1} - z_k)^2 + (x_{i+1} - x_i)^2}}, \quad \cos\theta = \frac{x_{i+1} - x_i}{\sqrt{(z_{k+1} - z_k)^2 + (x_{i+1} - x_i)^2}}$$

The total power dissipation $\dot{W}_t$ involving a metal-forming process is obtained by (Jia, 1998)

$$\dot{W}_t = \sum_{i=1}^{n} \left[\dot{W}_p(K_i) + \dot{W}_k(K_i) + \dot{W}_f(K_i) \right]$$

(8.9)

where i is the number of the UBET-elements. By minimizing the total power dissipation $\dot{W}_t$ with respect to the unknowns K_i (i.e., unknown velocities) using the "flexible polyhedron" optimization technique, the best velocity fields at each time step can be predicted and subsequently the minimum power dissipation $\dot{W}_{t,min}$ can be calculated at each time step using the best velocity fields.

The temperature module T-FORM and the material modeling techniques developed by Jia (1998) are integrated in each UBET numerical solution step, the effects of temperature rise, strain hardening, and strain-rate hardening on the material properties are automatically considered in the solution to the best velocity fields and the minimum power dissipation $\dot{W}_t$.

The minimum power dissipation $\dot{W}_{t,min}$ can be used to estimate the forming loads at each time step. For example, for extrusion, forging, and drawing processes, the forming load can be estimated by the following equation (Jia, 1998):

$$P = \frac{\dot{W}_{t,min}}{U_{tool}} \tag{8.10}$$

where P is the upper bound of the actual forming load and U_{tool} is the velocity of extrusion ram, forging die, or pulling speed of drawing process.

8.2.2 Analysis of Extrusion Process Using UBET—Case Study

A 3D double "I"-beam shear-die extrusion process as shown in Figure 8.5 is a challenging 3D problem which challenges most FEM codes with severe, local, plastic deformation, and resulting mesh break-down. The UBET method proposed by Jia (1998) has been successfully used to analyze this case. The UBET-mesh of the billet is shown in Figure 8.6. The optimization of velocity field is done by minimizing total power at each time increment. The velocity field and temperature field from last time increment is used as initial velocity field and initial temperature field for current time increment, respectively. Based on the optimized velocity field, the strain rate and temperature rise are calculated at each time increment. The effects of temperature rise and strain rate on the flow stress are considered in the calculation at each time step. The flow stress for each element is determined by the updated temperature and strain rate. The extrusion force, maximum and average strain rate, and maximum and average temperature at each time increment are predicted as shown in Figures 8.7–8.9.

The results obtained from the UBET were compared with the experimental results published in the literature (Laue and Stenger, 1988) and showed very good correlation. The validation shows that the UBET method has wide prediction capability in the analysis of extrusion processes and has a potential to become a convenient, simple, and economical method to obtain useful information applicable to the design and control of extrusion processes. In addition, compared to FEA and other analysis methods, the UBET proposed by Jia (1998) has the following advantages:

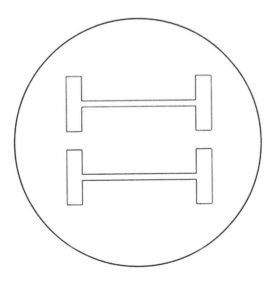

Figure 8.5 Geometry of die (Jia, 1998).

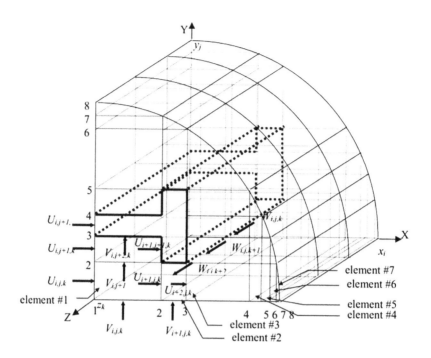

Figure 8.6 UBET-mesh (Jia, 1998).

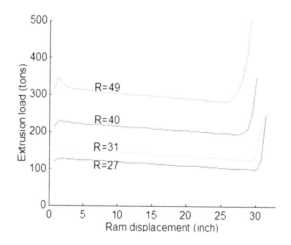

Figure 8.7 Extrusion loads at ram speed of 0.67 in./s (Jia, 1998).

(1) the UBET does not need to solve a huge equation system, whereas FEM must spend a lot of time solving the larger equation system for mid- or large-size problems, (2) the UBET can analyze the coupled temperature and stress problems quickly, whereas commercial FEA packages cannot, (3) the UBET does not have mesh break-down (or extremely distorted mesh) problems which challenge most FEA packages. This case study demonstrates that the UBET can be used to design, analyze, and optimize extrusion processes with the goal of reducing both thermal and mechanical energy, thus producing extrusion products in a sustainable way. More details of this case study can be found in Jia (1998).

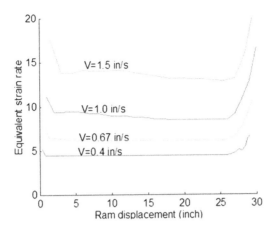

Figure 8.8 Equivalent strain rate at extrusion ratio, $R = 49.5$ (Jia, 1998).

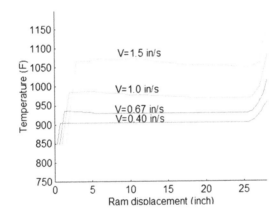

Figure 8.9 Temperature (°F) at extrusion ratio, $R = 49.5$ (Jia, 1998).

8.3 INTEGRATED ANALYSIS AND DESIGN TOWARD SUSTAINABLE ROLLING PROCESS

Rolling is a principal metal-forming technique and is usually a multipass process to achieve accumulated strains. Rolling process includes hot/cold sheet rolling, profile or shape rolling, and tube/pipe rolling. Sheet rolling or flat rolling is to reduce the cross-sectional area of the stock to produce strips, sheets, and plates. The rolled material generally elongates and spreads simultaneously. Shape rolling is a cold- or hot-forming process for reducing as well as shaping the cross-section of billet by passing it through a series of rotating sets of profiled rolls with appropriately shaped grooves. Tube rolling with or without mandrel is the process of reducing the cross-section and/or the diameter of a tube by passing it through powered rolls. Rolling consumes significant energy. Green rolling is one of the most important aspects in the sustainable material forming. Any energy efficiency improvement in rolling can produce a significant benefit to the environment and business. Many factors in rolling such as energy consumption in billet pre-heating, plastic deformation and parasitic energy loss; number of forming stages; material waste; and nonrecyclable/nondegradable materials used in forming processes are critically related to the sustainable rolling process. Rolling process must be well analyzed and well designed with the sustainability in mind.

As a multidisciplinary systematic approach, the integrated analysis and design method for material forming can be used to understand the process for sustainable manufacturing.

All metal-forming operations are aimed at obtaining a desired geometry based on tolerance limits and microstructural properties which may correspond to mechanical properties of the material. The optimal design of a rolling process consists of obtaining the desired shape/dimension and mechanical property without defects by appropriately setting and controlling the process parameters, which also satisfy the machine constraints. However, the analysis models commonly used in the rolling process typically rely on an iterative "correction" procedure which is similar to a one-factor-at-a-time experiment. The integrated analysis and design method is an efficient, powerful, and convenient tool to systematically approach to the analysis and design of various metal-forming processes and can generate new design concept. An integrated analysis and design method for material-forming processes was proposed by Jia (1998) for the robust design of metal-forming processes and the reduction of the number of design iterations required to complete forming process development throughout a synthetic utilization of matrix approach, Taguchi parameter design method, computer-based virtual experiments and physical experiments, Taguchi signal-to-noise (S/N) ratio analysis and analysis of variance (ANOVA), as well as their synthesis in a well-planned manner and, in turn, to produce a product which is well designed and controlled toward green rolling. The integrated analysis and design method considers the design of the whole material-forming process as opposed to a single parameter optimization. This method is generic and applicable to the design of all material-forming processes. It consists of four major stages: (1) Primary Design—Matrix Approach, (2) Detailed Design—Parameter Design, (3) Virtual and Physical Experiments, and (4) Result Analysis and Prediction of Optimum Condition. The flow chart of the method is shown in Figure 8.10 and the stages are explained in the following sections.

8.3.1 Stage 1: Primary Design—Matrix Approach

The goal of primary design is to define the design objectives based on various empirical rules and to select the main process design parameters. Before performing the detailed design, knowledge of the product and process under investigation is of prime importance for identifying the factors likely to influence the outcome. In order to compile a comprehensive list of the factors, the brainstorming for the quality characteristics and design parameters important to the product and process is a necessary step for determining the full range of factors to be investigated. As a powerful tool for brainstorming in the primary design, the Matrix Approach is utilized to study the relationships between the functional requirements and design parameters (or process variables), to establish the design matrix and to define and narrow the design space.

Both Design Matrix and Process Matrix studies are performed in the Primary Design. A Design Matrix maps the functional requirements and design parameters to identify the critical design parameters. A Process Matrix identifies the relation between process factors and critical design parameters. The integration of both matrix studies serves as the basis to select items to work on for the detailed design in the second stage—the parameter design.

8.3.1.1 Matrix Approach and Quality Function Deployment (QFD)

The QFD has been widely successfully employed by various manufactures in Japan since 1975 and in United States since 1984 (King, 1989, Phadke, 1989, Dehnad, 1989, American Supplier Institute, 1987, Hauser et al., 1988, Jia, 1995b). The QFD is an algorithmic design approach and a cross-functional tool. The QFD approach is a system for defining the design problem based on customer demands in terms of a conceptual map, which supports inter-functional planning and communications. The design problem is subsequently solved by a comprehensive prioritization scheme. In the QFD approach, the QFD matrix is a disciplined way of comparing two series of items and

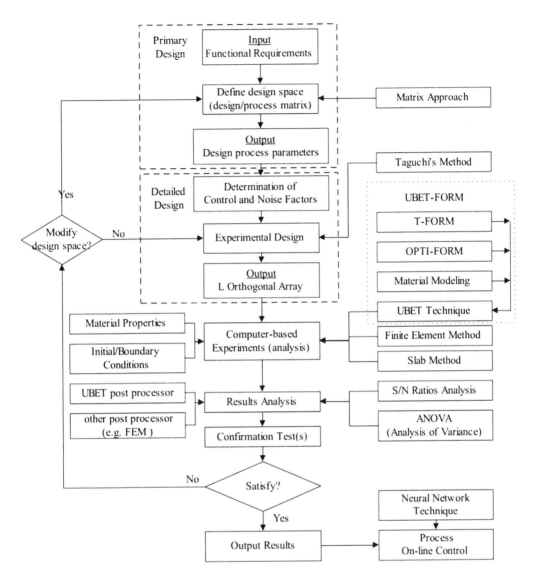

Figure 8.10 The flow chart of the integrated analysis design method for material-forming processes (Jia, 1998).

provides a logical, in-depth look at many of the critical aspects of any product/process. It virtually eliminates the need for redesign, especially on critical items (King, 1989). Matrix Approach stemming from the QFD is adopted in attempts to understand the relationships between functional requirements and design (process) parameters of material-forming processes, to identify main process design parameters and to define the design space. In Matrix Approach, there are two matrix charts (Jia, 1998): Design Matrix and Process Matrix. Each of them takes two groups of ideas and compares them against each other to decide if there are any correlations. The Design Matrix maps out in matrix form what the functions of the product require and how the material-forming process will meet that need. The design matrix chart compares functional requirements with design parameters and identifies strong, moderate, and possible correlations. The purpose of a Design Matrix is to develop the initial plan of how the functional requirements will be met based on the current process design and to find out which design parameters interact with others. The inputs include functional requirements and design parameters. The outputs include 3–6 key design parameters. A Process Matrix examines the correlations between the design parameters and the process factors with the most critical design parameters on the left side and the process factors on the top. The purpose is to identify which process factors are most related to the 3–6 critical design parameters.

8.3.1.2 Primary Design of Flat Rolling Using Matrix Approach—Case Study

A flat rolling process (Jia, 1998) has the initial thickness of 0.3 in. of sheet, which is reduced to 0.15 in. through two-pass reductions at room temperature. The material is the low carbon steel. The objective of the design is to determine the optimal process and operation parameters to achieve the desired thickness and properties without violating any constraints in the process and the robustness of the process for both economic and environmental sustainability.

In the flat rolling process, the functional (customer) requirements, the design parameters and the process setting and operation factors, as well as the constraints are defined as (1) The functional requirements: output thickness of sheet-metal product, flatness, mechanical properties, and surface finish; (2) The process design parameters: strain (hardening), strain rate (hardening), forming temperature, forming speed, rolling force, material properties of billet (choice of material) and reduction ratio; (3) The process operation and setting factors (including controllable and uncontrollable factors): roll diameter, roll length, roll speed, flattening of rolls, wear out of rolls (roll shape), deflection of the rolling mill (mill stiffness), forward/backward tensions, reduction distribution in two passes, friction (lubricant) condition, tolerance of the input (initial) thickness of sheet metal, preheating-temperature of sheet metal, ambient temperature, variation of incoming material property, batch-to-batch variation and process/piece to process/piece variation; (4) The constraints: maximum rolling force, maximum reduction ratio, maximum and minimum roll speed, maximum temperature of sheet metal and flat roll mill and maximum strain or strain rate.

The Design Matrix is used to relate the functional (customer) requirements and the process design parameters (engineering characteristics). The design matrix for the flat rolling process is shown in Table 8.1. Based on the Design Matrix, the first five important process design parameters—rolling force, reduction ratio, strain, incoming material properties, and forming speed—are selected for further study in the Process Matrix. The Process Matrix shows the relationship between the process design parameters and the process operation and setting factors, and examines their correlations to identify which process operation and setting factors are most related to the critical

Table 8.1   Design Matrix for the Cold Flat Rolling Process (Jia, 1998)

Customer (Functional) Requirements		D1	D2	D3	D4	D5	D6	D7
		Process Design Parameters						
Output thickness	5	#		*				*
Flatness	3			o			o	
Mechanical property	4	*	#	*	*	o	#	*
Surface finish	2			o			o	
Objective target values					13.5	90	50	50%
Design guideline's engineering competitive assessment		5	5	2	1	3	1	1
Technical importance	T	41	4	96	36	12	19	81
Rating	%	14	1	33	12	4	7	28
Importance value for process matrix		5	1	7	4	2	3	6

D1, strain (hardening); D2, strain rate (hardening); D3, rolling force; D4, material properties of billet (choice of material); D5, forming temperature (°F); D6, roll (forming) speed (rpm); D7, reduction ratio; *, strong (9); o, medium (3); #, week (1), and (blank) none (0).

design parameters. Based on the Process Matrix, the first eight important factors—Second-pass reduction ratio, roll radius, roll speed, forward tension, backward tension, initial thickness of sheet metal, friction condition, and the material properties of the billet—are selected as the main process parameters for the study in the Stage 2: Detailed Design—Parameter Design. The process matrix for the flat rolling process is shown in Table 8.2.

8.3.2 Stage 2: Detailed Design—Parameter Design

The goal of the detailed design stage is to determine the best combination of levels of controllable process factors, such that the product's functional characteristics are optimized and the effect of uncontrollable factors (noise) is minimized, using Taguchi's method. The parameter design provides keen insight into material-forming process behavior, isolates the critical process design variables that have the greatest impact, and communicates the sensitivity of product performance to the values of these variables. With this information, the process design parameters can be adjusted to maximize the product quality, which may be subjected to the broadest range of uncontrollable environmental factors. The parameter design is to remove or reduce the impact of the causes (the noise factors), instead of finding and eliminating causes of resulting in the variation of a product's functional characteristic, to achieve the robustness against noise factors. The parameter design is the key element to achieve high-quality, low-cost, and green material-forming process.

8.3.2.1 Taguchi Methodology and Parameter Design

"The quality of a product is the (minimum) loss imparted by the product to the society from the time the product is shipped."—Dr. Genichi Taguchi (Byrne and Taguchi, 1987). Taguchi views loss to society on a much broader scale. From an engineering standpoint, the losses of concern are those caused by a product's functional characteristic deviating from its desired target value.

Table 8.2 Process Matrix for the Cold Flat Rolling Process (Jia, 1998)

Design Parameters		Operation and Setting Factors												
		P1	P2	P3	P4	P5	P6	P7	P8	P9	P10	P11	P12	P13
Rolling force	5	*	*	*	o	*	#	*	o	#	*	*	#	*
Reduction ratio	4	*			#	*	#		*		o	o	#	
Strain	3	o			#	*	#	*	*	#	o	o	#	*
Material property	2				#	*			*	#	o	o		*
Forming speed	1	o		*				*			#	#	#	
Process objective/target value		6	24	50		25% / 15%			±5%	100	0–6	0–5		±10%
Importance	T	93	45	54	24	126	12	99	78	10	73	73	13	90
Rating	%	12	5	7	3	16	1	13	10	1	9	9	2	11
Rank		11	5	6	4	13	1	12	9	1	7	7	3	10

P1, roll radius (in.); P2, roll length (in.); P3, roll speed (rpm); P4, flattening of rolls; P5, 2nd-pass reduction (reduction design in two passes); P6, deflection of the rolling mill (mill stiffness); P7, friction (lubricant) condition; P8, tolerance of the input (initial) thickness of sheet; P9, preheating-temperature of sheet metal (°F); P10, forward tension (tons); P11, backward tension (tons); P12, batch/process to batch/process variation; P13, variation of incoming material property (ton/in.^{2}); *, strong (9); o, medium (3); # week (1), and (blank) none (0).

Therefore, the Taguchi Method attains quality by reducing the variation around the target and enhances quality in the design phase which involves two steps—optimizing the design of the product/process (system approach) and making the design insensitive to the influence of uncontrollable factors (robustness).

Traditional factorial design technique defines and investigates all possible conditions in an experiment involving multiple factors. In most situations, it is not practical due to time and costs. However, fractional factorial experiments investigate only a fraction of all the possible combinations. This approach saves considerable time and money, but it requires rigorous mathematical treatment both in the design of experiment and in the analysis of the results (Roy, 1990). Taguchi overcame the limitations of both factorial and fractional factorial experiments by a standardized design methodology—Taguchi Methodology, which minimizes variation rather than just optimizes a mean by adjusting and controlling the levels of factors, rather than by eliminating the causes of variation. The key element is the Parameter Design, in which product parameters on process factor levels are determined such that the product's functional characters is optimized and has minimal sensitivity to "noise". Unlike traditional design methods with "one-at-a-time" experimentation, Taguchi Methodology does "all-at-once" (systematic) experimentation. Taguchi Method includes two major parts: orthogonal arrays (OAs) in experimental design and S/N ratio in the result analysis. OAs are a representation of special experimental designs, which are, in fact, fractional factorial designs based on using symmetrical subsets of all the combinations of factor levels in the corresponding full factorials (Bendell et al., 1989). The S/N ratio takes into account the unpredictable element of the design and the mean response. The S/N analysis determines the most robust set of operating conditions from variations within the results and the effect of the noise factors on the functional requirements. The combination of OA arrays and S/N ratio produces consistency and reproducibility rarely found in any other statistical method (Roy, 1990). In addition, employing computational techniques to perform computer-based experiments with Taguchi parameter design delivers the greatest value for design guidance and savings in time and costs in conducting physical experiments, as experimental hardware in material-forming testing and development is very costly, which contributes to both environmental and business sustainability.

In a material-forming process, there are two types of factors: (1) controllable factors, which include all process design parameters and all process setting parameters, such as choice of material, dimension, configuration, forming temperature, and speed, etc., and (2) noise (uncontrollable) factors which are either difficult or impossible or expensive to control. Based on the number of control factors (parameters) and noise factor, an L OA is selected to lay out the experiments. The discovery of interactions between controllable factors and noise factors is the key to achieve robustness against noise. For designs with interactions, the *Triangular Table of Interaction* and *Linear Graphs* can be used to lay out experiments with interaction effects. For designs with mixed factor levels, the *Dummy Treatment* can be used to decompose a higher level column into lower-level column. More details can be found in Dehnad (1989), Byrne et al. (1987), Roy (1990), Ross (1988), and Clements (1991). The Parameter Design is used to (1) establish the best or the optimum condition for a product or a process, (2) estimate the contribution of individual factors, and (3) estimate the response under the optimum conditions.

8.3.2.2 Parameter Design of Flat Rolling Process—Case Study

In the parameter design, the first phase is to determine the response variables and controllable/uncontrollable factors. In the cold flat rolling process, the responses (i.e., outputs) are the dimensions and mechanical properties of the sheet-metal product. The desired output thickness is 0.15 in. Since the rolling force is often used as the control signal to control the flat rolling process, the rolling force is selected as a response variable in this study, instead of directly using the output thickness due to experimental limitation. To obtain the desired dimension and mechanical property, the rolling force should be properly controlled and minimized in order to reduce the energy dissipation for a green rolling practice.

The controllable factors (i.e., inputs) include all process design parameters and all process operation and setting factors in the flat rolling process. Based on the information provided in the Matrix Approach, the eight main factors are considered in the experimental design. Since the variation of initial thickness of sheet metal due to tolerance and the variation of the material property (flow stress) from batch-to-batch are difficult to control in the sheet rolling process, these two factors are considered noise factors. The other six factors, second-pass reduction, roll radius, roll speed, forward tension, backward tension, and lubricant (friction) condition, are considered controllable factors. Their levels are selected toward the edges of the range of manufacturability as shown in Table 8.3. Based on the number of controllable factors (parameters) and noise factors involved in the study, an L8 OA is selected to layout the experiments as shown in Table 8.4, which shows the inner and outer array and the organization of the experimental design procedure. Each controllable and noise factor is tested at two levels. It is assumed that the control factors behave linearly within the two prescribed levels for accurate characterization of the experimental design. The primary purpose of the outer array is merely to create noise during the experimentation to aid in the selection of controllable factor levels. The two noise levels are selected to represent their normal variation range. In addition, the interaction effect between the controllable factors is also taken into account in the experimental design of this case study. Based on the existing experience, one possible interaction between second-pass reduction and roll radius is assumed and assigned in Column 3 in the L8 OA based on the Linear graph for L8 (Roy, 1990).

Table 8.3 The Levels of the Controllable Factors and Noise Factors (Jia, 1998)

Type	Factor Description	Level 1	Level 2
Control	A: Reduction in second pass	25%	15%
Control	B: Roll radius (in.)	6	8.5
	C: Interaction A × B	-	-
Control	D: Forward tension (ton)	2	6
Control	E: Backward tension (ton)	0	4
Control	F: Roll speed (rpm)	40	60
Control	G: Friction coefficient (lubricate condition)	0.1	0.2
Noise	N1: Initial thickness (0.30% ± 5% tolerance, in.)	0.285	0.315
Noise	N2: Flow stress (13.5% ± 10% ton/in.^{2})	12.15	14.85

Table 8.4 L8 Orthogonal Array (Jia, 1998)

Outer Array							No.	1	2	3	4
2 Noise Factors							N2	1	2	2	1
Inner Array Seven Controllable Factors							N1	1	1	2	2
Run	A	B	C	D	E	F	G	Y1	Y2	Y3	Y4
1	1	1	1	1	1	1	1				
2	1	1	1	2	2	2	2				
3	1	2	2	1	1	2	2				
4	1	2	2	2	2	1	1				
5	2	1	2	1	2	1	2				
6	2	1	2	2	1	2	1				
7	2	2	1	1	2	2	1				
8	2	2	1	2	2	1	2				

8.3.3 Stage 3: Virtual and Physical Experiment

The third stage in the integrated analysis and design method is to conduct the experiments laid out in Table 8.4 by the parameter design. Depending on the availability and needs, virtual experiments (or computations) or physical experiments or their combinations may be employed to conduct experiments. Since the physical experiments of material-forming processes are usually very costly, such experiments should be replaced by computer-based virtual experiments, whenever possible. Physical testing is used only for confirmation of the predictions or when computer-based experiments are not applicable, for example, when the experiments must be conducted with such noisy factors as incoming material variation, batch-to-batch variation, the aging of the forming tool set, tool wear, process/piece-to-process/piece variation, etc. Using computer-based virtual experiments to replace physical experiments whenever possible will deliver a great savings in time and costs to achieve a sustainable computational-experiment.

In addition, whenever possible, the trial conditions should be run in a random order to avoid the influence of experiment setup. In some situations, repetition (simply continually repeating each run several times) or replication (repeating each run from setup several times) may be needed. A larger repetition sample will decrease the sampling error estimate, and a larger replication will decrease the error estimate due to settings. Usually, a large repetition is used for extremely "noisy" systems. Nowadays, one of the most popular computational tools used for compute-based experiments is the FEM.

8.3.3.1 FEM

As a powerful tool, the FEM has been extensively used effectively to solve the problems of material-forming processes. A large number of studies on the material-forming process using FEM have been performed by many researchers (Zienkiewicz et al., 1988, Cavaliere et al., 2001, Chenot et al., 2010, Soulami et al., 2014, Gunasekera et al., 1982, Kiuchi et al., 1991, Kobayashi et al., 1989, Yang et al., 1992). For example, the FEM has been used to investigate the die fill in a streamlined die extrusion (Jia and Gunasekera, 1995), the effects of the hydraulic pressure (residual strain and stresses) on the expansion of a thin-walled tube with longitudinal projections (Jia et al., 1996), and the metal flow in an extrusion with and without feeder plate (Holthe et al., 1992). 2D FEM was applied to investigate titanium tube extrusion (Udagawa et al., 1989) and the direct extrusion process of aluminum (Joun and Hwang, 1992). Shivpuri and Momin (1992) investigated curvature and twist problems in L-shape extrusion by modifying the die geometry (including entry angles and the die land) and simplifying the 3D problem into a 2D problem. Akiyama and Yamashita (1992) carried out axisymmetric rigid-plastic finite-element analyses and fundamental experiments in order to find the threshold value of the equivalent plastic strain to be given to the stainless steel billet for tube extrusion. Urbanski et al. (1992) used matrix FEM to predict the hardness distribution in the tubes draw on the cylindrical mandrel.

The FEM can be used easily to take into account of the complex geometries of the forming tools and products to accurately predict material-forming process parameters, such as forming load, and distributions of stress, strain, strain rate, displacement, velocity, temperature, etc. The FEM is based on the theory of elasticity and plasticity and the FE formulations used for metal forming usually include the elastic-plastic formulation and the rigid viscoplastic formulation. The details on the FEM theory and applications can be found from many publications and are not discussed in detail here.

The FEA for material-forming processes usually includes three steps, pre-processing, analysis, and post-processing. In the pre-processing step, an FE model is created to represent the forming process including billets, forming tools, and operation environments as necessary. The boundary conditions and the mechanical and thermal load conditions are then applied to the FE model, and

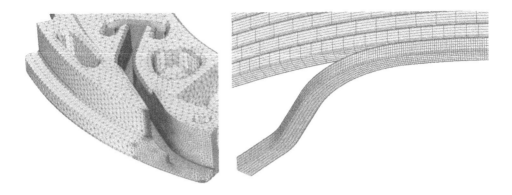

Figure 8.11   Mesh control of an engine component.

then, the material properties are applied to the billets and tools, if necessary. The knowledge of the material-forming process to be modeled is required to set up a good model for the simulation, such as the effective distribution of the mesh density (Figure 8.11 shows an example of the mesh control in an engine component FE model, which involves nonlinear materials, nonlinear geometry and nonlinear contacts), the reasonable simplification of the billet/tools geometry, and the proper approximation of the boundary/load conditions, to compromise the accuracy of the FE results and the computational time. In the analysis step, the model is submitted to run the simulation, which is the most complicated numerical computation process. For a complicated FEA, the convergence can be a challenging task and needs to be tuned with the knowledge and experience. Nowadays, running the simulation on a high-performance computing (HPC) system has been a common practice and the computational time has been significantly reduced to a fraction. In the post-processing step, the simulation results can be reviewed and analyzed for the decision-making on the design and control of the material-forming process. Many post-processing techniques, such as animation of the plastic deformation of the billet, load history, sectional plot, etc., can provide a great insight into the forming process being studied.

8.3.3.2 Virtual Experiments Using FEM—Case Study

The flat rolling trials for the experimental design procedure defined in Table 8.4 are carried out by the computer-based experiments. The experiments are therefore deterministic and are not replicated. A total of 32 computer-based experiments are conducted to estimate the rolling force. The outputs obtained from the computational trials are shown in Table 8.5 for the result analysis.

Table 8.5   Results (ton/in.) from the Flat Rolling Model (Jia, 1998)

Run	Y1	Y2	Y3	Y4
1	17.814	22.215	22.632	18.144
2	20.055	26.087	26.670	20.509
3	33.480	42.532	43.418	34.158
4	18.849	24.492	25.033	19.273
5	13.930	18.024	18.385	14.212
6	13.079	16.551	16.853	13.317
7	15.057	19.561	19.958	15.367
8	17.285	22.796	23.285	17.659

8.3.4 Stage 4: Result Analysis and Prediction of Optimum Condition

The integrated analysis and design method (Jia, 1998) uses two different routes to carry out the complete analysis—standard approach and *S/N* ratio analysis, to find the optimum condition, such as which factors contribute to the results and by how much and what will be the expected result at the optimum condition. The standard approach processes the results through main effect (factorial effects) and ANOVA analyses. The main effect is the difference between the two average effects of the factor at the two levels. The main effect indicates the general trend of the influence of the factors. Thus, the levels of the factors which are expected to produce the best results can be predicted. The optimum condition and the performance at the optimum condition can be determined by studying the main effects of each of the factors. Then, ANOVA is performed on the results to determine the percent contribution of each factor.

The *S/N* ratio is a performance measure to choose control levels that best cope with noise and provides the sensitivity of the output to the noise factors. In the Taguchi *S/N* ratio analysis, a high value of *S/N* implies that the signal is higher than the random effects of the noise factors. Product design or process operation consistent with highest *S/N* ratio always yields the optimum quality with minimum variance (Roy, 1990). Therefore, the best combination of levels of process parameters can be determined with the highest possible *S/N* ratio for the result, and thus, seeking for the highest possible *S/N* ratio for the result becomes the aim of any experiment. The *S/N* analysis is generally sufficient. Both the *S/N* ratio analysis and ANOVA are employed to analyze the results and to predict the optimum levels of process parameters.

Once the optimum condition is determined, a confirmation experiment is conducted to estimate the performance (response) at the combination of the best levels. The correlation between the predicted result and the result from the confirmation run will be established in statistical terms reflecting the level of confidence.

8.3.4.1 S/N Ratio Analysis

The *S/N* ratio is a key issue in the Taguchi method and is defined as follows (Dehnad, 1989):

$$S/N = -10 \, \text{Log}_{10}(\text{MSD}), \tag{8.11}$$

where MSD (mean squared deviation) is a statistical quantity reflecting the deviation from the target value of the quality characteristic, which is used to express how well a product performs the function. Three different *S/N* formulas can be derived as described below:

For *the smallest is best* quality characteristic (e.g., for minimization of total power in UBET),

$$\text{MSD} = \frac{1}{n} \sum_{i=1}^{n} y_i^2 \tag{8.12}$$

For *the nominal is best* quality characteristic (e.g., for a specific value problem, dimension),

$$\text{MSD} = \frac{1}{n} \sum_{i=1}^{n} (y_i - y_0) \tag{8.13}$$

For *the biggest is best* quality characteristic (e.g., for maximization problem, strength, yield),

$$\text{MSD} = \frac{1}{n} \sum_{i=1}^{n} \frac{1}{y_i^2} \tag{8.14}$$

where y_i is the result of experiment (observation), y_0 is the target value of result, and n is number of repetitions.

In the S/N ratio analysis, the value of MSD or greatest value of S/N represents a more desirable condition, as a high value of S/N implies that the signal is higher than the random effects of the noise factors. Therefore, the aim of any experiment is always to determine the highest possible S/N ratio for the result, and the S/N ratio is always interpreted the same way: *the larger the S/N ratio the better*. The S/N ratio analysis provides a guidance to the selection of the optimum level based on least variation around the target and, also, on the average value closest to the target. The S/N ratio calculation is based on data from all observations of a trial condition. The set of S/N ratios can then be considered as trial results without repetitions. Performing the S/N ratio analysis generally eliminates the need for investigating the specific interactions between controllable factors and noise factors.

8.3.4.2 ANOVA

Taguchi replaces the full factorial experiment with a lean, less expensive, faster, partial factorial experiment. Taguchi's design for the partial factorial is based on specially developed OAs. Since the partial experiment is only a sample of the full experiment, the analysis of the partial experiment must include an analysis of the confidence that can be placed in the results. A standard statistical technique called ANOVA can be used to provide a measure of confidence. ANOVA methods for Taguchi methodology are somewhat special and controversial. The technique does not directly analyze the data, but rather determines the variability (variance) of the data. Confidence is measured from the variance. Analysis provides the variance of controllable factors and noise factors. By understanding the source and magnitude of variance, robust process conditions for metal-forming processes can be predicted. More details about the mathematical development for ANOVA can be found in the literatures (Dehnad, 1989, Roy, 1990, Ross, 1988).

8.3.4.3 Result Analysis and Prediction of Optimal Process Parameters—Case Study

Figures 8.12 and 8.13 show the main effects for the average responses (S_m) and the S/N ratios, respectively, which clearly show the effect of each factor on the output, and thus, the significant factors and their optimum levels can be determined as, A2, B1, D1, E2, F1, and G1. Since the factors D, E, and F are not significant factors and their levels are determined by weighting the conditions. Based on the average responses, E2 and F1 are selected. D1 indicates the 2 tons forward tension, which is less than D2, 6 tons forward tension. For less equipment requirement, D1 is selected. Therefore, the optimum combination of the levels of the factors can be determined as shown in Table 8.6. The interaction effect diagram shown in Figure 8.14 indicates that lines A1 and A2 appear almost parallel. Hence, A and B interact slightly and are included in the optimum condition. Since the interaction between the second-pass reduction and the roll radius is not significant to the output, the minor interaction of A × B can be ignored in future study.

ANOVA is performed on the results to determine the percent contribution of individual factors and interaction effects to the variation of the output (rolling force), to differentiate significant from insignificant factors and to identify the relative influence of the factors and interactions by comparing their variances. ANOVA helps to determine which of the factors need control and which do not. Tables 8.7 and 8.8 show the ANOVA for the average responses (S_m) and the S/N ratios. The factor with the less than 5% contribution is pooled, and the pooled error is used to calculate for the percent contribution to the output from all of the pooled factors, errors, and missing factors. It can be seen that the factors A, B, and G make more than 60% contributions to the output. The factor A (second-pass reduction) has the most significant effect to the output with the 46% contribution to the

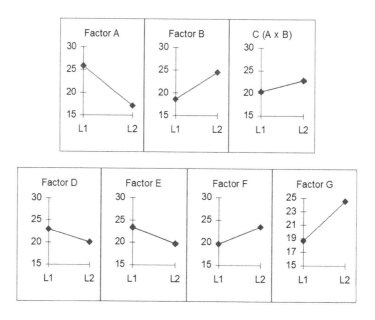

Figure 8.12 Main effects for the average responses (S_m) (ton/in.) (Jia, 1998).

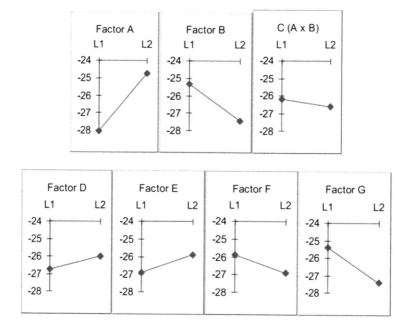

Figure 8.13 Main effects for the S/N ratios (dB) (Jia, 1998).

result based on the *S/N* ratio analysis and 33.7% based on the average response. If the variance ratio (*F*-ratio) is greater than the value in the *F*-table, this factor is significant to the rolling force and, in turn, significant to the output thickness of sheet metal, H2.

Based on the predicted optimum combination in the main effect analysis and the ANOVA analysis, the performance at the optimum condition is estimated only from the significant factors, and

Table 8.6   Optimum Combination of the Levels of the Factors (Jia, 1998)

Factor	Level	Value
A: Reduction in second pass	2	15%
B: Roll radius (in.)	1	6
C: Interaction A × B		
D: Forward tension (ton)	1	2
E: Backward tension (ton)	2	4
F: Roll speed (rpm)	1	40
G: Friction coefficient (lubricant condition)	1	0.1

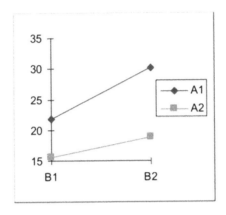

Figure 8.14   Interaction A × B (ton/in.) (Jia, 1998).

Table 8.7   ANOVA Table (S_m) (Pooled) (Jia, 1998)

Column	Factor	DOF	Sum of Squares	Variance	Variance Ratio	Percent Contribution
1	A	1	612.879	612.879	42.31	33.69%
2	B	1	274.535	274.535	18.95	14.64%
3	C	(1)	51.238	51.238		
4	D	(1)	69.308	69.308		
5	E	1	111.602	111.602	7.70	5.47%
6	F	1	122.553	122.553	8.46	6.09%
7	G	1	277.83	277.83	19.18	14.83%
All others/error		24	256.076			
Pooled error		26	376.621	14.485		25.28%
Total:		31	1776.02			100.0%

the pooled factors are not included in the estimate. The performance at the optimum condition and the contributions from the significant factors are estimated as shown in Table 8.9. The confidence interval (C.I.) represents the variation of the estimated result at the optimum, i.e., the mean result, lies between (m+C.I.) and (m−C.I.) at 90% confidence level.

Once the optimum condition is determined, confirmation experiment(s) is conducted to estimate the performance (response) at the optimum condition. Ten additional computer-based experiments (confirmation runs) were performed at the combination of the optimum levels with the randomly selected values for two noise factors and 10 data points were collected. The correlation between the

Table 8.8   ANOVA Table (S/N) (Pooled) (Jia, 1998)

Column	Factor	DOF	Sum of Squares	Variance	Variance Ratio	Percent Contribution
1	A	1	22.134	22.134	15.15	45.95%
2	B	1	8.823	8.823	6.04	16.36%
3	C	(1)	0.353	0.353		
4	D	(1)	1.108	1.108		
5	E	(1)	2.203	2.203		
6	F	(1)	2.174	2.174		
7	G	1	8.189	8.189	5.61	14.95%
All others/error		0	0			
Pooled error		4	5.838	1.461		22.73%
Total:		7	44.9898			100.0%

Table 8.9   Estimate of the Performance at the Predicted Optimum Condition (Jia, 1998)

Factor	Level	Contribution (S_m)	Contribution (S/N)
A	2	−4.376	1.663
B	1	−2.929	1.050
G	1	−2.947	1.012
Contribution from all factors (total)		−10.18	3.73 (dB)
Current grand average		21.58	−26.39 (dB)
Expected result at optimum condition		11.40 (ton/in.)	−22.67 (dB) (13.596 ton/in.)
Confidence interval (C.I.) at 90%		±2.81 (ton/in.)	±1.82 (dB)

predicted result and the results from the confirmation runs is established in statistical terms reflecting the level of confidence. The results from the confirmation runs are as follows:

Average rolling force: 13.10 (ton/in.)
S/N ratio: −22.39 (dB)

The results show that the average response is within C.I.: 11.40 ± 2.81 (ton/in.), and the S/N ratio is within C.I.: −22.67 ± 1.82(dB). The confirmation run at the optimum condition provides a satisfied result for the predicted optimum condition.

8.3.5 Summary

In summary, the integrated analysis and design method proceeds with material-forming design by first narrowing the design space using Matrix Approach, next performing parametric study to select the right processing parameters and their levels using the Taguchi Method, then conducting experiments using computations and analyzing results and predicting optimum processing parameters and their levels using S/N ratio and ANOVA. Finally, the confirmation test is used to determine the confidence level.

The integrated analysis and design method for material-forming processes has the following advantages (1) Systematically optimizing the development of material-forming processes; therefore, the need for redesign, especially on critical process parameters, may be reduced or even eliminated; (2) Driving whole design processes by functional requirements (customer demands); therefore, the objective is clear and consistent throughout overall forming process development; (3) Using robustness in design; therefore, the product's functional characteristic is optimized, and the effects of uncontrollable factors (i.e., "noise" factors) is minimized; (4) Applying off-line quality control in design phase; therefore, a nonoptimum design can be effectively prevented in the design stage, whereas traditional quality control or statistical process control usually cannot fully compensate for a bad design; (5) Minimizing the total number of experiments, especially for the need of physical tests; therefore, the time and cost in material-forming process development can be significantly reduced; (6) Removing or reducing the impact of the causes (the noise factors) by adjusting and controlling the levels of process parameters rather than by finding and eliminating causes resulting in the variation of a product's functional characteristic; and (7) Accelerating form-ing process development with parametric capabilities and sensitivity analysis which provide the greatest insight into how changes in process design variables will impact the reliability of forming process designs.

The principles and the fundamental characteristics of the integrated analysis and design method have been clearly demonstrated in the case study, which shows that the integrated analysis and design method have a potential to become an efficient, powerful, and convenient analysis and design tool for the systematic design of various material-forming processes for lower business costs, better product quality, and shorter time to market, which will make great contribution to the success of both economic sustainability and environmental sustainability.

8.4 NEURON NETWORK APPROACH TOWARD GREEN FORMING PROCESS CONTROL

The proper control of the material-forming process has become increasingly important with expanding global competition and growing sustainability requirement. Monitoring and control of material-forming processes play a major role in achieving quality goals and the sustainable material-forming process. There are several different control system design strategies ranging from simple, conventional open-loop to sophisticated hybrid control systems of conventional and artificial intel-ligent techniques, such as neural networks and expert systems. The conventional (open-loop and closed-loop) controls have been addressed by many researchers (Pietrzyk et al., 1991, Polakowski, et al., 1969, Bryant, 1973, Koot, 1973). The production of high-quality products on a reproducible basis can be achieved only through the use of automated control systems. Modern material-forming equipment, such as servo-hydraulic forge presses and section rolling mills, possess the necessary dynamic-response characteristics for implementing the process-control algorithm. Classical control methods employ a "signal-input/signal-output" approach which cannot effectively treat the nonlin-ear, time-varying multivariable nature of material-forming processes. However, the application of learning control techniques to complex material-forming process provides a flexible capability for designing and building adaptive, intelligent process controllers capable of improving control of the processes. When either analytical relationships for the process are not available or the process is too complicated to be represented by an analytical model, neural networks a most promising modeling tool for the use in analysis and control.

The neural network based online control strategy for material-forming processes proposed by Jia (1998) to realize the online learning and real-time optimizing control of material-forming processes includes two stages: *offline pre-tune* and *online learning*. The online learning provides the flexibil-ity to learn the control of an unknown, dynamically changing, multivariate process with arbitrary inputs and feedback data, while the offline learning provides the pre-tune of a learning control sys-tem to reduce start-up oscillation. The strategy of *offline pre-tune* and *online learning* makes itself distinguished from traditional control methods. The conventional "non-learning" control does not have the capability of *online learning* and *offline pre-tune*. In conventional "non-learning" control, it is usually assumed that accurate process and control models are available beforehand, but accu-rate and comprehensive process knowledge of complex nonlinear systems is rarely known *a priori*. The strategy of the *offline pre-tune* and *online learning* enables the developed neural networks to behave fast on-line learning to achieve a workable, economical and optimum solution of problem in less time, and in turn enables the precisely optimal process control toward a green forming process. As a case study, a neural network based online control for the flat rolling process is presented in this section to demonstrate the principles of the neural network based online control strategy.

8.4.1 Neural Network

A neural network is an information-processing system that has certain performance characteristics in common with biological neural networks. Neural networks have been widely used in the areas of pattern recognition, signal processing, vision, speech processing, forecasting, and modeling and robotics (Muller et al., 1991, Wasserman, 1993, Freeman, 1994, Mehra et al., 1992, Welstead, 1994, Wang et al., 1993, Davalo et al., 1991, Haykin, 1994, Kim et al., 2000, Bagheripoor et al., 2013). Information processing with neural networks consists of analyzing patterns of activity, with learned information stored as weights between node connections. Since knowledge is represented as numeric weights, the rules and reasoning process in neural networks are not readily explainable. Neural networks can solve problems which are difficult to be simulated using the logical, analytical techniques, and traditional approaches, where rules are not known or data are incomplete or noisy. Neural networks provide a novel approach for the interpretation of results from numerical models. A common characteristic is the ability to classify streams of input data without the explicit knowledge of rules and to use arbitrary patterns of weights to represent the features of the different categories. The ability of neural networks to serve as generic functional approximates makes them a useful tool for synthesizing such mechanisms. In that capacity, neural networks can realize nonlinear models with high-accuracy levels.

Back-propagation is a specific technique for implementing gradient descent in weight space for a multilayer network. When applied to multilayer networks, the back-propagation technique adjusts the weights in the direction opposite to the instantaneous error gradient. In its simplest form, back-propagation training begins by presenting an input pattern vector X to the network, sweeping forward through the system to generate a response vector O and computing the errors at each output. The next step involves sweeping the effects of the errors backward through the network to associate a square error derivative δ with each connection, computing a gradient from each δ and finally updating the weights of each connection based upon the corresponding gradient. A new pattern is then presented, and the process is repeated. The initial weight values are normally set to small random numbers. A two-layer back-propagation network architecture is shown in Figure 8.15. The training of a network by back-propagation involves three stages: the feed-forward of the input training pattern, the calculation and back-propagation of the associated error, and the adjustment of the weights (Dixit and Narayanan, 2013). After training, application of the net involves only the computations of the feed-forward phase. Even if training is slow, a trained net can produce its output very rapidly.

The net input to a neuron is expressed as follows:

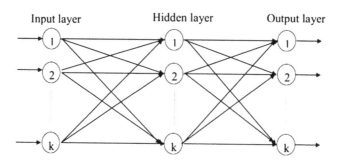

Figure 8.15   A two-layer back-propagation network architecture (Jia, 1998).

$$\text{Net}_i = \sum_{j=1}^{n} w_{ij} O_j + b_i \qquad (8.15)$$

The total error on the outputs when one training example is presented is calculated, i.e., the error (a function of the weights) to be minimized is as follows:

$$E = \frac{1}{2} \sum_i (t_i - O_i)^2 \qquad (8.16)$$

The back-propagation algorithm consists of carrying out a gradient descent minimization process on E. In general, each connection weight is modified following each presentation of example t using changes given by

$$\Delta w = -\eta \frac{\partial E}{\partial w} \qquad (8.17)$$

Thus, the updates for the weights to the output layer are given by

$$\Delta w_{ij} = -\eta \frac{\partial E}{\partial w_{ij}} = \eta (t_i - O_i) f'(\text{Net}_i) O_j \qquad (8.18)$$

and for the weights to the hidden layer,

$$\Delta w_{jk} = -\eta \frac{\partial E}{\partial w_{jk}} = \eta \left[\sum_i (t_i - O_i) f'(\text{Net}_i) w_{ij} \right] f'(O_j) O_k \qquad (8.19)$$

Back-propagation neural networks have been successfully applied in manufacturing, but the training process is a very complicated problem. The structure of a network can be modified by adjusting the number of neurons, as well as the number of connections per layer. An algorithm based on the second-derivative order information (Jia et al., 1996), which is derived from Optimal Brain Damage (OBD) technique (Cun et al., 1990), can be used to achieve a near-optimal network structure. The performance of the neural network in the learning errors, prediction errors, and training times can be significantly improved by the algorithm as demonstrated in the case study.

The OBD was proposed by Cun et al. (1990). They use information-theoretic ideas to derive a class of practical and nearly optimal schemes for adapting the size of a neural network. By removing unimportant weights from a network, several improvements can be expected: better generalization, fewer training examples required and improved speed of learning and/or classification. The basic idea is to use second-derivative information to make a tradeoff between network complexity and training set error. A simple strategy consists in deleting connections with small "saliency", i.e., those whose deletion will have the least effect on the training error. This method uses the second-derivative of the objective function with respect to the connections to compute the saliencies. The objective function can be expressed in a Taylor series. A perturbation δu of the connection vector will change the objective function by (Cun et al., 1990)

$$\delta E = \sum_i g_i \delta u_i + \frac{1}{2} \sum_i h_{ii} \delta u_i^2 + \frac{1}{2} \sum_{i \neq j} h_{ij} \delta u_i \delta u_j + O\left(\| \delta U \| \right)^3 \qquad (8.20)$$

The goal is to find a set of connections whose deletion will cause the least increase of E. The second derivatives back-propagated from last-layer to hidden layer is given by equation 8.25 and the second derivatives back-propagated from hidden layer to input layer is given by equation 8.26. Therefore, the saliencies for each connection can be obtained by equation 8.27. More detailed information can be found in Jia (1998). Thus, we have

$$h_{ij} = \frac{\partial^2 E}{\partial W_{ij}^2} = \frac{\partial^2 E}{\partial \mathrm{Net}_i^2} O_j^2 = \left[2O_i^2(1-O_i)^2 - 2(t_i-O_i)O_i(2O_i^2 - 3O_i + 1) \right] O_j^2 \tag{8.21}$$

$$h_{jk} = \frac{\partial^2 E}{\partial W_{jk}^2} = \frac{\partial^2 E}{\partial \mathrm{Net}_j^2} O_k^2 = \left[O_j^2(1-O_j)^2 \sum_i W_{ij}^2 \frac{\partial^2 E}{\partial \mathrm{Net}_i^2} \right.$$
$$\left. - O_j(2O_j^2 - 3O_j + 1) \sum_i \left(-(t_i - O_i)O_i(1-O_i)W_{ij} \right) \right] O_k^2 \tag{8.22}$$

$$S_{ab} = \frac{\left(\sum_{i=1}^n \dfrac{h_{ab}^i}{n} \right) w_{ab}^2}{2} \tag{8.23}$$

where w_{ab} is the weight between neuron a and neuron b, and i is the ith pattern. Deleting a connection is defined as setting it to 0 and freezing it there.

8.4.2 Neural Network Based Online Control of Flat Rolling Process—Case Study

As a case study, a neural network based online control for the flat rolling process was developed based on the back-propagation paradigm. The performance of this neural network was optimized by the algorithm developed based on the second-derivative order information (Jia et al., 1996, Gunasekera et al., 1998). This control strategy includes two stages: *offline pre-tune* and *online learning*. The *offline pre-tune* stage includes a four-step procedure (Jia, 1998): (1) Data Collection: the critical process parameters and constraints, and control variables or the design parameters and their ranges are identified as listed in Table 8.10. (2) Training the Neural Network: the neural network is trained with 4832 patterns containing eight inputs and one output (control signal). The first step is to determine the numbers of hidden layers and the numbers of neurons on each hidden layer. The second step is to determine the threshold. The system converges when the error for all of the patterns falls below the minimum error threshold, which is the sum of the root-mean-square (RMS). The threshold 0.0065 is set. The learning rate controls the rate of updating weights. The learning rates of 0.25 and 0.35 are set for the first 500,000 epochs of the output and the input layers, respectively. The momentum of 0.2 is set. The last step is that the trained neural network is tested with the testing data. Both the training and testing data are generated from the analytical model. The four initial neural networks are generated with the hidden neurons 10, 20, 30, and 40. The training and testing show that the neuron network with the hidden neuron 28 has the lowest error, which, therefore, is selected for further study using the second-derivative information. (3) Optimizing the Selected Neural Network Using the Second-Derivative Information: due to the nonlinear behavior of the problem, the network is optimized using the algorithm, the second-derivative information derived from OBD technique (Jia et al., 1996), to achieve a near-optimal neural network structure by removing unimportant weights from the network, improving generalization and speeding-up training. The algorithm reduces the size of the developed neural network by selecting deleting connections with very small saliencies ($S_{ab} \le 0.5$) each time. The performance of the developed neural

Table 8.10 Process Parameters and Their Ranges (Jia, 1998)

Parameters	Range
Billet thickness at annealed state (in.)	0.10–1.25
Billet thickness at entry (in.)	0.10–1.25
Billet thickness at exit (in.)	0.02–0.50
Roll speed (rpm)	40–70
Material flow stress (ton/in.^{2})	11.5–15.5
Forward tension (ton)	0.0–10.0
Backward tension (ton)	0.0–10.0
Coulomb's friction coefficient	0.05–0.15

network was significantly improved by this technique. The selected neural network has 28 hidden neurons, 8 inputs, 1 output, 1 bias, and total connections 281. The average prediction errors for training data were reduced seven times of connections using the OBD. When the reduction of connections reached at the maximum 37%–40% reduction of the original connections, the network can well predict the rolling force, even without the need of re-training the network. (4) Validation: the validation of the developed neural network is carried out in different phases to make sure that the results produced by the network match the validation data (target value) and fall within the given margin of the error. The results show that the prediction of the rolling force matches the target value very well over the billet thickness of 0.192–0.242 in., the friction coefficient of 0.07–0.13, and the roll speed of 46–55 rpm. The validation data are also used to study the mean and standard deviation of the prediction error for the developed neural network: the mean is 0.0449875 and the standard deviation is 0.0270691. All the prediction errors are within almost 5%. The validation provides a great deal of confidence in this methodology and their application in metal-forming processes. After the validation, the trained neural network can then be employed for the *on-line control* of flat rolling process as discussed below.

The second stage is *online learning*, which is the implementation stage of the control strategy. The proposed neural network based on-line, nonlinear system control for flat rolling process, as shown in Figure 8.16, does not require previous knowledge of the accurate models for the flat rolling process and is an on-line learning control. The goal of on-line learning is the real-time optimization of a large-scale nonlinear flat rolling process at minimal computational cost, since computational delays can have adverse effects on both rate of convergence and stability for on-line process optimization. On-line learning provides the flexibility to learn the control of an unknown, dynamically changing, multivariate process with arbitrary inputs and feedback data. The development of a learning-optimal control system will enable the flat rolling processes to high quality, productivity, and design flexibility. The assumed goal is to maintain the desired thickness of product (sheet metal) with the optimum operating conditions. The system uses the current state space to train the neural network, uses the required thickness h of product and dynamical thickness H at time $(t-1)$ of billet to predict the expected rolling force and uses the actual rolling force measured by sensor and predicted current rolling force to filter the effect of noises. Then the output is passed to the neural network controller to produce the control signal (e.g., roll speed, gap, and tensions) by comparing the actual rolling force with the desired rolling force needed to perform the real-time control of the process. Each parameter from the state space is checked by the expert system for material and machine constraints and limits, specifically for stability. The more detailed information can be found in Jia (1998).

8.4.3 Neural Network Expert System for Tailor Welded Blanks—Case Study

Tailor Welded Blanks (TWBs) consist of metal sheets of similar or dissimilar thicknesses, materials, and surface coatings welded in a single plane before forming. These are then formed

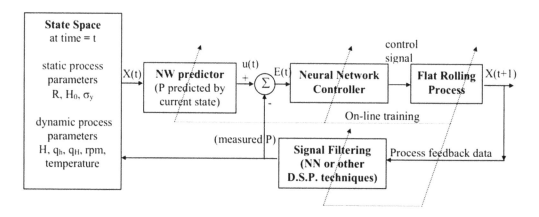

Figure 8.16   Neural network based on-line control for flat rolling process (Jia, 1998).

like unwelded blanks to manufacture components used predominantly in automotive industries. Appropriate tooling and forming conditions are required. Deck lids, bumper, car door inner panel, side frame rails are some of the applications of TWB in the automotive sector.

The advantages of using TWBs are as follows: (1) Part weight reduction and hence savings in fuel consumption, (2) efficient distribution of material thickness and properties resulting in part consolidation, (3) greater flexibility in component design, (4) re-usage of scrap materials, and (5) improved corrosion resistance and product quality. The final TWB product quality including its formability is critically influenced by thickness and material combinations of the sheets used, weld conditions like weld orientation, weld location, and weld properties, all affect in a synergistic fashion. Prior knowledge of appropriate thickness, strength combinations, weld line location and profile, number of welds, weld orientation and weld zone properties is essential for TWB design. Such optimum parameters can be predicted if an "expert system" is available for TWBs.

Here, another example of developing an "expert system" based on neural network (ANN) for welded blanks that can predict their tensile, deep drawing, forming behavior under varied base material and weld conditions is presented. In this case, as said before, to avoid experimental cost, the data required was obtained through FE model simulations only. The expert system design is depicted in Figure 8.17 (Veera Babu et al., 2009). All the three phases in the expert system have a design mode of operation and updating mode of operation.

In Phase 1, while the expert system is designed, a range of material properties and TWB conditions are defined within which neural network models are developed. In the usage mode, the user selects base material properties and TWB conditions within the chosen range for formability prediction. In this phase, user can select different material models like strain-hardening laws and yield theories to predict the forming behavior. In the design mode, the neural network models will be developed to predict the forming behavior using different material models. As a result, in the usage mode of the expert system, the user will have option of selecting desired material models to predict the forming characteristics.

Phase 2 involves selecting the forming behavior to be predicted for chosen base material and weld conditions. In the design mode, using standard formability tests, the tensile behavior, stamping behavior of welded blanks are simulated. Different category of industrial sheet parts will be simulated and expert system will be developed to predict their forming behavior. The global tensile properties of TWB including stress–strain curve will be monitored. Formability characteristics like forming limit curve, percentage thinning, dome height at failure, failure location will be predicted by available formability and deep drawability tests. It is also planned to develop expert system for predicting application specific formability results. In the usage mode, the user selects the type of test results that is required to be predicted.

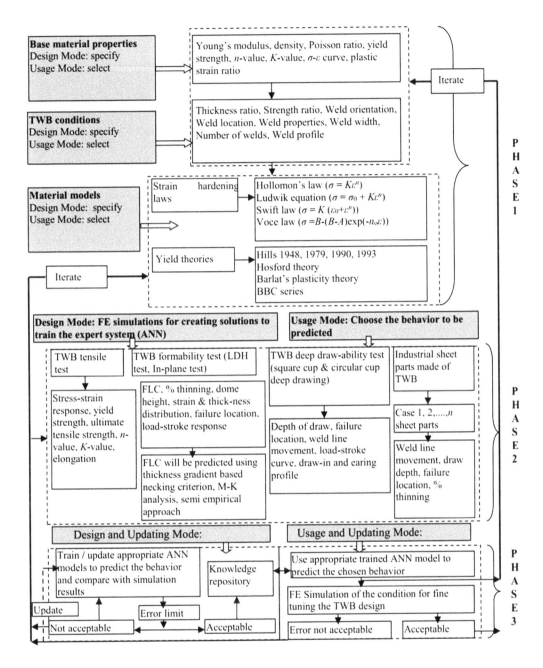

Figure 8.17 Expert system proposed for TWB forming. (With permission from Veera Babu et al., 2010.)

In Phase 3, the training, testing, usage, and updating the neural network predictions with simulation results are performed. In the design mode operation, various models are created and validated for predicting the forming behavior (computed in Phase 2) for various combination of material properties and TWB conditions and constitutive behavior (computed in Phase 1). In the usage mode, the user predicts the required forming behavior for an initially chosen material, TWB condition and constitutive behavior, until the error is within acceptable range.

Some representative predictions like the stress–strain behavior, draw-in profile, weld line movement, and draw depth during cup deep drawing, are presented as below. The tools required for tensile test, deep drawing test simulations, and neural network modeling details can be obtained from the earlier work (Veera Babu et al., 2009, Veera Babu et al., 2010, Dhumal et al., 2012, Dey et al., 2012). The comparison between expert system predictions (true stress–strain behavior) and simulation results are shown in Figure 8.18. The strain-hardening exponent (n) and strength coefficient (K) values obtained from neural network predictions are incorporated in the power law (hardening law), $\sigma = K\,\varepsilon^n$, for TWBs made of steel and aluminum alloy base materials for true stress–strain curve prediction. In Figure 8.19, the draw-in profiles during deep drawing are predicted for two different material combinations. Both the tensile behavior and deep drawing behavior are predicted with acceptable accuracy. In Dhumal et al.'s work, the drawn depth and weld line movement are predicted. Weld line movement is commonly observed during the forming of TWBs—the weld zone moves during stamping. The imbalance in the drawing resistance of the base sheets constituting TWBs leads to different levels of deformation or draw depths, but maintaining continuity, resulting in movement of the weld zone. In other words, the weld line movement is due to the heterogeneity in plastic deformation achieved in the two base metals during forming. The weld line movement is unwanted as it reduces formability. Sometimes tool design needs major modification incurring extra

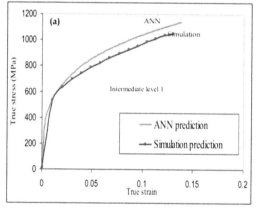 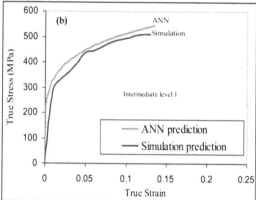

Figure 8.18   Prediction of true stress–strain behavior from ANN/expert system and comparison with FE simulation results (a) for Steel TWB and (b) for Al alloy TWB. (With permission from Veera Babu et al., 2009.)

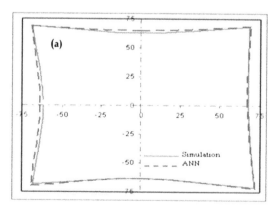 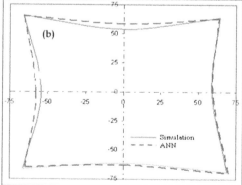

Figure 8.19   Prediction of cup draw-in behavior from ANN/expert system and comparison FE simulation results (a) for Steel TWB and (b) for Al alloy TWB. (With permission from Veera Babu et al., 2010.)

cost. Therefore, it is essential to predict the weld line movement during forming of TWBs and the expert system predictions are encouraging.

8.5 SUMMARY

As becoming a more and more fundamental requirement, the sustainability in material-forming industry has drawn a great attention to minimize the overall environmental impact under a life cycle perspective. The material-forming process must be designed, optimized, and controlled with environmental sustainability in mind as always. Huge research work and great effort has been done toward to this goal for the efficient use of materials and resources, the effective reduction of energy consumption, and minimum material waste in the material-forming process. In this great effort, computations have played a significant role for a multidisciplinary systematic approach, robust design, and optimum control of material-forming processes for environmental sustainability, economic sustainability, and social sustainability. To reveal the importance of the computations in material-forming sustainability, several major analytical methods, numerical methods, and virtual material-forming approaches and their applications have been discussed in detail in this chapter, to demonstrate that the computations not only reduces the energy consumption in physical tests, but also reduces the costs and time to thereby make the material-forming process sustainable in terms of both environmental and economic sustainability.

REFERENCES

Akiyama, M., Yamashita, M. 1992. Finite Element Analysis on the Optimum Cold Work for Stainless Steel. *NUMIFORM 92*, pp. 583–588.

American Supplier Institute, Inc. 1987. Quality Function Deployment, Center for Taguchi Methods.

Avitzur, B. 1968. *Metal Forming: Processes and Analysis*, McGraw-Hill, New York.

Bagheripoor, M., Bisadi, H. 2013. Application of artificial neural networks for the prediction of roll force and roll torque in hot strip rolling process. *Applied Mathematical Modelling*, 37(7), pp. 4593–4607.

Bendell, A., Disney, J., Pridmore, W.A. 1989. *Taguchi Methods: Application in World Industry*, Springer, Berlin Heidelberg.

Bryant, G.F. 1973. *Automation of Tandem Mills, Iron and Steel Institute*, London.

Byrne, D.M., Taguchi, S. 1987. The Taguchi Approach to Parameter Design. *40th Annual Quality Congress Transactions, ASQC*, Milwaukee, WI.

Cavaliere, M.A., Goldschmit, M.B., Dvorkin, E.N. 2001. Finite element analysis of steel rolling processes. *Computers & Structures*, 79(22–25), pp. 2075–2089.

Chenot, J.L., Fourment, L., Ducloux, R., Wey, E. 2010. Finite Element Modelling of Forging and Other Metal Forming Processes. *13th ESAFORM Conference on Material Forming*, Brescia, Italy. pp. 359–362.

Clements, R.B. 1991. *The Experimenter's Companion, A Guide and Reference to the Aspects of Research and Experimentation*, ASQC Quality Press, Milwaukee, WI.

Cramphorn, A.S., Bramley, A.N., Mcdermott, R.P. 1976. UBET Related Developments in Forging Analysis. *4th NAMRC*, pp. 80–86.

Cun, Y.L., Denker, J.S., Splla, S.A. 1990. Optimal Brain Damage. *Proceedings of the AIIE*, pp. 598–605.

Davalo, E., Naim, P. 1991. *Neural Networks*, Macmillan Press Ltd., Basingstoke.

Dehnad, K. 1989. *Quality Control, Robust Design, and Taguchi Method*, AT&T.

Dey, S.K., Narayanan, R.G., Kumar, G.S. 2012. Computing the tensile behavior of tailor welded blanks made of dual-phase steel by neural network-based expert system. *International Journal of Computer Integrated Manufacturing*, 25, pp. 158–176.

Dhumal, A.T., Narayanan, R.G., Kumar, G.S. 2012. Simulation based expert system to predict the deep drawing behaviour of tailor welded blanks. *International Journal of Modelling, Identification and Control*, 15, pp. 164–172.

Dixit, U.S., Narayanan, R.G. 2013. *Metal Forming Technology and Process Modelling*, McGraw Hill Education, New Delhi, pp. 1–568.

Falsafi, J., Demirci, E., Silberschmidt, V.V. 2016. Computational assessment of residual formability in sheet metal forming processes for sustainable recycling. *International Journal of Mechanical Sciences*, 119, pp. 187–196.

Freeman, J.A. 1994. *Simulating Neural Networks with Mathematics*, Addison-Wesley Publishing Company, Boston.

Gunasekera, J.S. 1989. *CAD/CAM of Dies*, Ellis Horwood Limited, UK.

Gunasekera, J.S., Hoshino, S. 1982. Analysis of extrusion of drawing of polygonal sections through strictly converging dies. *Journal of Engineering for Industry*, 104, pp. 38–45.

Gunasekera, J.S., Jia, Z. 1998. Analysis of Shear-Die Extrusion Process Using the Upper Bound Element Technique. *IDPT'98*, Berlin, Germany.

Gunasekera, J.S., Jia, Z., Malas, J., Rabelo, L.C. 1998. Development of a neural network model for a cold rolling process. *Engineering Applications of Artificial Intelligence*, 11(5), pp. 597–603.

Hauser, J.R., Clausing, D. 1988. The house of quality. Harvard Business Review, 66, p. 63.

Haykin, S. 1994. *Neural Networks, A Comprehensive Foundation*, Macmillan College Publishing Company, New York.

Holthe, K., Storen, S. 1992. Numerical Simulation of the Extrusion Process in a Series of Press Cycles. *NUMIFORM 92*, pp. 611–618.

Hosford, W.F. 2007. *Metal Forming, Mechanics and Metallurgy*, Cambridge University Press, Cambridge.

Hoshino, S., Gunasekera, J.S. 1985. An Upper Bound Solution for the Extrusion of Square Section from Round Bar Through Converging Dies. *Proceedings of the Twenty-First International Machine Tool Design and Research Conference*. pp. 97–105.

Hwang, B.C., Lee, H.I., Bae, W.B. 2003. A UBET analysis of the non-axisymmetric combined extrusion process. *Journal of Materials Processing Technology*, 139(1–3), pp. 547–552.

Jia, Z. 1995a. Optimization of Extrusion Process Using Quality Function Deployment & Taguchi's Methods. *Presented at the 3rd Annual American Society of Mechanical Engineer (ASME) GSTC Conference*, Kalamazoo, Michigan, USA.

Jia, Z. 1995b. Analytical Modeling of the Flat Rolling Process. *Research Report*, Mechanical Engineering, Ohio University.

Jia, Z. 1998. An Integrated Analysis and Design Method for Metal Forming Processes. *PhD dissertation*, Ohio University, USA.

Jia, Z., Gunasekera, J.S. 1995. Computer Simulation of Hollow Extrusion and Drawing Using 3D—FEM. *Proceedings of the 23rd North American Manufacturing Research Conference (NAMRC XXIII)*, Houghton, Michigan, USA, pp. 1–6.

Jia, Z., Gunasekera, J.S., Dehghani, M., Ali, A.F. 1996. Simulation of Hydraulic Expansion of Thin-Walled Tubes Using the Elastic-Plastic FEM. *Proceedings of the Third Biennial Joint Conference on Engineering Systems Design and Analysis (ESDA'96)*, Montpellier, France, pp. 169–174.

Jia, Z., Gunasekera, J.S., Malas, J.C. 1996. Application of Upper Bound Elemental Technique (UBET) for Aluminum Extrusion. *Proceedings of the 6th International Aluminum Extrusion Technology Conference (ET-96)*, Chicago, Illinois, USA, pp. 247–252.

Jia, Z, Gunasekera, J.S., Rabelo, L.C., Gunasekera M. 1996. Application of Neural network in the Analysis of Rolling Processes. Intelligent Engineering Systems through Artificial Neural Networks, Vol. 6. St. Louis, USA.

Joun, M. S., Hwang, S.M. 1992. Application of Finite Element Method to Process Optimal Design in Metal Extrusion. *NUMIFORM 92*, pp. 619–624.

Kim, D.J., Kim, B.M. 2000. Application of neural network and FEM for metal forming processes. *International Journal of Machine Tools and Manufacture*, 40(6), pp. 911–925.

King, B. 1989. *Better Design in Half the Time, Implementing QFD in America*, 3rd Edition, GOAL/QPC, USA.

Kiuchi, M. 1986. Complex simulation system of forging based on UBET. *Annals of the CIRP*, 25(1), pp. 147–150.

Kiuchi, M., Hsu, R. 1991. Numerical simulation of drawing of multi-cores clad rods and wires. *CIRP Annals 1990*, 39, pp. 271–274.

Kiuchi, M., Murata, Y. 1981. Study on application of UBET—Simulation of axisymmetric metal forming process. *Journal of JSTP*, 22(244), pp. 495–502.

Kobayashi, S. 1964. Upper-bound solutions of axisymmetric forming problems—I, II. *ASME Journal of Engineering for Industry*, pp. 122–126, 326–332.

Kobayashi, S., Oh, S.I., Altan, T. 1989. *Metal Forming and the Finite-Element Method*, Oxford University Press, Oxford.

Koot, L.W. 1973. Process Design Criteria for Cooling of A Cold Strip Mill. *Presented at ISI Meeting*, Amsterdam, Holland.

Kudo, H. 1960. Some analysis and experimental studies of axi-symmetric cold forging and extrusion—I. *International Journal of Mechanical Sciences*, 1, pp. 57–83.

Kudo, H. 1961. Some analysis and experimental studies of axi-symmetric cold forging and extrusion—II. *International Journal of Mechanical Sciences*, 3, pp. 91–117.

Laue, K., Stenger, H. 1988. *Extrusion: Processes, Machinery, Tooling*, American Society for Metals, Metals Park, OH.

Liu, Z. 2011. A Computational Tool for Industrial Sustainability Assessment and Decision Making. *ISEE*. www.aiche.org/conferences/aiche-annual-meeting/2011/proceeding/paper/480a-isee-computational-tool-industrial-sustainability-assessment-and-decision-making.

Mehra, P., Benjamin, W.W. 1992. *Artificial Neural Networks: Concepts and Theory*, IEEE Computer Society Press.

Muller, B., Reinhardt, J. 1991. *Neural Networks, an Introduction*, Springer-Verlag.

Phadke, M.S. 1989. *Quality Engineering Using Robust Design*, Prentice Hall, Englewood Cliffs, NJ.

Phogat, K.S., Shrama, A.K., Ranjan, R.K., Baipai, V.K. 2012. An Upper-bound solution for forging load of an elliptical disc. *Journal of mechanical Engineering Research*, 4(4), pp. 130–135.

Pietrzyk, M., Lenard, J.G. 1991. *Thermal-Mechanical Modeling of the Flat Rolling Process*, Springer-Verlag, Berlin.

Polakowski, N.H., Ready, D.M., Schmeissing, H.N. 1969. Principles of self control of product flatness in strip rolling mills. *Journal of Engineering for Industry*, 91(3), pp. 702–708.

Ross, P.J. 1988. *Taguchi Techniques for Quality Engineering*, McGraw-Hill Book Company, New York.

Roy, R.K. 1990. *A Primer on the Taguchi Method*, Van Nostrand Reinhold, New York

Shivpuri, R., Momin, S. 1992. Computer-aided design of dies to control dimensional quality of extruded shapes. *CIRP Annals 1990*, 41, pp. 275–279.

Soulami, A., Lavender, C.A., Paxton, D.M., Burkes, D.E. 2014. Rolling Process Modeling Report: Finite-Element Prediction of Roll-Separating Force and Rolling Defects. The U.S. Department of Energy. www.pnnl.gov/main/publications/external/technical_reports/PNNL-23313.pdf.

Udagawa, T., Kropp, E., Altan, T. 1989. Simulations of Titanium Tube Extrusion by FEM with Automated Remeshing Capability, Rept. #ERC/NSM-89-27, Engineering Research center for Net Shape Manufacturing, Columbus, OH.

Urbanski, S., Packo, M., Sadok, L., Kazanecki, J. 1992. Prediction of Hardness Distribution in Draw Tubes by Matrix Method. *NUMIFORM 92*, pp. 675–680.

Veera Babu, K., Narayanan, R.G., Kumar, G.S. 2009. An expert system based on artificial neural network for predicting the tensile behavior of tailor welded blanks. *Expert Systems with Applications*, 36, pp. 10683–10695

Veera Babu, K., Narayanan, R.G., Kumar, G.S. 2010. An expert system for predicting the deep drawing behavior of tailor welded blanks. *Expert Systems with Applications*, 37, pp. 7802–7812

Wang, J., Takefuji, Y. 1993. *Neural Networks in Design and Manufacturing*, World Scientific, Singapore.

Wasserman, P.D. 1993. *Advanced Methods in Neural Computing*, Van Nostrand Reinhold, New York.

Welstead, S.T. 1994. *Neural Network and Fuzzy Logic Applications in C/C++*, John Wiley & Sons, Inc, Hoboken, NJ.

Yang, D.Y., Kang, Y.S., Cho, J.R. 1992. Finite Element Analysis of Three-Dimensional Hot Extrusion of Sections Through Continuous Dies Considering Heat Transfer. *NUMIFORM 92*, pp. 687–694.

Yang, D.Y., Kim, J.H., Lim, C.K. 1985. An arbitrarily inclined triangular UBET element and its application to combined forging. *ASME Journal of Engineering for Industry*, 107, pp. 134–140.

Zienkiewicz, O.C., Taylor, R.L. 1988. *The Finite Element Method*, McGraw-Hill Book Co., London.

CHAPTER 9

Hot Stamping of Ultra-High-Strength Steel Parts

Ken-ichiro Mori
Toyohashi University of Technology

CONTENTS

9.1 INTRODUCTION

Although high-strength steel sheets are increasingly employed for automobile parts to reduce the weight and to improve the collision safety, cold stamping of ultra-high-strength steel sheets with a tensile strength >1 GPa is not easy. The forming load is large, the formability is low, the springback is large, the tool wear becomes remarkable, and seizure and galling occur for severe deformation. Thus, cold stamping of ultra-high-strength steel sheets >1.2 GPa in tensile strength is not practical, and hot stamping for producing ultra-high-strength steel parts having a tensile strength of 1.5 GPa is attractive.

Although magnesium and aluminum alloy sheets are conventionally warm- and hot-stamped to improve the formability, the main purpose of hot stamping is to produce ultra-high-strength steel parts, and the improvement of formability and the reduction in stamping load are secondary advantages. Hot stamping is a new forming approach for producing ultra-high-strength steel parts.

In hot stamping shown in Figure 9.1, a quenchable steel blank is heated to about 900°C in a furnace to be transformed into austenite, and then is transferred to a press for forming. The formed blank is quenched just after the end of forming by holding with dies at the bottom dead center for about 10 s, so-called die-quenching. Heating above the austenite temperature and rapid cooling by die-quenching are requisite to obtain a strength of about 1.5 GPa, i.e., the martensite transformation. Steel sheets are mainly coated with aluminum and zinc to prevent oxidation at high temperatures. In non-coated steel sheets, oxide scale of the formed parts is removed by shot blasting to complete

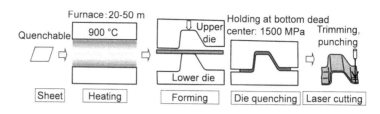

Figure 9.1   Hot stamping process of ultra-high-strength steel parts.

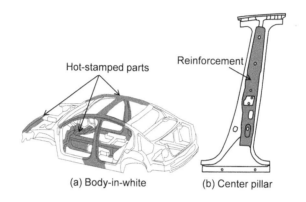

Figure 9.2   Application of hot-stamped parts to automobile parts.

welding and painting. Since the hot-stamped parts are too hard to be cold-sheared, laser cutting is generally employed to trim and punch the parts.

In order to improve the collision safety of automobiles, the application of hot stamping has been expanding, and body-in-white parts have been manufactured as shown in Figure 9.2. The hot-stamped parts having high strength are mainly used as reinforcements of cabin parts for protecting human bodies. However, the application is still limited to the reinforcements due to less ductility.

Hot stamping was first applied to the automobile parts by Saab AB of Sweden, and German automobile makers have begun full-scale production of hot stamping since the latter half of the 1990s. Hot stamping has expanded worldwide from the latter half of the 2000s. For example, 7% of the body-in-white for Volvo XC90 was launched in 2003 and 17% in 2006, 17% for Volkswagen Passat in 2006, 12% for BMW5 in 2009, and 12% for Volkswagen Golf VII in 2012. The percentage for Volvo XC90 launched in 2014 was raised to 40%. The number of hot-stamped parts rapidly increased and reached 250 million parts in 2014 (Habert, 2015).

9.2 PROPERTIES OF STEEL SHEETS

In cold stamping of high tensile strength steel sheets, the strength of the formed parts is dependent on that of the steel sheet, whereas hot-stamped parts are hardened by die-quenching. The strength of as-received steel sheets for hot stamping is considerably lower than that of the hot-stamped part. In hot stamping, conditions of heating and die-quenching are essential, and not the strength of the sheet.

The temperature history of the steel sheet in hot stamping is shown in Figure 9.3. The quenchable steel sheets are heated and kept in the furnace between 900°C and 950°C for about 2–5 min, are transferred to the press for several seconds, are die-quenched by holding for about 10 s at the

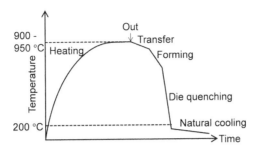

Figure 9.3 Temperature history of steel sheet in hot stamping.

bottom dead center of the press, and are naturally cooled after ejected from the dies <200°C. The steel sheet is transformed into a face-centered cubic crystal structure called austenite by heating >850°C, and is hardened by transforming into hard martensite in which carbon atoms are supersaturated into a body-centered cubic crystal structure of iron by rapid cooling. The temperature of the sheet just before forming is between 700°C and 800°C and the temperature at ejecting from the dies is <200°C of the martensite finish temperature. The temperature of the sheet is decreased by the partial contact with dies during forming, and is rapidly dropped by holding at the bottom dead center under a high pressure. This rapid drop leads to the martensite transformation.

The strength of quenchable steel sheets increases with increasing carbon content, whereas the ductility drops. In addition, the quenchability is changed by the amount of alloying elements, such as chromium, manganese, molybdenum, etc., the grain size of austenite, the cooling rate, and so on. In hot stamping, manganese–boron steel, 22MnB5, in which the quenchability is improved by adding elements such as manganese and boron, is generally used. As shown in Table 9.1, the carbon content of this steel is as low as 0.22 mass% to ensure toughness.

The tensile strengths for hot-rolled, cold-rolled, and annealed 22MnB5 sheets are about 590 MPa, about 780 MPa, and about 490–590 MPa, respectively, and that of hot-stamped parts is heightened to 1,500 MPa by die-quenching, two or three times higher. In hot stamping, the strength of the as-received sheets is not important, because the strength of the hot-stamped parts is dependent on die-quenching. Since hot stamping aims at reducing the weight of products, the thickness of the sheets is relatively small, between 0.8 and 2.2 mm. Sheets <1 mm in thickness are too soft to transfer to a press from a furnace and to move in a roller hearth furnace for continuous heating.

In hot stamping, the steel sheet is heated to about 900°C in the furnace and is formed under austenite. The flow stress curves were measured from a hot tensile test of an aluminum-coated 22MnB5 steel sheet (Lechler et al., 2008). In this test, the steel sheet was resistance-heated to 950°C to be transformed into austenite, and then was stretched just after air cooling to a desired temperature. At higher temperatures, work-hardening is not clear, and the temperature effect and strain-rate sensitivity are remarkable. The flow stress increases with decreasing temperature, and the heating effect is lost for a low temperature. It is important to shorten the transfer time from the furnace and the forming time.

The microstructures of the 22MnB5 steel sheet before and after die-quenching are shown in Figure 9.4. The as-received sheet has a mixed microstructure of ferrite and pearlite, is transformed into austenite by heating >850°C, and is transformed into elongated, dark, and fine martensite

Table 9.1   Chemical Components of Manganese–Boron Steel 22MnB5 (Mass%)

C	Mn	Cr	B
0.21–0.25	1.1–1.35	0.1–0.25	0.0015–0.004

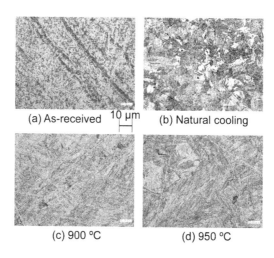

(a) As-received 10 μm   (b) Natural cooling

(c) 900 °C   (d) 950 °C

Figure 9.4   Microstructures of 22MnB5 steel sheet before and after die-quenching.

having 450–500 HV by die-quenching. The die-quenched parts are generally used without tempering as automobile parts. On the other hand, the air-cooled sheet has a mixed microstructure of pearlite and ferrite due to insufficient die-quenching, and the strength is not high. A high cooling rate in hot stamping is required to attain the high strength.

In the 22MnB5 steel sheet, the critical cooling rate causing martensite is 30°C/s (Nishibata and Kojima, 2010). As shown in Figure 9.5, the martensite transformation occurs above this cooling rate, and the hardness of the die-quenched parts is 450–500 HV. On the other hand, a mixed microstructure of ferrite and pearlite and a microstructure of bainite appear below this rate and the hardness is not high. When ferrite appears for a long transfer time from the furnace, the martensite transformation does not occur anymore. Although the cooling rate for die-quenching is not extremely high because of cooling with dies, the 22MnB5 steel sheet has a sufficient quenchability for a cooling rate of 30°C/s, and thus this sheet is generally employed for hot stamping. The cooling rate of die-quenching is lower than that of water and oil quenching, because the deforming shape, the wall thickness distribution, the oxide scale, etc., decrease the heat transfer to dies. Since the martensitic transformation starts from about 400°C and finishes around 200°C, the formed parts are held at the bottom dead center up to about 200°C.

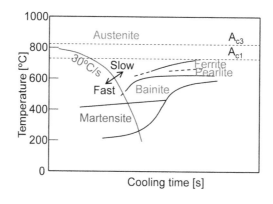

Figure 9.5   CCT diagram of 22MnB5 steel sheet heated to 950°C (Nishibata and Kojima, 2010).

9.3 OXIDATION AND PREVENTION

When a steel sheet is heated, oxide scale begins to appear from a temperature of 800°C, and the scale becomes remarkable around 900°C. It is necessary to remove the oxide scale from hot-stamped parts for subsequent welding and painting operations. In hot stamping, non-coated and coated sheets are employed. The hot-stamped part of the non-coated sheet heated in a furnace without controlling the atmosphere is shown in Figure 9.6. Much scale is formed on the surface of the formed part. The scale can be reduced by controlling atmosphere in a furnace.

For hot-stamped parts from the non-coated sheet, the oxide scale on the surface is removed by shot blasting treatment. The thickness of the oxide scale is about several μm, and the scale can be completely removed by shot blasting, and the weldability and paintability are almost the same as the cold-stamped parts. The oxide scale becomes large due to the contact with air during transfer from the furnace, and thus the reduction in transfer time is important for the prevention of not only temperature drop but also oxidation. In the removal of oxide scale by shot blasting, the wall thickness decreases by several percentages and the distortion of the formed part results.

In the coated sheets, oxide scale is not formed because the sheet is covered with a layer for protecting the oxidation, and thus shot blasting becomes unnecessary. The price of the coated sheets is higher than that of the non-coated sheets. Aluminum- and zinc-coated steel sheets are employed for hot stamping, and the aluminum-coated sheets are the most common. The aluminum coating layer contains about 10% of silicon and the melting point of this layer is about 600°C. As shown in Figure 9.7, aluminum reacts with iron and an intermetallic layer with a high melting point for protecting the oxidation is formed during heating. The concentration of aluminum in the intermetallic layer gradually decreases from the surface of the coated sheet, and that of iron increases. The thickness of the intermetallic layer is between 20 and 30 μm. The intermetallic layer has the function of not only prevention of the oxidation but also high corrosion resistance, weldability, and paintability.

The aluminum-coated steel sheet is generally heated at 900°C for about 5 min to generate the intermetallic layer. For a lower temperature and shorter heating time, aluminum is not sufficiently intermetallic, whereas excessive growth of the intermetallic layer is caused for long heating, and the oxidation protection, weldability, and paintability deteriorate. The size of the furnace for continuously heating the sheets becomes big for the heating time of 5 min. In order to reduce the size of the furnace, the heating time is shortened by controlling the heating history under a higher temperature.

Hot-dip galvanized and galvannealed steel sheets are employed as zinc-coated sheets. The galvannealed steel sheets are widely cold-stamped into automobile body panels, and a Zn–Fe alloy

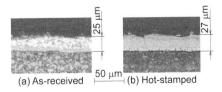

Figure 9.6 Hot bending of non-coated sheet heated in furnace without controlling atmosphere.

(a) As-received   50 μm   (b) Hot-stamped

Figure 9.7 Coating layer of aluminum-coated sheets.

layer is formed by reacting base steel with zinc for the hot-dip galvanized steel sheet. The thickness of the alloy layer for the galvannealed steel sheet is about 10 μm, and the layer grows to about 20 μm after heating. The formed layer containing zinc has high corrosion resistance, paintability, and spot resistance weldability. The hot-dip galvanized steel sheet can be alloyed for optimized heating conditions, and the heating time becomes long. Since the zinc-coated sheets have a risk of cracking due to liquid–metal induced embrittlement in the Fe-Zn systems, sufficient heating until forming the solid solution phase is required.

The hot-stamped parts from non-coated, aluminum-coated, and zinc-coated steel sheets are given in Figure 9.8. No scale occurs for the coated sheets and the amount of oxide scale for the non-coated sheet is suppressed by the atmosphere control. The aluminum-coated sheet is a glossy color of aluminum, and the color turns into a bluish black for alloying by heating. The hot-stamped part from the zinc-coated sheet is a brownish color. The features of the coatings used for oxide protection are summarized in Table 9.2.

9.4 STAMPING

Although the steel sheet is heated between 900°C and 950°C in hot stamping, the temperature decreases by about 100°C–150°C during transfer from the furnace to the dies. The relationship between the stamping load and the heating temperature in hat-shaped bending is illustrated in Figure 9.9. The bending load is reduced to about 1/3 to 1/4 of that for cold stamping, and thus high-strength parts can be produced for a small stamping load. The load reduction is effective in improving the tool life, whereas the high temperature of the heated sheet deteriorates the life. Cooling and coating of dies are useful.

In cold stamping, the formability largely depends on the material properties of the blank, while the temperature distribution in the deforming blank has a large effect on the formability in hot stamping. As shown in Figure 9.10, the temperature decreases around the regions in contact with the dies during forming, and the temperature in the regions not in contact with the dies is still high. The temperature is nonuniformly distributed in the deforming blank during forming. The temperatures

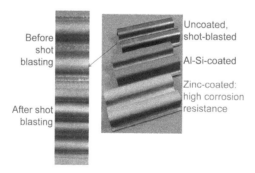

Figure 9.8 Hot-stamped parts from non-coated, aluminum-coated and zinc-coated steel sheets (courtesy of Aishin Takaoka Co., Ltd).

Table 9.2 Features of Coatings Used for Oxide Protection

Cases	Non-Coated	Aluminum-Coated	Zinc-Coated
Heating time	2–3 min	5–6 min	4–5 min
Tool wear	Large wear by oxidation	Predominantly adhesion	Predominantly abrasion
Scale removing	Shot blasting	Unnecessary	Unnecessary

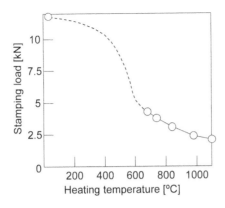

Figure 9.9 Relationship between stamping load and heating temperature in hat-shaped bending.

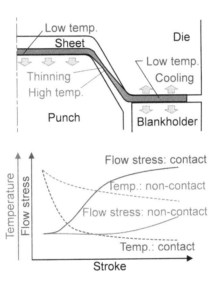

Figure 9.10 Effect of temperature distribution on formability in hot stamping.

in the non-contact and contact regions are high and low, respectively. The deformation in the non-contact region becomes large because of low flow stress, and thinning and rupture result due to the concentration of deformation. Although the ductility is heightened by heating, the deformation behavior is greatly influenced by this temperature distribution.

The drawn cups with a semi-spherical punch for different forming rates are shown in Figure 9.11 (Maeno et al., 2014). For hot stamping of the low forming rate, the fracture occurs, while no fracture occurs for cold stamping. The fracture results from the temperature distribution in the deforming blank shown in Figure 9.12 (Maeno et al., 2014). Even when the steel sheet is heated, the temperature falls after ejecting from the furnace, and the temperature becomes nonuniform. The temperature distribution becomes remarkable with decreasing forming rate, and thus quick hot stamping is desirable to minimize the temperature drop by the contact with dies. High stamping rates are possible for mechanical servo presses, whereas the stamping rate is low for conventional hydraulic presses. The forming limit diagram generally used for cold stamping is not proper because the effect of temperature distribution is not taken into consideration.

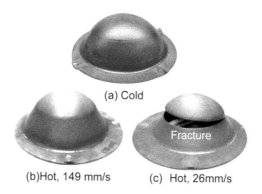

(a) Cold

(b)Hot, 149 mm/s   (c) Hot, 26mm/s

Figure 9.11   Drawn cups with semi-spherical punch for different forming rates (Maeno et al., 2014).

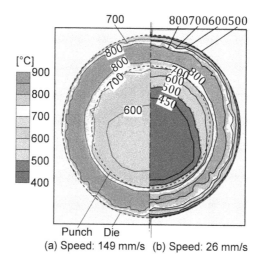

Figure 9.12   Temperature distribution in deforming blank (Maeno et al., 2014).

In cold stamping, the forming load increases with increasing strength of the steel sheet, and the springback induced by the elastic recovery in unloading becomes large as shown in Figure 9.13. In hot stamping, not only high-strength parts can be produced, but also almost no springback occur and the shape accuracy is very high.

The relationship between the springback and the holding time at the bottom dead center for hat-shaped bending is illustrated in Figure 9.14. The springback decreases with increasing holding time. Although formed parts are held at the bottom dead center for die-quenching, holding up to the martensite finish temperature of 200°C is required to prevent the springback because of the volume change induced by the martensite transformation. Holding up to the martensite finish temperature has the function of not only high strength but also the prevention of springback.

In cold stamping, the draw-type tools are generally employed, and the tension is applied with the sheet holders to reduce the springback as shown in Figure 9.15. In the draw-type tools, the blank is bent under sandwiching both edges between the upper and lower tools. On the other hand, the sidewalls and flanges are formed in the latter half with the punch and die. The temperature drop in the flange during bending for the form-type tools is smaller than that for the draw-type tools (Mori et al., 2016). The small drop is advantageous to hot stamping, and thus the form-type tools are mostly applied to hot stamping operations.

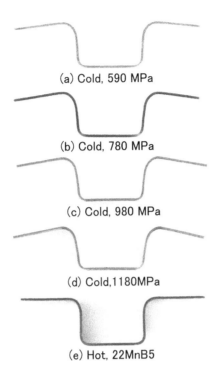

(a) Cold, 590 MPa

(b) Cold, 780 MPa

(c) Cold, 980 MPa

(d) Cold,1180MPa

(e) Hot, 22MnB5

Figure 9.13   Springback in cold stamping and hot stamping.

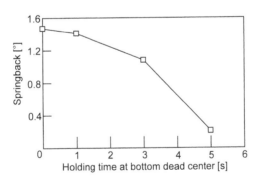

Figure 9.14   Relationship between springback and holding time at bottom dead center for hat-shaped bending.

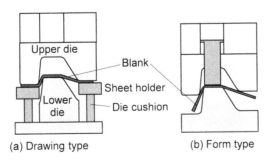

(a) Drawing type   (b) Form type

Figure 9.15   Draw- and form-type tools for hot stamping.

9.5 DIE-QUENCHING

Since high strength of the hot-stamped parts is obtained by die-quenching, the die-quenching behavior is very important. The relationship between the Vickers hardness and the holding time at the bottom dead center for hat-shaped bending is illustrated in Figure 9.16. As the holding time increases, the hardness increases, and then become constant >5 s. A certain holding time is required to attain the martensite transformation in die-quenching. Although the bent sheet used for Figure 9.16 is comparatively small, a holding time of about 10 s is necessary to large automobile parts.

Although formed parts are die-quenched by holding at the bottom dead center, the entire surface of the parts is not in contact with the dies, and thus the cooling rate of the non-contact regions becomes small. As shown in Figure 9.17, the thickness of the formed part is not uniform, and the corner and side wall are not in complete contact with the dies (Maeno et al., 2015). Low cooling rate brings about locally insufficient strength in the formed part. The holding force at the bottom dead center is considerably enlarged to increase the real contact area and the heat transfer to the dies. This leads to the increase in press capacity. The holding force and time are determined so as to quench the non-contact regions.

In hot stamping operations, the formed parts are cooled with dies during holding at the bottom dead center. The dies have water flowing channels for cooling and finally the heat is discharged by the cooling water. The design of the cooling channels is important because the cooling behavior is influenced by this design. To reduce the holding time at the bottom dead center, water is ejected from the tool surfaces just after reaching the bottom dead center and drained the water just after holding as shown in Figure 9.18 (Nomura et al., 2015).

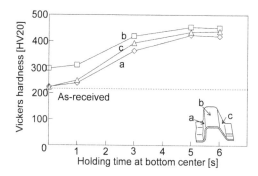

Figure 9.16   Relationship between Vickers hardness and holding time at bottom dead center for hat-shaped bending.

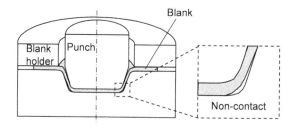

Figure 9.17   Contact between dies and formed part during holding at bottom dead center.

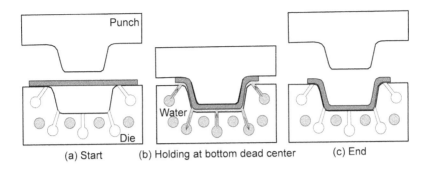

Figure 9.18 Direct water cooling for reduction in holding time at bottom dead center.

9.6 POST TREATMENTS

In hot stamping, the blank shape is estimated by developing the part shape, whereas the accuracy of this estimation is not very high because of plastic deformation of the blank. Regions requiring high dimensional accuracy are trimmed for hot-stamped parts. The hot-stamped parts are too hard to cold-trim. The tool life is considerably short and the trimmed parts have a risk of delayed fracture. Thus, high-precision laser cutting is generally used for trimming of the hot-stamped parts. The CO_2 laser, Nd:YAG laser, fiber laser, etc., are employed for the laser cutting processes, and the solid-state fiber laser has been mostly used. In the fiber laser, the intensity is high due to small focal diameter, the resistant to vibration is high and stable, the efficiency is high and the power consumption is low, whereas the equipment is expensive and the productivity is low.

In laser cutting, the production cost becomes relatively high. A large number of holes are made for body-in-white parts largely produced by hot stamping, and laser cutting of holes brings about a large cost increase. The laser cutting length of the hot-stamped parts is optimized by partially replacing laser cutting with blank development and hot shearing (Josefsson, 2011). The estimation of the blank shape by developing the part shape is improved by the finite-element simulation. In hot shearing, the formed part is trimmed and punched at high temperatures during holding at the bottom dead center, and the shearing load is low, while tooling becomes complicated. Laser cutting, hot shearing, and blank development are selected from a balance of dimensional accuracy and cost.

Although remarkable oxide scale is generated on the surface of hot-stamped parts from the non-coated steel sheets, the resistance spot weldability and paintability are enough improved by removing the scale using shot blasting. The shot-blasted parts are coated with rust preventive oil. The die-quenched non-coated and the 980 MPa ultra-high-strength steel sheets are spot-welded as shown in Figure 9.19. The oxide scale on the quenched non-coated sheet is removed with sandpaper.

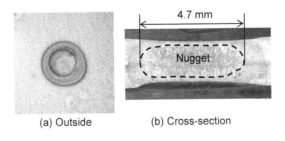

Figure 9.19 Spot resistance welding of die-quenched non-coated and 980 MPa ultra-high-strength steel sheets.

The nugget is formed in the weld and welding is sufficient. Because hot-stamped parts are mostly employed as reinforcements of body panels, hot-stamped parts are generally welded with cold-stamped parts.

9.7 FINITE-ELEMENT SIMULATION

The temperature of the steel blank heated in the furnace falls during the transfer, forming, holding at the bottom dead center as shown in Figure 9.20, and the temperature is nonuniformly distributed in the blank by partial contact with the dies. The temperature distribution in the blank affects the deformation behavior of the blank during forming. In addition, the dies have a nonuniform distribution of temperature by the heat transfer from the formed parts. If the temperature of dies is not low enough, the formed parts are not hardened, insufficient martensite transformation. The temperature, plastic deformation, and microstructure change are taken into consideration in simulation of hot stamping processes, and the finite-element methods are generally utilized for the simulation.

In the finite-element simulation, plastic deformation and temperature distributions in the forming blank and dies are calculated, and the microstructure is predicted from material models using the calculated strain, temperature, etc. The deformation, temperature and microstructure interact, and are calculated by exchanging obtained results in each time interval.

Although the temperature distribution, plastic deformation, etc., are calculated by the finite-element simulation, it is not easy to measure material properties used for calculation. The thermal conductivity, specific heat, density, etc., are functions of temperature, and are relatively measured as literature values. The heat transfer coefficient between the blank and die is a function of temperature, contact pressure and interface conditions, and the measurement is complicated. The temperature is greatly affected by the heat transfer coefficient. It is difficult to measure the flow stress curve of the sheet and the coefficient of friction which are functions of temperature. Expressions and material constants for predicting the microstructure are also problematic.

9.8 SMART HOT STAMPING

The present hot stamping processes still have the following drawbacks:

1. The equipment consisting of a furnace, press, laser cutting machine, etc., is large and expensive.
2. The productivity is low due to die-quenching and slow hydraulic presses, i.e., 2 or 3 shots per minute.
3. The applicable range is limited to reinforced parts of automobile cabins and bumpers.

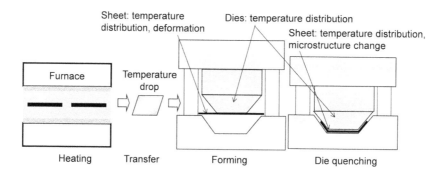

Figure 9.20   Finite-element simulation of temperature distribution, plastic deformation, and microstructure change.

4. For non-coated sheets, shot peening is required, and aluminum-coated sheets for preventing the oxidation at high temperatures are comparatively expensive.
5. Laser cutting of formed parts for trimming and punching becomes costly.

The heating approach of the steel sheets in hot stamping is crucial. Although the temperature drop in bulk workpieces used for the forging is comparatively small, that for the sheet metals after the ejection from the furnace is very rapid due to large surface area. In addition, oxide scale on the surface of the sheet is remarkable. The size of the furnace becomes large due to slow formation of the intermetallic compound on the surface of the aluminum-coated sheet during heating. For rapid laser heating, it is difficult to obtain a uniform temperature in the sheet due to local heating.

Mori et al. (2005) proposed a hot stamping process using rapid resistance heating to improve the productivity. The sheets are heated in only 2 s to 900°C required for quenching. Hot stamping using resistance heating and the hat-shaped sheets are shown in Figure 9.21. The oxide scale of the stamped parts is hardly generated due to rapid heating in comparison with furnace heating shown in Figure 9.6. Big furnaces are replaced with a power supply. The efficiency of resistance heating is higher than that of induction because of direct passage of current through sheets.

A smart hot stamping system using resistance hearting is illustrated in Figure 9.22. The big furnace and laser cutting can be omitted by resistance heating and hot trimming, respectively. The

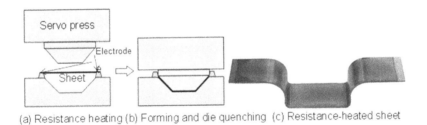

(a) Resistance heating (b) Forming and die quenching (c) Resistance-heated sheet

Figure 9.21 Hot stamping using resistance heating and hat-shaped sheets (Mori et al., 2005).

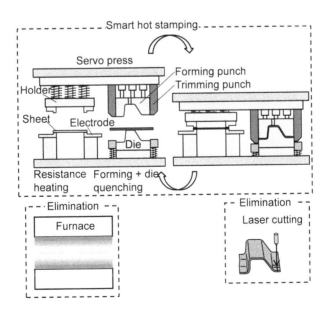

Figure 9.22 Smart hot stamping having resistance heating, servo press and hot trimming (Mori, 2015).

productivity is doubled by performing resistance heating and die-quenching simultaneously and the synchronization is facilitated by the servo press.

9.9 CONCLUSIONS

Hot stamping is an attractive process for producing ultra-high-strength steel parts, and the application to automobile parts rapidly expands. Since the hot stamping process is not yet fully established, new operations will be still developed. The improvement of productivity, the increase in product strength, compact processes, the improvement of accuracy of calculated results, application except for body-in-white parts, etc., are targets for hot stamping, and many challenging approaches have been tried.

REFERENCES

H. Habert, Products in Hot Stamped Boron Steel. *Capital Market Day*, October 6th, 2015.

P. Josefsson, Production Process Development for Hot Forming. *Proceeding of Tools and Technologies for Processing Ultra High Strength Materials TTP 2011*, Graz, (2011), 135–144.

J. Lechler, M. Merklein, M. Geiger, Determination of thermal and mechanical material properties of ultra-high strength steels for hot stamping, *Steel Research International*, 79(2) (2008), 98–104.

T. Maeno, K. Mori, M. Fujimoto, Improvements in productivity and formability by water and die quenching in hot stamping of ultra-high strength steel parts, *CIRP Annals—Manufacturing Technology*, 64(1) (2015), 281–284.

T. Maeno, K. Mori, T. Nagai, Improvement in formability by control of temperature in hot stamping of ultra-high strength steel parts, *CIRP Annals—Manufacturing Technology*, 63(1) (2014), 301–304.

K. Mori, *Smart Hot Stamping for Ultra-High Strength Steel Parts, 60 Excellent Inventions in Metal Forming*, (2015), 403–408, Springer-Verlag, Berlin Heidelberg.

K. Mori, T. Maeno, Y. Yanagita, Deep drawability and bendability in hot stamping of ultra-high strength steel parts, *Key Engineering Materials*, 716 (2016), 262–269.

K. Mori, S. Maki, Y. Tanaka, Warm and hot stamping of ultra high tensile strength steel sheets using resistance heating, *CIRP Annals—Manufacturing Technology*, 54(1) (2005), 209–212.

T. Nishibata, N. Kojima, Effect of quenching rate on hardness and microstructure of hot-stamped steel, *Tetsu-to-Hagané*, 96(6) (2010), 378–385.

N. Nomura, H. Fukuchi, A. Seto, Effect of High Cooling Rate on Shape Accuracy of Hot Stamped Parts. *Proceedings of 5th International Conference of Hot Sheet Metal Forming of High-Performance Steel*, Toronto, (2015), 549–557.

CHAPTER 10

Tubular Hydroforming and Hydropiercing

Nader Asnafi
Örebro University

CONTENTS

10.1 INTRODUCTION

Hydroforming of tubular components has been known by many other names such as bulge forming of tubes, hydrobulging, internal high-pressure forming, liquid forming of tubular components, etc. Different "hydroforming" methods have been reported earlier. Hydroforming or expansion in an open and in a closed tool (Dohmann and Hartl, 1996), free and die-bound hydroforming (Schäfer Hydroforming, 1996), low, high and sequenced pressure forming (Mason, 1996; VARI-FORM, 1996), and the Rolls–Royce method (Astrop, 1968) are some of the different methods tested, used, and reported so far.

In this chapter, *hydroforming* is defined as *a forming operation*, in which a tube is expanded radially while compressed axially by two cylinders, as shown in Figures 10.1 and 10.5. The purpose of axial compression is to feed in material to the expanding regions. Because of axial material feeding, the tube wall thinning (and thereby the risk of bursting) is reduced significantly.

T-branches/pieces, different types of tube items and water fittings, aircraft components, and automotive/vehicle parts are some of the components that are (or can be) manufactured by tube hydroforming. Figure 10.1 displays T-pieces made by hydroforming and how a T-piece is hydroformed. Exhaust manifolds, subframes, crash beams, and side members are some of the automotive parts that are made by hydroforming. Both the left- and the right-side members shown in Figure 10.2 were produced by hydroforming. These side members, which go from bumper to bumper, were developed and manufactured for an aluminum hybrid (car) body structure. The initial tubes were straight 6 m long extruded aluminum profiles, AA 6063-T4, with a circular cross-section. The outer diameter of the tube was 110 mm and the tube wall thickness was 2.5 mm. The straight tube was first cut to a length of 4,870 mm after which it was rotary-draw bent at ten different zones. The bent tube was then placed in a hydroforming tool. An internal pressure of 30–35 bar was built up. The tool was closed and the tube was hydroformed at 1,300 bar. The applied tool closure force (press force) was 9,600 metric tons. The front- and rear-end strokes were 18 and 32 mm, respectively, during the hydroforming operation (Asnafi et al., 2000, 2002).

The crash beam shown in Figure 10.3 was produced by tube bending and hydroforming. The straight steel tube, St37 Ti-NBK, had a circular cross-section with an outer diameter of 46 mm and a wall thickness of 2 mm. The straight tube was first rotary-draw bent at five zones, then placed in a hydroforming tool, and subsequently hydroformed at 1,300 bar in maximum internal pressure and with 50 mm in final stroke from each tube end (Asnafi et al., 2002).

Weight efficiency, reduced emissions, more sustainable products, and sustainable manufacturing processes and systems have been at the focus for many years within the automotive industry. These topics will be at the focus in the future too. Weight reduction is, for instance, of great value even for electrical vehicles. Hydroforming is an enabler in this context.

In a typical car, the body structure and hang-on parts account for 25% of the total weight. The largest contribution to this value comes from the load-carrying structure. Utilizing high, extra, and ultra-high strength/advanced steel grades, the body weight can be reduced by at least an additional 10%. Greater weight reduction can be accomplished using hydroforming.

Figure 10.4 shows that the manufacturing costs for each of the hydroformed side members in Figure 10.2 are reduced by 50%. Less material is used, the material and forming costs are reduced,

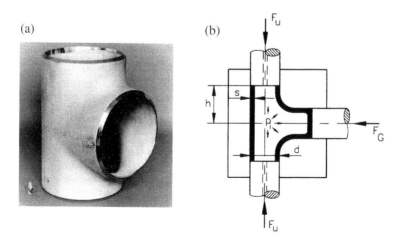

Figure 10.1 (a) T-pieces of steel made by hydroforming and (b) how these T-pieces are hydroformed. (Reprinted with permission from (a) Mücke, 1995, and (b) Dohmann and Hartl, 1996.)

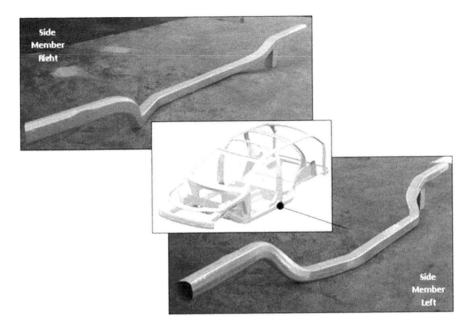

Figure 10.2 The (underbody) side members in an aluminum hybrid body structure. The initial tube, 2.5-mm-thick AA6063-T4 with 110 mm, was bent and hydroformed to make the side members shown in this figure. (Reprinted with permission from Asnafi et al., 2002.)

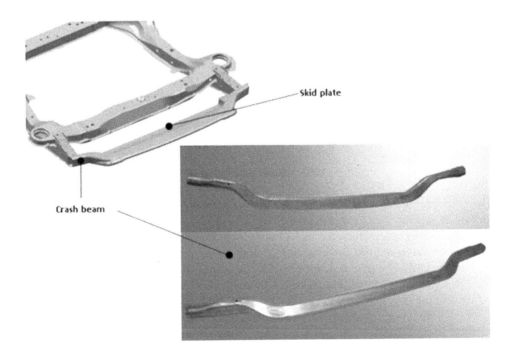

Figure 10.3 The Crash beam shown in this figure is made by tube bending and hydroforming. Initial Tube material = 2-mm-thick steel St37 Ti-NBK. (Reprinted with permission from Asnafi et al., 2002.)

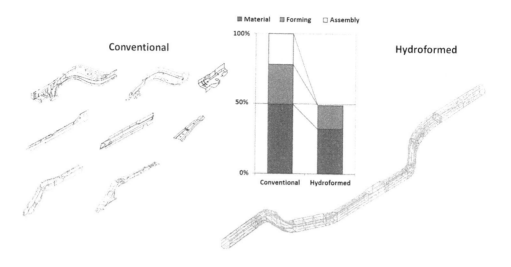

Figure 10.4 Each of the hydroformed side members in Figure 10.2 replaces eight stamped and assembly welded parts. The material and forming costs are reduced and the assembly costs vanish. The total costs for each side member is reduced by 50%, if hydroforming is selected as the manufacturing method.

and the assembly costs vanish, since the hydroformed side member replaces eight conventionally stamped and assembly welded parts. Since the hydroformed aluminum side member in Figures 10.2 and 10.4 replaces eight parts that are stamped in steel, the weight reduction is even greater.

In addition to the above-mentioned benefits, hydroformed components also have other advantages. A tubular beam (the hydroformed side member in Figures 10.2 and 10.4) displays, for instance, improved bending stiffness, reduced bending stress, higher torsional stiffness, and lower torsional stress compared to the stamped and assembly welded side member in Figure 10.4.

This chapter deals with the fundamentals of tube hydroforming, analytical models, and finite-element simulations for free forming, some sheet and tube material related issues, the procedure used in the design, manufacturing engineering, and prototyping of industrial, particularly automotive, components and how holes can be made in the component in the hydroforming process (hydropiercing). The readers interested in different sheet hydroforming methods are referred to Tolazzi (2010), Ocklund et al. (2002), Thuresson et al. (2003), and Asnafi (1999a).

10.2 FUNDAMENTALS

The principle of tube hydroforming is displayed in Figure 10.5. The hydroforming tool is mounted in a press. The tube is first filled with a liquid emulsion of a water-soluble material (concentration <5%), after which the tool is closed (by applying a press force). The tube is then forced to adopt the inner contour of the tool by application of an internal pressure and two axial forces. In some cases, the tube is formed by the increasing internal pressure only. In these cases, the axial forces are only so high that leakage is avoided. This means that the axial cylinders do not feed more material into the expansion zone. There are also cases in which the axial cylinders push more material into the expansion zone. In these cases, the tube is formed under the simultaneous action of the internal pressure and the axial forces. The tool is kept closed by applying a press force during hydroforming. The magnitude of this closing force is dependent upon the magnitude of the applied internal pressure (the higher the internal pressure, the larger the required closing force), as shown in Figure 10.5.

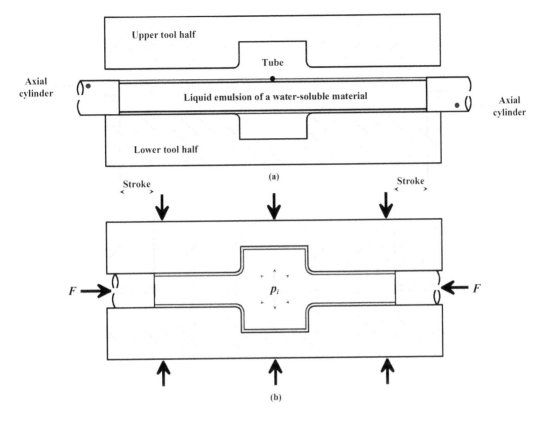

Figure 10.5 The principle of tube hydroforming: (a) original tube shape and (b) final tube shape (before unloading). Note the tool closing force acting on the lower and the upper tool halves. (Reprinted with permission from Asnafi et al., 2000.)

Hydroforming is either force-controlled (the axial forces are varied with the internal pressure) or stroke-controlled (the stroke is varied with the internal pressure), as shown in Figure 10.5. The failures encountered in tube hydroforming are displayed in Figure 10.6. The risk of *buckling* is greatest at the beginning of the hydroforming process, particularly if the hydroforming operation involves axial material feeding. If buckling occurs, it is not possible to continue the hydroforming process, since this process cannot be controlled any longer. The risk of buckling is dependent upon the so-called free tube length, l_f, the tube diameter, d_0, and the tube wall thickness, t_0 (see Figure 10.7). The free tube length, l_f, is not allowed to exceed two times the tube diameter, if $20 < d_0/t_0 < 45$. However, if $d_0/t_0 > 45$, the risk of buckling is very large (since the tube wall is very thin). Therefore, $l_f << 2d_0$. If $d_0/t_0 < 20$, the risk of buckling is very small (since the tube is very thick). The free tube length can therefore be allowed to exceed two times the tube diameter (see Figure 10.7).

Wrinkling, Figure 10.6, is not treated theoretically in this chapter. However, those interested in how wrinkling can be avoided are referred to Sauer et al. (1978), Geckeler (1928), and Asnafi (1997). *Fracture (bursting)*, Figure 10.6, will be addressed later on in this chapter. These limits (fracture and wrinkling) are displayed schematically in Figure 10.8.

The axial forces acting on the tube ends, Figure 10.5, must exceed a certain level to prevent leakage. This limit is also shown in Figure 10.8 (sealing). The deformation during hydroforming comprises an elastic and a plastic portion. The limit at which yielding occurs is, therefore, of great importance. The yielding limit is also shown in Figure 10.8. Once these limits (wrinkling, fracture,

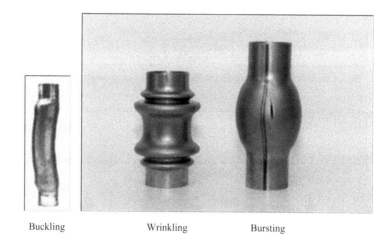

Buckling   Wrinkling   Bursting

Figure 10.6 Failures in tube hydroforming. (The picture showing buckling is reprinted with permission from Dohmann et al.,1996, and the picture showing wrinkling and bursting is reprinted with permission from Asnafi et al., 2000.)

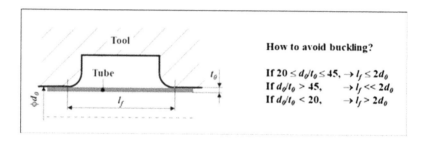

Figure 10.7 To avoid buckling, these rules should be followed during tool design and tube material selection. (Reprinted with permission from Asnafi, 1999.)

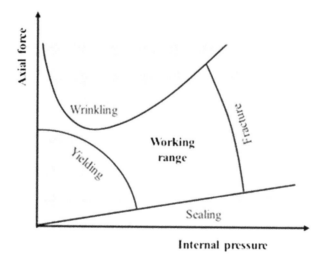

Figure 10.8 The limits and the working range in tube hydroforming. (Reprinted with permission from Asnafi, 1999.)

yielding and sealing) are determined, the working range can be established (see Figure 10.8). This working range is dependent on both tube material and tool parameters.

The hydroforming operation can be divided into two stages: *free forming* and *calibration*. This is illustrated in Figure 10.9. The part of the hydroforming operation, in which the tube expands without tool contact, is called *free forming*. Calibration starts as soon as tool contact is established (see Figure 10.9). During *calibration*, the tube is forced to conform to the inner contour of the tool by the internal pressure only. No additional material is fed into the expansion zone by the axial cylinders and the loading path during calibration (the path that describes how the axial force is to be varied with the internal pressure) is, therefore, parallel to the sealing limit (see Figure 10.9). (Note that Figure 10.9 shows an axisymmetrical hydroforming operation and not hydroforming of a T-branch.)

As mentioned above, the hydroforming operation is either force-controlled (the axial forces are varied with the internal pressure; Figure 10.5 and 10.9) or stroke-controlled (the stroke is varied with the internal pressure; Figures 10.5 and 10.10).

During free forming, the selected loading path determines the strain ratio at the top of the bulge—marked with an A in Figures 10.9 and 10.10. If the axial cylinders do not feed more material into the expansion zone, the applied axial force or the stroke provides only the sealing. In this case, the expansion is conducted at plane strain (see Figures 10.11 and 10.12). However, if the loading path lies on the borderline to wrinkling, the tube is deformed at pure shear as shown in Figures 10.11 and 10.12. (Note that Figures 10.11 and 10.12 show an axisymmetrical forming and not forming of a T-piece.)

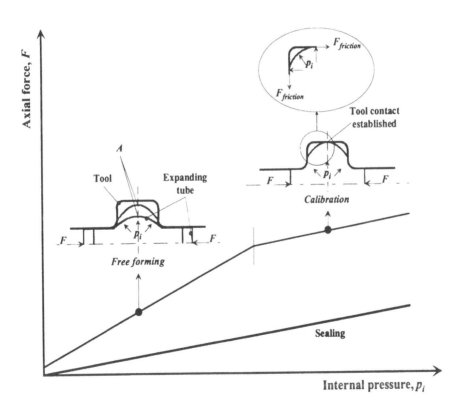

Figure 10.9   Tube hydroforming consists of free forming and calibration. This figure shows the so-called force-controlled tube hydroforming. (Reprinted with permission from Asnafi, 1999.)

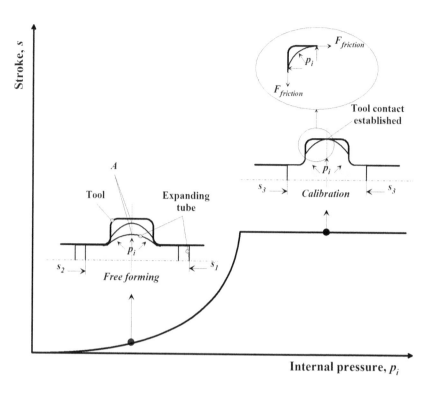

Figure 10.10   Tube hydroforming consists of free forming and calibration. This figure shows the so-called stroke-controlled tube hydroforming. (Reprinted with permission from Asnafi et al., 2000.)

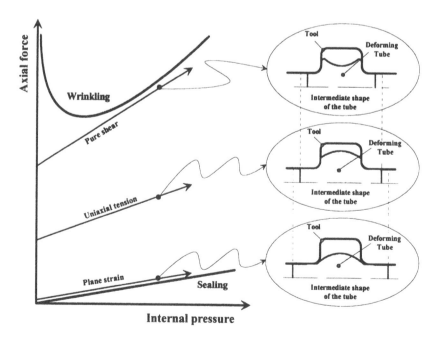

Figure 10.11   The selected loading path determines the deformation mode and the intermediate shape of the deforming tube. The figure shows the so-called force-controlled tube hydroforming. (Reprinted with permission from Asnafi, 1999.)

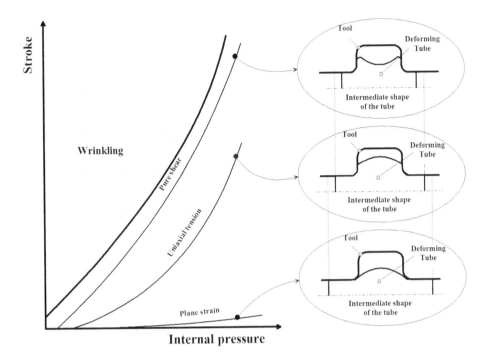

Figure 10.12 The selected loading path determines the deformation mode and the intermediate shape of the deforming tube. The figure shows the so-called stroke-controlled tube hydroforming. (Reprinted with permission from Asnafi et al., 2000.)

Wrinkles are, as shown in Figures 10.11 and 10.12, formed during free forming at the intake regions of the expansion zone, if the pure-shear path is selected. These wrinkles do not cause any problem and are straightened out during calibration. Comparing the intermediate shapes shown in Figures 10.11 and 10.12, one can see that the amount of material fed into the expansion zone during free forming is the greatest in the case of pure shear. The more material there is in the expansion zone as the calibration process starts, the less will be the wall thickness reduction. To minimize the wall thickness reduction, it is therefore recommended that the loading path during free forming be between the uniaxial-tension path and the pure-shear path (see Figures 10.11 and 10.12).

One of the aims of the *process design* is to determine how the tube is to be hydroformed (*selection of a loading path*), so that the wall thickness reduction is minimized without risking wrinkling or fracture. However, the loading path cannot be selected without considering the overall desired component shape. This is illustrated in Figure 10.13, which shows that the amount of material inflow varies dependent upon the position of the expansion zone in relation to the position of the axial cylinders. Assume that a zone at the base of the automotive subframe shown in Figure 10.13 is to be expanded. Since this zone lies far from the axial cylinders and is surrounded by two bends, the friction forces do not allow any material inflow into this expansion zone. Therefore, the expansion at this site can only occur at plane strain. Such a cross-section and longitudinal-section analysis must be carried out for different zones along the desired overall component shape. This means in turn that process design and *component design* are very closely related to each other. When designing the component, one has to know how much deformation the selected tube material can withstand at different deformation modes. (Regarding component design rules, the reader is also referred to Asnafi, 1997.)

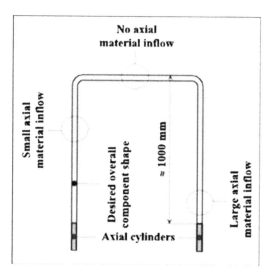

Figure 10.13 The desired overall component shape determines the appropriate loading path and the deformation modes prevailing at different zones. The figure shows the overall shape of an automotive subframe. (Reprinted with permission from Asnafi, 1999.)

10.3 ANALYTICAL MODELING OF FREE FORMING

10.3.1 Yielding

Consider a tube that is subject to an internal pressure, p_i, and compressive axial forces, F (Figure 10.14). For an element at the middle of this tube, the following equilibrium equation can be written (Asnafi, 1999):

$$\frac{\sigma_1}{\rho_1} + \frac{\sigma_2}{\rho_2} = \frac{p_i}{t_i} \tag{10.1}$$

and

$$\sigma_2 = \frac{p_i \rho_1}{2t_i} - \frac{F}{2\pi \rho_1 t_i} \tag{10.2}$$

von Mises yield criterion (plane stress) and the equivalent strain can be written as

$$\bar{\sigma} = \left(1 - \alpha + \alpha^2\right)^{\frac{1}{2}} \sigma_1 \tag{10.3}$$

and

$$\bar{\varepsilon} = \left[4\left(1 + \beta + \beta^2\right)/3\right]^{\frac{1}{2}} \varepsilon_1 \tag{10.4}$$

where

$$\alpha = \sigma_2/\sigma_1 \tag{10.5}$$

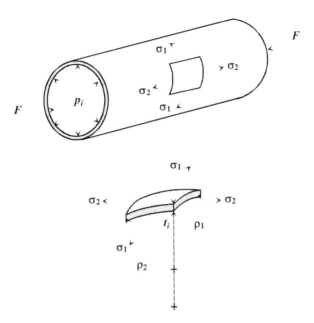

Figure 10.14 The stresses acting on an element at the *middle* of the tube. (Reprinted with permission from Asnafi, 1999.)

and

$$\beta = \varepsilon_2 / \varepsilon_1 \tag{10.6}$$

The tangential and radial strains, ε_1 and ε_3, can be denoted as

$$\varepsilon_1 = \ln\left(\rho_1 / \rho_0\right) \tag{10.7}$$

and

$$\varepsilon_3 = \ln\left(t_i / t_0\right) \tag{10.8}$$

where ρ_0 = initial tube radius, ρ_1 = instantaneous tube radius, t_0 = initial tube wall thickness, and t_i = instantaneous tube wall thickness.

The Levy–Mises flow rule yields (assuming volume constancy)

$$\alpha = \left(2\beta + 1\right) / \left(2 + \beta\right) \tag{10.9}$$

or

$$\beta = \left(2\alpha - 1\right) / \left(2 - \alpha\right) \tag{10.10}$$

Combining equations 10.1–10.3 and 10.5, one can write

$$p_i = \frac{\bar{\sigma} t_i}{\left(1 - \alpha + \alpha^2\right)^{1/2}} \cdot \left(\frac{1}{\rho_1} + \frac{\alpha}{\rho_2}\right) \tag{10.11}$$

and

$$F = p_i \pi \rho_i^2 \cdot \left(1 - \frac{2\alpha}{1 + \alpha\rho_1/\rho_2}\right) \qquad (10.12)$$

At the interface between elastic and plastic deformation, one can write or assume that

$$\rho_1 = (d_0 - t_0)/2 \qquad (10.13)$$

$$\rho_2 = \infty \qquad (10.14)$$

$$t_i = t_0 \qquad (10.15)$$

$$\bar{\sigma} = R_{p0.2} \qquad (10.16)$$

where d_0 = initial (outer) tube diameter and $R_{p0.2}$ = yield strength of the tube material.
    Substituting equations 10.13–10.16 into equations 10.11 and 10.12 yields

$$p_{iy} = \frac{2R_{p0.2}t_0}{(d_0 - t_0)\left(1 - \alpha + \alpha^2\right)^{\frac{1}{2}}} \qquad (10.17)$$

and

$$F_y = p_i \pi \cdot \frac{(d_0 - t_0)^2}{4} \cdot (1 - 2\alpha) \qquad (10.18)$$

In the analysis above, F is assumed to be equal to the forming force. To add the sealing and friction forces (which are not considered in the analysis above), assume that the hydroforming tool has the shape shown in Figure 10.15. Then, the forming force can be defined as

$$F_{forming} = F - F_{sealing} - F_{friction} \qquad (10.19)$$

The sealing force can be written as (Figure 10.15)

$$F_{sealing} = p_i \cdot \pi \cdot \left((d_0 - 2t_0)/2\right)^2 \qquad (10.20)$$

The friction force is determined by the normal stress, σ_N, the friction coefficient, μ, and the surface σ_N acts on (Figure 10.15)

$$F_{friction} = \mu \cdot \sigma_N \cdot \pi d_0 \cdot (l_0 - s) \cong \mu \cdot p_i \cdot \pi d_0 \cdot (l_0 - s) \qquad (10.21)$$

Note that $\sigma_N = p_i$, only if the tube is thin-walled. If $t_0/d_0 \leq 0.05$, the tube can theoretically be characterized as thin-walled.
    The friction force is, according to equation 10.21, strongly dependent upon the stroke, s, and the internal pressure, p_i (see also Figure 10.15). During the forming, $F_{friction}$ decreases with increasing s (more material is pushed into the expansion zone), while it increases with p_i (higher pressure is necessary to carry on with the plastic deformation of the tube). The combined effect of increasing s and p_i must, therefore, be considered.

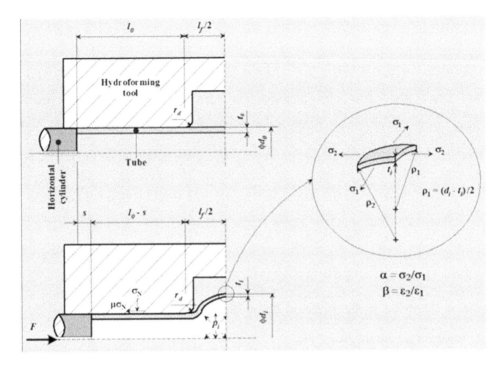

Figure 10.15 The hydroforming tool is assumed to have this shape. (Reprinted with permission from Asnafi, 1999.)

Substituting equation 10.19 into equation 10.12 and combining the result with equations 10.20 and 10.21 yields

$$F = p_i \pi \cdot \rho_1^2 \cdot \left(1 - \frac{2\alpha}{1 + \alpha \cdot \rho_1/\rho_2}\right) + p_i \pi \left((d_0 - 2t_0)/2\right)^2 + \mu \cdot p_i \cdot \pi d_0 \cdot (l_0 - s) \qquad (10.22)$$

At the yield limit, equations 10.13–10.16 are valid. Substituting these equations into equation 10.22 yields

$$F_y = p_i \cdot \pi \cdot \left(\left(\frac{(d_0 - t_0)}{2}\right)^2 \cdot (1 - 2\alpha) + \left(\frac{(d_0 - 2t_0)}{2}\right)^2 + \mu \cdot d_0 \cdot (l_0 - s_y)\right) \qquad (10.23)$$

To use equation 10.23, the magnitude of stroke, s, at the yield limit must also be calculated,

$$s_y = \Delta L^e/2 = (L_0 - L_y)/2 \qquad (10.24)$$

In equation 10.24, s_y = the stroke at the yield limit, L_0 = the initial tube length, and L_y = the tube length at the yield limit. One can write

$$\varepsilon_2^y = \ln(L_y/L_0) \qquad (10.25)$$

Therefore, equation 10.24 can be rewritten as

$$s_y = \frac{\Delta L^e}{2} = \frac{L_0 - L_0 e^{\varepsilon_2^y}}{2} = \frac{L_0}{2}\left(1 - e^{\varepsilon_2^y}\right) \tag{10.26}$$

where ε_2^y = axial strain at the yield limit. Based on equation 10.26, it is shown by Asnafi and Skogsgårdh (2000) that the following expressions can be used for calculation of the stroke at the yield limit:

$$s_y = -L_0\left(1 - e^{-vR_{p0.2}/E}\right)\beta \quad \text{at} -0.5 \leq \beta \leq 0 \tag{10.27}$$

and

$$s_y = \frac{L_0}{2}\left(1 - e^{-\frac{(v-\alpha)}{\sqrt{1-\alpha+\alpha^2}}\times\frac{R_{p0.2}}{E}}\right) \quad \text{at} -1 \leq \beta \leq -0.5 \tag{10.28}$$

Observe that the previously obtained expression for p_{iy} is still valid, equation 10.17.

Summarizing this section, equations 10.9, 10.10, 10.17, 10.23, 10.27, and 10.28 can be used to determine the yield limit in both force- and stroke-controlled hydroforming.

10.3.2 Plastic Deformation—$\rho_2 = \infty$

Assume that the tube expands in the fashion shown in Figure 10.16. This assumption means that $\rho_2 = \infty$ in Figures 10.14 and 10.15. With this assumption, equations 10.1 and 10.2 can be rewritten as

$$\frac{\sigma_1}{\rho_1} = \frac{p_i}{t_i} \tag{10.29}$$

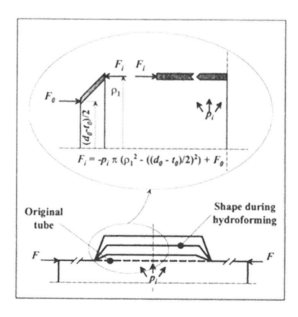

Figure 10.16 The expansion (the plastic deformation) is assumed to occur in this fashion ($\rho_2 = \infty$ in Figure 10.15). This assumption results in the force equilibrium shown in the upper part of the figure. (Reprinted with permission from Asnafi, 1999.)

and

$$\sigma_2 = \frac{p_i \rho_1}{2t_i} - \frac{F_i}{2\pi \rho_1 t_i} \tag{10.30}$$

Utilizing the force equilibrium in Figure 10.16, it can be shown that

$$F_i = -p_i \cdot \pi \cdot \left(\rho_1^2 - \left(\frac{d_0 - t_0}{2} \right)^2 \right) + F_0 \tag{10.31}$$

F_0 can be denoted as (Figures 10.14 and 10.16)

$$F_0 = F - F_{\text{sealing}} - F_{\text{friction}} \tag{10.32}$$

Combining equation 10.32 with equations 10.20 and 10.21 and substituting the result and equation 10.31 into equation 10.30 yield

$$\sigma_2 = \frac{p_i \rho_1}{2t_i} - \frac{1}{2\pi \rho_1 t_i} \left(\begin{array}{c} F - p_i \pi \left(\dfrac{(d_0 - 2t_0)}{2} \right)^2 \\[2ex] -p_i \pi \left(\rho_1^2 - \left(\dfrac{(d_0 - t_0)}{2} \right)^2 \right) \\[2ex] -\mu \cdot p_i \cdot \pi \cdot d_0 \cdot (l_0 - s) \end{array} \right) \tag{10.33}$$

Combining $\rho_1 = (d_i - t_i)/2$, in which d_i is the instantaneous tube diameter, with equation 10.29, one can obtain

$$p_i = \frac{2t_i \cdot \sigma_1}{d_i - t_i} \tag{10.34}$$

Substituting equations 10.5 and 10.34 into equation 10.33 yields

$$F = p_i \cdot \pi \cdot \left(\frac{1-\alpha}{2} \cdot (d_i - t_i)^2 - t_0 \cdot \frac{(2d_0 - 3t_0)}{4} + \mu \cdot d_0 \cdot (l_0 - s) \right) \tag{10.35}$$

Combining equations 10.3 and 10.34 yields

$$p_i = \frac{2t_i \bar{\sigma}}{(d_i - t_i)\sqrt{1 - \alpha + \alpha^2}} \tag{10.36}$$

Assume that the tube material obeys the Ludwik–Hollomon hardening relationship

$$\bar{\sigma} = K (\bar{\varepsilon})^n \tag{10.37}$$

Combining equations 10.4 and 10.37 with equation 10.36, one can obtain

$$p_i = \frac{2t_i K \varepsilon_1^n}{(d_i - t_i)\sqrt{1 - \alpha + \alpha^2}} \left(\sqrt{4(1 + \beta + \beta^2)/3} \right)^n \tag{10.38}$$

Substituting

$$\varepsilon_1 = \ln\left(\rho_1/\rho_0\right) = \ln\left[\left(d_i - t_i\right)/\left(d_0 - t_0\right)\right], \tag{10.39}$$

and equation 10.10 into equation 10.38 yields

$$p_i = \frac{2t_i}{d_i - t_i} K\left(\frac{2}{2 - \alpha}\right)^n \left(\sqrt{1 - \alpha + \alpha^2}\right)^{n-1} \left(\ln \frac{d_i - t_i}{d_0 - t_0}\right)^n \tag{10.40}$$

Assume now that

$$\varepsilon_1 + \varepsilon_2 + \varepsilon_3 = 0 \tag{10.41}$$

Combining equations 10.6–10.8 and $\rho_1 = \left(d_i - t_i\right)/2$ with equation 10.41, one can write

$$t_i = t_0 \cdot \left(\frac{d_i - t_i}{d_0 - t_0}\right)^{-(1+\beta)} \cong t_0 \cdot \left(\frac{d_i}{d_0}\right)^{-(1+\beta)} \tag{10.42}$$

Looking at equations 10.35, 10.40, and 10.42, one can see that all of the necessary parameters are known or can be calculated, except the instantaneous stroke, s. In the following, an expression is derived for s.

The initial tube material volume is as follows (Figure 10.17):

$$V_0 = \pi\left(d_0 t_0 - t_0^2\right) L_0, \tag{10.43}$$

where V_0 = initial tube material volume and L_0 = initial tube length.

The instantaneous tube material volume can be denoted as (Figure 10.17)

$$V_i = K_A\left(l_0 - s\right) + V_B + V_C, \tag{10.44}$$

where

$$K_A = 2\pi\left(d_0 t_0 - t_0^2\right), \tag{10.45}$$

$$V_B = 2\pi\left(d_i t_i - t_i^2\right) C, \tag{10.46}$$

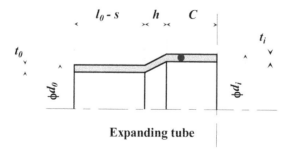

Expanding tube

Figure 10.17 The parameters used in derivation of an expression for the instantaneous stroke, s. (Reprinted with permission from Asnafi et al., 2000.)

and

$$V_C = \frac{\pi h}{12}\left[d_i^2 + d_i d_0 + d_0^2 - (d_i - 2t_i)^2 + (d_i - 2t_i)(d_0 - 2t_0) + (d_0 - 2t_0)^2\right]$$ (10.47)

Assuming volume constancy, equations 10.43–10.47 yield

$$s = l_0 - (V_0 - V_B - V_C)/K_A$$ (10.48)

Equations 10.35, 10.40, and 10.42–10.48 can be used to determine the force- or stroke-controlled loading path in the plastic region provided that the expansion occurs as displayed in Figure 10.16.

10.3.3 Plastic Deformation—$\rho_2 < \infty$

Assume now that the expansion occurs in the fashion shown in Figure 10.18. This assumption means that $\rho_2 < \infty$ in Figure 10.15. In this case, equations 10.1, 10.2, and 10.11 are still valid. Combining equation 10.4, $\rho_1 = (d_i - t_i)/2$, and equations 10.37 and 10.39 with equation 10.11, one can obtain

$$p_i = \left(\frac{2}{2-\alpha}\right)^n \cdot \left(\sqrt{1-\alpha+\alpha^2}\right)^{n-1} \cdot K \cdot t_i \cdot \left(\ln\frac{d_i - t_i}{d_0 - t_0}\right)^n \cdot \left(\frac{2}{d_i - t_i} + \frac{\alpha}{\rho_2}\right)$$ (10.49)

where

$$\rho_2 = \frac{(l_f - 2r_d)^2 + 4h^2}{8h}$$ (10.50)

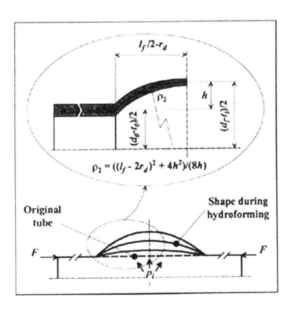

Figure 10.18 In this section, it is assumed that the expansion occurs in this fashion ($\rho_2 < \infty$ in Figure 10.15). (Reprinted with permission from Asnafi, 1999.)

h can, in turn, be denoted as follows:

$$h = \frac{(d_i - t_i) - (d_0 - t_0)}{2} \tag{10.51}$$

ρ_2 and h are represented in Figure 10.18.

Combining equations 10.12, 10.20, 10.21, 10.31, and 10.32 gives

$$F = p_i \cdot \pi \cdot \left[\left(1 - \frac{2\alpha\rho_2}{2\rho_2 + \alpha(d_i - t_i)} \right) \cdot \frac{(d_i - t_i)^2}{2} - t_0 \cdot \frac{(2d_0 - 3t_0)}{4} + \mu \cdot d_0 \cdot (l_0 - s) \right] \tag{10.52}$$

where ρ_2 is given by equation 10.50.

Utilizing Figure 10.18 and assuming volume constancy, it can easily be shown that

$$s = \frac{2h}{3} \cdot \left(\frac{l_f - 2r_d}{d_0 - t_0} + \frac{2h}{l_f - 2r_d} \right) \tag{10.53}$$

Equations 10.49–10.53 can be used to determine the force- or stroke-controlled loading path in the plastic region provided that the expansion occurs as displayed in Figure 10.18.

10.3.4 Significance of the Analytical Models of Free Forming

Table 10.1 displays the assumed tube and tool parameter values (see also Figures 10.15 and 10.17). These values and the mechanical properties shown in Table 10.2 are used in this section to examine the analytical models constructed above. *If not mentioned, the values given in* Table 10.1 *are used.*

Figure 10.19 is based on equations 10.10, 10.17, 10.20, 10.23, 10.27, 10.28, 10.35, 10.40, 10.42, and 10.48. Figures 10.20 and 10.21 are based on equations 10.10, 10.40, and 10.42. In all three figures, the material is hot-dip galvanized (HG/Z140) FeP06 (DX56) and the mechanical properties of the sheet are used (Table 10.2). Assume that there are no friction forces acting on the tube ($\mu = 0$). Assume too that the expansion occurs in the fashion shown in Figure 10.16 ($\rho_2 = \infty$). Then, Figure 10.19 displays the yield limit and the loading characteristics in hydroforming along different strain paths, β (stress modes (α)). In all of the cases shown in Figure 10.19, the axial force increases very rapidly at the end of free forming. This behavior is caused by instability. The free forming process becomes unstable when the internal pressure reaches a maximum.

Figure 10.20 displays the internal pressure versus the expansion, $((d_i - d_0)/d_0) \cdot 100$. As displayed in this figure, the internal pressure reaches a maximum at the end of free forming. At this point, the process becomes unstable. Figure 10.21 shows the instantaneous radius versus the internal pressure.

Table 10.1   Assumed Tube and Tool Parameter Values

Parameter	Designation (Figures 10.15 and 10.17)	Assumed Value
l_0	Initial contact length	215 mm
l_f	Free tube length	120 mm
r_d	Die profile radius	12.5 mm
d_0	Initial (outer) tube diameter	60 mm
v	Poisson's ratio	0.3
c	Length of the expansion zone	40 mm
h	Length of the transition zone	20 mm
L_0	Initial tube length	550 mm

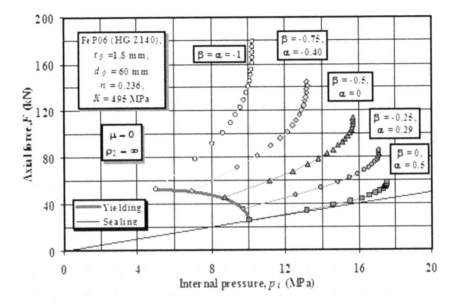

Figure 10.19 Yield limit and plastic deformation in hydroforming along different strain paths, β (stress modes (α)). $\rho_2 = \infty$ and $\mu = 0$. The figure is based on equations 10.10, 10.17, 10.20, 10.23, 10.27, 10.28, 10.35, 10.40, 10.42 and 10.48. See also Figures 10.15 and 10.16. (Reprinted with permission from Asnafi, 1999.)

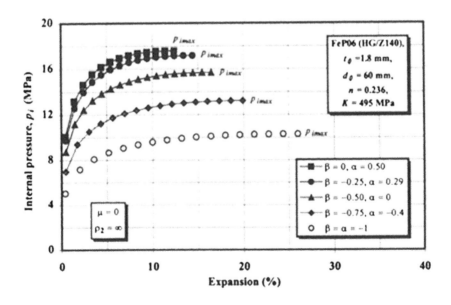

Figure 10.20 Internal pressure versus expansion, $((d_i - d_0)/d_0)\cdot 100$, at different strain modes, β (stress modes(α)). $\rho_2 = \infty$ and $\mu = 0$. The figure is based on equations 10.10, 10.40 and 10.42. See also Figures 10.15 and 10.16. (Reprinted with permission from Asnafi, 1999.)

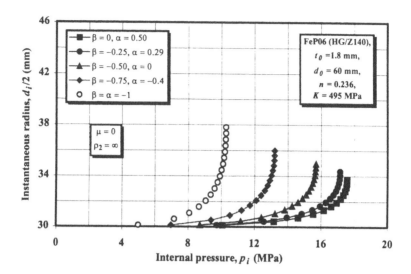

Figure 10.21   Instantaneous radius versus internal pressure. β and α are strain and stress modes, respectively. $\rho_2 = \infty$ and $\mu = 0$. The figure is based on equations 10.10, 10.40 and 10.42. See also Figures 10.15 and 10.16. (Reprinted with permission from Asnafi, 1999.)

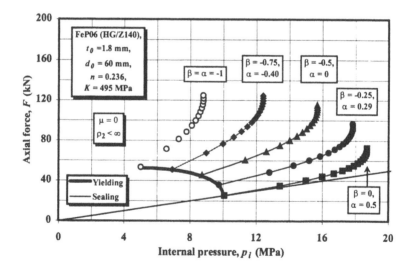

Figure 10.22   Yield limit and plastic deformation in hydroforming along different strain paths, β (stress modes (α)). $\rho_2 < \infty$ and $\mu = 0$. The figure is based on equations 10.10, 10.17, 10.20, 10.23, 10.27, 10.28, 10.49–10.53. See also Figures 10.15 and 10.16. (Reprinted with permission from Asnafi, 1999.)

This pressure reaches its maximum at different instantaneous radii, the latter being smallest at plane strain ($\beta = 0$) and largest at pure shear ($\beta = -1$).

Based on the assumption that $\rho_2 < \infty$ and $\mu = 0$ (equations 10.10, 10.17, 10.20, 10.23, 10.27, 10.28, and 10.49–10.53), Figure 10.22 shows the yield limit and plastic deformation in hydroforming along different strain paths, β (stress modes (α)). The material is hot-dip galvanized (HG/Z140) FeP06 (DX56) and the mechanical properties of the sheet are used (Table 10.2).

Table 10.2   Mechanical Properties of the Steel Sheets from Which the Tubes Were Made

Material	Orientation[a] (°)	t_0 (mm)	$R_{p0.2}$ (MPa)	R_m (MPa)	$A_{uniform}$ (%)	A_{80}[b] (%)	n	K (MPa)	r	E[c] (MPa)
FeP06 (HG/Z140)	90	1.80	141	284	25.3	48.6	0.236	495	1.97	205,000

Source: Data with permission from Asnafi (1999).
[a] With respect to the rolling direction.
[b] Gauge length = 80 mm.
[c] Determined by tensile testing.

A comparison of this assumption ($\rho_2 < \infty$) with the previous assumption ($\rho_2 = \infty$) shows that

- the same result is obtained at uniaxial tension ($\beta = -0.5$) regardless of the assumed intermediate tube shape ($\rho_2 = \infty$ or $\rho_2 < \infty$).
- instability occurs at higher internal pressures at $\beta = -0.75$ and $\beta = -1$, if it is assumed that $\rho_2 = \infty$. The same assumption leads, however, to lower instability pressures at $\beta = 0$ and $\beta = -0.25$.
- the instantaneous radius at instability is smaller at $\beta = 0$ and $\beta = -0.25$, if it is assumed that $\rho_2 = \infty$. The same assumption results, however, in a larger instantaneous radius at instability at $\beta = -0.75$ and $\beta = -1$.

It is assumed that the tube is made by bending and welding of rectangular sheets. The mechanical properties of the initial sheet and the tube are displayed in Tables 10.2 and 10.3, respectively. Using these properties, the yield limits and the plastic loading paths shown in Figure 10.23 are obtained. It is assumed in Figure 10.23 that the plastic deformation occurs at uniaxial tension ($\beta = -0.5$, $\alpha = 0$). This figure is, therefore, valid regardless of the value of ρ_2. Figure 10.23 is based on equations 10.17, 10.23, 10.27, and 10.28 combined with equations 10.35, 10.40, 10.42–10.48, or equations 10.49–10.53. It is, furthermore, assumed in Figure 10.23 that $\mu = 0$. As displayed in Figure 10.23, the difference between the yield limits is relatively large, while the difference between the plastic loading paths is relatively small.

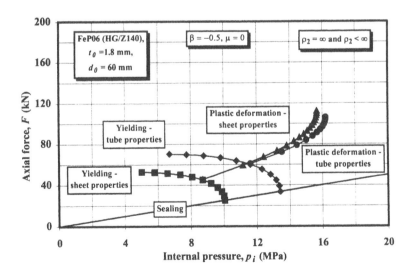

Figure 10.23   Sheet versus tube: the influence of used mechanical properties on the yield limit and the plastic loading path. See also Tables 10.2 and 10.3. $\mu = 0$ and $\beta = -0.5$ ($\alpha = 0$; uniaxial tension) during plastic deformation. The figure is valid regardless of the value of ρ_2. The figure is based on equations 10.17, 10.23, 10.27, and 10.28 combined with equations 10.35, 10.40, 10.42–10.48, or equations 10.49–10.53. (Reprinted with permission from Asnafi, 1999.)

Figure 10.24 shows how friction affects the yield limit and the plastic loading path. In this figure, uniaxial tension ($\beta = -0.5$, $\alpha = 0$) is assumed to be prevailing during the plastic deformation.

Figure 10.24 is, therefore, valid regardless of the value of ρ_2. This figure is based on equations 10.17, 10.23, 10.27, and 10.28 combined with equations 10.35, 10.40, 10.42–10.48, or equations 10.49–10.53. As exhibited in Figure 10.24, $\mu = 0.04$ requires higher axial forces than $\mu = 0$. However, the plastic loading paths are parallel (Figure 10.24).

It is assumed that μ is constant during the hydroforming operation (equation 10.21 and Figure 10.24). Assuming that equation 10.21 is valid, it can, however, be shown that μ is strongly dependent upon the internal pressure, feeding velocity (the rate at which the material is fed into the expansion zone), the tube material, the tube material surface roughness, the tool surface roughness, the lubricant, etc. In some cases, μ is constant. However, there are also cases where μ exhibits a drastic variation during the hydroforming. It is also worth noting that selecting a tube length that is as short as possible can reduce the influence of the friction forces (Figure 10.15).

10.3.5 Finite-Element Simulation of Free Forming

The CAD model of the tool was made in CATIA. The subsequent meshing was performed in DeltaMESH Stamping. The rest of the preprocessing was carried out in LS-Ingrid. The finite-element simulations were conducted using the explicit code LS-Dyna. In these simulations, the parallel version of LS-Dyna was used. The obtained results were studied in the postprocessor LS-Taurus (Asnafi and Skogsgårdh, 2000).

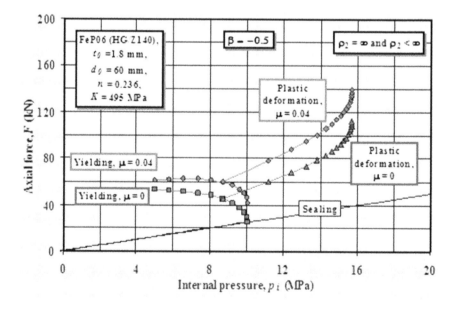

Figure 10.24 The influence of friction coefficient, μ, on the yield limit and the plastic loading path. $\beta = -0.5$ ($\alpha = 0$; uniaxial tension) during plastic deformation. The figure is valid regardless of the value of ρ_2. The figure is based on equations 10.17, 10.23, 10.27, and 10.28 combined with equations 10.35, 10.40, 10.42–10.48, or equations 10.49–10.53. (Reprinted with permission from Asnafi, 1999.)

The finite-element model consisted of the tool with 4,528 elements and a straight tube (tube diameter is 60 mm, tube length is 220 mm) with 6,512 elements (Figure 10.25). The tool was modeled as a rigid body. The used tube material model was transversely anisotropic elastic plastic. This latter model uses the yield criterion suggested by Hill in 1948. The friction coefficient was set at zero (Asnafi and Skogsgårdh, 2000).

The loading path (variation of the stroke with the internal pressure) was first calculated by the analytical model constructed above equations 10.9, 10.10, 10.40, and 10.42–10.48. This loading path was then used in the finite-element simulation (input to the finite-element simulation) (Asnafi and Skogsgårdh, 2000).

Figure 10.26 displays the yield limit and the loading paths (plastic deformation at different strain modes β) obtained analytically (equations 10.9, 10.10, 10.40, and 10.42–10.48) for material FeP06 and used as input to the finite-element simulations. The sheet properties, Table 10.2, were used in calculation of the values shown in Figure 10.26 (Asnafi and Skogsgårdh, 2000).

In the simulations, fracture was assumed to occur as soon as ε_2 at the pole (top of the bulge) did not grow in magnitude any longer (Asnafi and Skogsgårdh, 2000). As displayed in Figure 10.26, the plastic deformation is initiated at such small strokes that the position of the yield limit is theoretically irrelevant. This feature constitutes one of the major differences between stroke-controlled and force-controlled hydroforming. Compare Figure 10.26 with Figures 10.23 and 10.24 (with respect to the yield limit).

Looking at the loading paths in Figure 10.26, one can observe an asymptotic increase in the stroke at the end of forming (particularly at small strain ratios). This asymptotic increase in stroke is caused by instability. The free forming process becomes unstable, when the internal pressure reaches a maximum (refer also Figure 10.20).

Figure 10.27 shows the major true strain versus the minor true strain (at the top of the bulge marked with an A in Figure 10.10) obtained by finite-element simulations. If the legend (which shows the analytically desired β) is compared to the plotted finite-element values in Figure 10.27, one can see that the finite-element simulation predicts smaller strain ratios than the analytical model, particularly at the beginning of the hydroforming operation. One can also note that the strain paths attained by finite-element simulations are not linear, Figure 10.27.

Figure 10.28 displays the shape of the expansion zone at different internal pressures. This figure is obtained by finite-element simulation. The (analytically) selected strain path (desired strain path)

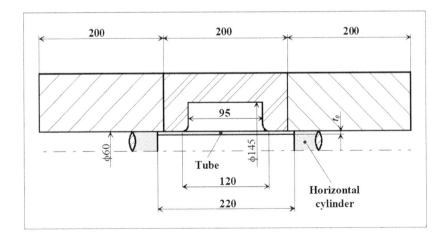

Figure 10.25 These tool and tube values were used in the simulations. (Reprinted with permission from Asnafi et al., 2000.)

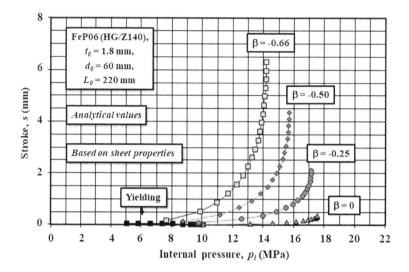

Figure 10.26 Stroke versus internal pressure. The yield limit is calculated in accordance to equations 10.9, 10.10, 10.17, 10.24, 10.27, and 10.28. The loading paths (plastic deformation) are computed by equations 10.9, 10.10, 10.40, and 10.42–10.48. Material = hot-dip galvanized (HG-Z140) FeP06. Sheet properties were used, Table 10.2. β is strain ratio. t_0, d_0, and L_0 are the initial tube wall thickness, tube diameter, and tube length, respectively. (Reprinted with permission from Asnafi et al., 2000.)

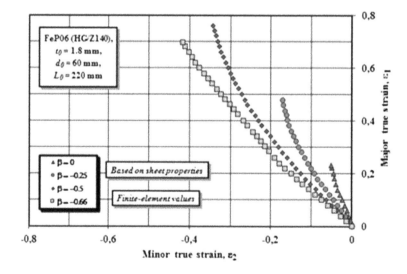

Figure 10.27 Major true strain versus minor true strain (at the top of the bulge). The figure shows the obtained finite-element values. Material = hot-dip galvanized (HG-Z140) FeP06. Sheet properties, Table 10.2, were used in the simulations. β is the analytically targeted strain ratio. t_0, d_0, and L_0 are the initial tube wall thickness, tube diameter, and tube length, respectively. (Reprinted with permission from Asnafi et al., 2000.)

was uniaxial tension ($\beta = -0.5$). The intermediate shape of the tube corresponds, as exhibited in Figure 10.28, to that assumed in Figure 10.16 from 0 up to 18 MPa in internal pressure. However, a drastic expansion occurs from 18 to 20.5 MPa. This drastic expansion results in an intermediate shape that differs from that assumed in Figure 10.16. This drastic expansion corresponds better to

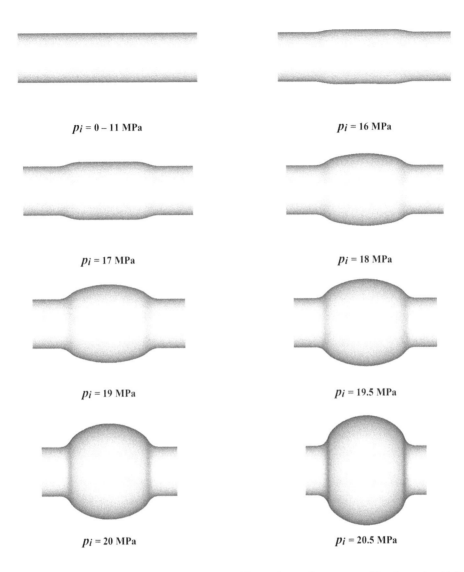

$p_i = 0 - 11$ MPa

$p_i = 16$ MPa

$p_i = 17$ MPa

$p_i = 18$ MPa

$p_i = 19$ MPa

$p_i = 19.5$ MPa

$p_i = 20$ MPa

$p_i = 20.5$ MPa

Figure 10.28 The shape of the expansion zone at different internal pressures. The figure is obtained by finite-element simulation. Material is hot-dip galvanized (HG-Z140) FeP06 (sheet properties, Table 10.2, were used). The loading path used in the finite-element simulation was calculated analytically, using equations 10.9, 10.10, 10.40 and 10.42–10.48. The (analytically) selected strain path = uniaxial tension ($\beta = -0.5$). The fracture is, according to the finite-element model, predicted to occur at $p_i = 20.5$ MPa. (Reprinted with permission from Asnafi et al., 2000.)

Figure 10.18 and leads also to a strain path change. This explains the nonlinear strain path obtained for (the analytically targeted) $\beta = -0.5$ plotted in Figure 10.27.

10.4 MATERIALS

Tables 10.2–10.4 display the mechanical properties of the materials at the focus in this chapter. In these tables, t_0 = sheet or tube wall thickness, $R_{p0.2}$ = yield strength, R_m = ultimate tensile strength, $A_{uniform}$ = uniform elongation, A_{80} (A_{50} or A_{60}) = total elongation (the value stands for the gauge length in tensile testing), n = strain-hardening exponent, K = strength coefficient, r = plastic strain

Table 10.3   Mechanical Properties of the Steel Tubes

Material	Orientation[a] (°)	t_0 (mm)	$R_{p0.2}$ (MPa)	R_m (MPa)	$A_{uniform}$ (%)	A_{60}[c] (%)	n	K (MPa)	r[d]	E[e] (MPa)
FeP06 (HG/Z140)	0[b]	1.80	188	293	25.8	44.0	0.168	447	1.00	205,000

Source: Data with permission from Asnafi (1999).
Note: Initial outer tube diameter = 60 mm.
[a] Longitudinal direction (along the tube).
[b] With respect to the rolling direction of the original sheet.
[c] Gauge length = 60 mm.
[d] Assumed value.
[e] Determined by tensile testing.

Table 10.4   Mechanical Properties of the Extruded Aluminum Profile

Material	Orientation[a] (°)	t_0 (mm)	$R_{p0.2}$ (MPa)	R_m (MPa)	$A_{uniform}$ (%)	A_{50}[b] (%)	n	K (MPa)	r	E[c] (MPa)
AA6063-T4	0	2.5	76	159	16.9	21.2	0.259	295	0.517	69,000
	45	2.5	97	170	18.4	22.1	0.212	258	0.247	69,000
	90	2.5	88	162	21.1	27.6	0.215	264	2.53	69,000

Source: Data from Asnafi et al. (2003).
Note: Initial outer tube diameter = 110 mm.
[a] Testing direction with respect to the extrusion direction.
[b] Gauge length = 50 mm.
[c] Determined by tensile testing.

ratio, and E = Young's modulus. It is assumed that the materials in these tables obey the Ludwik–Hollomon hardening relationship (equation 10.37).

Tube materials are manufactured in different manners. The steel tube in Table 10.3 was made by bending and laser welding of rectangular sheets. These sheets have the properties shown in Table 10.2. The longitudinal direction of the rectangular sheet coincided with the rolling direction. On a tube made in this fashion, the deformation direction during hydroforming is perpendicular to the rolling direction of the initial sheet. Table 10.2 displays, therefore, the mechanical properties of the initial sheet in the 90° direction.

To determine the mechanical properties of the steel tubes, tensile specimens were cut by water-jet from the tubes and tested. These specimens were cut along the longitudinal direction of the tube. In the case of tubes made from sheets, this means that the testing direction coincided with the rolling direction of the initial sheet. Table 10.3 shows the mechanical properties of the steel tube. The r-value was not determined when testing these specimens. Note, therefore, that Table 10.3 shows the assumed r-value.

The aluminum tube AA6063-T4 in Table 10.4 was extruded. The tensile specimen cut 45° and 90° to the extrusion direction were straightened out before tensile testing. These specimens were therefore pre-deformed approximately 2.3% prior to tensile testing. The grain size varied between 56 and 77 μm in the extrusion direction and between 48 and 61 μm perpendicularly to the extrusion direction.

Figure 10.29 displays the forming limit curves of 1.5 and 2.5-mm thick tubes (outer diameter = 70 mm) of AA6063-T4 determined by hydroforming (Asnafi et al., 2000). These curves are established by testing the tube material in a tool set of the type shown in Figure 10.30.

10.5 INDUSTRIAL TUBE HYDROFORMING

As mentioned above, hydroforming is in this chapter defined as a forming operation, in which a tube is expanded radially while compressed axially by two cylinders (Figure 10.5). The purpose

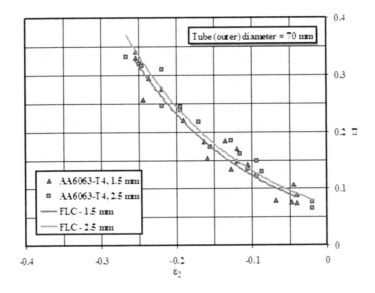

Figure 10.29 The forming limit curves of 1.5 and 2.5-mm thick tubes (outer diameter = 70 mm) of AA6063-T4 determined by hydroforming. (Reprinted with permission from Asnafi et al., 2000.)

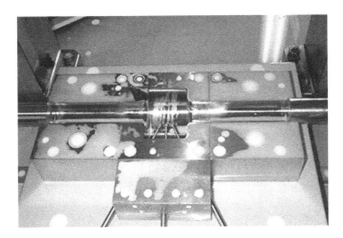

Figure 10.30 The tool used for determination of the forming limit curve of tube materials with an outer diameter of 60 by tubular hydroforming. The figure shows lower tool set, the axial cylinders, and the three sensors used to measure the incremental change in axial tube curvature during hydroforming. (Reprinted with permission from Asnafi and Andersson, 2001.)

of axial compression is to feed in material to the expanding regions to reduce the tube wall thinning during the expansion (and thereby the risk of bursting) significantly. This axial compression is however possible in case the component is straight and relatively short.

In other words, the amount of material inflow varies dependent upon the position of the expansion zone in relation to the position of the axial cylinders. The obtained strain ratio, β (equation 10.6), is therefore dependent upon the position of the expansion zone in relation to the position of the axial cylinders. $\beta = 0$ (plane strain) in an expansion zone at the base of the automotive subframe shown in Figure 10.13. In an expansion zone near the axial cylinders in Figure 10.13, it is however possible to obtain $\beta < -0.5$ by feeding material into this zone.

Except the position of the expansion zone, the overall component shape plays also a significant role. In many cases, the tube needs to be bent before hydroforming. This means that the tube is pre-deformed (pre-strained) in some areas before it is deformed further during the hydroforming. The hydroformability of the tube material is, in other words, dependent upon whether the tube is deformed (pre-strained) before hydroforming.

Figure 10.31, which shows a roof rail, constitutes another example. The position of the expansion zone in relation to the position of tube ends (position of the axial cylinders) determines the possible strain ratio, β. The overall shape of the component determines the need of preforming (axial and radial bending) which in turn affects how much the tube can be deformed by hydroforming.

Within the engineering and automotive industry, hydroforming of tubular blanks may therefore comprise (Figure 10.31):

1. Tube bending
2. Preforming (radial bending) by tool closure. The preforming (tool closure) can/might be conducted after a low internal pressure is built up.
3. Hydroforming
4. Post-processing (normally end trimming and/or hole-making)

Component design, manufacturing engineering, and prototyping comprise therefore a cross-section and a longitudinal-section analysis. Formability limits can be avoided by a proper component design. This analysis is addressed in the next section.

Formability limit curve determined by tube hydroforming shown in Figure 10.29 needs to be used in the cross- and longitudinal-section analysis and as failure criterion in FE simulations.

In general, and as rules of thumb, the following can be mentioned and need to be considered:

• If/when it is possible to accomplish a large axial material inflow, then the maximum perimeter enlargement (expansion) $\gg A_{\text{uniform}}$ (see also Tables 10.2–10.4).
• If/when only a small axial material inflow is possible, then the maximum perimeter enlargement (expansion) $= A_{\text{uniform}}$ (see also Tables 10.2–10.4).
• If/when no axial material inflow is possible, then the maximum perimeter enlargement (expansion) $= 0.5 \cdot A_{\text{uniform}}$ (see also Tables 10.2–10.4).

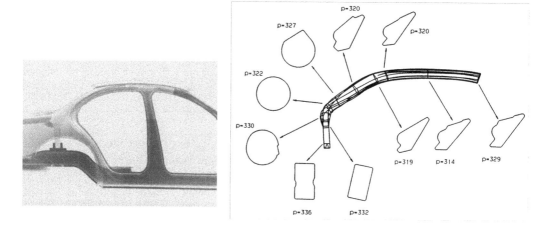

Figure 10.31   Hydroformed side roof rail. The shape of the side roof rail varies in several zones according to its function. p = perimeter in mm. (Reprinted with permission from Final Report of Ultra Light Steel Auto Body Consortium (ULSAB), *Porsche Engineering Services Inc.*, 1996.)

The required internal pressure is dependent on the material properties, the tube wall thickness, and the minimum internal component radius (see Figure 10.32). The values given in this figure can be considered as guiding indicators or "rules of thumb". The higher the internal pressure during hydroforming, the larger force (press force) is required to keep the hydroforming tool closed (Figure 10.5).

10.5.1 Component Design, Manufacturing Engineering, and Prototyping

In the following sections, component design, manufacturing engineering, and prototyping will be discussed using two (actually three) developed automotive side members.

10.5.1.1 The Swan Neck

Figure 10.33 shows an automotive side member, a so-called Swan Neck. Tube bending and hydroforming were used to make this component. The starting tube was a straight aluminum extrusion AA6063-T4 with a circular cross-section. The initial tube diameter and wall thickness were 107 and 2.5 mm. The tube diameter was however increased to 110 mm after completion of the analysis described below. This straight tube was first bent and then placed in the hydroforming tool.

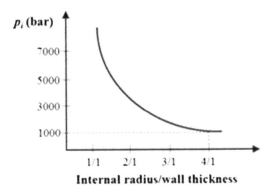

Figure 10.32   The required internal pressure is dependent on the material properties, the tube wall thickness, and the minimum internal component radius.

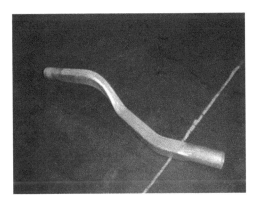

Figure 10.33   The car body side member (Swan Neck) made by bending and hydroforming. The initial tube was a straight extruded aluminum profile AA6063-T4 with a circular cross-section.

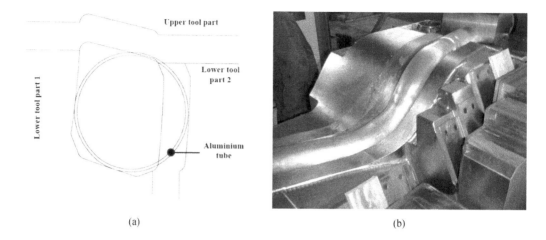

(a)   (b)

Figure 10.34   The side member Swan Neck was made in this hydroforming tool: (a) a section of the hydroforming tool (both when the tool is open and as it is closed) and a section of the bent tube placed
in the open hydroforming tool, (b) the bent tube is placed in the tool prior to tool closure and
hydroforming.

Figure 10.34 shows a section of the bent tube and the hydroforming tool (both when the tool
is open and as it is closed). After placing the bent tube in the tool (Figure 10.34), the tube was
hydroformed.

The whole forming procedure comprised the following:

   I. The straight tube was bent axially at five positions.
  II. The bent tube was placed in the hydroforming tool.
 III. A preforming internal pressure of 30–35 bar was built up.
  IV. The tool was closed and the press force (closing force) increased to 1,800 metric tons.
   V. The internal pressure was increased to 950 bar. The final maximum stroke was 57 mm from the right
      tube end and 2.5 mm from the left tube end in Figure 10.33.

To obtain the Swan Neck in Figure 10.33 using the procedure and load path described in I–V
above, a cross- and longitudinal-section analysis and finite-element simulations were first conducted.

Figure 10.35 shows some of the results of the conducted cross- and longitudinal-section analysis. The figure displays the results of the perimeter measurements conducted at 18 sections on a bent
tube, a preformed tube and a bent, preformed, and hydroformed tube.

The cross- and longitudinal-section analysis is of great importance. This analysis enables avoidance of wrinkling and bursting or severe thinning. It is of substantial value in obtaining good shape
accuracy (minimization of the springback). The connection between this analysis and shape accuracy is addressed further below.

The whole process was also modeled by finite-element simulation using the explicit code
LS-Dyna. The simulation results were used to identify the critical zones (the zones at which the risk
of fracture was high) where tool adjustments could be regarded as necessary. Figure 10.36 shows the
effective plastic strains after tube bending and hydroforming predicted by finite-element simulation.

As the result of the analysis described above, the initial tube diameter was increased from 107
to 110 mm. After this diameter increase, the Swan Neck in Figure 10.33 was made by the procedure and loading path described in I–V above. New measurements and analyses were conducted.
Figure 10.37 displays some of the strain measurement results obtained in this analysis. The analysis
led to approval of the Swan Neck in Figure 10.33.

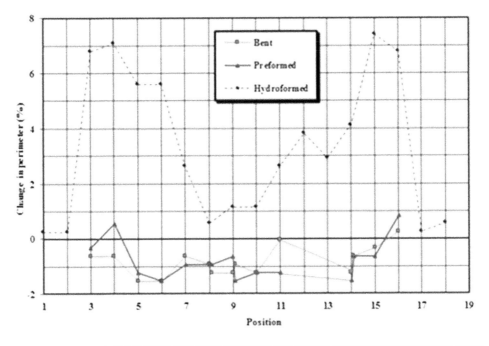

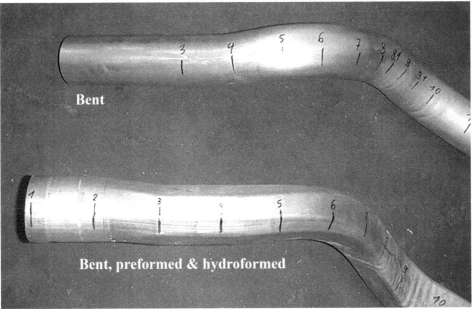

Figure 10.35   The cross- and longitudinal-section analysis conducted for the automotive side member Swan Neck. The photo shows a bent tube and a bent, preformed, and hydroformed tube. The perimeter has been determined at the positions indicated on the tubes.

It is worth noting that the major portion of the cross- and longitudinal-section analysis can be conducted already in the design stage and the finite-element simulation of the forming operation can be carried out latest during the manufacturing engineering phase in the product creation process.

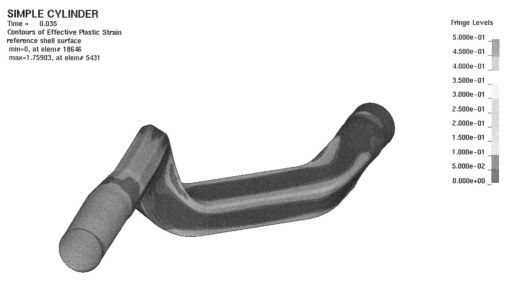

SIMPLE CYLINDER
Time =   0.035
Contours of Effective Plastic Strain
reference shell surface
min=0, at elem# 18646
max=1.75903, at elem# 5431

Fringe Levels

5.000e-01
4.500e-01
4.000e-01
3.500e-01
3.000e-01
2.500e-01
2.000e-01
1.500e-01
1.000e-01
5.000e-02
0.000e+00

Figure 10.36   The effective plastic strains obtained by finite-element simulation of the bending, preforming, and hydroforming of the side member Swan Neck. See also Figures 10.33–10.35.

10.5.1.2 Side Members in the Underbody Structure

In this section, the manufacturing engineering and prototyping of the Side Member Right and Side Member Left depicted in Figure 10.2 are used to illustrate hydroforming further. This section deals specifically with some issues related to tube bending prior to hydroforming, preforming, required minimum deformation at each component section obtained in hydroforming, geometrical spread after hydroforming, and the impact of hydroforming tool separation.

The Swan Neck in Figure 10.33 is very similar (almost identical) to the front end of Side Member Left in Figure 10.2. The contents of the previous section are, in order words, also valid even in this section both qualitatively and quantitatively.

The final forming procedure (for both side members in Figure 10.2) comprised the following (Asnafi et al., 2000, 2002):

I. The straight tube was rotary-draw bent axially at ten zones.
II. The bent tube was placed in the hydroforming tool.
III. A preforming internal pressure of 30–35 bar was built up.
IV. The tool was closed and the press force (closing force) increased to 9,600 metric tons.
V. The internal pressure was increased to 1,300 bar. The front and rear end strokes were 18 and 32 mm, respectively, during the hydroforming operation.

Figure 10.38a displays the principle of rotary-draw bending with and without application of an axial force. The magnitude of the applied axial force has a significant effect on the strains during tube bending. As shown in Figure 10.38b, the maximum strain on the outer surface of the bent region decreases dramatically with increasing axial force. However, the strain on the inner surface increases significantly with increasing axial force. Note that the maximum strain on the outer surface is tensile, while the maximum strain on the inner surface is compressive. Figure 10.38b signifies that if a sufficiently high axial force is applied at the tube end, while the tube is rotary-draw bent, the amount of thinning on the outer surface is decreased significantly and the tube thickens intensely on the inner surface.

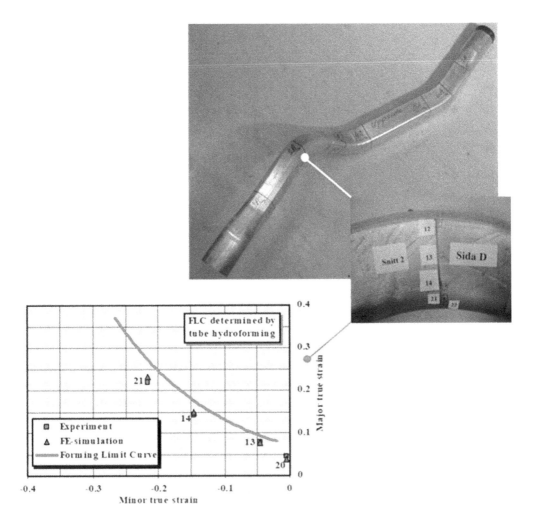

Figure 10.37   Strain measurements at positions 13, 14, 20 and 21 at Section 2 (Snitt 2) on Side D (Sida D) on the bent, preformed, and hydroformed Swan Neck.

Avoidance of severe thinning is considered as more important than occurrence of thickening. It is therefore important to consider rotary-draw bending with application of an axial force.

In the case of Side Members Right and Left in Figure 10.2, fracture occurred during the hydroforming both in a tube that was rotary-draw bent with application of an axial force and a tube that was only rotary-draw bent. In both workpieces (tubes), the crack was obtained in the same section but at different positions, Figure 10.38c. If a rotary-draw bent tube was lubricated better before hydroforming, no cracks could however be observed during hydroforming. Due to the tight time schedule, all of the tubes used to make Side Member Right and Side Member Left in Figure 10.2 were rotary-draw bent without applying an axial force.

How the bent tube "lies" (is positioned) in the hydroforming tool before hydroforming plays an important role. As many cross-sections as possible or at least the fracture critical cross-sections should be in contact with the bottom cavity prior to hydroforming. The bending program needs therefore to be "optimized".

The spread after bending may have a large impact on the spread after hydroforming, especially if high internal pressure cannot be applied in hydroforming.

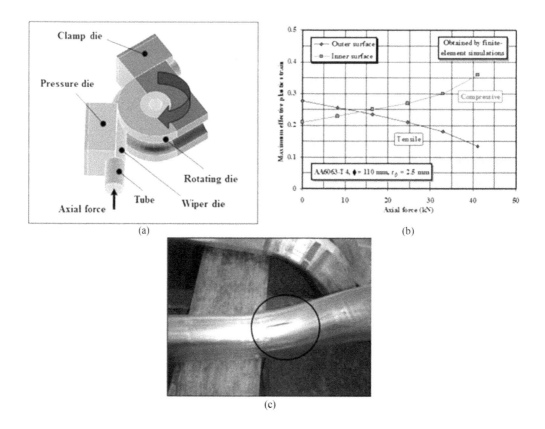

Figure 10.38 (a) The principle of rotary-draw bending with and without application of an axial force. (b) Rotary-draw bending with and without application of an axial force: the maximum effective plastic strain versus the applied axial force. (c) In this case, a crack was obtained during hydroforming of a tube that was rotary-draw bent with an axial force. (Reprinted with permission from Asnafi et al., 2000.)

After bending the tubes that were to be hydroformed to Side Member Left and Right, the spread was measured in the fixture shown in Figure 10.39. Dial gauges were mounted on this fixture to measure the "shape" of the bent tubes and the hydroformed side members.

The maximum spread was approximately 8 mm after bending. This spread

- was very significant, if the bent tubes were hydroformed with 1,100 bars in internal pressure. 72% of the tubes with large deviations after bending also exhibited unacceptably large deviations after hydroforming.
- did not exhibit any measurable impact, if the bent tubes were hydroformed with 1,300 bars in internal pressure.

Trials with three different lubricants showed that the following lubrication gave the best results: All of the bent tubes were first wrapped up in one layer of plastic film ("deep-drawing" film) in the bent zones. Then the whole bent tube was lubricated with *Iloform BWN 180*, which is a mineral oil-containing lubricant, before hydroforming. This selection was based on trials with three different lubricants (Asnafi et al., 2003).

Figure 10.40 displays the lower hydroforming tool for Side Member Right. Tool closure with and without application of a preforming (assisting) internal pressure were tested. Large buckles were

Figure 10.39 Both the bent tubes and the hydroformed side members were measured in this fixture. Dial gauges were mounted on the fixture. These gauges measured the "shape" of the bent tubes and the hydroformed side members.

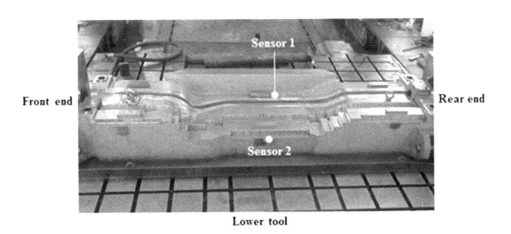

Lower tool

Figure 10.40 The Lower hydroforming tool for Side Member Right. The indicated sensors were used to measure the tool separation.

formed, as the tool was closed without an assisting internal pressure (Figure 10.41a). Traces of these buckles could be observed on the hydroformed side members.

Tool closure with application of a preforming internal pressure resulted, however, in a shape that was very close to the final, Figure 10.41b. With one exception, no buckles could furthermore be detected on the hydroformed side member provided that the tool was closed with an assisting

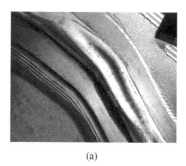
(a)

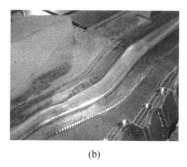
(b)

Figure 10.41   The shape of the workpiece (the tube) when the tool is closed (a) without application of a preforming internal pressure and (b) with application of a preforming internal pressure. (Reprinted with permission from Asnafi et al., 2003.)

internal pressure. Tool closure with an assisting internal pressure was therefore used in all cases. The mathematical expressions in Sections 10.3.1–10.3.3 were used to find the magnitude of this assisting (preforming) internal pressure.

The conducted cross- and longitudinal-section analysis showed that the average change in cross-section perimeter varied between −0.2% and +6% initially. Finite-element simulations predicted that some cross-sections would be critical (due to fracture). The same cross-sections were also critical in practice. At these cross-sections, the magnitude of the plastic deformation was too large and the risk of fracture very high. Rubber pads (thickness = 3 mm) were taped on the tube before hydroforming and putty (filler) was used to find the maximum attainable deformation level at these critical sections (Figure 10.42). At these sections, the circumference was then decreased by TIG-welding (and grinding and polishing) in the tool.

It was found during the course of the work that the cross-sections, at which the magnitude of the plastic deformation was too small, affected the dimensional accuracy and stability of the side members negatively. At such sections, the perimeter was therefore enlarged by carving out "bananas" in the tool, Figure 10.43. (The left tool was modified in exactly the same fashion as the right.)

The average change in cross-section perimeter was therefore altered to +2% and +5% due to the actions that the cross- and longitudinal-section analysis led to. The maximum average deformation needed to be reduced to 5% to avoid fracture and at least +2% in cross-sectional deformation was required to attain dimensional accuracy and stability.

The hydroforming tool is mounted in a press that is used to keep the tool closed during hydroforming. This closing force exerted by the press (the press capacity) must exceed the internal pressure

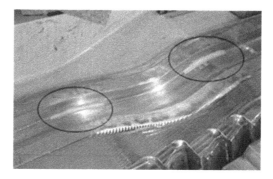

Figure 10.42   Where required, the radii were increased with the aid of putty (filler). (Reprinted with permission from Asnafi et al., 2003.)

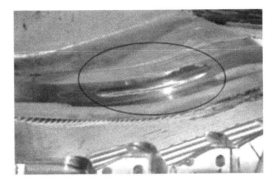

Figure 10.43 Carving out "bananas" in the tool enlarged the perimeter. (Reprinted with permission from Asnafi et al., 2003.)

used during hydroforming multiplied with the surface that this pressure is acting on. Therefore, the maximum internal pressure needed to obtain acceptable results in hydroforming times the die surface this pressure is acting on cannot be allowed to exceed the press capacity. If the press capacity is exceeded, a tool separation will be observed (the tool "yawns").

Using the sensors in Figure 10.40, the tool separation was measured. Figures 10.44 and 10.45 show the measured sensor values, the definition of tool separation, and a schematic description of changes in the cross-section caused by this tool separation. It is of great significance to minimize the consequence of such a tool separation (Asnafi et al., 2000) or avoid it completely.

During post-processing, the hydroformed side members (both the right and the left) were placed in specially devised transport fixtures and transported to another workshop in another city. At these

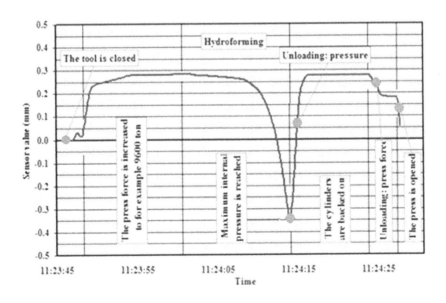

Figure 10.44 Tool (hydroforming tool) separation: sensor values and process description. (Reprinted with permission from Asnafi et al., 2000.)

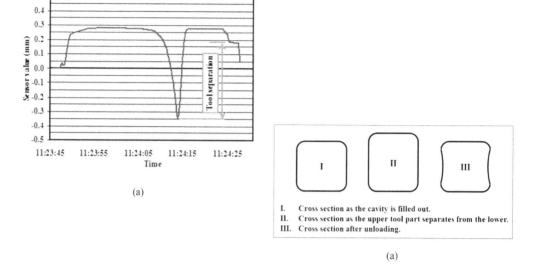

I.   Cross section as the cavity is filled out.
II. Cross section as the upper tool part separates from the lower.
III. Cross section after unloading.

(a)

Figure 10.45 Hydroforming: (a) the definition of tool separation and (b) schematic description of changes in the cross-section caused by tool separation. (Reprinted with permission from Asnafi et al., 2000.)

Table 10.5   The Change in Maximum Spread during Post-Processing

The Measurements Were Conducted...	Max. Spread (mm)
... after hydroforming in city 1[a]	0.57
... upon arrival—city 2[b]	0.57
... after heat treatment—city 2[b]	0.62
... after end-cutting and hole-making—city 2[b]	0.72
... after washing—city 2[b]	0.74
... at Volvo Cars—city 3[c]	0.80

Source: Data from Nader Asnafi et al. (2000).
[a] The fixture shown in Figure 10.39 was used.
[b] The fixture was an exact copy of that shown in Figure 10.39.
[c] Performed in a coordinate-measuring machine.

workshops, the side members were moved to heat-treatment fixtures and artificially aged to the temper T6. After artificial aging, end-cutting, and hole-making, the side members were washed, placed in transport fixtures, and transported to Volvo Cars in Göteborg. At Volvo Cars, the side members were subject to a thorough measurement program in a coordinate-measuring machine. Table 10.5 displays how the spread changed during the post-processing. The spread, an important approval criterion, needs to be minimized and made traceable.

10.6 HYDROPIERCING

Hole-making is needed in many cases. To avoid adding separate process steps (a separate hole-making step) to the manufacturing procedure, the preference is to hydropierce the hole/holes. The hole/holes are, in other words, made in the tool used to hydroform the part directly after hydroforming and prior to unloading.

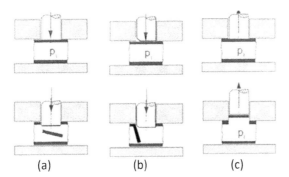

Figure 10.46 The tested hydropiercing methods: (a) hydropiercing inwards; (b) hydropiercing by folding the "scrap piece" into the tube cavity; (c) hydropiercing outwards. (Reprinted with permission from Asnafi et al., 2003.)

Figure 10.46 displays different hydropiercing methods—hydropiercing inwards, outwards and by folding the "scrap" piece into the tube cavity. Making the hole (hydropiercing) is however not sufficient. The hole must also have the intended/required quality. Both of these topics (hydropiercing and hole quality) are addressed in this section.

10.6.1 Theoretical Treatment

Figure 10.47 displays the nomenclature for hydropiercing. In hydropiercing inwards, the punch is used to make the hole and the inside pressure acts as a counterpunch. In hydropiercing outwards, the hole is made by the inside pressure while the punch acts as a counterpunch. For details, see Figure 10.47, where F = punch force, p_i = inside pressure, c = clearance between the punch and the tool, l_c = cut line, and t = tube wall thickness. In common mechanical punching, the punch force can be expressed as follows:

$$F = k_s \cdot A_{cl} = k_s \cdot l_c \cdot t = 0.6 \cdot R_m \cdot l_c \cdot t \qquad (10.54)$$

where k_s = shear strength of the tube material and A_{cl} = cut-line area. It is assumed here that the shear strength is 60% of the ultimate tensile strength, R_m.

In hydropiercing a *circular hole* inwards, the exerted punch force must both cut the hole and counteract the inside pressure. Force equilibrium (Figure 10.47) and equation 10.54 yield (Asnafi et al., 2003)

$$F = p_i \cdot \pi d_p^2 \big/ 4 + 0.6 \cdot R_m \cdot \pi d_p \cdot t \qquad (10.55)$$

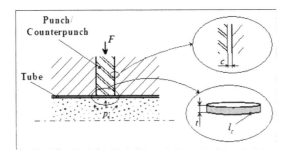

Figure 10.47 Nomenclature for hydropiercing. (Reprinted with permission from Asnafi et al., 2003.)

where d_p is the punch/hole diameter. c, clearance between the punch and the tool in Figure 10.47, is thereby neglected.

In hydropiercing an *"oval"* *hole* inwards, force equilibrium (Figure 10.47) and equation 10.54 yield (Asnafi et al., 2003)

$$F = p_i \left(\pi d_p^2 / 4 + h_c \cdot d_p \right) + 0.6 R_m \left(\pi d_p + 2h_c \right) \cdot t \qquad (10.56)$$

where h_c and the shape of the "oval" hole are as shown in Figure 10.48.

In hydropiercing outwards, the inside pressure cuts the hole. For a *circular hole*, the force equilibrium (Figure 10.47) and equation 10.54 yield (the clearance c in Figure 10.47 is neglected) (Asnafi et al., 2003)

$$p_i = \frac{0.6 \cdot R_m \cdot \pi d_p \cdot t}{\pi \cdot d_p^2 / 4} = 2.4 \cdot R_m \cdot t / d_p \qquad (10.57)$$

which is only valid, if the counterpunch force is smaller than $p_i \cdot \pi \cdot d_p^2 / 4$ (Figure 10.47).

For the *oval hole* (Figure 10.48) and hydropiercing outwards, it can be shown that (Asnafi et al., 2003)

$$p_i = 0.6 R_m \left(\pi d_p + 19.6 \right) \cdot t / \left(\pi d_p^2 / 4 + 9.8 d_p \right) \qquad (10.58)$$

Finite-element simulation can be used to model hydropiercing. The description below is based on finite-element simulations, in which an axisymmetric 2D solid model was used to simulate inward and outward hydropiercing of the round holes. To simulate the hydropiercing by folding the "scrap piece" inwards, it was however not possible to use an axisymmetric model. In this case, a 2D-plane strain model was used, since the run-time for a 3D model would be too long. The code LS-DYNA was used in all cases. The clearance c, Figure 10.47, was not neglected in the simulations (c was neglected in the analytical modeling above). The tube material properties in Table 10.4 were used in these simulations.

Figure 10.49 concerns **hydropiercing inwards** and displays the calculated hydropiercing force required for a circular hole (30.2 mm in diameter) versus the internal pressure. The analytical values in this figure are based on equation 10.55.

The pressure required to make the same circular hole ($\phi = 30.2$ mm) by **hydropiercing outwards** is 327.8 bar according to equation 10.57. FE-simulations predict, however, that this pressure should be 290 bar.

The FE-simulations show that high-quality holes demand high internal pressures (see, e.g., Figures 10.50 and 10.51). Figure 10.50 concerns hydropiercing inwards and shows how the inside pressure affects the hole quality (described by the parameters A, B, and C). Figure 10.51 concerns

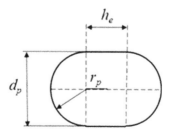

Figure 10.48   The shape of the "oval" hole.

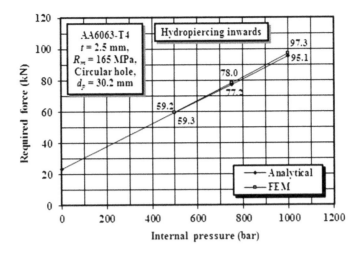

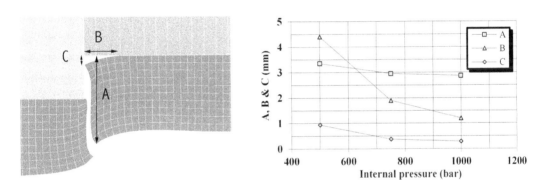

Figure 10.49 Hydropiercing inwards: the calculated hydropiercing force required for a circular hole (hole diameter = 30.2 mm) versus the internal pressure. The figure shows both analytical (equation 10.55) and FE-values. (Reprinted with permission from Asnafi et al., 2003.)

Figure 10.50 Hydropiercing inwards: The influence of the internal pressure on the hole quality. The figure is based on FE-simulations. Note too that a high internal pressure requires a high hydropiercing force (see equation 10.55). (Reprinted with permission from Asnafi et al., 2003.)

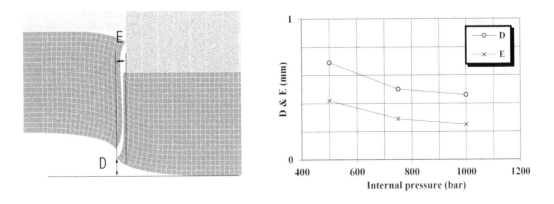

Figure 10.51 Hydropiercing outwards: The influence of the internal pressure on the hole quality. The figure is based on FE-simulations. (Reprinted with permission from Asnafi et al., 2003.)

hydropiercing outwards (the applied internal pressure makes the hole). It is based on FE simulations and depicts the influence of the internal pressure on the hole quality (measured by the parameters D and E).

For hydropiercing by folding the "scrap piece" into the tube cavity (Figure 10.46b), different tool (punch and punch cavity) geometries were tested using FE-simulations. Figure 10.52 displays some of the tested geometries. The conducted simulations showed that a flat-bottomed punch with a profile radius of 5 mm would significantly decrease the risk of fracture in the encircled zone in Figure 10.52. The simulations also showed that a friction coefficient (μ) of 0.1 was too large for this type of hydropiercing ($\mu = 0.1$ in all other simulations). In this case, μ had to be set at 0.01 in order to avoid fracture.

10.6.2 Hydropiercing in Practice

In the practical test, straight 1,110-mm-long tubes of (extruded) AA6063-T4 were used. The tubes were circular in cross-section. The outer diameter and wall thickness of the tube were 107 and 2.5 mm, respectively. Table 10.4 displays the mechanical properties of this tube material. The round tube was first placed in the hydroforming tool, after which the tool was closed. The tube was then hydroformed at 1,300 bar in internal pressure, the stroke being ca 2 mm at both tube ends (see also Figure 10.56). Hydroforming changed the shape of the cross-section. Prior to unloading, three holes were hydropierced in the tube. Figure 10.53 shows how the shape of the cross-section was changed by hydroforming and the hydropierced holes. Figure 10.54 displays the hydroforming tool and the position of the punches and counterpunches (holes). As shown in Figure 10.55, hole 1 is the "oval" hole while holes 2 and 3 are circular with a diameter of 30.2 mm (punch/die).

All of the tubes were first hydroformed at 1,300 bar, after which the inside pressure was kept at the same level, increased or decreased prior to hydropiercing. Figure 10.56 displays a case, in which the inside pressure was decreased prior to hydropiercing. Refer also Table 10.7.

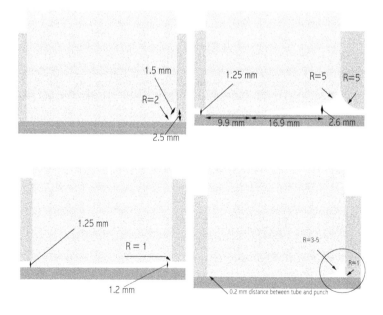

Figure 10.52 Hydropiercing by folding the "scrap" piece into the tube cavity: some of the tool (punch and punch cavity) geometries tested by FE-simulations. The figure shows the initial positions. (Reprinted with permission from Asnafi et al., 2003.)

Figure 10.53 A part that is hydroformed and hydropierced before unloading. The initial tube was extruded aluminum profile, AA6063-T4, circular in cross-section, 107 mm in diameter and 2.5 mm in wall thickness. (Reprinted with permission from Asnafi et al., 2003.)

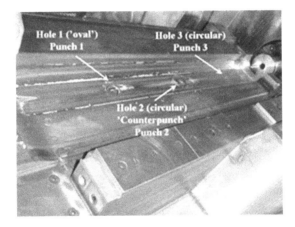

Figure 10.54 The lower part of the hydroforming and hydropiercing tool.

Hydropiercing could be conducted successfully in all of these cases:

- Alternative I: From the beginning of the hydroforming/hydropiercing operation, the punches/counterpunch were positioned, so that they coincided with the inner surface of the hydroforming tool. This is illustrated in Figure 10.57.
- Alternative II: From the beginning of the hydroforming/hydropiercing operation, punches 1 and 3 were backed and punch 2 (counterpunch) was moved forward according to Figure 10.58. This approach resulted in creation of "embossments" on the tube before the hydropiercing was carried out. The values of $x_1 - x_3$ used in the tests are shown in Table 10.6.
- Alternative III: From the beginning of the hydroforming/hydropiercing operation, punches 1 and 3 were moved forward 0.4 and 0.6 mm, respectively (see Figure 10.59). In these experiments, Teflon was sprayed on punches 1 and 3 prior to hydroforming and hydropiercing. No other lubricant (*Gleitmo* and *Iloform BWN 180* were also tested) could separate the "scrap" piece from the hydropiercing punches.

Table 10.7 summarizes the successfully conducted tests and the figures below show how the holes look like. Note that each test was repeated ten times, i.e., 10 beams were hydroformed and hydropierced with the same parameters.

The deflection/elevation at the hole edge and the hole diameter were measured in a coordinate-measuring machine and by a micrometer, respectively. As displayed in Figure 10.55, the diameter

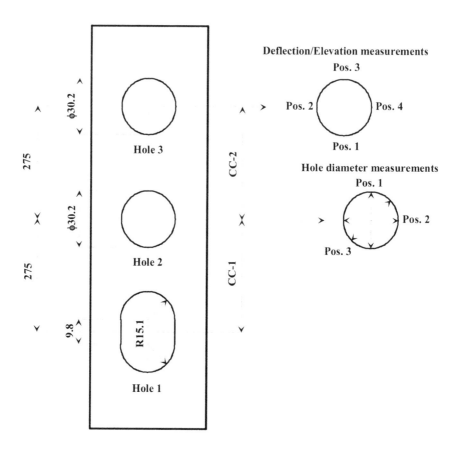

Figure 10.55 The positions at which the hole diameter and the deflection/elevation at the hole edge were measured. The digits displayed in the figure are the target values.

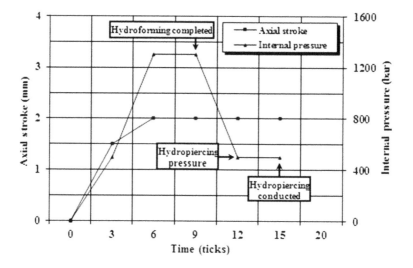

Figure 10.56 All of the tested tubes were first hydroformed at 1,300 bar, after which the internal pressure was kept at the same level, decreased or increased prior to hydropiercing. See also Table 10.7. (Reprinted with permission from Asnafi et al., 2003.)

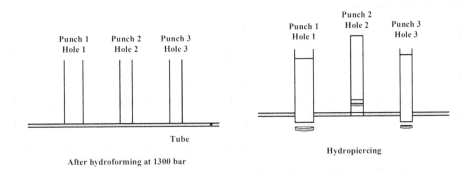

Figure 10.57 Positioning of the hydropiercing punches/counterpunch according to alternative I. (Reprinted with permission from Asnafi et al., 2003.)

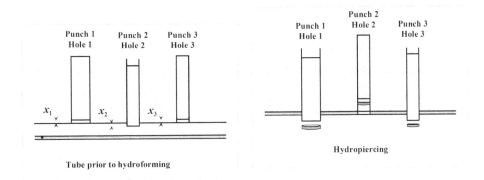

Figure 10.58 Positioning of the hydropiercing punches/counterpunch according to alternative II. See also Table 10.6. (Reprinted with permission from Asnafi et al., 2003.)

Table 10.6 The Values of x_1–x_3 Used in Hydropiercing according to Alternative II, Figure 10.58

Test Designation		
T9	x_1 (mm)	0.2
	x_2 (mm)	0.2
	x_3 (mm)	0.2
T10	x_1 (mm)	0.4
	x_2 (mm)	0.4
	x_3 (mm)	0.4

Source: Data from Asnafi et al. (2003).

of the circular holes (holes 2 and 3) was measured at three positions. As far as the "oval" hole (hole 1) is concerned, the hole "diameter" was measured at positions 1 and 2, Figure 10.55. The hole diameters were measured according to the following:

- Holes made by hydropiercing inwards (holes 1 and 3, Figures 10.60 and 10.61): The *upper diameter* was measured 0.5 mm below the upper edge. The *lower diameter* is defined as the smallest diameter measured below *the upper diameter*.
- Holes made by hydropiercing outwards (hole 2, Figures 10.62 and 10.63): only the smallest measured diameter, *lower diameter*, was registered.

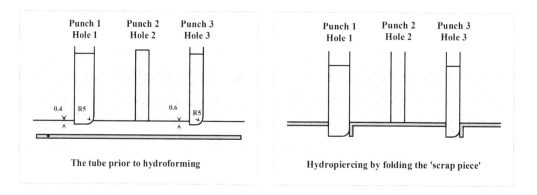

Figure 10.59 Hydropiercing according to alternative III. (Reprinted with permission from Asnafi et al., 2003.)

Table 10.7 The Conducted Practical Tests

Test Designation	Hydropiercing according to Alternative	Hydropiercing Pressure (bar)	Hole/ Holes	No. of Beams Made
T1	I	500	All 3	10
T2	I	750	All 3	10
T3	I	1,000	All 3	10
T4	I	1,100	All 3	10
T5	I	1,200	All 3	10
T6	I	1,300	All 3	10
T7	I	1,400	All 3	10
T9	II	1,300	All 3	10
T10	II	1,300	All 3	10
T12	III	750	1 and 3	10
T13	III	1,300	1 and 3	10

Source: Data from Asnafi et al. (2003).
Note: All of the beams were first hydroformed at 1,300 bar. See also Figure 10.56. For the different hydropiercing alternatives, see Figures 10.57–10.59. For alternative II, see also Table 10.6.

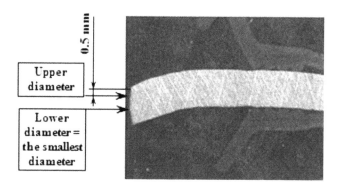

Figure 10.60 Hole edge after hydropiercing inwards (alternative I, holes 1 and 3 in Figure 10.57) at 500 bar. The figure also shows how upper and lower hole diameters are defined.

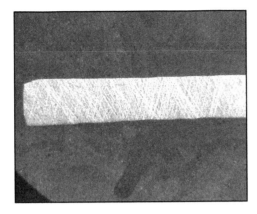

Figure 10.61   Hole edge after hydropiercing inwards (alternative I, holes 1 and 3 in Figure 10.57) at 1,400 bar.

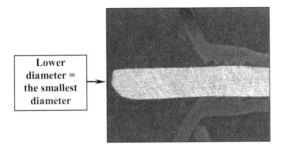

Figure 10.62   Hole edge after hydropiercing outwards (alternative I, hole 2 in Figure 10.57) at 500 bar. The figure also shows how lower hole diameter is defined.

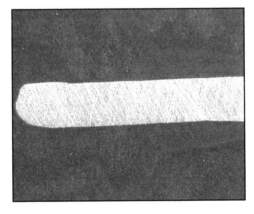

Figure 10.63   Hole edge after hydropiercing outwards (alternative I, hole 2 in Figure 10.57) at 1,400 bar.

Figure 10.64 displays a cross-section of a circular hole (hole 3) made by folding the "scrap" piece into the tube cavity. This hole was made by a flat-bottomed punch with a profile radius of 5 mm at 1,300 bar in hydropiercing pressure. As mentioned above, Teflon was the only lubricant that could separate the "scrap" piece from the hydropiercing punches and result in a successful hydropiercing.

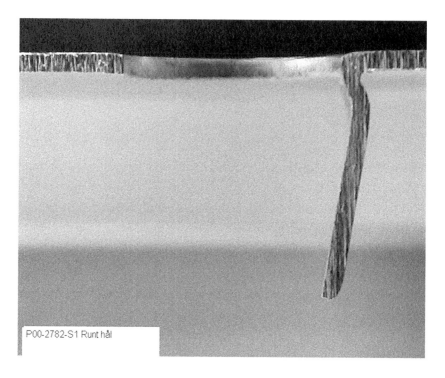

Figure 10.64   Hydropiercing by folding the "scrap" piece into the tube cavity. Hydropiercing pressure = 1,300
bar. The figure shows the circular hole (hole 3, alternative III, Figure 10.59). (Figures 10.60–10.64 are reprinted
with permission from Asnafi et al., 2003.)

10.6.3 Hole Quality

As mentioned before, ten beams were made per set of parameters (Table 10.7). The diameters
of the circular holes were, furthermore, measured at three positions (Figure 10.55). All of these
measured values were averaged. Table 10.8 displays the average values obtained for all of the manufactured beams. In Table 10.8, deflection is given as a negative value, while elevation is displayed as
a positive value. Compare Table 10.8 with Table 10.7.

The following are some of the interesting features found in Table 10.8:

* A deflection is measured at the edges of the holes made by hydropiercing inwards (holes 1 and 3,
  tests T1–T7 and T12–T13). This deflection decreases in magnitude with increasing hydropiercing
  pressure.
* A small elevation is measured at the edges of the hole made by hydropiercing outwards (hole 2).
  This small elevation increases with increasing hydropiercing pressure up to 1,200 bar.
* Hole 2 is made by hydropiercing outwards. The measured diameter for this hole is in all cases
  smaller than the target diameter (30.2 mm). This should be considered as a confirmation of the prediction made by FE simulations (Figure 10.51). This means that the diameter of the counterpunch
  cavity must be larger than the target diameter of the hole. FE simulations and/or practical tests are
  required to find the optimum diameter of the counterpunch cavity.

Deflection/elevation, lower diameter (the smallest diameter), and *center distance* are the most
important parameters, as far as the hole quality is concerned (see Table 10.8). Assuming that the
largest (in magnitude) acceptable deflection is 0.2 mm, Table 10.8 shows that hydropiercing inwards
at ≥1,300 bar yield the best hole quality.

Table 10.8 The Average Values Obtained on All Beams

Test Designation	Hydropiercing according to Alternative	Hydropiercing Pressure (bar)	Deflection/Elevation Hole 1 (mm)	Hole 2 (mm)	Hole 3 (mm)	Upper Diameter Hole 1[a] Pos. 1 (mm)	Pos. 2 (mm)	Hole 3 (mm)	Lower Diameter Hole 1 Pos. 1 (mm)	Pos. 2 (mm)	Hole 2 (mm)	Hole 3 (mm)
					Target	40.0	30.2	30.2	40.0	30.2	30.2	30.2
T1[b]	I	500	−0.372	0.009	−0.702	40.16	31.16	30.25	39.78	29.90	29.67	30.12
T2[b]	I	750	−0.451	0.010	−0.556	40.10	30.35	30.27	40.01	30.12	29.61	30.22
T3[b]	I	1,000	−0.411	0.017	−0.354	40.06	30.26	30.26	40.03	30.21	29.60	30.24
T4[b]	I	1,100	−0.388	0.021	−0.296	40.04	30.24	30.25	40.03	30.22	29.57	30.23
T5[b]	I	1,200	−0.299	0.024	−0.222	40.05	30.24	30.25	40.03	30.21	29.54	30.24
T6[b]	I	1,300	−0.201	0.023	−0.158	40.05	30.23	30.26	40.03	30.21	29.49	30.24
T7[b]	I	1,400	−0.178	0.021	−0.131	40.04	30.24	30.24	40.03	30.21	29.44	30.23
T9[c]	II	1,300	−0.174	0.018	−0.184	40.12	30.23	30.29	40.10	30.21	29.51	30.27
T10[d]	II	1,300	−0.210	0.014	−0.233	40.10	30.22	30.28	40.09	30.20	29.51	30.26
T12	III	750	−0.730		−1.321				40.00	30.18		30.22
T13	III	1,300	−0.208		−0.195				40.02	30.18		30.23

Source: Data from Asnafi et al. (2003).
Note: Compare with Table 10.7.
[a] Hole 1 is "oval", while holes 2 and 3 are circular, Figure 10.55.
[b] Both CC-1 and CC-2 were 275.0 mm.
[c] CC-1 = 275.1 mm and CC-2 = 275.0 mm.
[d] CC-1 = 275.1 mm and CC-2 = 275.1 mm.

10.7 SUMMARY

As displayed in Figure 10.4, a hydroformed tubular component replaces eight conventionally stamped sheet metal parts. The material utilization degree in stamping of each of these eight components is less than 70%. The material utilization degree for the hydroformed variant is more than 90%. In tube hydroforming, the material thickness can be reduced due to the inherent strength of the tube material. The weakest points of a stamped and joined subassembly are the zones at the spot welds. The solution for the stamped and joint subassemblies is often to start with a thicker sheet material to ensure that the weak points will be acceptable. This weakness vanishes completely, in case a tubular base material is used. Tubular hydroforming enables a further reduction of the weight or more weight-efficient components. Less material is in other words used in tube hydroforming and the material cost is therefore reduced, as displayed in Figure 10.4.

Each of the eight stamped parts in Figure 10.4 requires a tool and die set. The number of tool sets is therefore reduced, in the case of tube hydroforming. These stamped parts need to be assembled by joining (spot welding). This is not needed, in the case of tube hydroforming and the assembly cost is therefore reduced to zero, as displayed in Figure 10.4. The cross-section shape can be varied, which also enables a better utilization of the package room, as shown in Figure 10.31.

The number of process stages is reduced significantly by tube hydroforming. Compare, for instance, all of the process stages required to make a side member by stamping and joining eight stamped sheet metal parts with the hydroformed tubular variant shown in Figure 10.4.

In tube hydroforming and hydropiercing, the whole shape including all of the holes is made in the same die cavity in one stage. There is no "springback", due to the high applied pressure. The repeatability is excellent. All these features lead to better quality for hydroformed components.

Tube hydroforming (and hydropiercing) is in other words a more sustainable manufacturing method (less material usage, fewer parts, fewer process steps, no subassembly joining, and better quality). Except more sustainable manufacturing processes, it enables also weight efficiency, reduced emissions, and more sustainable products within the automotive industry. These topics will be at the focus also in the future. Weight reduction is, for instance, of great value even for electrical vehicles.

The benefits mentioned above are not the only advantages gained by selecting the hydroforming solution. The hydroformed parts exhibit also benefits, when in service. A hydroformed tubular component displays, for instance, improved bending and torsional stiffness (and lower bending and torsional stresses).

REFERENCES

Asnafi, N. 1997. Hydroformability of Extra High Strength Steels in Structural Tubular Applications—An Analysis Based on Literature Survey, Report No. IM–3521. Swedish Institute for Metals Research (SIMR/Swerea KIMAB), Stockholm.

Asnafi, N. 1999a. On stretch and shrink flanging of sheet aluminium by fluid forming. *J. Mater. Process. Technol.* 96:198–214.

Asnafi, N. 1999b. Analytical modelling of tube hydroforming. *Thin-Walled Struct.* 34:295–330.

Asnafi, N. 2000. Tube bending and hydroforming. *Svetsaren/Welder—A Weld. Rev.* 2:34–35.

Asnafi, N., Nilsson, T., and Lassl, G. 2000. Automotive tube bending and tubular hydroforming with extruded aluminium profiles, Paper No. 2000-01-2670. In *Proceedings of the International Body Engineering Conference & Exposition (IBEC2000)*, October 3–5, Detroit, MI.

Asnafi, N. and Skogsgårdh, A. 2000. Theoretical and experimental analysis of stroke-controlled tube hydroforming. *Mater. Sci. Eng.* A279:95–110.

Asnafi, N. and Andersson, R. 2001. Tubular hydroforming viewed from the Swedish perspective. *Tube and Pie Journal*, March.

Asnafi, N., Ocklund, J., and Lassl, G. 2002. Tubular hydroforming of side members and crash beams—A study from the perspective of Volvo Cars. In *Proceedings of the 22nd Biennial Congress of the International Deep Drawing Research Group (IDDRG)*, May 20–22, Nagoya, Japan, pp. 289–298.

Asnafi, N., Lassl, G., Olsson, B., and Nilsson, T. 2003. Theoretical and experimental analysis of hydropiercing, Paper No. JSAE20037165/SAE2003-01-2884. In *Proceedings of the International Body Engineering Conference & Exposition 2003 (IBEC2003)*, Oct 27–29, Chiba, Japan.

Asnafi, N., Nilsson, T., and Lassl, G. 2003. Tubular hydroforming of automotive side members with extruded aluminium profiles. *J. Mater. Process. Technol.* 142:93–101.

Astrop, A.W. 1968. The Rolls-Royce hydro bulging metal forming technique. *Machinery and Prod. Eng.* 18:604–609.

Dohmann, F. and Hartl, C. 1996. Hydroforming—a method to manufacture light-weight parts. *J. Mater. Process. Technol.* 60:669–676.

Geckeler, J.W. 1928. Plastisches Knicken der Wandung von Hohlzylindern und einige andere Faltungserscheinungen an Schalen und Blechen. *Zeitschrift für angewandte Mathematik und Mechanik*, Band 8(Heft 5. Oktober):341–351.

Mason, M. 1996. Tube hydroforming—Advancements using sequenced forming pressures. In *Proceedings of a Symposium/Conference on Innovations in Hydroforming Technology*, held by *Tube and Pipe Association International*, Sept. 25, Nashville.

Mücke, K. 1995. Innen-Hochdruck-Umformen in der Serienfertigung. *Blech Rohre Profile* 42(1):17–20.

Ocklund, J., Asnafi, N., Amino, H., and Maki, T. 2002. Hydromechanical forming of trunk lid outer. In *Proceedings of the 22nd Biennial Congress of the International Deep Drawing Research Group (IDDRG)*, May 20–22, Nagoya, Japan, pp. 279–288.

Sauer, W.J., Gotera, A., Robb, F., and Huang, P. 1978. Free bulge forming of tubes under internal pressure and axial compression. In *Proceedings of the Sixth North American Manufacturing Research Conference (NAMRC-6)*. Gainsville, FL, pp. 228–235.

Schäfer Hydroforming GmbH & Co. 1996. *Technological Guide Lines for ASE-Components.* Schäfer Hydroforming GmbH & Co, Wilnsdorf.

Thuresson, G., Servin, O., Bencan, D., Asnafi, N., Jönsson, M., Hellgren, K., and Johannisson, T. 2003. Pre-stretching, forming, trimming, flanging and calibration of hood outer conducted in the same cycle in a fluid cell (QUINTUS®) press. In *Proceedings of the Annual Conference of the International Deep Drawing Research Group (IDDRG)*, May 11–14, Bled, Slovenia.

Tolazzi, M. 2010. Hydroforming applications in automotive: a review. *Int. J. Mater. Form.* 3(Suppl 1):307–310.

Ultra Light Steel Auto Body Consortium (ULSAB). 1996. Final report 8-16-96, *Porsche Engineering Services Inc.*

VARI-FORM—Global leader in hydroforming technology for tomorrow's automotive requirements. 1996. Information brochure from VARI-FORM, Telford, England.

CHAPTER 11

Some Recent Developments in Microforming

Ming Yang and Tetsuhide Shimizu
Tokyo Metropolitan University

Tsuyoshi Furushima
The University of Tokyo

Tomomi Shiratori
Komatsuseiki Kosakusho Co. Ltd.

CONTENTS

11.1 INTRODUCTION

Micro metal forming is considered to manufacture parts in size of millimeter or submillimeter for the audio–video (AV) or information technological (IT) devices, and also to manufacture micro functional devices, such as MEMS (Micro Mechanical Electro System) for mechanical sensors and actuators, and for bioassays and medical devices, as an alternative process for mass production. Metallic materials have high strength, ductility, and conductivity, and the metal forming adapts to manufacture 3D structures, which are important for the devices in chemical and biological applications. The investigations performed on metal forming showed that the micro metal forming could be a new technology for fabrication of MEMS in low cost and large quantities.

On the other hand, accuracy of the processes and the occurrence of size effects due to scaling-down will be significant in micro metal forming. When the grain size is approaching the material thickness, individual grain significantly affects the deformation behavior of the material [1]. The inhomogeneous deformation occurs and it leads to the large strain concentration at the fracture region which induces formability decreasing. Furthermore, ratio of surface area vs. volume increases with a decrease in its scale [2]. As a result, tribological properties become more important in comparison with that in macro scale. The dies bear higher stress during process than macro process due to size effect. Development of equipment, dies, and processes for overcoming the issues due to the scaling-down is an important factor in the micro metal forming.

Scale effects in deformation of material and process tribology during metal forming were investigated [1–3]. The authors have been working on development of a novel micro-metal-forming technology by combining micro press-forming and automatically assembling in a progressive die, and applying it to a fabrication of micro parts and MEMS [4,5]. Micro parts in several hundreds of micrometers and a micro pump in several millimeters were made by using this microforming technology.

Several investigations including development of novel processes including combining micro press-forming and automatically assembling in a progressive die for fabrication of micro devices, heat-assisted forming, dieless tube forming, and coating system for micro-die adapted to microforming are introduced in this chapter. Furthermore, a micro device, metallic pump fabricated by using micro metal forming, is also introduced.

11.2 DEVELOPMENT OF NEW PROCESSES FOR MICROFORMING

11.2.1 Development of Desktop Servo Driven Precise Press Machine and Progressive Die for Precise Forming and Assembly of Micro Components

11.2.1.1 Introduction

Press-forming is one of the most significant metal-forming processes for fabrication of complicated parts in a press machine. Most of the parts used in AV and IT devices are fabricated by the press-forming. The feature of these parts became smaller and smaller in the recent years. Furthermore, MEMS and biochips consist of multiparts constructing a system or a unit with complicated structure. Fabrication of parts/units with feature of submillimeters using press-forming could be important in the next few years [4]. It is important to develop a proper press-forming system for meeting these needs. In this section, the author will introduce a micro fabrication system consisting of a desktop servo-press and precise progressive dies, in which not only forming but also assembly of micro components into a unit part or device will be accomplished.

11.2.1.2 Desktop Servo-Press Machine and Micro Fabrication System

A miniature press machine with desktop size was developed for press-forming and assembly of micro parts. Figure 11.1 shows a photo of the machine and its specifications. The press is actuated by a servomotor and controlled precisely. It is possible to supply materials simultaneously in different directions during the process.

A micro fabrication system was proposed by using press-forming to form micro parts in multisteps, and then to assemble parts into a unit in a progressive die. Figure 11.2 shows the concept

Type: S-23 desk-top servo press machine
Max. Force 23kN. Max. speed 500spm.
Max Stroke:15mm.
Slide Adjustment 15mm. Die height: 100mm.
Dimensions: W340 x D330 x H545

Figure 11.1 Miniature desktop servo-press machine with its specification. (With permission from Yang et al., 2007, Springer Nature.)

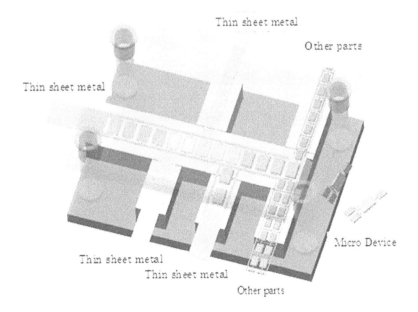

Figure 11.2 Schematic configuration of press-forming and in-process assembly in a progressive die. (With permission from Yang et al., 2007, Springer Nature.)

of the micro fabrication system. Several kinds of materials are supplied into the progressive die by feeders, and formed simultaneously in the progressive die in several steps and then, assembled together in the same die. As a result, a unit part could be formed as an output of the press-forming. In order to establish the micro fabrication system, accuracy of each element of dies in dimensions and of alignment is challengeable. The main issues are listed as follows:

- Establishing new methods for fabrication and evaluation of die features smaller than 20 μm;
- Keeping errors in dimension and allocation of the elements smaller than 1 μm;
- Treatment of die surface for protection from wearing and degradation and for processes without lubricant;
- Removing burrs larger than several microns before assembly process;
- Miniaturization and digitization of press machine for reducing errors in vertical and horizontal positions during the processes;

Methodologies on fabrication of micro-die in size of several ten micrometers and design of the press-forming process with the assembly system will be discussed.

11.2.1.3 Production of Micro Components and Units with the Fabrication System

A two-step punching process was carried out for fabrication of a micro gear with a central hole by using the micro fabrication system. In this case, alignment of elements of the die for each process step, and feeding and positioning of material at each step will be very important. Figure 11.3 shows the schematic of the gear with specification and of the progressive dies. The gear was successfully fabricated and the error in concentric circles was about 3 μm. In this case, the process rate was 60 spm.

Fabrication of a unit part was also carried out using the micro fabrication system. A unit part with three components was designed as a sample shown in Figure 11.4. Two movable components and a base plate with several pins for fixing the movable components are press-formed and assembled together in the same progressive die. The movable components are joined by dowel pins on the base plate and rotational around the pins. The process includes punching, bending, imprinting, pinning, and so on. For this purpose, a 5-step process for punching the shape of functional elements, respectively, a 10-step process for punching and forming the base plate, and a 4-step process for joining and assembly were performed. Figure 11.5 shows the products fabricated by the process.

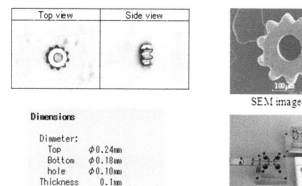

Figure 11.3   Fabricated micro gear and the dies for the press-forming. (With permission from Yang et al., 2007, Springer Nature.)

Unit parts image   Extended image

Die for micro parts stamping and assembly

Figure 11.4 Design picture of press-forming dies and unit part. (With permission from Yang et al., 2007, Springer Nature.)

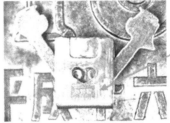

Marking H0.5mm × W1.0mm

Figure 11.5 Unit part on a coin produced by the press-forming and auto assembly system. (Dimensions: W2.4 mm × H3.0 mm × D0.5 mm, Material: Case: SUS304 t = 0.2 mm, moving parts: SUS304CSP t = 0.1 mm.)

The results show that the unit part was successfully fabricated by using the press-forming in a progressive die. One of the most important issues for fabrication of micro functional devices, such as MEMS, biochips, and so on, is the manipulation and assembly of the parts. The automatic manipulation and assembly in the progressive die manifests very higher potential for production of micro functional devices by using micro press-forming.

11.2.2 Contact Heating System for Improving Formability in Microforming

11.2.2.1 Introduction

Several investigations were carried out to indicate the advantages of heating-assisted microforming for improving and formability of workpiece and accuracy of the product. Egerer et al. carried out micro upsetting tests with material of CnZn15 at different temperatures (from 20°C to 400°C) using tools with heat element. The results showed the reduction in scatter of flow stress with an increase in the temperatures [6]. Wolfsberg et al. proposed laser-assisted microforming with sapphire tools and showed that the formability was increased by elevated temperatures [7]. Peng et al. investigated with FE simulations for electrical and laser heating-assisted blanking. The results showed an increase in the aspect ratio (thickness/width) and a decrease in the blanking force at elevated temperatures [8].

In this section, a high-energy-assisted microforming system is introduced. This combines heating and vibration to improve the formability and accuracy by using servo-press system with a resistance heating system embedded dies. Micro-deep drawing and micro-forging were carried out to evaluate the performance of the high-energy assistance and effect on the formability of material and accuracy of the products.

11.2.2.2 Equipment for Heat-Assisted Microforming

A desktop servo-press system with a resistance heating system embedded dies was developed. Figure 11.6a shows the experimental apparatus. By using servo motor to the press machine, either simple clank motion or complicated motions, such as step motion, vibration motion, are available, and the motion can be arbitrarily controlled by a computer. In the case of vibration motion, frequency of the vibration motion can achieve to 200 Hz. The heat-assisted microforming system with resistance heating was designed to satisfy low energy consumption and high forming accuracy to heat only the workpiece material and to limit the heat transferring to tools. In upper die, there are a punch and a blank holder. Two contact probes as electrodes are inserted in blank holder, as shown in Figure 11.6b. Since there are springs inside the contact probes, the forming and heating steps can be carried out at the same time. In lower die, there are dies at the center and two guide pins which carries out the positioning the blank (workpieces). The blank cut off the fringe of forming area for charging electrical current preferentially. Then the forming area achieves high temperatures. Nonconductive ceramics were used for the punch, the die, and the blank holder, to prevent from the heating loss and electrical leak. For experimental data, there are load cells on the punch and a radiation thermometer under the die in the equipment. The temperature was controlled real-time by the same controller.

During the forming process, temperature was measured and fed back to the computer to control the output of the power supply. Figure 11.7 shows an example of temperature control during the process. After the power was supplied, temperature of the workpiece rises to the preset temperature in a few seconds. While the tool contacted on the heated material, the temperature of material decreased

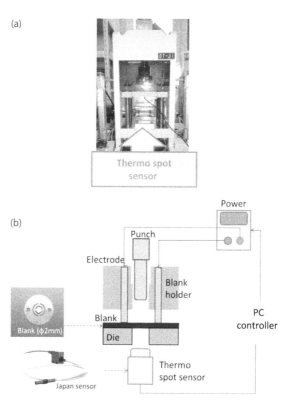

Figure 11.6 High-energy-assisted microforming system with a desktop servo-press machine (a) and a resistance heating system embedded dies (b). (With permission from Tanabe, 2012.)

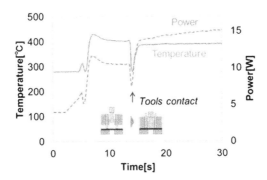

Figure 11.7 Temperature control during forming process. (With permission from Tanabe, 2012.)

by about 200°C temporarily, but returned to the preset temperature again within 1–2 s. It is seen that this system is very effective to heat the workpiece to over 400°C rapidly with very small amount of energy and keep the temperature steady during the contact with dies.

11.2.2.3 Microforming with Localized Heating

In order to verify the influence on forming characteristic, micro-deep drawing and forging using thin foil of pure Ti were performed. Deformation properties and forming limitation were evaluated. Since Ti has lower formability at room temperature, heat assistance was expected to improve the formability. Deep drawing test of micro cylinder cup with a diameter of 1 mm was performed. Pure Ti foil with thickness of 20 μm was utilized as workpiece. Since the recrystallization temperature of titanium was about 450°C, the temperature was raised to 600°C in maximum in order to verify the influence of recrystallization characteristics on the deformation behavior and the drawing limit. Figure 11.8 shows experimental results on relation of drawing limit and temperature. The height of the drawn cup increased from 0.2 to 0.55 mm with the increase in temperature. Figure 11.9 shows the fracture position of the micro cup. The fracture position is changing to the die shoulder from the punch shoulder with the increase in temperature. This indicates that the stress state and deformation properties change with the increases in temperature. In order to verify the influence of heating effect on homogeneity of deformation, the micro cup drawn to the same height for various temperatures was performed and surface of the drawn cup was observed microscopically. Figure 11.10 shows a microscopic observation of surface of the drawn for different temperatures. It is evident that slip lines were observed for room temperature at 300°C, but not for 600°C. It is known that Ti has few slip systems for plastic deformation in room temperature, and the slip system could increase

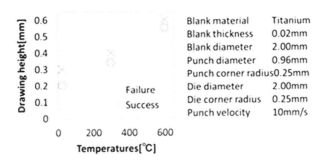

Figure 11.8 Relationship between drawing height and temperature. (With permission from Tanabe, 2012.)

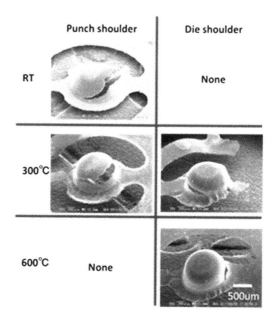

Figure 11.9   Fracture area at elevated temperatures. (With permission from Tanabe, 2012.)

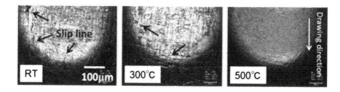

Figure 11.10   Images of surface of a drawn cup. (With permission from Tanabe, 2012.)

according to temperature raise, especially higher than the recrystallization temperature. As a result, the surface of drawn cup for higher temperature is much smoother than that for room temperature. The experimental result shows that the micro-deep drawing assisted with heating is effective in improvement in ductility and homogeneity, and in improvement in drawn limit and in reduction of surface roughness of drawn cup.

Forging assisted by heating and vibrating was performed for improvement of the formability and quality of product since the forging with vibration was very effective to enhance deformation on the surface [9]. In this experiment, frequency of the vibration was found to be 16 Hz. Figure 11.11 shows the effect of temperature, forming speed, and vibration on surface roughness. The surface roughness decreased more significantly for higher temperature with faster forming rate and vibration assistance., meaning that higher energy could be concentrated on the very surface of the workpiece to enhance the localized deformation on the surface. Figure 11.12 shows the effect of heating and vibration on the surface roughness in microscale and nanoscale. With only in heating, the surface roughness was decreased to 50% in the nanoscale and 54% in the microscale, and 29% in the microscale and 38% in nanoscale by combining heating and vibration. The surface roughness was increased more significantly by applying higher density of vibration energy since higher frequency was more effective for reduction of the further surface roughness, as shown in previous study [9].

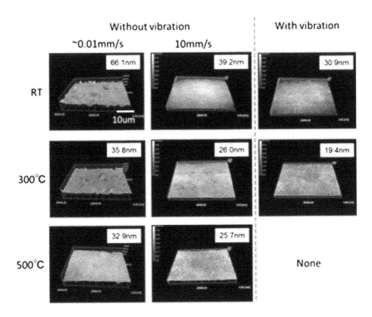

Figure 11.11 Surface images of titanium after heat and vibration-assisted micro-forging. (With permission from Tanabe, 2012.)

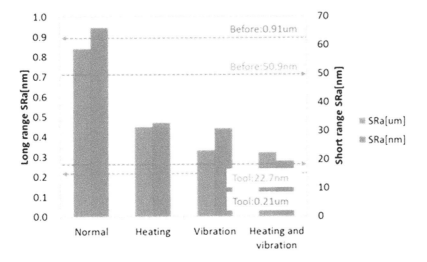

Figure 11.12 Summary of effect of heating and vibration on surface roughness. (With permission from Tanabe, 2012.)

11.2.3 Dieless Forming of Micro Tube

11.2.3.1 Introduction

Microtube which is one of the important micro components is used in various applications such as a painless injection needle and stent, heat exchanger, and micro nozzle. The micro tubes have the functional advantages that are not only weight reduction but also fluid such as gas and liquid can flow inside the tube. As a method of fabricating microtubes, a conventional cold die drawing process has been used. However, it is difficult to scale down the conventional process into a microscale due to the difficulty of manufacturing and handling dies and tools such as mandrel and plug, the size effect of friction resistance. In addition, the increase in frictional resistance between the tool and the microtube due to the size effect with miniaturization causes a decrease in the forming limit. Thus, the development of new drawing technologies for fabrication of microtubes without tools such as dies, plugs, and mandrels has become necessary. For these problems, a dieless forming process is presented in this section. The original heat-assisted dieless drawing process has been developed by Weiss and Kot [10]. The dieless forming by the use of local heating of metal wires, bars, and tubes is to replace the conventional die drawing which are subjected to tensile deformation. This technique can leads to a great reduction of materials in a single pass by the local heating and cooling approach compared with conventional die drawing. This section describes dieless forming technique and its peripheral technologies for fabrication of microtubes.

11.2.3.2 Fundamental Principle of Dieless Drawing

A dieless drawing is a unique deformation process without the need of conventional dies and tools. The superplastic dieless drawing is realized by combination of local heating, tensile deformation, and continuous movement for superplastic materials. Continuous dieless drawing process is shown in Figure 11.13. The local heating devices are fixed and the metal tube moves from left to right through the heating zone. Induction heating [11,12], laser heating [13,14], and direct resistance heating [4] were used for heating source of the dieless drawing process. Concurrently, the workpiece is subjected to tension by the difference in speed between pulling V_1 and feeding V_2. The relationship between the reduction in area R and the speeds V_1 and V_2 is follows:

$$R = 1 - A_2/A_1 = 1 - V_2/V_1 \qquad (11.1)$$

where A_1 and A_2 are original and deformed cross-sectional area, respectively.

11.2.3.3 Fabrication of Microtubes

Figure 11.14 shows the circular microtubes fabricated by the dieless drawing process. A superplastic material Zn–22Al alloy circular microtube with $D = 191\,\mu m$ and $d = 90\,\mu m$ can be obtained, as shown in Figure 11.14 [11]. In addition, the dieless drawing process was applied not only to circular tube but also on noncircular tubes. Figure 11.15 shows the noncircular microtubes fabricated by the dieless drawing process [12]. A noncircular micro tube, which has inner square tubes with a $335\,\mu m$ side and an outer rectangular tube of $533\,\mu m \times 923\,\mu m$, was fabricated successfully without closing of inside of tube and getting distorted cross-sectional shape. Figure 11.16 shows the effect of reduction in area, initial shape, and material on diameter ratio d/D for circular tubes and similarity ratio x/a for noncircular tubes during the superplastic dieless drawing [11,12,15]. The results show

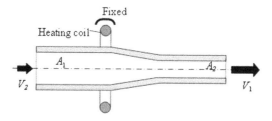

Figure 11.13 Schematic illustration of continuous dieless drawing process.

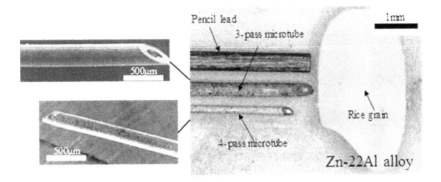

Figure 11.14 Photographs of circular microtubes fabricated by superplastic dieless drawing. (With permission from Furushima and Manabe, 2007, Elsevier.)

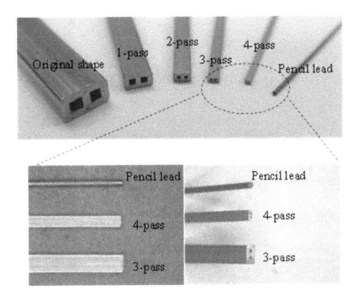

Figure 11.15 Noncircular Zn–22Al alloy microtubes fabricated by superplastic dieless drawing process. (With permission from Furushima et al., 2014, Elsevier.)

that the diameter ratio d/D and similarity ratio x/d are kept at a constant value as well as initial value of original shape. In other words, the geometrical similarity law in the cross-section for circular and noncircular tube was satisfied during superplastic dieless drawing without depending on the material and its original shape. These results show that the dieless drawing is effective not only in circular tubes but also in making noncircular tubes.

11.2.4 Die Surface Treatment for Improving Tribological Properties

11.2.4.1 Introduction

Interfacial behavior between forming dies and work materials has a significant effect on formability and its accuracy in microforming, since the surface area-to-volume ratio greatly increases with the miniaturization of process dimensions. In particular, one of the key technologies playing a role for achieving industrial requirements is enhancement of tool life [16]. In practice, the sticking

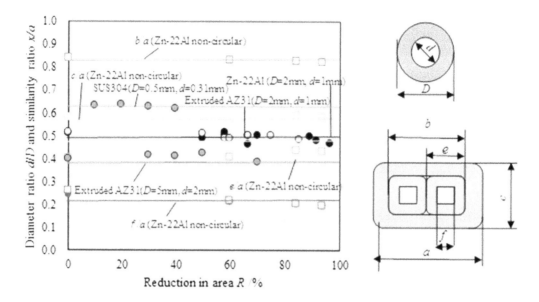

Figure 11.16   Geometrical similarity law in cross-section in superplastic dieless drawing.

of the work material and the wear of the micro-die surface with increasing the amount of product number results in the remarkable decrease of its forming accuracy [17]. In manufacturing site for micro-punching process, for example, it is said that the 10,000 shots of the process is the limitation to keep the quality of produced holes with accuracy of submicrometer range [18]. Furthermore, since the time for precision alignment of the positioning of die set with every interval of tool life accounts for a large part of the total operation time, the enhancement of the tool life makes a significant role to decrease the production cost. Therefore, the development of the high performance micro-dies which realizes the high wear toughness and anti-adhesion of the work material is strongly required. In addition to the general problem of tool life enhancement, the process design for the microforming conflicts with the issue of so-called scale effects [19]. The fundamental research on the scale effect of tribological behavior in microforming has been performed worldwide [20]. The general understanding for the scale effect of friction, which is well described with a lubricant pocket model by Engel [20], is the low effect of lubricant in microscale forming. Furthermore, from the standpoint of dirt handling, contamination of products, and unstable formability, the lubricant-free microforming process is a strong demand from the industry. This section takes a brief overview of the die surface treatment toward the final goal of the achievement of lubricant-free microforming.

11.2.4.2 Application of Hard Film Coatings

Activities of the application of die surface treatment, especially focused on hard film coatings targeting on microforming, are gradually increasing on the background of rapid development of hard film coatings for cutting tools [21]. Hanada et al. fabricated a micro-die utilizing the chemical vapor deposition (CVD) diamond coating [22]. Fuentes et al. investigated the adhesion properties of pure aluminum on the PVD-coated carbide forming-tools for microforming [23]. The authors also showed the anti-adhesion properties of DC magnetron sputtered WC–C-coated carbides. The diamond-like carbon (DLC) films are originally applied to micro-extrusion die by Yang et al. by using pulse plasma CVD method [24]. The studies on the tribological properties of DLC coated micro-tools were followed by Takatsuji et al. and Krishnan et al. [25,26] for micro-extrusion process and Gong et al., Hu et al., and Wang et al. [27–29] for micro-sheet metal-forming process. The tool life endurance test was firstly conducted by Fujimoto et al. [30]. The ionized physical vapored DLC film was coated on the micro-bending die and it shows the high wear resistance during the 50,000 shots of micro-bending tests due to the pre-treatment of the high-energy ion beam irradiation for finishing die-surfaces. Furthermore, a nanolaminated DLC is deposited by radio frequency (RF) sputtering and it bears to the 10,000 shots of the severe contact of micro-bending with ironing of stainless steel metal foils [31].

11.2.4.3 Micro-Surface Texturing

In spite of these developments above, the problems in low adhesion strength of DLC films with substrate has been remained as a technical challenge to endure under the high impact surface pressure in forming process. As an approach to enhance the adhesion properties of DLC films, surface texturing by segmentation of DLC films, which is possible to decrease inner stress and applied strain in deposited films, was proposed by Aoki et al. [32]. They demonstrated the twice longer endurance for micro-textured DLC films compared with the nontextured DLC films. In addition to this anti-delamination effect, surface texturing has further advantages in tribological properties under dry friction, which is the function of wear debris entrapment effect in the trench or valley of the textured geometries [33]. Figure 11.17a,b shows surface images of micro-textured DLC films deposited by ionized physical vapor deposition (I-PVD) using masking with metallic mesh and Figure 11.18 shows evolutions of the coefficient of friction (COF) during the 100,000 times rotations of ball-on-disk type friction tests for the nontextured and the micro-textured

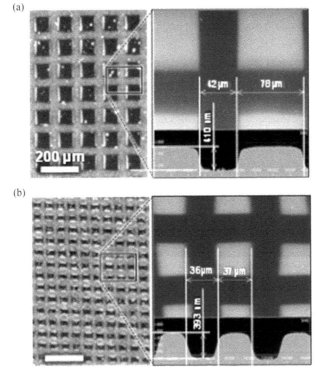

Figure 11.17 DLC films deposited with the mesh with different grid-interval. (With permission from Shimizu et al., 2014, Elsevier.)

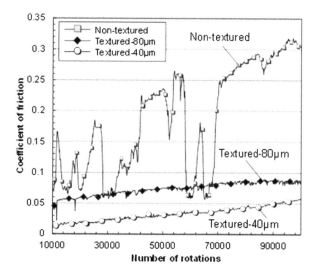

Figure 11.18   Evolution of COF of nontextured and micro-textured DLC films under dry sliding friction during ball-on-disk friction test. (Conditions: sliding speed: 100 mm/s, initial Hertzian pressure: 1.2 GPa.) (With permission from Shimizu et al., 2014, Elsevier.)

DLC films [34]. While the high COF with relatively large deviation is shown for the nontextured films, textured DLC films demonstrate the stable evolution with low COF. Thus, micro-textured structure plays an important role to prevent the wear debris trapping in the area of contact and to promote its ejection.

  a. 80-μm-grid-interval
  b. 40-μm-grid-interval

11.2.4.4 Deposition of High Performance Films on Complex Geometry

Another issue in the conventional hard film deposition process in microforming die is the difficulties in the deposition of high-performance films on the inner wall of closed shaped micro-dies with submillimeter scale holes. In order to enhance the tool life as in industrial scale, thin films with antiadhesive and high-wear toughness are required to deposit uniformly on the three-dimensional complicated shape structure. In view of the uniformity of the nonflat surface, although the CVD process has larger advantages than PVD process, the film properties and substrate temperature would not be applicable for micro-die manufacturing. As for the PVD process, the control of the directionality and energy is achieved by ionization of the vapored species [35]. One of the emerging PVD processes, which can obtain a high degree of ionization, is high-power impulse magnetron sputtering (HIPIMS) technology [36]. HIPIMS deposited TiAlN films at inner wall of 1 mm scale tiny holes demonstrated the highly dense microstructure with fine smooth surface and relatively high hardness of more than 35 GPa even on the inner walls of the small hole structure [37]. The applicability of the HIPIMS deposition films for microforming die application was successfully demonstrated through the micro-circular drawing process (Figure 11.19) [38]. Lower forming load was obtained in the process with the HIPIMS-deposited micro-die. The fine smooth and high hardness properties at inner wall of the micro-dies contributed to the antiadhesive performance during the micro-circular drawing of stainless steel foils. This high performance of the coatings on the inner wall of the HIPIMS deposition film will be expected to contribute to the further enhancement of the tool life in microforming.

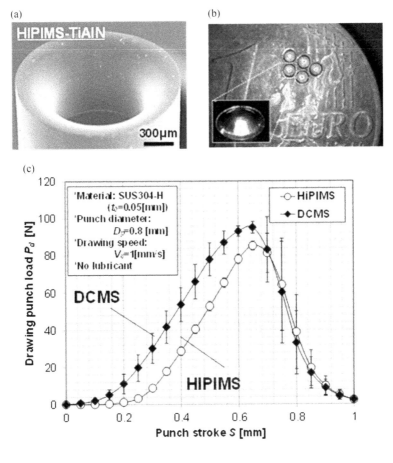

Figure 11.19 Applicability of HIPIMS deposited TiAlN films in micro-circular drawing process. (With permission from Shimizu et al., 2018, Elsevier.)

a. SEM image of microdrawing die with TiAlN films deposited by HiPIMS
b. Appearance of tiny domical shape parts produced by micro-deep drawing
c. Comparison of punch load-punch stroke curves between the TiAlN-coated micro-die by conventional dc magnetron sputtering and HIPIMS

11.3 DEVELOPMENT OF MICRO DEVICES WITH FINE-GRAIN MATERIALS

11.3.1 Development of Highly Precise Tooling System for Micro Stamping

11.3.1.1 Introduction

Stamping and shearing were one of the important processes for plastic deformation. A pair of punch and die presses the material, and then material will break with fractured surface. The sheared surface in the hole was configured as rollover, burnished surface, fractured surface, and burrs, as shown in Figure 11.20. Clearance between punch and die was determined by ratio of the material thickness. The clearance at the conventional setting condition in stamping of SUS304 was recommended from 10% to 15% of material thickness [39]. However, there are some problems. If the

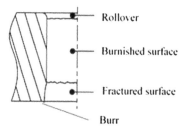

Figure 11.20 Sheared surface in cross-section of hole.

material thickness decreases until 10 μm, allowance clearance was only 1 μm at 10% of material thickness. In general, accuracy of the tooling parts and the die system has been considered about 1 μm. Thus, clearance cannot be set beyond precision of tooling and positioning accuracy. Especially, adjusting tool position has been adopted by analog processes such as mechanical grinding and polishing. This section reports the newly developed digital tool positioning system and effectiveness of adoption of the digital adjusting process.

11.3.1.2 Development of Digital Clearance Adjusting Systems

Figure 11.21 shows the conventional analog process of adjusting die position. Figure 11.21b has a misalignment between punch and die. In conventional process, the left side of die was ground until same amount of misalignment (Figure 11.21c). After that metal foil at same amount of misalignment was set into the right side of the die (Figure 11.21d). Then, misalignment gets resolved in the die. These analog-adjusting processes take about 1 h.

Figure 11.22 shows developed nanometric digital clearance adjusting systems. This system adopts nanometric X–Y stage and it is set under of the die. These stages are controlled by linear encoder and movement of stage was monitored by laser displacement meter at 1-nm resolution (Keyence SI-F01). The nanometric X–Y stage can move the stage at 5-nm resolution in 40-μm range. All die system plates are made by hardened SKD11. Guide post was adopted roller guide that was made by AGATHON. Thus, the stripper plate and punch plate were able to move in vertical direction by keeping zero clearance. If the system is used, the clearance adjusting time decreases from 1 h to only 15 min. The productivity of clearance adjustment was improved four times than conventional analog process. Furthermore, this digital adjusting system brings effectiveness of standardization.

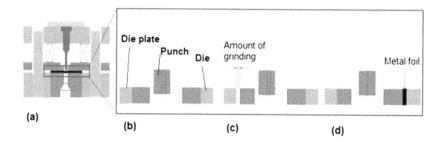

Figure 11.21 Conventional analog clearance adjusting method: (a) construction of die system, (b) before adjusting, (c) grinding of die, and (d) after grinding.

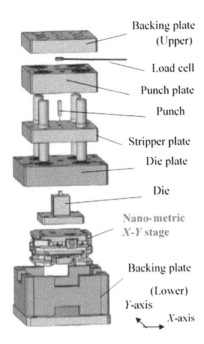

Figure 11.22   Structure of digital clearance adjusting system.

11.3.2 Development of a Micro Metal Pump by Stamping and Diffusion Bonding

11.3.2.1 Introduction

The silicon process has been adopted, in general, in MEMS manufacturing process. In the field of micro pump, a silicon pump has been developed [39] and a drug delivery system is one of the practical areas. However, there are some problems with a silicon pump. For example, it is known that the material and the processes costs are high and the limited ductility determines the flow rate. Thus, the manufacturing process of a micro pump with stainless steel has been developed by using etching and diffusion bonding processes [40]. However, a stainless steel micro pump shows diffluent problems. Firstly, each part of the pump needs higher micro-meter accuracy and the parts have to be bonded together without leakage and deformation. To overcome these problems, the micro-stamping process has been converted into etching process in the forming process of sheet parts, and a new low-temperature diffusion bonding processes were developed. As a result, these newly developed technologies can fulfill the demand for mass production of three-dimensional structured metal MEMS.

11.3.2.2 Structure of Diaphragm Pump

Figure 11.23 shows the structure of the diaphragm pump and a schematic view of the driving pump. In this pump, the diaphragm is raised and depressed by adding positive and negative voltages to a piezoelectric element periodically, and the flow rate is regulated by two valves. Thickness of the valves is only 10 µm. The valve is an important component for suction and discharging. Flatness of pump affects leakage performance of pump. Therefore, it is considered that the dimensional accuracy of valves determines the pump performance. Figure 11.24 shows the enlarged drawing of the valve. The valve is suspended by three U-shaped springs in 120-µm width.

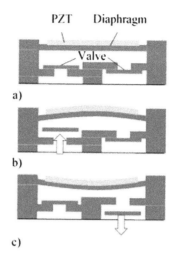

Figure 11.23 Structure and schematic view of pump: (a) structure of diaphragm pump, (b) add negative voltage, and (c) add positive voltage.

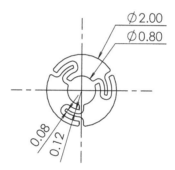

Figure 11.24 Valve shape.

11.3.2.3 Zero Clearance Die Systems for Stamping the Metal Foil Valve

While stamping three U-shaped 10 μm thickness valve, zero clearance between the punch and the die is required. It is a quite difficult task to create such die systems. Figure 11.25 shows the developed zero clearance die. This die adopts newly developed zero clearance punch, round fine centering systems, and special structure for nanometric accuracy. There are three manufacturing steps for creating the zero clearance punch. Figure 11.26 shows the shaving process of punch. NAK80 ($Hv = 450$) and ultra-fine-cemented carbide are used for this punch and this die, respectively. At the first step, roughly three U-shaped punch shape is cut by pico-second laser. After that, laser cut punch is made by shaving into the die. The cutting width is set at 10 μm. In the final step, the burr of shaving one is removed by pico-second laser.

Figure 11.27 shows the rough shaped punch after pico-second laser cutting in (a) after shaving, and burr removing punch shape in (b). The clear U-shape appears in after shaving case and burr removing punch. Figure 11.28 [41] shows the SEM images of stamped three U-shaped valves using SUS304 metal foils. Three U-shapes are processed with only one shot stamping. There are stable shared surface at rollover side and fractured surface side. The flatness of valve has been secured and there are almost no burrs at fractured surface. Therefore, it is considered that developed

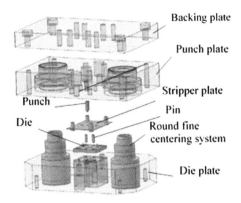

Figure 11.25 Construction of zero clearance die system.

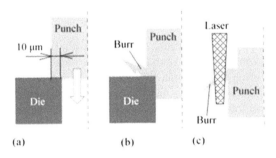

Figure 11.26 Shaving process of punch: (a) before shaving, (b) shaving, and (c) laser deburring.

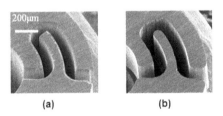

Figure 11.27 Enlarged SEM view at tip of U-shape punch (a) after laser cutting and (b) after shaving and laser deburring.

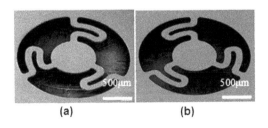

Figure 11.28 SEM image of stamped U-shape valve (a) rollover side and (b) fractured surface side.

zero-clearance die systems can fulfill the demands for mass production of metal foil stamping with irregular shapes.

11.3.2.4 Low-Temperature Diffusion Bonding

At the conventional SUS304 diffusion process, using diffusion bonding temperature should be at least 800°C. The reason is that oxygen and carbon in the passive film are halved at this temperature [42]. If using a fine-grained SUS304, diffusion bonding temperature decreases at 700°C [43]. Figure 11.29 [41] shows the 700°C low-temperature diffusion bonding result at eight-layered diaphragm pump using fine-grained SUS304. Gaps between each sheet are small and the boundary of bonding surface is stable. The stable boundary could improve the reliability of pump and enhances the safety performance. Furthermore, the material with fine-grain can improve the flow rate of the pump, because of its higher yield stress. The grain size can control with adjustment of diffusion bonding time.

11.4 SUMMARY

Several new technologies on microforming are introduced. These technologies are aimed to overcome issues in the microforming due to the size effects while scaling-down. A novel process combining micro press-forming and automatically assembling in a progressive die is effective for fabrication of micro components precisely and for fabrication of micro devices. A heat-assisted forming system can improve formability and accuracy of the parts with low energy consumption.

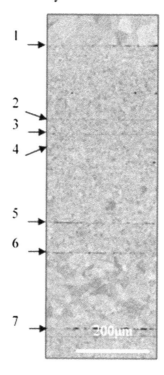

Figure 11.29 Result of 700°C low-temperature diffusion bonding at eight-layered diaphragm pump.

A dieless tube forming system can produce micro tubes without dies, and a coating system for micro-die adapts to improve tribological properties without any lubricants. Furthermore, a process of fabricating a micro device, metallic pump by using precise press-forming system, is introduced. By these developments, micro metal forming becomes more practical for manufacturing micro parts and micro functional devices, such as MEMS for low-cost sensors and actuators, and for bio-assays and medical devices.

REFERENCES

1. U. Engel, R. Eckstein, 2002, Microforming-from basic research to its realization, *Journal of Materials Processing Technology*, 35, 125–126.
2. F. Vollertsen, Z. Hu, H.S. Niehoff, C. Theiler, 2004, State of the art in micro forming and investigations in-to micro deep drawing, *Journal of Materials Processing Technology*, 151, 70–79.
3. Y. Saotome, 2003, Technology for micro parts and die process in nano, *Journal of Tooling Engineering*, 43–13, 84–89.
4. M. Yang, 2004, Fabrication of MEMS using micro metal forming process, *Proceedings of ICNFT*, Bremen, pp. 135–140.
5. M. Yang, K. Manabe, K. Ito, 2007, Micro press forming and assembling of micro parts in a progressive die, *Journal of Mechanical Science and Technology*, 21, 1338–1343.
6. E. Egerer, U. Engel, 2004, Process characterization and material flow in microforming at elevated temperatures, *Journal of Manufacturing Processes*, 6(1), 1–6.
7. J.P. Wulfsberg, M. Terzi, 2007, Investigation of laser heating in microforming applying sapphire tools, *Annals of the CIRP*, 56(1), 321–326.
8. X. Peng, Y. Qin, 2006, Coupled electrical-thermal-mechanical analysis for electrical/laser heating assisted blanking, *Chinese Journal of Mechanical Engineering*, 19(2), 294–299.
9. H. Tanabe, Development of heat assisted micro forming system and evaluation of forming characteristic, Master thesis, Tokyo Metropolitan University, Hachioji, 2012.
10. V. Weiss, R.A. Kot, 1969, Dieless wire drawing with transformation plasticity, *Wire Journal*, 9, 182–189.
11. T. Furushima, K. Manabe, 2007, Experimental study on multi-pass dieless drawing process of superplastic Zn–22%Al alloy microtubes, *Journal of Materials Processing Technology*, 187–188, 236–240.
12. T. Furushima, K. Manabe, 2010, Fabrication of AZ31 magnesium alloy fine tubes by dieless drawing process, *Journal of the Japan Society for Technology of Plasticity*, 51, 990–992.
13. T. Furushima, Y. Imagawa, S. Furusawa, K. Manabe, 2014, Deformation profile in rotary laser dieless drawing process for metal microtubes, *Procedia Engineering*, 81, 700–705.
14. Y. Li, N.R. Quick, A. Kar, 2002, Thermomechanical effects in laser microprocessing for dieless metal wire drawing. *Journal of Laser Applications*, 14, 91–99.
15. P. Tiernan, M.T. Hillery, 2004, Dieless wire drawing—An experimental and numerical analysis, *Journal of Materials Processing Technology*, 155–156, 1178–1183.
16. T. Furushima, A. Shirasaki, K. Manabe, 2014, Fabrication of noncircular multicore microtubes by superplastic dieless drawing process, *Journal of Materials Processing Technology*, 214, 29–35.
17. T. Shimizu, S. Iwaoka, K. Manabe, Y. Teranishi, K. Morikawa, 2012, Tribological properties of anodized pure titanium foils in micro-deep drawing with ironing, *Proceedings of 15th AMPT, (CD-ROM)*.
18. I. Aoki, M. Sasada, 2008, Micro shearing process, *Journal of The Japan Society for Technology of Plasticity*, 49–570, 619–623 (in Japanese).
19. F. Vollertsen, 2008, Categories of size effects, *Production and Engineering: Research and Development*, 2–4, 377–383.
20. U. Engel, 2006, Tribology in microforming, *Wear*, 260, 265–273.
21. K.-D. Bouzakis, N. Michailidis, G. Skordaris, E. Bouzakis, D. Biermann, R. M'Saoubi, 2012, Cutting with coated tools: Coating technologies, characterization methods and performance optimization, *CIRP Annals Manufacturing Technology*, 61, 703–723.
22. K. Hanada, L. Zhang, M. Mayuzumi, T. Sano, 2003, Fabrication of diamond dies for microforming, *Diamond and Related Materials*, 12, 757–761.

23. G.G. Fuentes, M.J. Diaz de Cerio, R. Rodriguez, J.C. Avelar-Batista, E. Spain, J. Housden, Y. Qin, 2006, Investigation on the sliding of aluminium thin foils against PVD-coated carbide forming-tools during micro-forming, *Journal of Materials Processing Technology*, 177, 644–648.

24. X.D. Yang, T. Saito, Y. Nakamura, Y. Kondo, N. Ohtake, 2004, Mechanical properties of DLC films prepared inside of micro-holes by pulsed plasma CVD, *Diamond and Related Materials*, 13, 1984–1988.

25. N. Takatsuji, K. Dohda, T. Makino, T. Yoshimura, 2007, Friction behavior in aluminum micro-extrusion, *Proceedings of ICTMP*, pp. 157–162.

26. N. Krishnan, J. Cao, K. Dohda, 2007, Study of the size effects on friction conditions in microextrusion— Part 1: Microextrusion experiments and analysis, *ASME Journal of Manufacturing Science Engineering*, 129(4), 669–676.

27. F. Gong, B. Guo, C. Wang, D. Shan, 2011, Micro deep drawing of micro cups by using DLC film coated blank holders and dies, *Diamond and Related Materials*, 20, 196–200.

28. Z. Hu, A. Schubnov, F. Vollertsen, 2012, Tribological behaviour of DLC-films and their application in micro deep drawing, *Journal of Materials Processing Technology*, 212, 647–652.

29. C. Wang, B. Guo, D. Shan, X. Bai, 2013, Tribological behaviors of DLC film deposited on female die used in strip drawing, *Journal of Materials Processing Technology*, 213, 323–329.

30. K. Fujimoto, M. Yang, M. Hotta, H. Koyama, S. Nakano, K. Morikawa, J. Cairney, 2006, Fabrication of dies in micro-scale for micro sheet metal forming, *Journal of Materials Processing Technology*, 177, 639–643.

31. T. Aizawa, K. Itoh, E. Iwamura, 2010, Nano-laminated DLC coating for dry micro-stamping, *Steel Research International*, 81(9), 1169–1172.

32. Y. Aoki, N. Ohtake, 2004, Tribological properties of segment-structured diamond-like carbon films, *Tribology International*, 37, 941–947.

33. N.P. Suh, M. Mosleh, P.S. Howard, 1994, Control of friction, *Wear*, 175(1–2), 151–158.

34. T. Shimizu, T. Kakegawa, M. Yang, 2014, Micro-texturing of DLC thin film coatings and its tribological performance under dry sliding friction for microforming operation, *Procedia Engineering*, 81, 1884–1889.

35. S.M. Rossnagel, 1995, Directional and preferential sputtering-based physical vapor deposition, *Thin Solid Films*, 263, 1–12.

36. V. Kouznetsov, K. Macak, J.M. Schneider, U. Helmersson, I. Petrov, 1999, A novel pulsed magnetron sputter technique utilizing very high target power densities, *Surface and Coatings Technology*, 122, 290–293.

37. T. Shimizu, H. Komiya, Y. Teranishi, K. Morikawa, H. Nagasaka, M. Yang, 2017, Pressure dependence of (Ti, Al)N film growth on inner walls of small holes in high-power impulse magnetron sputtering, *Thin Solid Films*, 624, 189–196.

38. T. Shimizu, H. Komiya, T. Watanabe, Y. Teranishi, K. Morikawa, H. Nagasaka, K. Morikawa, M. Yang, 2014, HIPIMS deposition of TiAlN films on inner wall of micro-dies and its applicability in micro-sheet metal forming, *Surface and Coating Technology*, 250, 44–51.

39. www.emft.fraunhofer.de/en/mediacenter/press-briefings/2015-11-16_smallest-micropump.html.

40. www.kikuchiseisakusho.co.jp/mechatro2/images/MMP_20150817_ENG.pdf.

41. T. Shiratori, S. Nakano, M. Katoh, Y. Suzuki, T. Aihara, M. Yang, 2016, Development of metal MEMS manufacturing technologies using pierced metal foil and diffusion bonding process at low temperature, *4M/IWMF2016 The Global Conference on Micro Manufacture*, September 13–15, 2016, Denmark, pp. 209–212.

42. O. Ohashi, S. Suga, 1992, Effect of surface composition on diffusion welding in stainless steel, *Journal of the Japan Institute of Metals*, 56(5), 579–585.

43. M. Katoh, N. Sato, T. Shiratori, Y. Suzuki, 2017, Reduction of diffusion bonding temperature with recrystallization at austenitic stainless steel, *ISIJ International*, 57(5), 883–887.

Mechanical Joining Processes

V. Satheeshkumar
National Institute of Technology Tiruchirappalli

R. Ganesh Narayanan
IIT Guwahati

CONTENTS

12.1 INTRODUCTION

Mechanical joining processes are known for joining similar and dissimilar components in which no metallurgical bonding takes place. In these processes, parts are joined by mechanical joining methods either temporarily or permanently depending on the application. Some of the typical examples of mechanical joining processes include mechanical fastening, clinching, joining by forming or plastic deformation, stapling, nailing, stitching, and joining by fits. The mechanical joining processes play a key role in pressure vessel fabrication, truss construction, and in aircraft, automobile, and ship building sectors. The major advantages of mechanical joining processes include no heat energy required for joining, no generation of fumes, no requirement of flux and filler rod, joint can be repaired, capable of joining multi-material construction, and ability to join dissimilar materials like metal to glass, metal to polymers, and thicker parts. On the other hand, the major disadvantage of the mechanical joining process is of stress concentration at the location of holes in the components for fastening. The main features of the mechanical joining processes that help us in waste management and developing sustainable structures include design for disassembly, light-weight structure construction, deconstruction, and recyclability. In this chapter, various mechanical joining processes and the joining processes that offer mechanical interlocking like adhesive bonding are briefly discussed.

12.2 THREADED FASTENERS

Temporary joining of components is performed by mechanical fasteners like threaded fasteners, pins, and clips. A few studies in relation to threaded fasteners predominantly used in industries are discussed in this section. Figure 12.1 shows the examples of joining components using threaded fasteners like screws or bolts and nut combinations. Screw joints are not meant for thinner sheets as these do not provide sufficient load-bearing length. Therefore, other joining methods such as flow drill screws, collar forming, or spring nuts are used to join thinner sheets. Bolted joints are used for joining of thicker and large sectioned components. Threaded fasteners are easy for disassembly and during remanufacture. Figure 12.1 also shows that bolted joints are applicable in tensile and shear modes of loading (Martinsen et al., 2015).

McComb and Tehrani (2015) demonstrated the effect of sheet metal screws for resisting transverse shear in a polymer composite concrete enhanced with a steel sheet deck as shown in Figure 12.2. Results showed that there is a significant difference in the ultimate deflection of enhanced deck with fasteners as compared with without fasteners cases. Nearly 100% increase in strength was observed with the addition of the fasteners.

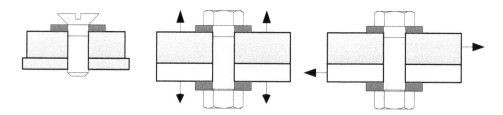

Figure 12.1   Temporary mechanical fastening methods such as screw joint, bolted joint in shear, and bolted joint in tension. (With permission from Martinsen et al., 2015, Elsevier.)

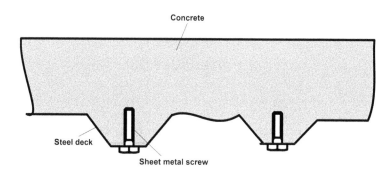

Figure 12.2   Fastening of sheet metal screws. (With permission from McComb et al., 2015, Elsevier.)

There is also evidence that the type of fastener head influences the joint performance. In the study carried out by Qin et al. (2013), the joint performance of a double-lap joint fabricated with a protruding head bolt and countersink bolt was compared. The joint made by the countersink bolt shows about 32% lesser yield strength and larger damage around the outer plate hole on bolt head side as compared with the protruding head joint. The fastener effect on the final strength is very small. Intricate details indicate that the clearance between the bolt and the plates plays a dominant role in determining the joint failure.

12.3 PERMANENT MECHANICAL JOINING PROCESSES

In this section, permanent mechanical joining processes like mechanical fastening processes, clinching, joining by forming methods, joining by fits, and adhesive bonding are discussed.

12.3.1 Permanent Mechanical Fastening Processes

Traditional fusion welding processes are not capable of joining multimaterial construction of dissimilar materials. Using mechanical fasteners like bolt and nut combinations or rivets for the purpose is one of the alternative methods. Conventional bolting or riveting is a two-step joining method that includes pre-drilling and fastening. One-step mechanical fastening processes, such as self-pierce riveting (SPR), clinching, friction riveting (FR), friction stir blind riveting (FSBR), friction self-piercing riveting (FSPR), rotation friction drilling riveting (RFDR), and rotation friction pressing riveting (RFPR) have been developed and applied to join similar and dissimilar materials. Though riveting is used for joining high-strength metal alloys, mechanical fastening of the light-weight alloys like aluminum and magnesium alloys and their joining to polymeric materials has many challenges (Min et al., 2015a; Padhy et al., 2018).

12.3.1.1 Solid and Blind Riveting

In conventional solid and blind riveting processes, two steps—hole making in the component and riveting—are responsible for joining (Figure 12.3). In solid riveting, after the rivet is inserted into the hole, it is plastically deformed at the opposite end to complete joint fabrication. In blind riveting, after insertion of the rivet head, the mandrel within the rivet is pulled out to plastically deform the rivet head in the bottom end to make the joint. The requirement of a preexisting hole in the solid and blind riveting processes increases manufacturing time and decreases fatigue life (Trimble et al., 2016). However, the performance of such joints is at par with the joints made by solid-state welding

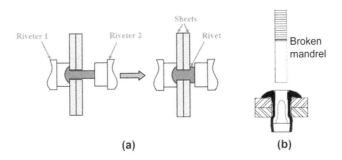

Figure 12.3 Riveting processes: (a) solid riveting. (With permission from Li et al., 2017, Elsevier.) (b) Blind riveting. (With permission from Min et al., 2015a, Elsevier.)

methods and adhesive-bonded joints. Electromagnetic riveting (EMR) is also possible as depicted by Li et al. (2017).

Laforte and Lebel (2018) developed a riveting process to fasten a composite laminate using thermoplastic composite rivet blank. In this process, the rivet placed into the joint hole is heated using resistance heating and then formed by applying force, as shown in Figure 12.4. The riveting process can be performed at different temperatures. At elevated temperature (say at 450°C), the properties of composite get affected. This study recommended that thermoplastic could be used in riveting at temperatures higher than the glass transition temperature of epoxy without damaging carbon fiber/epoxy laminates.

Min et al. (2015b) proposed a modified one-step blind riveting method for joining dissimilar materials. The conventional rivet is modified to have an inclined inward tip (Figure 12.5) to reduce the penetration force. During joining, the usual method described for blind riveting is followed with the modified rivet rotating at a high speed of 9,000 rpm (Figure 12.6). The fastening process was demonstrated by fabricating lap joints made of AA6111-T4 and AA6022-T4 with feed rates lower than 480 mm/min. It was observed that both the torque and force increase with an increase in feed rate during the rivet penetration. There is about 20% increase in ultimate tensile load when using modified rivet as compared with using conventional rivet.

Kato et al. (2001) proposed a short pipe rivet and a wavy ribbon rivet for joining metal sections as shown in Figure 12.7. In pipe riveting, the rivet is forcefully inserted into the workpieces by the application of force as shown in Figure 12.7a,b. Here ends of the rivet undergo slight flaring to achieve compact joint. The performance is found to be encouraging as observed from in-plane and out-of-plane tests. The optimized inclination for pipe rivet ends is 45°–60°. The wavy ribbon

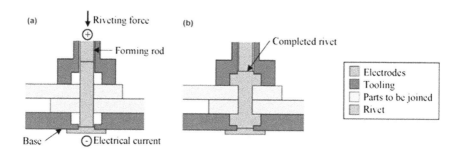

Figure 12.4   Thermoplastic composite riveting process: (a) joint assembly before forming of rivet head and (b) joint formed after forming the rivet head. (With permission from Laforte and Lebel, 2018, Elsevier.)

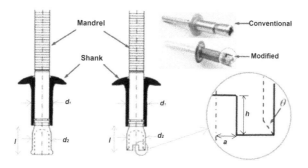

Figure 12.5   Schematic of conventional rivet and modified rivet. (With permission from Min et al., 2015b, Elsevier.)

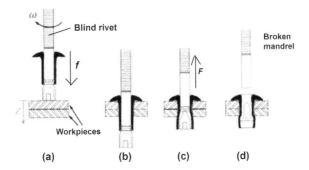

Figure 12.6 Stages of single-sided blind riveting process: (a) positioning of rivet, (b) rivet tip cutting and penetration into the workpieces with a high rotational speed, (c) mandrel pulled back to join the workpieces, and (d) excess mandrel portion is cut. (With permission from Min et al., 2015b, Elsevier.)

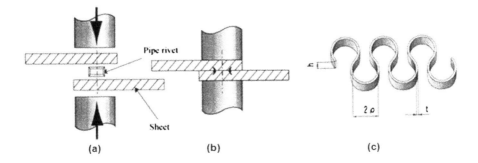

Figure 12.7 Schematic of (a) initial stage of joining by pipe rivet, (b) fabricated joint using pipe rivet, and (c) wavy ribbon rivet. (With permission from Kato et al., 2001, Elsevier.)

rivet (Figure 12.7c) is used for a continuous, long interface joining purpose. The wave rivet joining strength is almost same as that of pipe rivet joining.

Apart from conventional rivets, some special rivets have been developed in order to increase joint strength, interference percentage, and other mechanical performance (Mucha, 2013). Mucha fabricated a solid rivet for joining sheet materials using SPR. Application of electromagnetic repulsive force is another way to fabricate rivet joints. Li et al. (2017) discussed the mechanical properties and fatigue behavior of EMR lap joints and compared to regular pressure riveting (RPR). In EMR, the powerful repulsive force is generated between the two magnetic fields and the force is utilized on the punch to strike the rivets. The fatigue performance of EMR is either equal or 2–3 times larger than that of RPR. This is due to better and uniform interference fit. The shear load is slightly better in case of EMR. Moreover, the joining time per riveting is 1 ms for EMR as compared with 2 min in case of RPR.

12.3.1.2 SPR

SPR is a one-step mechanical joining process which is used for joining metallic sheets, multi-sheet material combination, and plastics. Steel sheets can be joined with aluminum sheets, copper sheets, magnesium alloys, plastics, etc., for a wide range of industry applications. There is a large difference in mechanical and physical properties between these materials making fusion welding very difficult. A considerable process optimization is required. In such situations, SPR is a better choice. In this process, a metallic hollow rivet is used to mechanically interlock the sheets by controlled plastic deformation. The joining has no metallurgical bonding. A predrilled hole is not required as observed in other spot joining methods. The rivet is driven through the upper sheet and the tip of the rivet is flared to form a hook profile and join the upper and lower sheets as shown in Figure 12.8. The high-strength boron steel is used to make rivet with 500 HV hardness. Often zinc plating is done to prevent corrosion (Mori et al., 2013).

Van Hall et al. (2018) improved the flaring performance of rivet material, 10B37, by incorporating a decarburized ferrite surface layer through a modified quench and temper process. The results showed that the process of decarburization shows threefold improvement in flaring performance than the untreated case. Although SPR is suitable for joining dissimilar materials, their ductility plays a vital role in deciding the joint fabrication without cracks occurring during joining. FSPR is a suitable joining technology, which combines the SPR with friction stir spot welding (FSSW). Modeling studies of Ma et al. (2016) reveals that during FSPR the temperature increases in the upper sheet until piercing and then decreases once it is completed. At the same time, the bottom sheet experiences increasing temperature throughout the process. Such higher temperature in the severely deformed region of the lower magnesium sheet improves the rivetability, although its ductility is less (Ma et al., 2016). A solid rivet can be used instead of tubular rivet as suggested by Mucha (2013). This rivet is used in the production of spot welds for thin-walled car body parts.

12.3.1.3 FR

Blaga et al. (2015) introduced FR (FricRiveting) technique and demonstrated the feasibility of riveting glass fiber-reinforced polymer (FRP)/titanium/aluminum structure for structural engineering applications. The principle of FR includes friction-based heat generation by the metallic rivet, melting the polymer, followed by rivet tip plastic deformation by providing axial force as shown schematically in Figure 12.9. The peak temperature ranges from 450°C to 600°C. FricRiveted joints

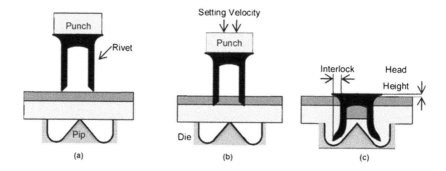

Figure 12.8 SPR stages (a) locating tools and sheets, (b) deformation and piercing of upper sheet, and (c) deformation of lower sheet and formation of joint. (With permission from Van Hall et al., 2018, Elsevier.)

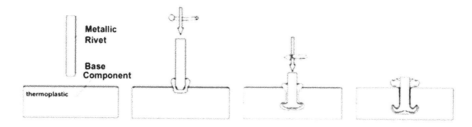

Figure 12.9 Schematic of the friction riveting process: (a) positioning of rivet and component, (b) inserting the rivet into the polymer, (c) forming of rivet, and (d) joint formation. (With permission from Blaga et al., 2015, Elsevier.)

have benefits like high strength, short joining cycles, and no complex surface preparation require-ment, yielding a joint strength 20% larger than bolted joints. Borba et al. (2018) proposed direct-FR technology that is suitable for joining polymer composites using titanium (Ti–6Al–4V) rivets. Micro-mechanical interlocking within the constituent materials along with macro-mechanical rivet locking are responsible for improved joint strength as compared with bolted joints.

12.3.1.4 FSBR

In the FSBR process, a blind rivet is rotated at about 3,000–12,000 rpm and pressed against the upper sheet to be joined. The frictional heat generated from the contact established between the rotating rivet and the sheet, softens the workpiece, allowing the rivet to plunge through the two sheets. After that, the inner mandrel is pulled upwards to deform the bottom end of the rivet to com-plete the joint fabrication. The exposed region of the mandrel is removed. A pre-existing hole is not required for FSBR, and one-sided access is sufficient to join the sheets. It is applicable for dissimilar materials as well (Trimble et al., 2016).

12.3.1.5 RFPR

Duan et al. (2014) proposed RFPR method for AZ31 alloy. In RFPR operation, Figure 12.10, friction heat is generated between the rivet and sheets to be joined with a plug rotating at high speed. This softens the sheet materials and enables the rivet to be drilled into the sheets under reduced force. Once the rivet is fully inserted, its rotation is halted. The plug inside the rivet is immediately pressed into the shank of the rivet by a bottom punch establishing a mechanical interlock between the rivet and the sheets by rivet expansion. The joint produced by RFPR shows superior shear strength and fatigue property as compared with SPR. The punch pressure should be optimized to reduce the unwanted compressive residual stress and strain in the specimens.

12.3.2 Clinching

Mechanical clinching (or press joining) is used for joining similar and dissimilar sheet materials like SPR. This process is successfully used in automobile industry (Mucha et al., 2011). In mechani-cal clinching, sheets are joined by local cold forming with a punch and die, but without a permanent rivet inserted into the joint (Figure 12.11). Two metal sheets are placed on the upper surface of a die containing a shaped groove. The punch is then driven into the sheets. When the punch moves downward, the upper sheet is spread into the lower sheet and through an annular gap at the groove of the die, and the two layers are mechanically locked to create a joint (Gao and Budde, 1994; Xing et al., 2015).

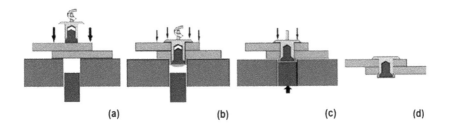

(a)   (b)   (c)   (d)

Figure 12.10   Stages in RFPR process: (a) positioning of rivet, (b) penetration of rivet by rotating and axial force, (c) rotation stop and pressing plug into the shank, and (d) completion of riveting. (With permission from Duan et al. 2014, Elsevier.)

Figure 12.11   Schematic of mechanical clinching of sheets. (With permission from Xing et al., 2015, Elsevier.)

The strength of the clinched joint is influenced by the degree of interlock and the thickness of the upper sheet at the neck. Tool profile, either round or square, also decides the joint strength (Mori et al., 2013). The advantages of mechanical clinching are short joining time, high tool life, simple and affordable equipment, useful for joining coated sheets that are difficult to join by resistance spot welding (RSW). The disadvantages include double-side access required for the joint, bulges and indents in the joint, and relatively large joining force. Neck fracture or button failure are the two major modes of failure observed while testing clinched joints.

While several research contributions are seen in conventional clinching, Lambiase (2015) investigated the influence of preheating of sheets (or warm clinching) on the material flow and mechanical behavior of clinched joints made of Al6xxx sheets. The employment of preheating allows using deeper dies which produce larger material flow and larger interlocks as compared to those produced with shallow dies. Lambiase and Di Ilio (2015) investigated the warm clinching of AA5053 and polystyrene sheets. At suitable preheating, joint forms without polymer cracking, while excessive preheating causes softening of polymer and weaker joints. Moreover, the overall joint fabrication time is about 2 s. Hydroclincing and hydro SPR are also possible for spot joining of metallic structures (Neugebauer et al., 2008). The bottom side of the joint which usually contains a die is replaced by hydrostatic pressure making the process effective in joining tubular structures. An existing hydroforming machine can be used to provide hydrostatic pressure like in conventional tube/sheet hydroforming.

In addition, heated flat dies have been used to reduce the preheating time for the low ductile materials like magnesium alloy sheets (Mori et al., 2013). Gerstmann and Awiszus (2014) reported developments in flat-clinching. The sheets to be joined are mounted on a planar anvil. The blankholder clamps the sheets and the punch moves down to deform the sheet such that joining happens within the total sheet thickness eliminating bulges and indents created in conventional clinching.

There are other varieties of clinching process, namely friction-assisted clinching introduced by Lambiase and Paoletti (2018) for joining aluminum sheets with polymer laminates, two-step clinching that includes clinching followed by upsetting (Chen et al., 2016), and single-side dieless clinching in which a groove in the blankholder contributes in joint formation (Atia and Jain, 2017) are notable. These are simple clinching methods that have potential applications in automotive industries. Joining titanium sheets with aluminum alloy sheets and copper sheets are possible by extensible die clinching process (He et al., 2015).

Hole clinching is a suitable process for joining less ductile materials like composites with ductile metallic materials as shown in Figure 12.12a. In this process, the upper sheet is forged against the lower sheet with a hole and the joint formation occurs by the extruded upper sheet through the hole. There is an anvil cavity at the bottom that forms a joint head. To improve the mechanical stability, water tightness, and corrosion resistance of bolted and riveted joints, adhesive bonding is used additionally (Jahn et al., 2016). Lee et al. (2017) proposed a spring assisted die for hole clinching for joining reinforced plastic and ductile materials. In this method, a spring die supported by a

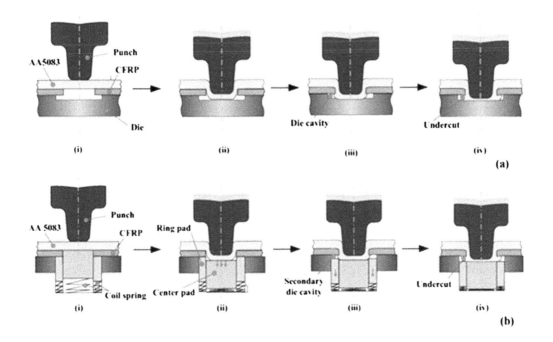

Figure 12.12 Process stages for hole clinching: (a) with conventional die, (b) with spring die. [Steps: (i) positioning of the sheets and punch above die, (ii) forming of the upper sheet, (iii) extrusion and spreading of the upper sheet to make an undercut, and (iv) final cross-section of the joint.] (With permission from Lee et al., 2017, Elsevier.)

coil spring at the bottom is used to improve the sheet formability and reduces damage to the laminate (Figure 12.12b). The spring die reduced the damage accumulation by about 50%.

12.3.3 Joining by Forming

In joining by forming, workpieces are mechanically joined by controlled plastic deformation. Joining by forming processes such as hydroforming, electromagnetic forming, incremental forming, crimping, and hemming are discussed in this subsection.

12.3.3.1 Interference Fit and Form-Fit

The tubular and cylindrical parts are joined either by interference-fit or form-fit. An interference-fit joint is created based on the difference in the elastic recovery between the parts to be joined. The interference pressure between the parts after deformation thus results in by either expansion or compression of the parts to be joined. For example, in Figure 12.13, after aligning the parts, the

Figure 12.13 Interference-fit joint made by expansion. (With permission from Mori et al., 2013, Elsevier.)

forming pressure is applied up to a critical value in order to expand the internal surface of the tube by plastic deformation. Once the internal pressure is removed, both the parts undergo elastic recovery, but differently, fabricating a joint. The joint strength in this case is predominantly governed by Coulomb's law of friction. In the case of form-fit joints, circumferential grooves are produced in one of the parts to be joined (Figure 12.14). The applied internal pressure deforms the tube into the groove of the mandrel to make a joint. The tube can be deformed by the application of electromagnetic force. The rectangular, triangular, or circular grooves are used for joining. Out of these, the rectangular grooves yield stronger joints (Mori et al., 2013). The forming pressure can be induced by fluid pressure like in hydroforming to fabricate interference-fit and form-fit joints (Mori et al., 2013).

12.3.3.2 Joining by Shear Spinning

Shear spinning is used to manufacture multi-material structures. In this process, in the first step, the deformable structure is formed to create undercuts using a split mandrel. In the second step, a ring is positioned with another mandrel between the undercuts. Another shear spinning yields form-fit joint as shown in Figure 12.15. Rotary swaging can be used to achieve the same compound structures (Mori et al., 2013).

12.3.3.3 Joining by Crimping and Hemming

Mechanical crimping and hydraulic crimping are the two different methods available. Mechanical crimping is used to fabricate form-fit or interference-fit joint using segmented tools called crimping dies. In the crimping process, the tube is plastically deformed to fill a groove, which is machined/contained on a solid cylinder. In the case of hydraulic crimping process, forming pressure is applied on an elastomer, which in turn causes plastic deformation of the outer tubular part causing an interference-fit (Shirgaokar et al., 2004; Mori et al., 2013).

Hemming is essentially a sheet bending operation in which a sheet edge is bent 180° such that it clamps another sheet edge to form a sandwich-like arrangement. It has application in automotive, electrical appliances, and food industries. Sometimes warm hemming is done for materials like magnesium alloys. Springback, roll-in, roll-out, sheet fracture, and recoil are some of the defects

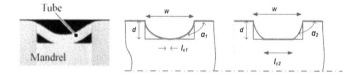

Figure 12.14   Form-fit joint made by electromagnetic force. (With permission from Mori et al., 2013, Elsevier.)

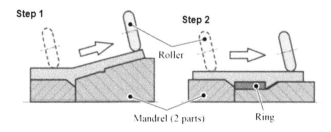

Figure 12.15   Form-fit joint made by shear spinning. (With permission from Mori et al., 2013, Elsevier.)

observed during hemming (Muderrisoglu et al., 1996). Seaming is another process to make water-tight or airtight joints.

12.4 JOINING OF TUBES BY END FORMING

Tubes are assembled by various mechanical joining and metallurgical joining methods such as fastening, crimping, end forming, welding, adhesive bonding, and brazing (Figure 12.16). The mechanical joining methods help in joining tubes made from dissimilar materials, which otherwise is difficult to join by fusion welding (Alves et al., 2017a).

Alves et al. (2017a) proposed a single stroke tube joining operation that includes expansion to produce two adjacent counterfacing surfaces and compression beading to lock the tubes. These tubes can be of dissimilar grades/materials—ferrous steel to stainless steel (Figure 12.17). They also proposed a two-stroke variance of the same process for joining the carbon steel tube to aluminum tube. In another attempt, Alves et al. (2017b) proposed a new tube joining method that has two operations—tube reduction in one of the tubes and rounding of the other tube, and finally, creating mechanical interlocking between these two ends. Alves et al. (2017c) presented another end forming method in which the tube wall is deformed in the thickness direction to create features that can clamp sheets making a stiff joint. In this method, the sheet is un-deformed until joining occurs. Such end forming techniques can also be used to make inclined joints, between a tube and a sheet, by compression beading (Goncalves et al., 2014).

Agrawal and Narayanan (2017) proposed and demonstrated a simple method of joining a tube to a sheet through end curling as shown in Figure 12.18 with the aim of developing a sustainable manufacturing process by eliminating environmental pollution caused by fusion welding processes. They studied the effect of many crucial process and tool conditions on the joint quality. Finally, they concluded that the die groove radius, friction conditions, and tube support length influence the quality of the joints significantly. Figure 12.19 shows some examples of the successful and unsuccessful joints made by the process. Narayanan (2018) proposed a novel method of joining an aluminum rod to an aluminum sheet by material accumulation process. The main advantage is that the process is performed in a lathe machine, and the sheet is undeformed. A rigid tool is used to plastically deform the rod to create the first accumulation. Then the sheet is placed and second accumulation

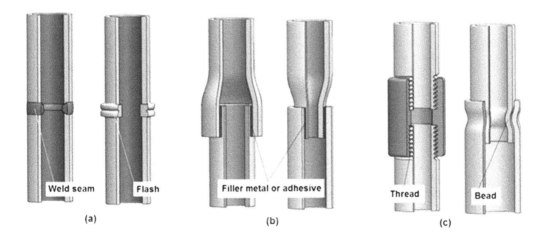

Figure 12.16 Joining tubes at their ends by: (a) welding, (b) adhesive bonding or brazing, and (c) mechanical fastening and crimping. (With permission from Alves et al., 2017b, Elsevier.)

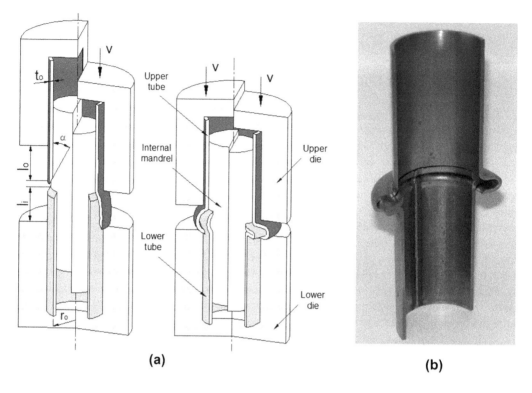

(a)   **(b)**

Figure 12.17 Joining tubes by end forming: (a) process schematic and (b) sectioned view of the joint made between carbon steel and stainless steel tubes. (With permission from Alves et al., 2017a, Elsevier.)

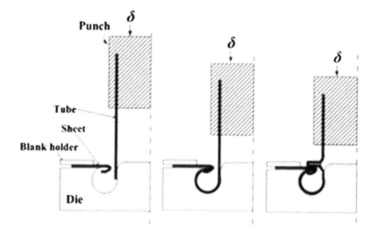

Figure 12.18 Schematic of joining a tube to a sheet using end curling. (With permission from Agrawal and Narayanan, 2017, Elsevier.)

is done to rigidly clamp the sheet between the two accumulations making the joint as presented in Figure 12.20.

Rotary swaging of tubes (Zhang et al., 2014) to form a joint and joining of two plates arranged perpendicular to each other by plastic deformation (like mortise-tenon joint) (Silva et al., 2018) are the other methods that are affordable and are potentially applicable in industries. The pull-out

Figure 12.19   Joints fabricated by end curling process (a) and (b) are successful, and (c) is unsuccessful. (With permission from Agrawal and Narayanan, 2017, Elsevier.)

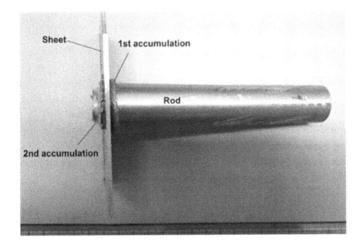

Figure 12.20   Joining a rod to a sheet by material accumulation (Narayanan, 2018).

destructive testing of the mortise-tenon joint revealed that the safe maximum tensile force is about 12 kN, while its applicability in shear deformation mode is restricted.

12.5 ADHESIVE BONDING

The adhesive bonding technology is used in manufacturing sectors in order to eliminate stress concentrations at the location of joints caused by the conventional mechanical fastening methods. It also eliminates thermal stresses generated by fusion welding. The advantages of using adhesive-bonded blanks in the manufacturing sectors are usage of lightweight materials for reducing fuel consumption, utilization of different mechanical properties, surface attributes and loading conditions, joining materials having different thickness levels and properties toward cost reduction without compromising the quality, and component design flexibility (Pocius, 2002). The adhesive bonding is mainly used in the construction of aircraft and automotive body structures. Curing (or hardening) occurs within the bulk of the adhesive, and adhesion occurs at the interface between adhesive and substrate. Mechanical interlocking, diffusion, adsorption, chemical bonding, electro-static, and wetting theory are the adhesion theories proposed (Adams, 2005).

The performance of adhesive-bonded joints depends on various parameters such as bonding methods, surface preparation, adhesive and adherend properties, geometrical parameters such as adhesive thickness, overlap length, stacking sequence, ply angle, and fillet, moisture, temperature, and humidity. The defects like voids, soft spots, micro bubbles, and micro-cracks occur in the adhesive during curing. The joint strength decreases as the adhesive thickness increases, and the fracture energy of the joint increases with increase in adhesive thickness for a ductile adhesive, but it is opposite for a brittle adhesive. The ductility of the adhesive increases with an increase in adhesive thickness (Budhe et al., 2017). Liu et al. (2012) investigated the influence of interfacial adhesion strength between skin sheet and core polymer on the formability of AA5052/polyethylene/AA5052 sandwich sheets. The forming limit of the sandwich sheet increases with an increase in interfacial adhesion strength. Satheeshkumar and Narayanan (2014) carried out a study on the influence of adhesive properties on the formability of adhesive-bonded steel sheets. The structural adhesive properties were varied by changing the hardener to resin ratios. There is a significant improvement in the ductility of adhesive-bonded sheets and limit strains in the base material constituents which is due to the improved plasticity of adhesive with hardener rich formulation and presence of good interface bond between adhesive and base materials. Kim et al. (2009) applied aluminum sandwich sheets with polypropylene core for an automotive hood part with the aim of reducing weight. They generated the basic formability data by conducting in-plane stretching and out-of-plane stretching tests. The fabricated sandwich hood was 65% lighter than the steel sheet for the same bending stiffness and 30% lighter than an aluminum alloy sheet. They proposed 17% as limit in-plane strain condition below which the sandwich can be deformed for making the hood. Surface treatment is vital in improving the bonding ability of sheets, specifically aluminum sheets. From Correia et al. (2018) work, it is revealed that such surface treatments are adhesive specific. In this study, sulfuric acid anodizing is good for AF 163-2 adhesive, while boric–sulfuric acid anodizing is good for EA 9658 AERO adhesive. However, both the adhesives are used to join aluminum sheets. It is worth noting the review done by Parashar and Mertiny (2012) toward finding solutions and challenges associated with adhesive bonding of FRP pipe sections.

12.6 PERFORMANCE OF MECHANICAL JOINING PROCESSES

A few comparison studies in relation to mechanical joining processes are discussed in this section. Sun et al. (2007) showed that the fatigue strength of SPR joints is 100% higher than that of the RSW joints made of 2-mm-thick 5182-O sheet. Xing et al. (2015) compared the joining performance of joints made by SPR and clinching of copper alloy H62. The SPR joints performed much better than clinched joint as presented in Table 12.1.

Mori et al. (2012) compared the static and fatigue strengths of joints made by clinching, SPR and RSW made of AA5052-H34 sheets. SPR produced strongest joint both in static and fatigue testing conditions. Table 12.2 summarizes the results obtained by three joining processes from static cross-tension test.

Table 12.1   Comparison of Results between Joints Made by SPR and Clinching of Copper Alloy H62 (Xing et al., 2015)

Joining Method	Mean Peak Load (kN)	Displacement at Failure (mm)	Energy Absorption (J)
SPR	5.39	3.88	15.30
Clinching	3.74	1.49	3.50

Table 12.2   Comparison of Results Obtained by Three Joining Processes from Static Cross-Tension Test (Mori et al., 2012)

Joining Method	Mean Peak Load (kN) (Approx.)	Displacement at Failure (mm) (Approx.)
Clinching	1.5	7
RSW	2.3	13
SPR	2.8	16

Moroni et al. (2010) evaluated the performance of hybrid weld-, rivet-, or clinch-bonded joints and compared with simple adhesive, spot-welded, riveted, or clinched joints. The weld-bonded joints perform better with improved strength, stiffness, and energy absorption as compared with simple spot-welded joints. The contribution of adhesive bonding plays a dominant role in hybrid-fastened-bonded joints than in the hybrid weld-bonded joints.

12.7 SUMMARY

In this chapter, some of the mechanical joining processes that include different mechanical fastening, clinching, joining by forming, joining by fits, and the adhesive bonding that imparts mechanical interlocking are described with available research contributions. Most of the mechanical joining processes are advantageous than conventional fusion and solid-state welding processes when environmental issues are concerned. Dissimilar materials are joined successfully by the mechanical joining processes. Most of the mechanical joining processes discussed in this chapter have been used in real-time shop floor applications successfully. Such joining methods are at par or better with respect to the existing welding processes in terms of mechanical performance.

REFERENCES

Adams, R.D. 2005. *Adhesive Bonding-Science, Technology and Applications*, Woodhead Publishing Limited and Press LLC, UK.
Agrawal, A.K., Narayanan, R.G. 2017. Joining of a tube to a sheet through end curling, *Journal of Materials Processing Technology* 246: 291–304.
Alves, L.M., Nielsen, C.V., Silva, C.M.A., Martins, P.A.F. 2017a. Joining end-to-end tubing of dissimilar materials by forming, *International Journal of Pressure Vessels and Piping* 149: 24–32.
Alves, L.M., Silva, C.M.A., Martins, P.A.F. 2017b. Joining of tubes by internal mechanical locking, *Journal of Materials Processing Technology* 242: 196–204.
Alves, L.M., Gameiro, J., Silva, C.M.A., Martins, P.A.F. 2017c. Sheet-bulk forming of tubes for joining applications, *Journal of Materials Processing Technology* 240: 154–161.
Atia, M.K.S., Jain, M.K. 2017. Die-less clinching process and joint strength of AA7075 aluminum joints, *Thin-Walled Structures* 120: 421–431.
Blaga, L., dos Santos, J.F., Bancila, R., Amancio-Filho, S.T. 2015. Friction riveting (FricRiveting) as a new joining technique in GFRP lightweight bridge construction, *Construction and Building Materials* 80: 167–179.
Borba, N.Z., Blaga, L., dos Santos, J.F., Amancio-Filho, S.T. 2018. Direct-friction riveting of polymer composite laminates for aircraft applications, *Materials Letters* 215: 31–34.
Budhe, S., Banea, M.D., de Barros, S., da Silva, L.F.M. 2017. An updated review of adhesively bonded joints in composite materials, *International Journal of Adhesion and Adhesives* 72: 30–42.
Chen, C., Zhao, S., Cui, M., Han, X., Fan, S. 2016. Mechanical properties of the two-steps clinched joint with a clinch-rivet, *Journal of Materials Processing Technology* 237: 361–370.
Correia, S., Anes, V., Reis, L. 2018. Effect of surface treatment on adhesively bonded aluminium-aluminium joints regarding aeronautical structures, *Engineering Failure Analysis* 84: 34–45.
Duan, H., Han, G., Wang, M., Zhang, X., Liu, Z., Liu, Z. 2014. Rotation friction pressing riveting of AZ31 magnesium alloy sheet, *Materials and Design* 54: 414–424.
Gao, S., Budde, L. 1994. Mechanism of mechanical press joining, *International Journal of Machine Tools and Manufacture* 34(5): 641–657.
Gerstmann, T., Awiszus, B. 2014. Recent developments in flat-clinching, *Computational Materials Science* 81: 39–44.
Goncalves, A., Alves, L.M., Martins, P.A.F. 2014. Tube joining by asymmetric plastic instability, *Journal of Materials Processing Technology* 214: 132–140.
Van Hall, S.N., Findley, K.O., Freis, A.K. 2018. Improved self-pierce rivet performance through intentional decarburization, *Journal of Materials Processing Technology* 251: 350–359.

He, X., Zhang, Y., Xing, B., Gu, F., Ball, A. 2015. Mechanical properties of extensible die clinched joints in titanium sheet materials, *Materials and Design* 71: 26–35.

Jahn, J., Weeber, M., Boehner, J., Steinhilper, R. 2016. Assessment strategies for composite-metal joining technologies—A review, *Procedia CIRP* 50: 689–694.

Kato, K., Okamoto, M., Yasuhara, T. 2001. Method of joining sheets by using new type rivets, *Journal of Materials Processing Technology* 111: 198–203.

Kim, K.J., Rhee, M.H., Choi, B.I., Kim, C.W., Sung, C.W., Han, C.P., Kang, K.W., Won, S.T. 2009. Development of application technique of aluminum sandwich sheets for automotive hood, *International Journal of Precision Engineering and Manufacturing* 10: 71–75.

Laforte, L.P., Lebel, L.L. 2018. Thermal analysis and degradation of properties in carbon fiber/epoxy laminate riveting at high temperatures, *Polymer Testing* 67: 205–212.

Lambiase, F. 2015. Clinch joining of heat-treatable aluminum AA6082-T6 alloy under warm conditions, *Journal of Materials Processing Technology* 225: 421–432.

Lambiase, F., Di Ilio, A. 2015. Mechanical clinching of metal–polymer joints, *Journal of Materials Processing Technology* 215: 12–19.

Lambiase, F., Paoletti, A. 2018. Friction-assisted clinching of aluminum and CFRP sheets, *Journal of Manufacturing Processes* 31: 812–822.

Lee, C.J., Kim, B.M., Kang, B.S., Song, W.J., Ko, D.C. 2017. Improvement of joinability in a hole clinching process with aluminum alloy and carbon fiber reinforced plastic using a spring die, *Composite Structures* 173: 58–69.

Li, G., Jiang, H., Zhang, X., Cui, J. 2017. Mechanical properties and fatigue behavior of electromagnetic riveted lap joints influenced by shear loading, *Journal of Manufacturing Processes* 26: 226–239.

Liu, J.G., Wei, L., Wang, J. 2012. Influence of interfacial adhesion strength on formability of AA5052/polyethylene/AA5052 sandwich sheet, *Transactions of Nonferrous Metals Society of China* 22: s395–s401.

Ma, Y., Li, Y., Hu, W., Lou, M., Lin, Z. 2016. Modeling of friction self-piercing riveting of aluminum to magnesium, *Journal of Manufacturing Science and Engineering* 138: 061007-1–061007-9.

Martinsen, K., Hu, S.J., Carlson, B.E. 2015. Joining of dissimilar materials, *CIRP Annals: Manufacturing Technology* 64: 679–699.

McComb, C., Tehrani, F.M. 2015. Enhancement of shear transfer in composite deck with mechanical fasteners, *Engineering Structures* 88: 251–261.

Min, J., Li, Y., Li, J., Carlson, B.E., Lin, J. 2015a. Mechanics in frictional penetration with a blind rivet, *Journal of Materials Processing Technology* 222: 268–279.

Min, J., Li, Y., Carlson, B.E., Hu, S.J., Li, J., Lin, J. 2015b. A new single-sided blind riveting method for joining dissimilar materials, *CIRP Annals: Manufacturing Technology* 64: 13–16.

Mori, K.I., Bay, N., Fratini, L., Micari, F., Tekkaya, A.E. 2013. Joining by plastic deformation, *CIRP Annals: Manufacturing Technology* 62: 673–694.

Mori, K., Abe, Y., Kato, T. 2012. Mechanism of superiority of fatigue strength for aluminium alloy sheets joined by mechanical clinching and self-pierce riveting, *Journal of Materials Processing Technology* 212: 1900–1905.

Moroni, F., Pirondi, A., Kleiner, F. 2010. Experimental analysis and comparison of the strength of simple and hybrid structural joints, *International Journal of Adhesion and Adhesives* 30: 367–379.

Mucha, J., Kascak, L., Spisak, E. 2011. Joining the car-body sheets using clinching process with various thickness and mechanical property arrangements, *Archives of Civil and Mechanical Engineering* 11(1): 135–148.

Mucha, J. 2013. The effect of material properties and joining process parameters on behavior of self-pierce riveting joints made with the solid rivet, *Materials and Design* 52: 932–946.

Muderrisoglu, A., Murata, M., Ahmetoglua, M.A., Kinzel, G., Altan, T. 1996. Bending, flanging, and hemming of aluminum sheet-an experimental study, *Journal of Materials Processing Technology* 59: 10–17.

Narayanan, R.G., 2018. A novel method of joining a rod to a sheet by end deformation: A preliminary experimental study, *International Journal of Precision Engineering and Manufacturing* 19: 773–779.

Neugebauer, R., Mauermann, R., Grützner, R. 2008. Hydrojoining, *International Journal of Material Forming* 1(1): 1303–1306.

Padhy, G.K., Wu, C.S., Gao, S. 2018. Friction stir based welding and processing technologies—processes, parameters, microstructures and applications: A review, *Journal of Materials Science and Technology* 34: 1–38.

Parashar, A., Mertiny, P. 2012. Adhesively bonded composite tubular joints: Review, *International Journal of Adhesion and Adhesives* 38: 58–68.

Pocius, A.V. 2002. *Adhesion and Adhesive Technology–An Introduction*, 2nd Edition, Hanser, Germany.

Qin, T., Zhao, L., Zhang, J. 2013. Fastener effects on mechanical behaviors of double-lap composite joints, *Composite Structures* 100: 413–423.

Satheeshkumar, V., Narayanan, R.G. 2014. Investigation on the influence of adhesive properties on the formability of adhesive bonded steel sheets, *Proceedings of the Institution of Mechanical Engineers, Part C: Journal of Mechanical Engineering Science* 228(3): 405–425.

Shirgaokar, M., Cho, H., Ngaile, G., Altan, T., Yu, J.H., Balconi, J., Rentfrow, R., Worrell, W.J. 2004. Optimization of mechanical crimping to assemble tubular components, *Journal of Materials Processing Technology* 146: 35–43.

Silva, C.M.A., Bragança, I.M.F., Alves, L.M., Martins, P.A.F. 2018. Two-stage joining of sheets perpendicular to one another by sheet-bulk forming, *Journal of Materials Processing Technology* 253: 109–120.

Sun, X., Stephens, E.V., Khaleel, M.A. 2007 Fatigue behaviors of self-piercing rivets joining similar and dissimilar sheet metals, *International Journal of Fatigue* 29(2): 370–386.

Trimble, A.Z., Yammamoto, B., Li, J. 2016. An inexpensive, portable machine to facilitate testing and characterization of the friction stir blind riveting process, *Journal of Manufacturing Science and Engineering* 138: 095001-1–095001-8.

Xing, B., He, X., Wang, Y., Yang, H., Deng, C. 2015 Study of mechanical properties for copper alloy H62 sheets joined by self-piercing riveting and clinching, *Journal of Materials Processing Technology* 216: 28–36.

Zhang, Q., Jin, K., Mu, D. 2014. Tube/tube joining technology by using rotary swaging forming method, *Journal of Materials Processing Technology* 214: 2085–2094.

CHAPTER **13**

Hybrid Joining Processes

V. Satheeshkumar
National Institute of Technology Tiruchirappalli

R. Ganesh Narayanan
IIT Guwahati

CONTENTS

13.1 INTRODUCTION

Material joining is used for assembling raw materials or fabricated components. Either a permanent joint or a temporary joint is made depending on the application. Mechanical fastening, welding, adhesive bonding, brazing, and soldering are the widely used joining processes. Each joining process has its own advantages. At the same time, there are some disadvantages as well. Holes in the components for riveting and fastening process result in stress concentration, the electrode consumption, fume evolution, difficulty in welding of thicker objects, and higher power consumption in the welding processes, and debonding and delamination issues in the adhesive bonds are some of the problems faced in conventional welding. These problems can be sorted out by combining the advantages of a few joining methods. Such methods are termed as hybrid joining methods. For example, (1) the delamination issue in adhesive bonds can be eliminated by spot welding and self-pierced riveting of adhesive-bonded parts, and (2) laser–metal inert gas (MIG) hybrid welding technology can be utilized for better gap bridge ability between two parts, deeper penetration, welding larger thickness of samples, higher welding speed, less power consumption, and better metallurgical properties. Sometimes hybrid solid-state welding process can also be used to avoid fume and emissions. Hybrid joining methods use the advantages of a few joining methods, and improve the process and materials toward sustainable manufacturing. These methods help to resolve the problems or to overcome the limitations existing in the traditional joining processes as well. In this chapter, various hybrid joining methods are presented with their importance presented from the available literature. In particular, hybrid fusion welding, hybrid solid-state welding, hybrid resistance welding, hybrid adhesive bonding-spot joining, hybrid brazing and soldering, and hybrid weld surfacing are discussed.

13.2 HYBRID FUSION WELDING PROCESSES

In this section, the hybrid fusion mainly focuses on welding processes such as laser–MIG hybrid welding, laser–tungsten inert gas (TIG) hybrid welding, TIG–MIG hybrid welding, and hybrid plasma–arc welding.

13.2.1 Laser–MIG Hybrid Welding

The laser–MIG hybrid welding technique has been developed to combine the advantages of both the processes in order to meet the welding industry needs. Arc welding is beneficial as it is inexpensive and uses available energy source. It has a considerable effect in changing microstructure by acceptable melting of the groove. On the other hand, laser welding has high welding rate and produces deeper welds. As a result, in the hybrid variety, arc stability and thermal efficiency improve because of the presence of plasma. Figure 13.1 shows the schematic of laser–MIG hybrid welding. The two sources involved in hybrid laser–MIG process act complementarily, and hence, faster welding, deep bead penetration associated with the formation of keyhole, and high energy concentration achieved by the laser beam are observed. Gap bridging and cost-effectiveness are the other advantages of the process. The type of laser source (Nd:YAG, CO_2, fiber laser), arc welding/laser-welding parameters, shielding gas composition, filler wire composition, number of laser welding/arc welding sources, and location of energy sources are some of the variables identified during the welding process, which can be optimized for sustainable welding.

Different types of laser sources have been used with MIG welding process. The combination of a high energy CO_2 laser with the MIG arc welding has been demonstrated and analyzed by Wieschemann et al. (2003). AH 36 steel, a high-strength structural steel used in ship construction,

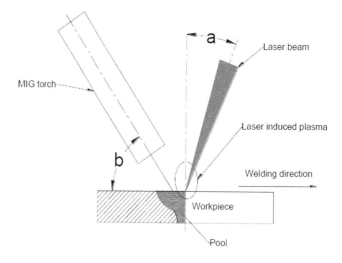

Figure 13.1 Laser–MIG hybrid welding. (With permission from Liu et al., 2016b, Elsevier.)

of 12-mm and 20-mm thicknesses is welded by the hybrid welding successfully with a 20-kW laser. Welding of the 12-mm thick plate is possible without using backing plate with a groove width of 0.5 mm. Successful welds are achieved by following the sets of conditions such as 15.9 kW and 1.8 m/min, and 9 kW and 1.2 m/min.

Tee-joints in the form of longitudinal stringers are used in ships to improve the stiffness of the plate. Such joints are fabricated successfully by hybrid laser–MIG process. The hybrid process ensures good mechanical strength by providing welds on the either side of the web in the form of reinforcement and the presence of filler does not reduce the welding rate. Good Tee-joints are made with a 6-kW laser at a welding rate of 1.23 m/min.

Li et al. (2010) carried out a study on coupling of Nd:YAG laser beam and a MIG arc with the spectrum of the plasma. The results clearly show that the hybrid variety is advantageous than MIG. It can be seen from their results that the laser hybrid welding causes the energy to concentrate at the center of the arc. The temperature of laser–MIG hybrid plasma, which is about 15,000 K, is 400–800 K higher than MIG in different selected surfaces. The laser–MIG hybrid plasma has higher electronic density, 5.0×10^{23} m^{-3}, than the electronic density of MIG, which is about 4.0×10^{23} m^{-3} at the center of the arc.

In general, the wavelength of Nd:YAG laser, 1.064 μm, is ten times smaller than the CO_2 wavelength of 10.64 μm, which is suitable for absorbing by most metals (Ready, 1971). However, the small wavelength of Nd:YAG laser inhibits its ability to be absorbed by many other materials such as wood, polymers, fabrics, and so on. Moreover, Nd:YAG laser is expensive and costly to maintain than CO_2 laser. Now it is important to choose suitable laser source in the hybrid joining processes aiming at cost-effectiveness, and efficient output toward sustainable manufacturing. In this context, the CO_2 laser–MIG hybrid welding can be utilized for welding activities because of good power-to-cost ratio. Tani et al. (2007) studied the effect of shielding gas composition (He–Ar–O_2) and flow rate on the stability of the process and on the weld bead dimensions during hybrid CO_2 laser–MIG welding of 304 stainless steel sheet. The results show that 30%–40% of He content and flow rate between 10 and 30 L/min yield a good process feasibility and is cost-effective as well. Furthermore, Liu et al. (2013) reported the importance of arc characteristics, droplet diameter, and droplet transfer behavior during laser–metal active gas (MAG) hybrid welding of 7-mm-thick high-strength steel plates. A 5-kW laser is used with 30%He+70%Ar shielding gas flowing at the rate of 25 L/min. For arc, 20%CO_2+80%Ar shielding gas with flow rate of 16 L/min is used. Globular transfer and spray transfer are observed depending on the arc power. This shows that selection of welding parameters decides the arc power, and finally sustainable welding is ensured.

Welding wire composition is also important in deciding the weld strength and sustainability in welding. Yan et al. (2009) reported that the joint efficiency improved to 78% when ER2319 (Si: 0.2; Cu: 5.8–6.8; Mg: 0.02; Mn: 0.2–0.4; Fe: 0.3; Ti: 0.2; Zn: 0.1) wire is used as compared to 69% in the case of ER4043 welding wires (Si: 4.5–6.0; Cu: 0.3; Mg: 0.05; Mn: 0.05; Fe: 0.8; Ti: 0.2; Zn: 0.1; Al: remaining) while welding 2A12 aluminum alloy using CO_2 laser–MIG hybrid welding.

Fiber laser is another type of laser in which an optical fiber doped with rare-earth elements such as erbium, neodymium, ytterbium, thulium, praseodymium, holmium, and dysprosium acts as an active gain medium. Briefly, fiber laser light is created by diodes. The fiber optic cable channels and amplifies the light like data transfer. The fiber laser light is 200% more efficient than traditional CO_2 laser, with no expensive optical mirrors, and its delivery is simple. Fiber laser is also used for welding purposes and has own merits in hybrid joining processes. Zhang et al. (2016c) carried out joining of A6061 material that is phosphate anodized (pretreated) with carbon fiber-reinforced plastic (CFRP) using fiber laser characterized by 2,000 W maximum power, 1,070 nm wave length, 300 mm focal length, and 0.2 mm focal spot diameter. The pretreatment strengthens the joint between CFRP and A6061, and the shear strength improves eight times as compared with the case without pretreatment. The highest shear strength reaches up to 42 MPa at 10–20 min anodizing time.

The pretreated surface improves the wettability of CFRP on A6061, resulting in a stronger bond or mechanical anchoring effect at the interface.

In the fiber laser–GMA hybrid welding of double-side welded T-joints made of K36D ship building steel, Wahba et al. (2015) found that argon-rich shielding gas could be replaced by 100% CO_2 gas for sustainable and inexpensive welding operation. In the welding process using 100% CO_2 gas, deeper penetration and reduced porosity are achieved, though spatter formation and weld surface irregularities are seen. In hybrid welding, the spattering is reduced through buried-arc transfer by controlling arc voltage, arc leading arrangement and wire extension, and high-quality joints can be made. The spatter generation could also be reduced by controlling the relative distances between the two heat sources in the X- and Y-directions and by producing a proper profile of the arc cavity which traps spatters.

Superalloys that are generally difficult to get welded by conventional fusion welding can be welded successfully by hybrid joining. Superalloy GH909 10-mm-thick material has been welded by laser (ytterbium fiber)–MIG hybrid welding (Liu et al., 2016b). A 4-kW laser power has been used and a gap of 1 mm or 3 mm is maintained between arc and laser. The result shows that the weld joints with a desirable weld profile that is of wine-glass shape has been achieved by optimized process conditions. This has yielded acceptable joint tensile strength of about 633 MPa which is nearly 77% of the base metal.

The efficiency of hybrid welding is compared with variations of conventional fusion welding. Some of the variations of traditional welding are at par with the hybrid welding methods. For example, Zhang et al. (2016b) compared the double-sided MIG (DMIG) welding and single-pass laser–MIG hybrid joining processes during welding of 4-mm thick T2 copper plates. Figure 13.2a,b shows the schematic of DMIG weld and laser–MIG hybrid weld. Although the performance of both the welding methods is comparable, the hybrid welding exhibits higher efficiency, lower heat input, smaller fusion zone (FZ) and heat-affected zone (HAZ), and finer grains in the FZ and HAZ as compared with DMIG. As a result, the tensile strength of the joint made by hybrid welding has reached 74% of that of base metal, as opposed to 69% in the case of the DMIG joints. On the other hand, defects like pore and undercut are prevalent in hybrid joints. The electrical conductivity and thermal diffusion coefficient are larger for hybrid joints. Both the joints show enhanced ductility as compared with the base metal. The tensile test data are summarized in Table 13.1.

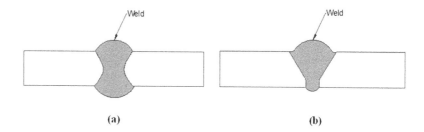

(a)   (b)

Figure 13.2 Schematic of weld made from (a) DMIG welding and (b) laser–MIG hybrid welding. (Zhang et al., 2016b)

Table 13.1 Tensile Properties of Joints Fabricated from DMIG and Laser–MIG Hybrid Welding of Cu Plates (Zhang et al., 2016b)

Cases	Ultimate Tensile Strength (UTS) (MPa)	Displacement Till UTS (mm)	Displacement at Failure (mm)
Base metal	275	1.5	7.5
DMIG	180	10.5	12.5
Hybrid	210	8.6	10.8

Generally, in the laser–MIG hybrid welding processes, laser beam is traversed prior to the MIG torch. However, it can be traversed after MIG torch as well (Figure 13.3). As suggested by Liu et al. (2016a), the effect of modifications in the location of energy sources is to improve the tensile strength of the joint. This has been demonstrated with a 4-kW Nd:YAG laser–MAG hybrid welding of high-strength steel. The arc–laser hybrid welding (ALHW) produced 1,035 MPa strong joint, while it is 875 MPa in the case of laser–arc hybrid welding (LAHW). However, the impact energy of ALHW joint, 13 J, is lower than LAHW joint, which is about 17 J. The weld bead appears to be cone-shaped and cocktail cup-shaped under ALHW and LAHW configurations, respectively.

Laser–twin-arc hybrid welding which includes Nd:YAG laser welding and twin-arc welding (Figure 13.4) is another hybrid variety (Gu et al., 2013). In this scheme, the main function of laser is to provide a conductive and stable plasma channel for the arcs. This influences the arc shape, slower droplet transfer, reducing resistivity, and stabilizing arc. The laser–twin-arc hybrid weld exhibits

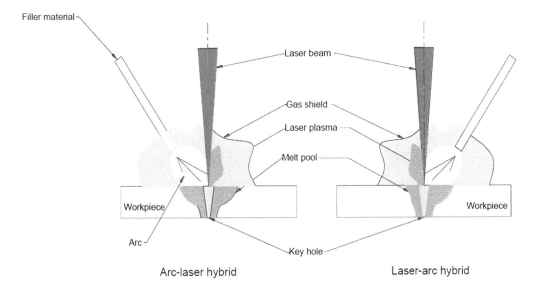

Figure 13.3 Laser and arc arrangement in laser–MAG hybrid welding. (With permission from Liu et al., 2016a, Elsevier.)

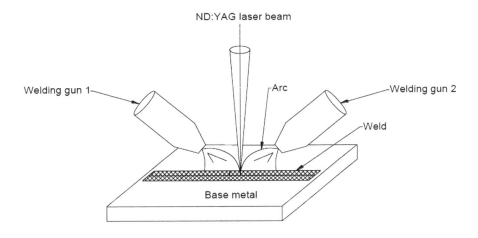

Figure 13.4 Laser–twin-arc hybrid welding. (Gu et al., 2013)

deeper penetration and wider HAZ than in the twin-arc weld alone while welding Q235A steel plate of 6-mm thickness. Hence, there are possibilities of such newer welding schemes improving the sustainability of the joining processes by improving the quality of welds.

Dual-beam laser welding is another scheme for improving the sustainability of welded joints. Ma et al. (2017) carried out a comparative study on welding stability between single- and dual-beam laser welding of 2-mm-thick stainless steel sheet with filler wire. The dual-beam laser is like hybrid of two laser beams (Figure 13.5). Two different configurations (Figure 13.6) of dual laser beam are followed. The transfer process and welding stability are improved by stabilizing the molten pool dynamics in side-by-side configuration as compared with single-beam laser welding with filler wire. The side-by-side configuration showed its superiority on improving the welding quality.

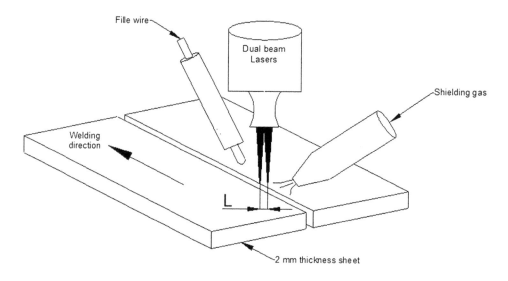

Figure 13.5   Dual-beam laser welding (laser–laser hybrid). (With permission from Ma et al., 2017, Elsevier.)

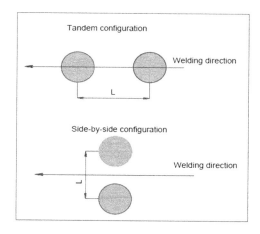

Figure 13.6   Configurations of dual-beam laser welding. (With permission from Ma et al., 2017, Elsevier.)

13.2.2 Laser–TIG Hybrid Welding

In the case of laser–TIG hybrid welding processes, arc generated by conventional TIG welding process is combined with the laser beam to join materials. Figure 13.7 shows a schematic of laser–TIG hybrid welding. Like laser–MIG welding, there are various parameters that can be optimized to have sustainable welding and defect free joint. A few examples are given here. Chen et al. (2017) investigated the relationship between laser keyhole characteristics and the porosity formation during pulsed laser–GTA welding of AZ31B Mg alloy. They have concluded that by controlling the laser keyhole behavior, porosities can be minimized in the weld region. A large depth-to-width ratio of the laser keyhole increases porosity generation. Two solutions, (1) keeping the keyhole outlet in open state for a longer time and (2) proper overlap of the adjacent two laser keyholes, were identified to restrict the porosity. Qiao et al. (2017) addressed welding of 4-mm-thick AA7N01P-T4 using fiber laser-variable polarity tungsten inert gas (VPTIG) hybrid welding with filler wire with the effect of natural aging and reinforcement in the weld joint. The natural aging increases the quantity of precipitates along the grain boundaries enriched with Zn and Mg and the dislocation density in the joint. The fatigue limit of joints without reinforcement is higher than the case with reinforcement, i.e., 160–115 MPa. Moreover, weld zone has faster fatigue crack growth rate as compared with base metal followed by HAZ.

A variant of laser–TIG hybrid welding is laser–micro TIG hybrid welding. This is similar to conventional laser–TIG hybrid welding, except that both the laser-beam and TIG arc work in pulsed mode. The process can be performed either in pulsed laser–micro TIG welding, if the laser welding is the leading process, or in pulsed (micro) TIG–laser welding, if the TIG arc is the leading source as shown in Figure 13.8. The pulse shapes and frequency phase shifting provide window for sustainable joining. Birdeanu et al. (2012) conducted a study using bead on plate trials on a 3-mm-thick austenitic stainless steel and revealed that the pulsed TIG–laser hybrid welding performs better than the other one in establishing the process stability. TIG frequency values which led to the stable welding process are of the same order as the laser pulse repetition rates.

13.2.3 Arc–Plasma–Ultrasonic Hybrid Welding

In this section, hybrid welding including arc–arc (like TIG–MIG), plasma–arc, and ultrasonic–arc combinations for welding variety of base materials are discussed. In TIG–MIG hybrid welding,

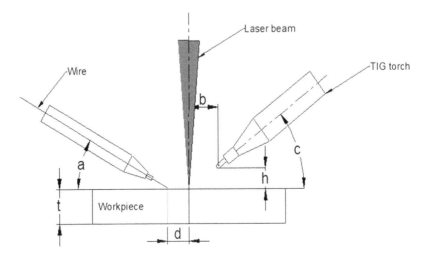

Figure 13.7   Schematic of Laser–TIG hybrid welding. (With permission from Qiao et al., 2017, Elsevier.)

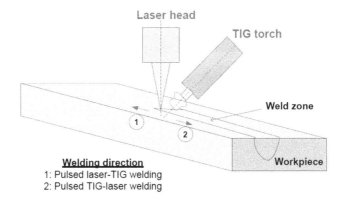

Figure 13.8 Schematic of pulsed laser–micro TIG welding. (Birdeanu et al., 2012)

two arc sources are used for welding, while the plasma–arc and ultrasonic–arc hybrid welding are shown in Figures 13.9 and 13.10. The plasma–arc hybrid welding combines both plasma–arc welding and MIG welding in a single torch. As shown in Figure 13.9, the plasma–arc surrounds the MIG arc. Since plasma–arc surrounds MIG wire and weld pool, it helps in preheating and post-heating of base metal resulting in higher deposition rate and reducing porosity. In ultrasonic–arc (like GTAW) hybrid welding, the arc source is located ahead of the rotating sonotrode so that it helps in preheating the sheet material to enhance the weldability during ultrasonic welding. The sonotrode vibrates at a frequency of about 20 kHz. In order to improve gripping of sheets, the sonotrode tip contains serrations. The current, axial force, and welding speed can be varied to improve the sustainability of welding (Figure 13.10).

TIG–MIG hybrid welding of AZ31B Mg alloy and ferritic stainless steel with Cu as intermediate layer performed by Ding et al. (2015) and MIG–TIG double-sided arc welding–brazing (DSAWB) of AA5052 and a low-carbon steel by Ye et al. (2017) are some of the recent examples in the arc–arc hybrid category. In the first case, since Mg and Fe elements do not interact with each other, Cu was introduced as an intermediate layer that can interact with both of them. The Cu layer thickness plays a vital role in developing the joint metallurgy. Cu foils of 0.02 mm and 0.1 mm thicknesses were used for the analyses. There is about 47% shear strength improvement while using 0.1-mm-thick Cu

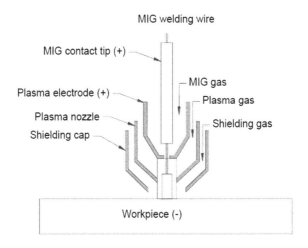

Figure 13.9 Plasma–MIG hybrid welding. (With permission from Lee et al., 2015, Elsevier.)

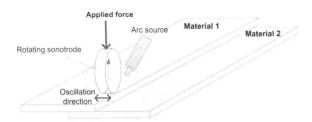

Figure 13.10   Ultrasonic–arc (GTAW) hybrid welding. (Dai et al., 2015)

interlayer as compared with 0.02-mm Cu (Ding et al., 2015). In the case of DSAWB, two merits are observed: (1) Fe_2Al_5 intermetallic layer thickness is restricted to about 2.03 µm as compared with conventional MIG welds and (2) the maximum tensile strength of the joint reaches about 148 MPa in DSAWB, while it is only 56 MPa in traditional MIG weld. The strength of weld generated by DSAWB is about 70% of AA5052.

In the case of plasma–arc hybrid (PAH) welding process (Figure 13.9), the surface defects in the weld region can be minimized by optimizing plasma current. Lee et al. (2015) showed that smut (a black spot considered as a defect) and undercut are generated by increasing the plasma current. Three categories of smut—smut outside the weld bead, smut spots on the weld bead, and smut inside the weld bead—are formed with increase in plasma current. The amount of smut formed is less than that observed in MIG weld. Such defects deteriorate the weld zone properties and hence the optimization of plasma current plays a vital role in these conditions. Ono et al. (2009) observed that plasma–MIG hybrid welding has the advantage of reducing the fumes and spatter, and producing better weld appearance while welding an Al alloy. As a result, the PAH welding method can be considered as a candidate for sustainable joining.

AZ31B Mg alloy and Al 6061 alloy can be welded successfully by ultrasonic–GTAW hybrid welding (Figure 13.10). Here, the arc is used as a preheating source. The tensile-shear strength of the joint increases with GTAW current up to a maximum limit and then decreases with a further increase in GTAW current (Dai et al., 2015). The maximum lap shear strength achieved in this study is 1 kN at 30 A GTAW current, which is approximately 40% of strength of AZ31B Mg alloy. Wen et al. (2015) developed a device for superimposing vibration on the workpiece in both horizontal and vertical directions during TIG welding of AZ31 sheets. The device has a maximum power output of 2 kW at a frequency of 15 kHz. The average grain diameter within the joints decreases from 49.4 µm, when welded without vibration, to about 34 µm, when welded under varying vibration conditions. Increasing the amplitude of the vibration results in a significant increase in the hardness, tensile strength, elongation, and microstructure refinement of the joint.

Generally, the effect of arc–plasma–ultrasonic hybrid welding is to improve the mechanical strength and microstructure of the joint as compared with the conventional ones. Joining of dissimilar materials becomes efficient when the hybrid method is followed. Lesser defect formation and good weld bead appearance are few other merits of the hybrid welding. Such advantages are vital for sustainable joining.

13.3 HYBRID SOLID-STATE WELDING PROCESSES

Like fusion welding, solid-state welding methods can also be combined either with one of the fusion welding or solid-state welding methods for joining. Use of solid-state welding methods minimizes the use of harmful consumables and fluxes, thus improving the sustainability of the process and product. The hybrid variety helps in enhancing the joint quality. In this context, friction stir-based methods like friction stir welding, friction welding, friction stir spot welding, friction riveting, diffusion bonding, and other methods like magnetic pulse welding are used in various industrial applications (Martinsen et al., 2015). Some of the hybrid varieties and benefits toward sustainable manufacturing are discussed in this section.

MIG–FSP (MIG–friction stir processing) is a hybrid welding process shown in Figure 13.11. In a particular, FSP has been carried out in MIG welded (T-joint) plates made of 6-mm-thick Al 6082-T651 for modifications in tensile and fatigue properties (Costa et al., 2014). The MIG-welded region showed lesser hardness and strength as compared with base material. Not much improvement in strength of FSPed and MIG welded region was observed, but about 5% improvement in ductility was observed. Moreover, FSP improves fatigue strength of MIG welded region due to grain refinement and elimination of defects like porosity and lack of wetting. Stress concentration is reduced due to increased toe radius.

Induction heating-based FSW, i-FSW, as shown in Figure 13.12, uses FSW tool heated by induction for welding of metals and nonmetals. The conventional FSW tool is surrounded by an induction coil. The tool gets heated by the generation of eddy current when alternating electric current is applied to the induction coil. The induction heating of tool enables the plastic material to soften in a short time and helps in easier stirring. FSW joints were made on HDPE (Vijendra and Sharma, 2015). The strength of the joint is governed by the ductile nature of the material at high pin temperature and turbulence, and stirring action at low pin temperature. The hardness at the stir zone decreased with an increase in tool-pin temperature due to the ductile nature of the joints. Pin temperature of 45°C and a rotational speed of 2,000 rpm yielded joints with optimum strength.

Electric current based FSW (EFSW) is another hybrid solid-state welding technique. In this process, electric current flows from the FSW tool to the sheet through contact interfaces (Figure 13.13). The total heat generated in EFSW is a combination of heat generated from FSW and by electric resistance, where electric resistance is given by I^2Rt. Luo et al. (2014) used EFSW for welding of

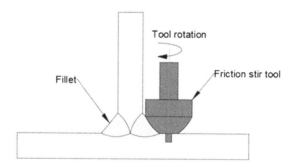

Figure 13.11   MIG–FSP hybrid welding and weld-tool position during FSP. (With permission from Costa et al., 2014, Elsevier.)

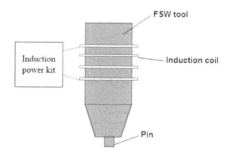

Figure 13.12   Schematic of i-FSW tool arrangement. (Vijendra and Sharma, 2015)

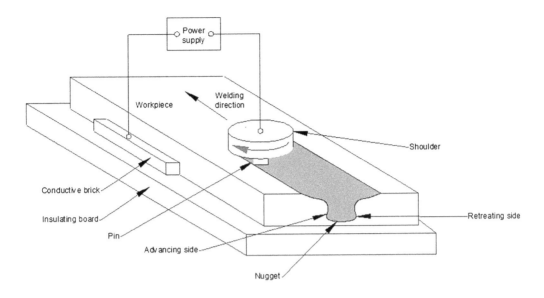

Figure 13.13 Schematic of EFSW. (Luo et al., 2014)

AZ31B and Al 7075 sheets. In the case of AZ31B, the resistant heat source promoted significant grain refinement, and hardness improvement in the weld nugget zone (NZ). In the case of Al 7075, the grain size in the NZ and HAZ increased slightly with the increase in electric current intensity. EFSW can be used for joining high-strength alloys like stainless steel and carbon steel.

FSW assisted by external heat by an arc source (like GTAW torch) in front of FSW tool, and externally cooled FSW after the FSW tool are the two hybrid varieties suggested by Mehta and Badheka (2017) (Figure 13.14a,b). The external cooling is done by compressed air or water. Other than FSW parameters, the arc welding parameters play a vital role in preheating. The strength of the Cu–Al 6061 joint was significantly affected by the preheating currents of 40, 80, and 120 A, with a

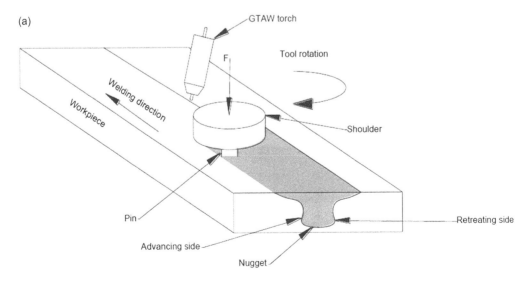

Figure 13.14 (a) TIG-assisted FSW. (Mehta and Badheka, 2017) (b) External cooling assisted FSW. (Mehta and Badheka, 2017)

(Continued)

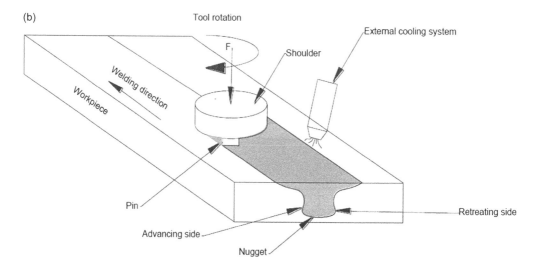

Figure 13.14 (CONTINUED)   (a) TIG-assisted FSW. (Mehta and Badheka, 2017) (b) External cooling assisted FSW. (Mehta and Badheka, 2017)

slight increase in strength at 40 A as compared with normal FSW, while it decreased from 80 to 120 A. On the other hand, cooling-assisted FSW by water provided stronger joint with 158 MPa as tensile strength. The formation of intermetallic compounds like $CuAl_2$ and Cu_9Al_4 increased with increase in preheating current, while it has significantly decreased by increasing the cooling effect with water.

Hybrid friction diffusion bonding (HFDB) (Figure 13.15) is another hybrid welding method in which a tube can be welded to another thinner or thicker tube. The conical feature plays a significant role in the stirring action at the end of the tube made in the insertion phase during the welding phase. Dethlefs et al. (2014) applied the hybrid method to fabricate tube-to-tube-sheet connections for AA5xxx aluminum coil-wound heat exchangers. Leak tests revealed an acceptable leak rate of less than 10^{-9} mbar/s. The pull-out strength of the joint was found to be about 80% of the strength of the base metal.

In order to improve the mechanical properties of the FSW joint, the second-phase particles are added to the stir zone as reinforcement, to constitute a particle-reinforced metal matrix composites. In conventional FSW, the particle distribution is nonuniform in the stir zone. In order to overcome

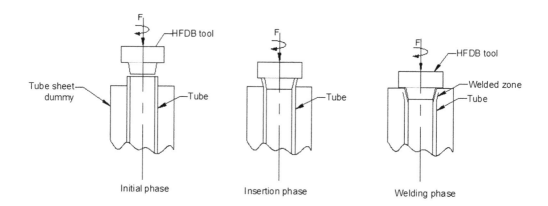

Figure 13.15   Hybrid friction diffusion bonding. (With permission from Dethlefs et al., 2014, Elsevier.)

these problems, Fouladi and Abbasi (2017) performed friction stir vibration welding (FSVW) of Al 5052 alloy workpieces in which vibration is done normal to the weld line during FSW, while SiO$_2$ particles are incorporated into the weld. Homogeneous distribution of particles is witnessed during FSVW, and homogeneity improves with increase in vibrating frequency resulting in sustainable joining. Optimum vibration frequency of about 33 cycles/s imparts superior mechanical properties of welded specimens.

13.4 HYBRID MECHANICAL JOINING PROCESSES

Hybrid mechanical joining processes are fumeless, flux less, and filler rod less joining process in which plastic deformation causes joining of materials with or without utilizing rivets or bits. A specific combination of joining process may be used to join materials that are difficult to weld while yielding stronger joints. Clinching, self-pierced riveting, and joining by end forming are some of the methods in this category. Although there can be few hybrid combinations, Friction-based Injection Clinching Joining technique (F-ICJ), and hybrid resistance spot clinching process are presented here. In F-ICJ process (Figure 13.16), thermoplastic–metal pair can be joined by heating and deforming a polymer stud within a pre-existing hole such that it mechanically clinches the sheet pair at the end of the operation (Abibe et al., 2016). In this operation, during friction phase, because of the stirring action of the rotating tool, the polymer stud melts. During the process, the stud deforms just above the metal sheet (upper sheet) to form an anchor. After this process, in the consolidation phase, axial force is provided by the tool on the joint surface without rotation to avoid dimensional recovery. At the end of the process, the tool is retracted. The joint made between polyetherimide (PEI) and Al 6082-T6 is characterized by ultimate load of about 1,420 N in lap shear test and 430 N in cross-tensile tests, which are almost equal to what is achieved in ultrasonic staking joints. On the other hand, the F-ICJ process demands longer joining cycles, 7.5 s, as compared with ultrasonic staking (or joining), 3 s.

A novel combination of resistance spot welding and mechanical clinching resulted in "hybrid resistance spot welding clinching" process (Zhang et al., 2017). Joining of Al 5052 sheets showed that under lower heat input, the load-bearing capacity of the resistance spot clinching joints is

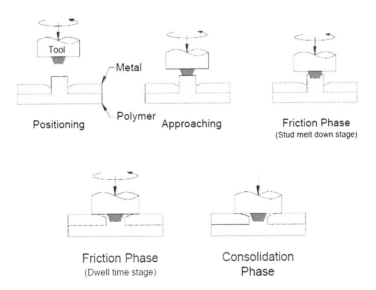

Figure 13.16 F-ICJ process and stages. (With permission from Abibe et al., 2016, Elsevier.)

superior to that of the traditional resistance spot welding joints. This demonstrates that the hybrid variety is energy efficient and produces stronger joints making the process sustainable.

13.5 HYBRID JOINING WITH ADHESIVE BONDING

The adhesive bonding technology is mostly used for fabricating tailor-made blanks for lightweight applications in the manufacturing sector. The adhesive bonds are integrated with various fusion welding, solid-state welding, resistance welding, and mechanical joining methods to form hybrid joining methods. For instance, laser welding, resistance spot welding (RSW), self-pierced riveting (SPR), clinching, FSW, friction bit joining (FBJ), embossing, and bolt joints can be combined with adhesive bonding for joint formation. Some of the typical characteristics of such hybrid joining methods are discussed below.

In the case of laser spot weld-adhesive bonding, the joint made between mild steel sheets showed increased ductility (about 20%–30%) and breaking load as compared to the traditional adhesive-bonded joint and laser spot weld joint during tensile-shear tests and peel tests (Tao et al., 2014). In another case, the joint made by hybrid adhesive-RSW performed better than hot spot welding without adhesive bonding. The hybrid joints showed two times higher load-carrying ability and six times higher energy absorption ability as compared to spot welding without adhesive boding (Sadowski et al., 2014). Similar attempts made by Costa et al. (2015) on interstitial free steel revealed that the hybrid RSW–adhesive bond joint performed better than adhesive-bonded joint, but the joint quality deteriorated when compared with RSW joint because of the damage caused in the adhesive during spot welding and possible contamination in the nugget.

SPR–adhesive bonding and clinching–adhesive bonding are the hybrid varieties that use mechanical joining for sustainable joining. The SPR–adhesive bond hybrid joints between carbon fiber and Al 2024-T6 endured larger peak load of 4.5 kN in untreated condition, while about 5 kN in treated condition when compared with SPR joints having 3.4 kN (Franco et al., 2013). The adhesive-bonded joints performed either equally or poorly as compared to the hybrid joints. On the other hand, the energy absorption ability of SPR, 10.12 J, and 21 J in two different cases is either much larger or almost equal to that of the hybrid variety. The clinched–bonded hybrid joint between Cu and steel sheets can be fabricated in two ways: clinching before adhesive curing and clinching after curing as suggested by Balawender et al. (2012). The experimental data showed slight advantage of clinching before curing when compared with the case after curing during joint separation tests.

FBJ is another solid-state spot welding method. FBJ is more suitable for joining lightweight metal and nonmetals to high-strength steels. Initially, a rotating, consumable bit is driven by a tool through upper material toward lower material. The driven bit forms a plasticized region at the interfaces due to frictional heat and pressure during the joining phase. In the final stage, tool is withdrawn leaving the consumable bit in the joint. Lim et al. (2015) employed FBJ to spot weld the adhesive-bonded joint, between Al 7075-T6 and dual phase (DP) 980 steel, under different corrosion conditions. The accelerated cyclic corrosion tests were carried out on the joints made by FBJ and the hybrid process. The joint strength of FBJ specimen significantly decreases with increase in corrosion cycles, while weld-bonded specimens maintained more than 80% of their actual strength. The presence of adhesive compensates the mechanical integrity of the joints in the presence of a corrosive medium.

The presence of hook defect in the overlap joint fabricated by FSW process is a defect that should be minimized. The defect acts as crack initiation and reduces the static and fatigue strength. The hybrid FSW and adhesive bonding was proposed to minimize the hook defect in FSW process (Braga et al., 2015). The hybrid joining method and adhesive bonding show higher lap strength and ductility than the joints made by FSW only in the case of Al 6082-T6 base sheet. It is also found that a single-pass hybrid joint with continuous adhesive layer reduces distortion levels than joints made by FSW.

Huang et al. (2013) proposed an adhesive bonding–embossing hybrid joining process for joining thin metallic sheet and CFRP sheets. In this method, CFRP sheets and metal sheets are bonded by adhesive and embossed. This embossing process eliminates cutting of continuous carbon fibers and stress concentration caused when conventional method such as bonding and rivet joining are used. The maximum tensile-shear load, slip displacement, and static absorption energy of the adhesively bonded joints increased by 69%, 48%, and 165%, respectively, when hybrid joining is used. Some of the main factors that affect joint quality include the mechanical anchor effect of the embossed pit, the expansion of the adhesive area, the concentration of adhesive at the edge of the pit, and the heating procedure.

13.6 HYBRID BRAZING–SOLDERING–WELDING PROCESSES

This section focuses mainly on some important hybrid brazing–soldering–welding processes. Joining steel to aluminum, or copper, is of great interest to research communities to use merits of both the materials. For such applications, the use of solid stage welding is restricted because of longer welding time and lesser joint types. The use of fusion welding is a simplest solution, but its use is limited by high heat input resulting in unacceptable intermetallic compound (IMC) formation. Hence, a hybrid brazing–welding joining (fusion welding type) that releases lesser heat input has been introduced to join steel to aluminum. For example, hybrid EBW–brazing, shown in Figure 13.17, was carried out to join SS 304 to commercial pure aluminum (CP-Al) with Al_5Si filler wire under different beam current levels (Wang et al., 2017). The results indicate that beam current has a significant effect on the thickness of IMC layer, weld appearance, and tensile strength of joints. The ultimate tensile strength of the joint was found to be 93 MPa at an optimum welding condition, which is about 83% equal to that of CP-Al.

A hybrid MIG welding–brazing process was developed to join aluminum alloy and stainless steel with auxiliary TIG arc (Zhang et al., 2014). The auxiliary TIG improved the wettability of molten metal on the entire surface of steel resulting in a sound brazing joint. About 51% increase, from 97 to 147 MPa, in the tensile strength of joint was achieved while using auxiliary TIG arc as compared to conventional joining without auxiliary TIG. The friction stir spot welding (FSSW) has been modified to a hybrid variety "friction stir spot brazing (FSSB)" (Figure 13.18) by Zhang et al. (2016a) to join Cu plates. In this process, a pinless tool is used to eliminate keyhole, and the

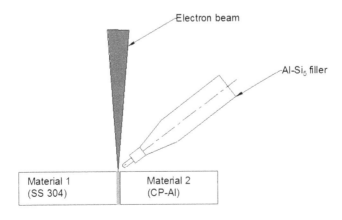

Figure 13.17 Hybrid EBW–brazing. (Wang et al., 2017, Elsevier.)

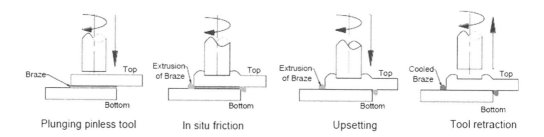

Figure 13.18 FSSB process. (With permission from Zhang et al., 2016a, Elsevier.)

preplaced braze between two base metals to eliminate hook formation and lack of mixing. Zn foil was used as the braze material for experiments on Cu plates. The braze material improves the metallurgical action with the help of thermo-mechanical changes initiated by a rigid tool. Such a hybrid process enhances the wettability beyond the shoulder in air and reduces some of the defects making the process sustainable. A solder material can also be used instead of braze material.

Friction stir soldering (FSS) was carried out by Ebrahimian and Kokabi (2017) with the aid of prior electroplating to fabricate graphite–copper lap joints with the aim of studying rotational speed influence on mechanical behavior of the joint. Sn–37Pb solder foil was placed between a copper sheet and a copper-coated graphite substrate. A rotating pinless tool was plunged into the copper sheet and allowed to travel along the interface edge. The adhesion strength of the copper coating on the graphite substrate increased from 4 to 13.3 MPa when the plating current density is decreased from 0.1 to 0.01 A/cm^2. At 20-mm/min traverse speed, with an increase in rotational speed from 300 to 800 rpm, the void volume percentage decreased from 14% to 4%, the IMC layer thickness increased from 3.8 to 8 μm, and the fracture load decreased by about 600 N. Hence, it is evident that the braze and solder materials enhance the sustainability of the joint by improving the strength and reducing the IMC formation during welding of the dissimilar materials at optimized conditions.

13.7 SUMMARY

In this chapter, some of the hybrid joining processes that include the hybrid varieties of fusion welding, solid-state welding, adhesive bonding, mechanical joining, brazing–soldering are described with available research contributions. Most of the hybrid varieties enhance the strength of the joint, reduce welding defects, and improve the joint microstructures, as compared with the original nonhybrid types, resulting in sustainable joining. Dissimilar materials are welded successfully by the hybrid welding processes. The hybrid welding methods include larger parametric set for optimization as they combine merits of the existing methods. This means that there is more scope for improving the efficiency of the process toward sustainability. Numerical simulations help in such optimization minimizing the experiments and resources.

REFERENCES

Abibe, A.B., Sonego, M., dos Santos, J.F., Canto, L.B., Amancio-Filho, S.T. 2016. On the feasibility of a friction-based staking joining method for polymer–metal hybrid structures, *Materials and Design* 92: 632–642.
Balawender, T., Sadowski, T., Golewski, P. 2012. Numerical analysis and experiments of the clinch-bonded joint subjected to uniaxial tension, *Computational Materials Science* 64: 270–272.

Birdeanu, A.V., Ciuca, C., Puicea, A. 2012. Pulsed LASER-(micro) TIG hybrid welding: Process characteristics, *Journal of Materials Processing Technology* 212: 890–902.

Braga, D.F.O., de Sousa, L.M.C., Infante, V., et al. 2015. Aluminum friction stir weld bonding, *Procedia Engineering* 114: 223–231.

Chen, M., Xu, J., Xin, L., Zhao, Z., Wu, F., Ma, S., Zhang, Y. 2017. Effect of keyhole characteristics on porosity formation during pulsed laser-GTA hybrid welding of AZ31B magnesium alloy, *Optics and Lasers in Engineering* 93: 139–145.

Costa, J.D.M., Jesus, J.S., Loureiro, A., Ferreira, J.A.M., Borrego, L.P. 2014. Fatigue life improvement of MIG welded aluminium T-joints by friction stir processing, *International Journal of Fatigue* 61: 244–254.

Costa, H.R.M., Reis, J.M.L., Souza, J.P.B., et al. 2015. Experimental investigation of the mechanical behaviour of spot welding–adhesives joints, *Composite Structures* 133: 847–852.

Dai, X., Zhang, H., Liu, J., Feng, J. 2015. Microstructure and properties of Mg/Al joint welded by gas tungsten arc welding-assisted hybrid ultrasonic seam welding, *Materials and Design* 77: 65–71.

Dethlefs, A., Roos, A., dos Santos, J.F., Wimmer, G. 2014. Hybrid friction diffusion bonding of aluminium tube-to-tube-sheet connections in coil-wound heat exchangers, *Materials and Design* 60: 7–12.

Ding, M., Liu, S.S., Zheng, Y., Wang, Y.C., Li, H., Xing, W.Q., Yu, X.Y., Dong, P. 2015. TIG–MIG hybrid welding of ferritic stainless steels and magnesium alloys with Cu interlayer of different thickness, *Materials and Design* 88: 375–383.

Ebrahimian, A., Kokabi, A.H. 2017. Friction stir soldering: A novel route to produce graphite-copper dissimilar joints, *Materials and Design* 116: 599–608.

Fouladi, S., Abbasi, M. 2017. The effect of friction stir vibration welding process on characteristics of SiO_2 incorporated joint, *Journal of Materials Processing Technology* 243: 23–30.

Franco, G.D., Fratini, L., Pasta, A. 2013. Analysis of the mechanical performance of hybrid (SPR/bonded) single-lap joints between CFRP panels and aluminum blanks, *International Journal of Adhesion and Adhesives* 41: 24–32.

Gu, X., Li, H., Yang, L., Gao, Y. 2013. Coupling mechanism of laser and arcs of laser-twin-arc hybrid welding and its effect on welding process, *Optics and Laser Technology* 48: 246–253.

Huang, Z., Sugiyama, S., Yanagimoto, J. 2013. Hybrid joining process for carbon fiber reinforced thermosetting plastic and metallic thin sheets by chemical bonding and plastic deformation, *Journal of Materials Processing Technology*, 213: 1864–1874.

Lee, H.K., Park, S.H., Kang, C.Y. 2015. Effect of plasma current on surface defects of plasma-MIG welding in cryogenic aluminum alloys, *Journal of Materials Processing Technology* 223: 203–215.

Li, Z., Wang, W., Wang, X., Li, H. 2010. A study of the radiation of a Nd:YAG laser–MIG hybrid plasma, *Optics and Laser Technology* 42: 132–140.

Lim, Y.C., Squires, L., Pan, T.Y., et al. 2015. Study of mechanical joint strength of aluminum alloy 7075-T6 and dual phase steel 980 welded by friction bit joining and weld-bonding under corrosion medium, *Materials and Design* 69: 37–43.

Liu, S., Liu, F., Xu, C., Zhang, H. 2013. Experimental investigation on arc characteristic and droplet transfer in CO_2 laser–metal arc gas (MAG) hybrid welding, *International Journal of Heat and Mass Transfer* 62: 604–611.

Liu, S., Li, Y., Liu, F., Zhang, H., Ding, H. 2016a. Effects of relative positioning of energy sources on weld integrity for hybrid laser arc welding, *Optics and Lasers in Engineering* 81: 87–96.

Liu, T., Yan, F., Liu, S., Li, R., Wang, C., Hu, X. 2016b. Microstructure and mechanical properties of laser-arc hybrid welding joint of GH909 alloy, *Optics and Laser Technology* 80: 56–66.

Luo, J., Chen, W., Fu, G. 2014. Hybrid-heat effects on electrical-current aided friction stir welding of steel, and Al and Mg alloys, *Journal of Materials Processing Technology* 214: 3002–3012.

Ma, G., Li, L., Chen, Y. 2017. Effects of beam configurations on wire melting and transfer behaviors in dual beam laser welding with filler wire, *Optics and Laser Technology* 91: 138–148.

Martinsen, K., Hu, S.J., Carlson, B.E. 2015. Joining of dissimilar materials, *CIRP Annals - Manufacturing Technology* 64: 679–699.

Mehta, K.P., Badheka, V.J. 2017. Hybrid approaches of assisted heating and cooling for friction stir welding of copper to aluminum joints, *Journal of Materials Processing Technology* 239: 336–345.

Ono, K., Liu, Z., Era, T., Uezono, T., Ueyama, T., Tanaka, M., Nakata, K. 2009. Development of a plasma MIG welding system for aluminium, *Welding International* 23: 805–809.

Qiao, J.N., Lu, J.X., Wu, S.K. 2017. Fatigue cracking characteristics of fiber Laser-VPTIG hybrid butt welded 7N01P-T4 aluminum alloy, *International Journal of Fatigue* 98: 32–40.

Ready, J.F. 1971. *Effects of High Power Laser Radiation*, New York: Academic Press.

Sadowski, T., Golewski, P., Knec, M. 2014. Experimental investigation and numerical modelling of spot welding–adhesive joints response, *Composite Structures* 112: 66–77.

Tani, G., Campana, G., Fortunato, A., Ascari, A. 2007. The influence of shielding gas in hybrid LASER–MIG welding, *Applied Surface Science* 253: 8050–8053.

Tao, W., Ma, Y., Chen, Y., Li, L., Wang, M. 2014. The influence of adhesive viscosity and elastic modulus on laser spot weld bonding process, *International Journal of Adhesion and Adhesives* 51: 111–116.

Vijendra, B., Sharma, A. 2015. Induction heated tool assisted friction-stir welding (i-FSW): A novel hybrid process for joining of thermoplastics, *Journal of Manufacturing Processes* 20: 234–244.

Wahba, M., Mizutani, M., Katayama, S. 2015. Hybrid welding with fiber laser and CO_2 gas shielded arc, *Journal of Materials Processing Technology* 221: 146–153.

Wang, T., Zhang, Y., Li, X., et al. 2017. Influence of beam current on microstructures and mechanical properties of electron beam welding-brazed aluminum-steel joints with an Al5Si filler wire, *Vacuum* 141: 281–287.

Wen, T., Liu, S.Y., Chen, S., Liu, L.T., Yang, C. 2015. Influence of high frequency vibration on microstructure and mechanical properties of TIG welding joints of AZ31 magnesium alloy, *Transactions of Nonferrous Metals Society of China* 25: 397–404.

Wieschemann, A., Kelle, H., Dilthey, D. 2003. Hybrid-welding and the HyDRA MAG+LASER processes in ship building, *Welding International* 17: 761–766.

Yan, J., Zeng, X., Gao, M., Lai, J., Lin, T. 2009. Effect of welding wires on microstructure and mechanical properties of 2A12 aluminum alloy in CO_2 laser-MIG hybrid welding, *Applied Surface Science* 255: 7307–7313.

Ye, Z., Huang, J., Gao, W., Zhang, Y., Cheng, Z., Chen, S., Yang, J. 2017. Microstructure and mechanical properties of 5052 aluminum alloy/mild steel butt joint achieved by MIG-TIG double-sided arc welding-brazing, *Materials and Design* 123: 69–79.

Zhang, H.T., Liu, J.H., Feng, J.C. 2014. Effect of auxiliary TIG arc on formation and microstructures of aluminum alloy/stainless steel joints made by MIG welding–brazing process, *Transactions of Nonferrous Metals Society of China* 24: 2831–2838.

Zhang, G., Zhang, L., Kang, C., et al. 2016a. Development of friction stir spot brazing (FSSB), *Materials and Design* 94: 502–514.

Zhang, L.J., Bai, Q.L., Ning, J., Wang, A., Yang, J.N., Yin, X.Q., Zhang, J.X. 2016b. A comparative study on the microstructure and properties of copper joint between MIG welding and laser-MIG hybrid welding, *Materials and Design* 110: 35–50.

Zhang, Z., Shan, J.G., Tan, X.H., Zhang, J. 2016c. Effect of anodizing pre-treatment on laser joining CFRP to aluminum alloy A6061, *International Journal of Adhesion and Adhesives* 70: 142–151.

Zhang, Y., Shan, H., Li, Y., Guo, J., Luo, Z., Ma, C.Y. 2017. Joining aluminum alloy 5052 sheets via novel hybrid resistance spot clinching process, *Materials and Design* 118: 36–43.

CHAPTER **14**

Green Lubricants and Lubrication

Sayanti Ghosh, Lalit Kumar, Vivek Rathore, Shivanand M. Pai, and Bharat L. Newalkar
Bharat Petroleum Corporation Limited

CONTENTS

14.1 INTRODUCTION

Lubricants are complex products consisting of 70%–99% base oils admixed with additives that modify the natural properties of the fluid to meet its intended requirements. The main advantage of lubricants (as shown in Table 14.1) is to facilitate the relative movement between surfaces by minimizing friction, wear and prevent overheating and corrosion (Lansdown, 2004).

Based on these features, the American Petroleum Institute (API) established a base oil classification (API Standard, 2002) as given in Table 14.2.

Owing to the global environmental imbalance, there are some critical issues related to the area of lubrication, as large quantities of petroleum or mineral oil-based industrial lubricants on disposal or break down are a potential threat to the environment. This led to two basic approaches for dealing with environmental safety of lubricants: the first focuses on ways to eliminate the disposal of lubricants into the environment and the other defines the use environmentally safe products in environment-sensitive applications. As a result, terms like biodegradable, eco-friendly, and green have been echoing in industrial sectors and become a selling point for lubricant manufacturers. Along with legislative compliance, one of the reasons for this recent green initiative is the growing awareness and demand to use more environmentally safe products. Thus, use of biodegradable synthetic lubricants is gaining momentum worldwide (Jaina and Suhanea, 2013). Nevertheless, one of the biggest challenges is to develop eco-friendly and biodegradable base stocks that could replace the existing mineral oil base stocks without compromising on the equipment functionality (Rensselar, 2011).

In general, three types of base fluids viz., mineral oils, vegetable oils (VO), and synthetic esters find application in formulation of lubricants as shown in Table 14.3.

Presently, high-performing commercial lubricants are complex materials composed of a lubricant base stock formulated with an additive package for desired property improvement. Conventional lubricant systems are highly diverse, ranging mainly from common lube oils (nonaqueous liquids) to oil-in-water emulsions (e.g., used in water-miscible cutting fluids), water-in-oil emulsions (as in metal-forming), oil-in-oil emulsions (applied in metalworking), water-based solutions (applied in chip-forming metalworking operations), greases, and solid lubricants.

However, the perfect lubricant selection depends on the specifications of targeted application and its suitable disposal in the environment. Hence, the aforementioned factors led to the development

Table 14.1   Merits of Lubrication

- Reduction of friction
- Wear and corrosion protection
- Shock cushioning
- Heat transfer
- Seal action
- Insulation
- Production efficiency
- Component lifetime (equipment durability)
- Energy economy

Table 14.2   API—Classification of Base Oils

Group	Saturate (wt%)	Sulfur (wt%)	Viscosity Index
I	<90 and/or	>0.03	>80 to <120
II	≥90 and	≤0.03	≥80 to ≤120
III	≥90	≤0.03	≥120
IV	All poly alpha olefins (PAOs)		
V	All base stocks not included in Groups I–IV		

Table 14.3   Properties of Different Types of Lubricant Base Stocks

Properties	Mineral Oil	Vegetable Oils	Synthetic Esters
Density at 20°C (kg/m³)	880	940	930
Viscosity index	100	200	220
Pour point (°C)	−15	−20 to +10	−60 to −20
Cold flow properties	Good	Poor	Very Good
Biodegradability (CEC)%	10–30	70–100	10–100
Oxidation stability	Good	Poor	Good
Hydrolytic stability	Good	Poor	Fair

Table 14.4   Critical Factors for a Green Lubricant

- Type of base stock
- Biodegradability
- Nontoxicity
- Energy efficiency
- Ease of maintenance
- Service life
- Waste generation
- Safety

of new class of lubricants commonly known as "green lubricants." Boyde (2002) described green lubricants as environmentally benign lubricants that degrade naturally, thereby reducing harmful emission of greenhouse gases (GHGs), toxic waste streams into air, water, and land. These lubricants are formulated with renewable oils in majority, readily biodegradable, and free from heavy metals or other toxic by-products. In addition, it also offers an advantageous edge when formulated with VO by exhibiting improved lubricity with reduced friction and wear, a high-viscosity index and flash points, respectively, for improved safety. Thus, new generation bio/synthetic-based environmentally friendly lubricants serve dual purposes: these lubricants are environmentally safe and are alternatives to petroleum base stocks. However, lubricant selection with a greener edge is complex and depends on a number of critical factors as shown in Table 14.4.

Typically, VO and synthetic ester-based lubricants offer excellent features with respect to high-viscosity index, high flash point, and high degree of biodegradability. Also, fatty-acid-based ester lubricants offer a platform to develop base stocks with varied viscosities depending on the required applications. However, VO-based base stocks demonstrate poor oxidation stabilities and cold flow properties.

In view of the above, research and developmental efforts have been made in the past two decades toward the selection and designing of technically superior VO and polyol-ester-based lubricant base stocks for industrial applications. Accordingly, the current chapter describes benefits of biodegradable base stocks and further attempts to bring out the greener advances made in synthesis methodologies for the industrial applications of electrical transformer oils, refrigeration oils, metal-working coolants, food-grade lubricants, VO-derived diesel lubricity improvers, and VO-based greases.

14.2 LUBRICANT BASE STOCKS

Typically, lubricant base stocks are classified into three categories as described in Table 14.3 (Rudnick, 2006). Mineral oil-based base stocks are the most common lubricating oils utilized due to their availability, cost, and compatibility with many engineering materials. However, due to their poor biodegradability there is a paradigm shift toward biodegradable vegetable and synthetic ester-based base stocks.

Table 14.5 Biodegradability (%) of Different Lubricant Base Stocks (Buenemann et al., 2003)

S. No.	Base Stocks	Biodegradability, % (as per CEC-L-33-A-93)
1	Mineral oil	20–40
2	Vegetable oil	90–98
3	Esters	75–100
4	Polyol esters	70–100

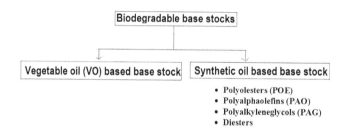

Figure 14.1   Carboxylate ester, R and R' denotes any alkyl or aryl group.

```
                    ┌─────────────────────────┐
                    │ Biodegradable base stocks │
                    └─────────────────────────┘
                          │
            ┌─────────────┴──────────────┐
            ↓                            ↓
┌──────────────────────────────┐  ┌────────────────────────────┐
│ Vegetable oil (VO) based base stock │  │ Synthetic oil based base stock │
└──────────────────────────────┘  └────────────────────────────┘
                                      • Polyolesters (POE)
                                      • Polyalphaolefins (PAO)
                                      • Polyalkyleneglycols (PAG)
                                      • Diesters
```

Figure 14.2   Classification of biodegradable base stocks.

The degree of biodegradability for the said base stocks is typically measured by standard test protocols as defined under CEC-L-33-A-9 (Batters, 2005), OECD301B, and ASTM D6384 (Buenemann et al., 2003). Estimated biodegradability of various base stocks is compiled in Table 14.5.

From the above information on biodegradability (%), it is evident that VO and ester-based base stocks are preferred choice for development of eco-friendly biodegradable lubricants. This is mainly due to the presence of ester groups $-O-C=O$ (Figure 14.1) which aid in biodegradation.

Accordingly, the major biodegradable ester-based lubricant base stocks can be classified as per the following scheme as shown in Figure 14.2.

14.2.1 VO Base Stock Based Lubricants

VOs are the potential source of green lubricants owing to their biodegradability, sustainability, and excellent lubricity (Adhvaryu and Erhan, 2002; Biresaw et al., 2002, 2003; Erhan et al., 2006; Fox and Stachowiak, 2007; Lawal et al., 2012). Typically, VO-based base stock consists of a triglyceride moiety which primarily comprises of esters of fatty acids ($C_{12}-C_{22}$) linked to the backbone of a glycerol molecule (a shown in Figure 14.3), unsaturated bonds ($-C=C-$) and hydroxyl groups based on the composition of fatty acids.

Most of the VOs contain distinct areas of polar and nonpolar groups in the same molecule. The polar groups induce amphiphilic nature in the VO that enables it to be used as a boundary lubricant (Adhvaryu et al., 2004; Allawzi et al., 1998; Asadauskas et al., 1997; Biresaw et al., 2003; Erhan et al., 2006; Fox and Stachowiak, 2007; Lawal et al., 2012). The VO has merits of high flash or low volatility, high-viscosity index, and high solubility owing to the presence of high molecular weight triacylglycerol molecule and polarity of ester groups (Adhvaryu et al., 2005; Biswas et al., 2005; Erhan et al., 2008; Jayadas and Nair, 2006; Salimon et al., 2012; Sharma et al., 2006). The triglyceride moiety of the VO accounts for the inherent disabilities of being used as a lubricant. The unsaturation present in fatty acids is responsible for poor cold flow properties (Asadauskas and Erhan, 1999; Bartz, 1998; Choi et al., 1997; Kassfeldt and Goran, 1997; Rhee et al., 1995; Wu et al., 2000) and thermal oxidative stability. Predominately, poor oxidation depends on the degree of unsaturation levels present in the fatty acids and is induced through free-radical mechanism. Moser and Erhan (2007) described the standard representation of auto-oxidation mechanism of VO, as shown in Figure 14.4

Typically, free-radical mechanism is initiated from the methylene group (near to the C=C double bond) formed by removal of a hydrogen atom followed by reaction with oxygen to form the peroxy-radical. Thereafter, the peroxy-radical attacks another triglyceride molecule to remove a hydrogen atom to form a hydroperoxide and another free radical which in turn propagates the oxidation mechanism (Hwang and Erhan, 2007). Ruger et al. (2002) inferred that degradation leads to an increase in viscosity of the oil that in turn limits the usefulness of VO as base stock. Even prolonged exposure of VO to low temperature (−10°C to 0°C) induces turbidity and solidifies the oil (Bantchev et al., 2009; Hagemann, 1988). Hwang and Erhan (2006) studied the importance of transforming alkene groups (−C=C−) of the VO to other stable functional groups to improve the oxidation stability and cold flow properties.

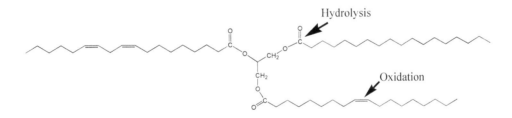

Figure 14.3 Molecular structure of a typical triglyceride moiety.

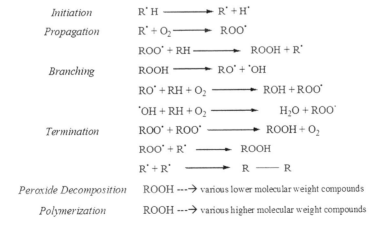

Figure 14.4 Auto-oxidation mechanism of vegetable oils.

14.2.1.1 VO with High-Oleic Content

VOs are triglycerides containing esters of glycerol and three straight chained fatty acids as shown in Figure 14.3. The two key variables of the straight chained fatty acids are the number of double bonds (–C=C–) and the chain length (C_n). In general, a longer chain length results in a higher melting point and viscosity whereas more double bonds correspond to lower melting points and viscosity, and poor thermo-oxidative stability (Garces et al., 2011). Monounsaturated fatty acids, such as oleic acid, have been found to maintain a good balance of low melting point, with good thermo-oxidative stability and viscosity (Nagendramma and Kaul, 2012; Wagner et al., 2001a). Therefore, feed stocks with high-oleic (HO) acid contents are generally preferred due to its perfor-mance and biodegradability. In recent years, HO varieties of rapeseed, sunflower, and soybean with oleic acid levels of up to 94% have been developed (Cole et al., 1998; Fick, 1988; Stoutjesdijk, 2000). Lawate and Lal (1995) reported HO palm oil-based esters with excellent pour point and lubricat-ing properties. For example, trimethylolpropane (TMP) trioleate displays excellent low-temperature (Lal, 1994) properties (PP −45°C). The properties of TMP trioleate with HO content (>90%) is shown in Table 14.6. TMP esters of oleic acid are widely applied for hydraulic fluids.

Recently, special techniques viz., selective breeding and gene manipulation have been evolved for cropping of seeds that produce HO oils (Duvick, 1996).

14.2.2 Synthetic Base Stocks

Synthetic lubricants are obtained through a synthetic reaction route of well-defined chemical compounds. Synthetic oils are generally superior to mineral oils in terms of:

- Better oxidation stability
- Higher Viscosity Index (VI)
- Low PP
- High level of biodegradability

This class of lubricant is classified as Gr. V lubricant base stocks (Table 14.2). The following sections describe different types of synthetic esters.

14.2.2.1 Polyol ester (POE)

Polyol esters are a class of synthetic lubricants which are derived from fatty acid and polyols. These esters are formed by esterifying polyhydric alcohols, also called *polyols*, like neopentylglycol (NPG), TMP, pentaerythritol (PE), and dipentaerythritol (DiPE), with carboxylic acids (as shown in Figure 14.5).

POEs form a broad range of base stocks with properties depending greatly on the chemical struc-ture. In general, they have excellent low-temperature properties with tunable viscosity index (VI). In principle, ester properties can be fine-tuned to fit desired applications by varying carbon chain length and degree of branching of acid moiety (Pettersson, 2007). For example, bis(*n*-octyl)adipate

Table 14.6 Properties of High-Oleic (90%+) TMP Trioleate

Property	Value
VI	190
PP (°C)	−45
Flash point (°C)	300
Oxidation stability	Good
Biodegradation (CEC%)	95

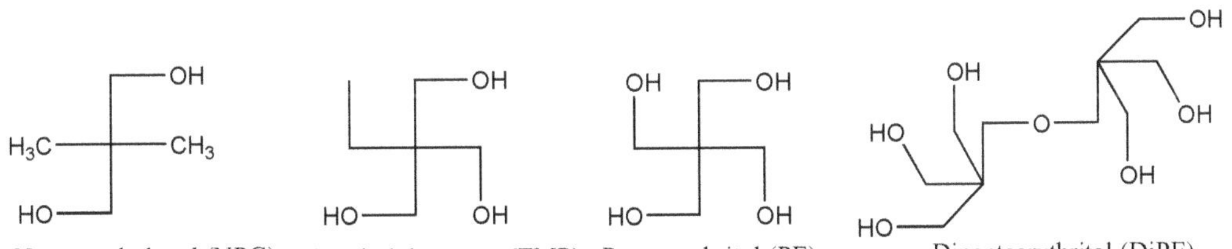

Neopentyl glycol (NPG) Trimethylolpropane (TMP) Pentaerythritol (PE) Dipentaerythritol (DiPE)

Figure 14.5 Different types of polyols.

Table 14.7 Selected Properties for Synthetic Ester Structures

Polyol Ester	VI	PP (°C)
Bis(*n*-octyl)adipate	218	4
Bis(2-ethylhexyl)adipate	123	–78
TMP C8–C10	140	–53
TMP oleate	181	–54
NPG C16–C18b	158	–24
NPG Iso C18b	148	–42

and bis(2-ethylhexyl)adipate just differ in branching (Table 14.7); however, their VI and pour point have a significant difference.

Likewise, the fatty-acid chain length of TMP oleate increases viscosity and VI as compared to TMP C8–C10. Apart from VI and cold flow properties, polyol ester-based lubricant base stocks have excellent thermal stability due to the absence of beta-hydrogen in polyols employed for preparation of the base stocks.

Polyol esters are of great interest for formulation of environmentally adapted lubricants (EALs) due to their nontoxic nature and generally show excellent biodegradability. EALs (Pal and Singhal, 2000) have been repeatedly heralded as one of the few future growth segments of the lubricant industry. Presently, they have been applied in the areas of hydraulic fluids, engine oils, circulation oils, and greases, etc.

14.2.2.2 Polyalphaolefins (PAO) Based Lubricants

PAOs are also widely used commercial semi-synthetic lubricants which are obtained from polymerization reactions of linear alpha olefins (LAOs), such as 1-decene (Shubkin, 1993). They are used as base fluids in many high-performance lubricants, synthetic motor oils, low-temperature hydraulic fluids, etc. There are many different types of PAO oligomers, from PAO2 to PAO100, where the number denotes the kinematic viscosity (in cSt) at 100°C. Rudnick (2006) reported that low polarity PAOs can lead to seal shrinkage and loss of elasticity, making them good blending stocks. Oxidative stability of antioxidant containing PAO is comparable to mineral oil-based lubricants.

14.2.2.3 Polyalkylene Glycols (PAG) Based Lubricants

Conventionally PAGs are derived from co-polymerization of ethylene oxide (EO) and propylene oxide (PO) (Lawford, 2009). PAGs provide performance benefits in terms of excellent friction control, good low-temperature properties, and high-viscosity index. Copolymers of EO and PO with a broad range of molecular weights and viscosities are the only major water-soluble lubricant base oil available today. This unique feature provides functional advantages for fire-resistant hydraulic

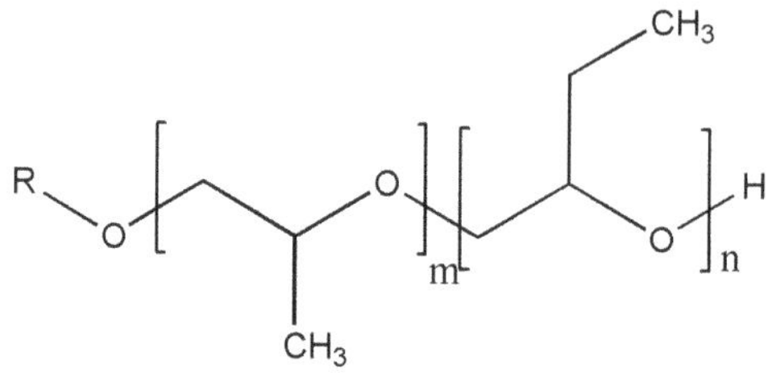

Figure 14.6   Structure of oil-soluble PAG.

fluids, metal-working fluids, and textile lubricants, etc. Instead of having hydrocarbon backbone, the molecules have alternating ether linkages, which results in high polarity of the material.

Greaves (2011) explored a new range of mineral oil-soluble PAGs (OSPs). The design of OSPs can be accomplished by synthesizing downstream derivatives of butylene oxide (BO) to form polybutoxylate homopolymers or by synthesizing copolymers of PO and BO (Figure 14.6). OSP base oils have higher VIs than mineral oils and low PPs. Synthetic OSPs provide opportunities to formulators to upgrade hydrocarbon oils and solve current lubrication challenges by leveraging the inherent benefits of PAGs. OSPs can be used as primary base oil, co-base oil, and as a performance-enhancing additive.

PAGs are used in applications such as gears, bearings, and compressors for hydrocarbon gases. PAGs have been used extensively as rotary screw compressor lubricants and may be applied as engine oils.

14.2.2.4 Diesters Based Lubricants

Synthetic dicarboxylic acid esters or diesters (as shown in Figure 14.7) are obtained from esterification of dicarboxylic acids (such as adipic acid) with hydroxylated alcohols. Their high molecular weight addresses volatility issues and ester linkages provide high solubilizing property. They demonstrate very good low temperature performance. Despite the above facts, diesters demand is continuously increasing as base stocks for biodegradable hydraulic fluids. Their blend with PAOs offers excellent lubricants properties in the synthetic segments that possess specially compressor and turbines lubricants (McTurk, 1953).

14.2.2.5 Epoxidized Soyabean Oil (ESO) Based Lubricants

Synthetic lubricant base stocks are prepared from ESO to be used alone or with polyalphaolefin (PAO). Adhvaryu and Erhan (2002) reported ESO as promising intermediates for improving the low-temperature properties and oxidative stabilities by blending VOs with diluents like PAO and oleates. The epoxy ring of the ESO is readily functionalized and gives rise to value-added products. The ring opening reaction of ESO followed by esterification of the resulting hydroxyl groups form lubricant candidates with enhanced low-temperature stability and high VI.

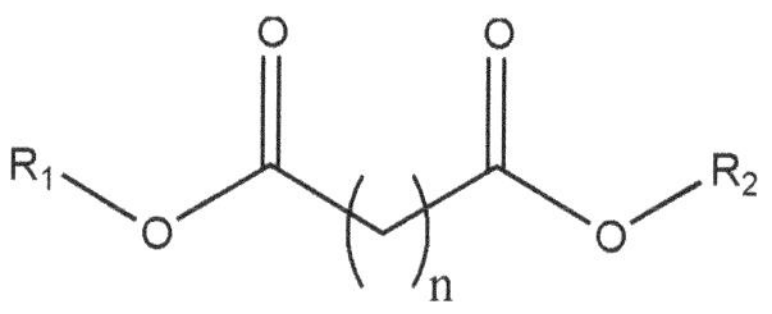

Figure 14.7   A diester.

14.2.2.6 Emerging Trends: Ionic Liquids (IL) Based Lubricants

ILs is an emerging class of environment friendly lubricants that have started making a mark into the lubricant market for specific applications. A typical IL comprises of an organic cation grouped with a variety of organic or inorganic anions. Liu et al. (2006) investigated the distinct and valuable characteristics of ILS including enhanced thermal stability, low melting point, and negligible vapor pressure. The application of ionic liquids as lubricants was initially reported in the year 1961, where fluoride-based molten salts (i.e., LiF and BeF_2) were subjected to high-temperature (650°C–815°C) ranges. The low melting forms of usual molten salts or the room temperature ionic liquids (RTILs) were first evaluated as synthetic lubricating fluids (Ye et al., 2001). Yao et al. (2009) reported key application of ILs as lubricants, base oils, additives, and thin films. ILs when employed as base oils have offered good tribological performance for steel and ceramic-based sliding pairs (Jimenez et al., 2006; Liu et al., 2002; Lu et al., 2004; Mu et al., 2004, 2005; Omotowa et al., 2004; Phillips and Zabinski, 2004; Reich et al., 2003; Wang et al., 2004; Yu et al., 2006). Fluoride anion-based ILs like tetrafluoroborate (BF_4^-) or hexafluorophosphate (BF_6^-) display superior tribological properties under very high pressures and elevated temperatures (Minami et al., 2010; Shah et al., 2009). Although the chemical structure of the cationic and anionic substituents of an IL can vary greatly, the hydro-phili/phobicity property of ILs can be altered by varying the nature of anions and chain length which further improves their thermal and oxidative stability (Minami, 2009). Ohtani et al. (2008) extensively explored the imidazolium ion-based ILs owing to the high thermal stability of the imidazole rings. Furthermore, the chain length on the imidazolium cation can be readily fine-tuned. In addition to the improved thermo-oxidative stability obtained from hydrophobic anions, a decrease in stability is observed with more hydrophobic cations. Jimâenez and Bermâudez (2007) investigated the importance of longer alkyl chains in ILs which aids in demonstrating exceptional tribological properties at wide temperature ranges (–30°C to 200°C).

14.3 SYNTHESIS METHODOLOGIES FOR BIODEGRADABLE BASE STOCKS

In view of the aforementioned lubricant base stocks, various synthesis methodologies have been evolved in last two decades to alter the properties of base stocks desired for various applications. These methodologies are reviewed in the following sections.

14.3.1 Synthesis of Vegetable Oil-Based Base Stocks

In order to use VO-based lubricants, numerous approaches have been adopted to enhance the cold flow and oxidative properties of VOs. Esterification/transesterification, hydrogenation, epoxidation, and oligomerization are few of the methodologies that have been employed to boost the performance and stability of VOs for their direct applications. The chemical modifications mentioned above are discussed extensively in the following sections.

14.3.1.1 Esterification/Transesterification

Transesterification of triglyceride backbone of the VO is the most ubiquitous method for altering the carboxyl group of the fatty-acid chain. Transesterification reaction is catalyzed in the presence of acidic or basic catalysts and leads to the separation of esters (Wagner et al., 2001b). The lubricant properties can be fine-tuned by the choice of various alcohols and acids. The transesterification of VO with short-chain alcohols like methanol or ethanol leads to the formation of biodiesel, while bio-lubricants are obtained when higher alcohols (C6–C8) and polyols are used for transesterification. Ghazi et al. (2010) illustrated few examples of transesterification and esterification products as shown in Figures 14.8 and 14.9.

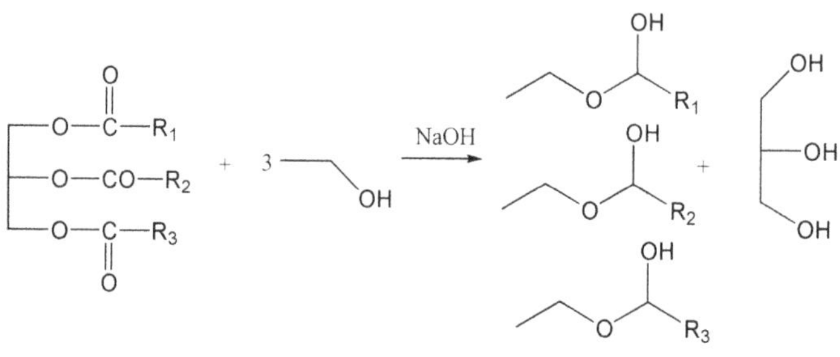

Figure 14.8 Synthesis of methyl/ethyl esters via transesterification of vegetable oil.

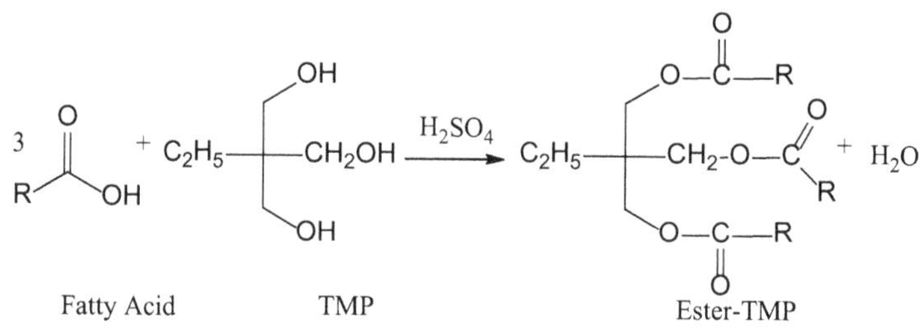

Figure 14.9 A typical esterification reaction.

It is worthwhile to note that such approaches upgrade the thermo-oxidative stability, cold flow, and tribological properties of a modified VO.

14.3.1.2 Partial/Selective Hydrogenation

Choo et al. (2001) described hydrogenation of VO as one of the essential methodologies to transform the physical characteristics of the oil which in turn improves its oxidative and degradation stability. By selective hydrogenation, the polyunsaturated fatty acids can be changed into monounsaturated fatty acids. Natural fats and oils contain polyunsaturated fatty acids such as linoleic and linolenic acids that degrades the aging stability of the oil. Selective hydrogenation aids in transforming the easily oxidizable compounds to more stable compounds. This significantly improves the aging behavior of the oils which proves critical for their use as lubricants. However, during the hydrogenation, the monounsaturated fatty acids formed as products may isomerize to form the cis- and trans-acids. Cis-isomer remains in liquid form at ambient temperature as compared to trans-isomer. The selective hydrogenation of VOs must have minimum of 80% cis-oleic acid to meet the industrial needs of bio-lubricants (Nohair et al., 2005). Selectively hydrogenated VOs into high-oleic content oils find application as electric insulation fluid in transformers due to their increased thermo-oxidative stability and less corrosive nature.

14.3.1.3 Estolides

Estolides are oligomeric esters, obtained when the carboxylic acid functionality of one fatty acid attaches to the site of unsaturation of another fatty acid. Cermak and Isbell (2000) reported that secondary ester linkages obtained during an estolide formation are more stable toward hydrolysis, making estolides a promising biolubricant. The extent of oligomerization of the molecule is represented by estolide number (EN) (Isbell, 2011). Cermak and Isbell (2001) also investigated a number

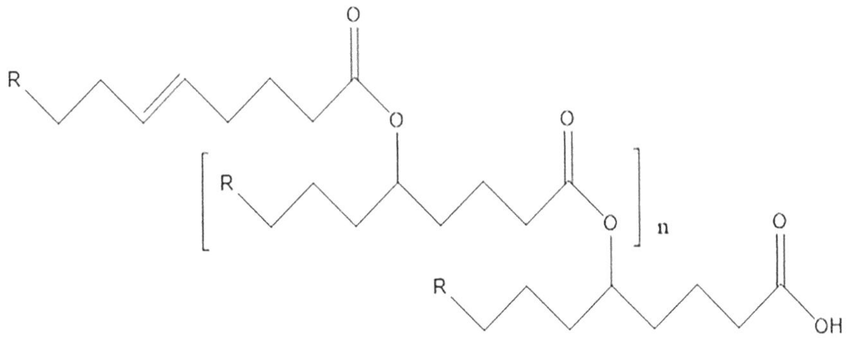

Figure 14.10 Estolide structure.

of saturated estolides, where oleic acid and saturated fatty acids ranging from butyric to stearic acid are treated with 0.4 equivalents of perchloric acid around 45°C–55°C to generate complex estolides. The EN varies with reaction temperature and change in saturated fatty acids. The shorter chain saturated fatty acids provide estolides with higher degree of oligomerization (EN = 3.3) than long-chain fatty acids (EN = 1.4). These oleic estolides when formulated with a small amount of oxidative stability package show better oxidative stability as compared to petroleum and VO-based base stocks. Complex estolides are formed by intermolecular reaction of C18 fatty acids wherein one or more double bonds react with each other at temperatures of about 210°C–250°C in the presence of layered aluminosilicate catalysts (e.g., montmorillonite) forming a complex mixture of C36 dicarboxylic acids (dimeric fatty acids), C54 trimer fatty acids, and C18 monomer fatty acids (Baumann et al., 1988; Koster et al., 1998; Zoebelein, 1992). These oleic estolides can be esterified with 2-ethylhexanol to obtain base stocks with excellent low-temperature property (PP) (as shown in Figure 14.10) (Cermak and Isbell, 2002). The viscosity index ranged from 122 to 155 for the free acid estolides, while the estolide 2-ethylhexyl esters had slightly higher indices which ranged from 172 to 196. These new estolide esters demonstrated superior low-temperature properties and are more suitable as a base stock for biodegradable lubricants than current commercial materials. Estolides have a variety of potential applications as lubricants, greases, printing inks, cosmetics, and surfactants.

14.3.1.4 Epoxidation

Uosukainen et al. (1998) reported epoxidation as an extensively used chemical methodology to upgrade the lubricity and oxidative stability of lubricating oil. A typical industrial route for epoxidation of VO is based on *in situ* epoxidation, wherein a peracid generated from reaction of acetic/formic acid and hydrogen peroxide in the presence of strong mineral acids further reacts with the unsaturated part of the VO to produce an epoxide (as shown in Figure 14.11) (Petrovic et al., 2002). Epoxidation can perform in the presence of a homogeneous (Greenspan and MacKellar, 1948;

Figure 14.11 An epoxidation reaction.

Figure 14.12 Schematic representation of synthesis of POE-based base stock.

Rangarajan et al., 1995; Swern, 1971) or a heterogeneous catalyst (Bu et al., 2000). The degree of epoxidation is commonly estimated by the *oxirane oxygen content* value.

Epoxide oil comprises of a three-atom ring system which is more reactive to form other various chemical compounds via ring opening. The ring opening of epoxy group and subsequent derivatization of epoxy carbons lead to improved oxidative and thermal stability.

14.3.2 Synthesis of Polyol (POE) Based Base Stocks

Polyol esters are synthesized by esterification of polyols with monocarboxylic acids in the presence of acid catalyst and azeotrophic solvent for efficient removal of water. The polyols used are DiPE, PE, TMP, and NPG. Usually, the acid is used in 1.5–2.5 molar excess of the amount of polyol used for completion of reaction to ensure high hydrolytic stability as per application requirements (as shown in Figure 14.12).

The reactions are performed in the presence of homogeneous catalysts like *p*-toluene sulfonic acid and methane sulfonic acid. During esterification reactions, water by-product is continuously removed for reaction completion. The reaction is generally performed at a temperature range of 180°C–240°C. After the reaction, the excess acid and the catalyst are removed by water wash and alkali treatment, respectively.

14.4 INDUSTRIAL APPLICATION OF VO-BASED AND SYNTHETIC OIL-BASED LUBRICANTS

The last decade has witnessed the replacement for mineral oil-based lubricants with synthetic polyol-ester-based lubricants for refrigeration applications. This was in accordance with the Kyoto and Montreal protocols to eliminate chemicals that are capable of ozone depletion and global warming. The polyol esters are compatible with HFC refrigerants which have zero ozone depleting potential. Owing to severe environmental concerns posed by mineral oil-based lubricants, the lubricant industry is gradually shifting to biodegradable or green lubricant-based base stocks for a wide range of industrial applications. Some of the applications in the area of transformer oils, refrigeration oils, metalworking fluids, VO-based diesel lubricity improver, and VO-based greases have been discussed in the following section.

14.4.1 Transformer Oils

Electrical transformers are usually packed with dielectric fluids which offer electrical insulation to transformer core and windings to remove the heat. This further protects the cellulosic paper insulation from probable attack of air and moisture (Darwin et al., 2010). Over the past decades, polychlorinated biphenyl (PCBs)-based mineral oils have been frequently used to fill in transformers until when PCB is recognized as a toxic and constant pollutant by the Toxic Substance Act. Environmental Protection Agency (EPA) regulations soon followed and prohibited the production of PCB and its use in transformers (McShane, 1998). The gradual elimination of PCB from transformers compelled the researchers to look for alternatives that are inflammable and environmentally acceptable. Hence, ester-based oils for transformer application have been developed. Fernandez

Table 14.8 Comparison between Mineral Oil-Based and Ester Oil-Based Transformer Oils

Parameter	Mineral Oil	Ester-Based Oil
Sludge formation	Forms sludge inside the transformer	No sludge formation
Cellulose insulation	Prone to damage	No damage
Biodegradability	Nonbiodegradable	Biodegradable
Toxicity	Toxic	Nontoxic
Service life	Poor	Good
Flash point	150°C	330°C

et al. (2013) studied the application of polyol and natural ester-based dielectric fluids as transformer oils for their excellent fire safety and environmental characteristics. The merits of ester-based oils vis-a-vis mineral oils are compared in Table 14.8.

VO or natural ester-based transformers are sustainable, safe, and eco-friendly alternative. [Native Advertisement] Natural esters for electrical applications are commonly obtained from soya, rapeseed, and sunflower oils. Polyol esters are thermo-chemically stable, readily biodegradable, and nontoxic in nature (Al-Amin et al., 2013). Their high flash points with improved fire-resistant characteristics make them safer (Gardiner et al., 1975). In comparison to mineral oils, polyol esters can absorb huge amount of water; it removes water from transformer insulation paper that leads to less paper degradation and extended transformer service life (Oomen et al., 1998). Bashi et al. (2006) reported that polyol esters with superior low-temperature fluidity suitable for subzero conditions can be designed by modifying the polyols or mixture of polyols with suitable aliphatic acids. Owing to aforementioned benefits, polyol esters are sorted to be the ideal choice for liquid-filled transformers. With increasing environmental concerns and need for improved performance, there is a rising trend of polyol esters in transformer industry (Hof et al., 2008).

14.4.2 Refrigeration Oils

Lubricants play a vital role in the domestic and industrial refrigeration applications. An ideal refrigeration lubricant must lubricate internal parts, remove heat generated during compression, act as a fluid seal, and reduce energy requirements. One of the main factors in the selection of refrigeration oil is refrigerant miscibility/solubility. The basic ingredients of a refrigerant include ethyl/methyl ethers, ammonia, carbon dioxide, sulfur dioxide, methyl chloride, ethyl chloride, and light petroleum distillates, which made the refrigerants flammable and unsafe. The search for safer refrigerants started in 1930s with chlorinated fluorocarbons (CFCs) followed by hydrochlorofluorocarbons (HCFCs) and HFCs. However, these refrigerants led to ozone layer depletion and global warming. Stein and Mead (1993) projected the global trend for replacement and phase out of the refrigerants as given in Table 14.9.

Further research led to the development of hydrofluorinated olefins (HFOs). Presently, HFO-1234yf (2,3,3,3-tetrafluoroprop-1-ene) is the potential candidate of low global warming refrigerant in mobile air conditioning (Short and Rajewski, 1995). New low GWP refrigerant blends are under development for use in commercial and industrial refrigeration (Council Directive 70/156/EEC,

Table 14.9 Historical and Projected Progression of Refrigerants

1996	CFCs banned, replacement by HFCs and HCFCs
2010	Complete replacement by HFCs
2011–2017	Usage of HFOs
2020	US-R22 phase out
2030	HCFCs phase out

Table 14.10   Comparison between Mineral Oil and POE-Based Refrigeration Oils

Properties	Mineral Oil	POE Oil
Lubrication	Poor	Excellent
Thermal stability	Poor	Good
Miscibility in HFC refrigerants	Less miscible	More miscible
Biodegradability	Nonbiodegradable	Highly biodegradable

2006). There is some renewed interest in carbon dioxide and hydrocarbons for commercial use but installation costs and safety issues have limited their application.

The shift from CFCs to HFOs has led to the changes in the lubricant base stocks used in combination with the refrigerants. The lubricants used with CFCs and HCFCs were based on refined naphthenic base oils. The attempts to use mineral oils with the HFCs met with limited success. The alkylated benzenes were reported for better stability at operational temperatures with favorable viscosities for CFC and HCFC refrigerants. Sasaki et al. (2003) revealed that alkylated benzene-based compressor oils are not compatible with chlorine-free HFC refrigerants such as R134a, R404A, and R407C. A comparison of the properties of the mineral oils and synthetic oils used in refrigeration is given in Table 14.10.

Accordingly, synthetic polyol ester lubricants that are compatible with these refrigerants have replaced alkylated benzenes in compressor systems. Other synthetic lubricants like polyalkylene glycols (PAGs) are commonly used in automotive applications. The synthetic oils, polyol esters, and PAGs are compatible with the HFCs and HFOs due to its polar nature and hence have better solubility with the refrigerants. The POE-based synthetic oils are hydrophilic than mineral oils and therefore need to be handled more carefully.

To overcome the issue of hydrolytic stability of POE-based oils, polyvinylether (PVE)-based synthetic refrigeration oils have been developed (Ciantar et al., 2001). PVE being nonhydrolytic eliminates the use of filter dryers in refrigeration equipments, thereby aids in cutting costs. Due to its flexible polymeric characteristics, properties required for refrigeration application can be optimally modified.

14.4.3 Metal Working Fluids

The present market is dominated with mineral oil-based metal-working fluids. Sokovic and Mijanovic (2001) reported that mineral oil-based cutting fluid formulations are nonbiodegradable hydrocarbon mixtures and contain toxic substances such as metals, sulfur, and aromatic compounds. Moreover, they possess low flash points, low VI, and less shear stability. Owing to the technical limitations, cutting fluids cause serious health and environmental hazards. The major health hazards associated with use of mineral oil-based metal-working fluids led to dermatitis (skin diseases) primarily caused due to skin coming in contact with nitrosamines, heavy metal compounds, and bactericides: respiratory problems due to inhalation of toxic mists and exposure to oil fags and fumes (Wilfried, 1998). It is worthwhile to note that about 80% of all occupational diseases of operators are reported from application of metal-working fluids (Abdalla et al., 2007). Also, inappropriate disposal of cutting fluids resulted into serious environmental impact leading to soil and water pollution (Alves and Gomes de Oliveira, 2008).

In view of the demerits of mineral oil-based metal-working fluids, natural esters (VOs and derivatives) and some synthetic ester-based metal-working fluids are being developed to meet the challenges faced with the mineral oils. The properties of natural esters, synthetic esters, and mineral oil-based cutting oils are compared in Table 14.11.

Jacob et al. (2004) concluded that use of VO-based cutting fluids can mitigate the skin problems and inhalation of toxic mists in the working environment.

Table 14.11   Comparison of Natural and Synthetic Esters with Mineral Oil-Based Base
             Stocks

Properties	Natural Ester	Synthetic Ester	Mineral Oils
VI	High	High	low
Flash point	High	High	low
Evaporation rate	Low	Low	high
Wetting tendency	High	High	low
Toxicity	Nil	Nil	high
Bio-degradability	High	High	poor
Shear stability	High	High	low

Typically, cutting fluids are designed to enhance lubrication between moving metal parts and remove the heat generated due to friction during cutting operations. Because of their lubrication and cooling characteristics, it promotes extended tool life by reducing wear, thermal deformation, and improves surface finish. A metal-cutting fluid serves as a dual function of lubrication and coolant. It is formulated with either oil/water or both oil and water as base material and additives such as corrosion inhibitors, de-foamers, emulsifiers, metal deactivators, and antioxidants. Furthermore, mineral oil-based cutting fluid possess low flash point resulting in high evaporation rates and losses, whereas these limitations can be handled with VO-based metal-working fluids.

14.4.4 VO Based Lubricity Improvers for Euro IV/VI Diesel

Over the last two decades, strict government mandates and regulations have forced the refiners to bring down the level of sulfur in fuels thereby encouraging the implementation of green and eco-friendly fuels. Hydroprocessing is one such process to de-sulfurize fuel and reduce aromatic levels. Minimizing these elements can affect the inherent lubricity of the fuel. Typically, a diesel engine depends on the fuel to lubricate many of the moving parts of fuel injection system. Low lubricity fuel can lead to fuel pump wear and engine failure. To meet the lubricity requirements of many diesel fuel specifications, lubricity improvers are therefore frequently required to prevent engine wear and reduce maintenance costs. Most commercial lubricity additives (Claydon, 2014) are produced from long-chain carboxylic acids. They are of two types acidic or neutral. If the acidic head group is not reacted, the product is classed as acidic. If the head group is reacted with an alkanol amine then the chemistry would be an amide and alternatively reaction with polyhydric alcohols, among others, would produce an ester. Both esters and amides are classed as neutral lubricity additives. An example of lubricity additive chemistry is shown in Figure 14.13.

The polar group that is attracted to metal surfaces causes the additive to form a thin surface film. The film acts as a boundary lubricant when two metal surfaces come in contact. Monoacids are more effective; therefore, lower concentrations are used (10–50 ppm). Because esters and amides are less polar, they require a higher concentration range from 50 to 250 ppm.

14.4.5 Vegetable Oil-Based Greases

Grease is a semisolid lubricant consisting of emulsified soap with mineral or VO. Greases are applied infrequent lubrication applications. They act as sealants to prevent ingress of water and incompressible materials. Currently, worldwide trend is to use eco-friendly lubricating greases in view of environmental concerns on biodegradability raised on mineral oil-based greases (Panchal et al., 2015). Presently, these biodegradable greases have been formulated from modified palm oil as base oil and lithium soap as thickener. Such palm-grease formulations are applied in agriculture, forestry, and coastal areas. Dwivedi and Sapre (2002) described the preparation of VO-based

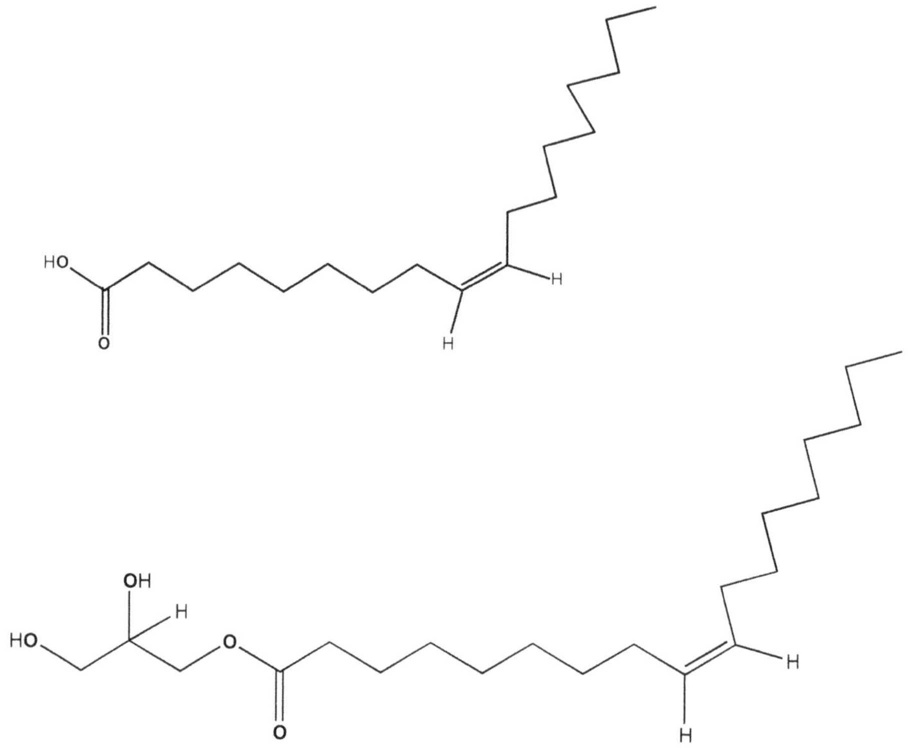

Figure 14.13 Examples of long-chain carboxylic acid and their ester used as commercial lubricity additives.

greases using saponified castor oil as base oil and lithium or sodium soaps as thickener. Currently, their application is limited due to its performance and high production cost (Bower, 1984).

14.4.6 Lubricants for Food-Grade Applications

In food industries, lubricants must comply with strict protocols. Food-grade lubricant categories are defined on the basis of its likelihood to be in contact with food. The Food and Drug Administration (FDA) classified original food-grade lubricant as H1, H2, and H3 grade lubricants; the terminology prevails and is presently used in food-grade lubricant markets (Williamson, 2003). The approval and registration of a new lubricant into one of these categories depends on the ingredients used in the formulation. The three food-grade lubricant categories are described in the following section.

14.4.6.1 H1 Grade Lubricants

These food-grade lubricants are normally used in food processing environments where there is some possibility of incidental food contact. Lubricant formulations may only be composed of one or more approved base stocks, additives and thickeners (if grease) listed in 21 CFR 178.3750.

14.4.6.2 H2 Grade Lubricants

These lubricants are used on equipment and machine parts in locations where there is no possibility that the lubricant or lubricated surface is in contact with food. Since there is no the risk involved for any contact with food, thus H2 type lubricants does not contain defined list of acceptable ingredients. Also, the ingredients must not include substances that are carcinogens, mutagens, teratogens, or mineral acids.

14.4.6.3 H3 Grade Lubricants

They are also known as soluble or edible oil. They are used to clean and prevent rust on hooks, trolleys, and similar equipment.

The US FDA approvals are based on the various FDA Codes mentioned in Title 21 that dictate approval (NSF Guidelines, 2004) for ingredients used in production of food-grade lubricants which may have incidental contact with food. These are listed below:

- 21 CFR 178.3570—Allows ingredients for the manufacture of H1 lubricants
- 21 CFR 178.3620—White oil as a component of nonfood articles intended for use in food
- 21 CFR 172.878—USP mineral oil for direct contact with food
- 21 CFR 172.882—Synthetic iso-paraffinic hydrocarbons
- 21 CFR 182—Substances generally recognized as safe

14.4.6.3.1 Acceptable Food-Grade Base Stocks

In addition to H1 or H2 approved lubricant, base stocks can be either mineral or synthetic. Mineral oils used in H1 food-grade lubricants are either technical white mineral or USP-type white mineral oils. These are highly refined and are colorless, tasteless, and odorless base stocks. Technically, white oils meet the regulations specified in 21 CFR 178.3620.

Synthetic H1 lubricant base stocks are often polyalphaolefins (PAO). Compared to white mineral oils, PAO have significantly greater oxidation stability and great range of operating temperatures. Another approved H1 synthetic base stock is polyalkylene glycols (PAG).

14.5 FUTURE DRIVERS

As the environmental standards are tightening year by year worldwide, the use of green lubricants is envisaged to increase rapidly. The anticipated growing demand is expected to drive development of cost effective biodegradable lubricants which would add impetus for deriving new synthesis methodologies, and exploring new feedstocks such as biomass to prepare suitable eco-friendly base stocks.

ACKNOWLEDGMENTS

The authors are thankful to the management of Bharat Petroleum Corporation Limited for granting permission to publish the present chapter.

List of Abbreviations

API   American Petroleum Institute
ASTM   The American Society for Testing and Materials
CEC   Coordinating European Council
CFC   Chlorinated fluorocarbons
CFR   Code of Federal Regulations

DiPE	Dipentaerythritol
EAL	Environmentally adapted lubricants
EN	Estolide number
EO	Ethylene oxide
EPA	Environmental Protection Agency
ESO	Epoxidized soyabean oil
FDA	Food and Drug Administration
GHG	Greenhouse gases
GWP	Global warming potential
HCFC	Hydrochlorofluorocarbon
HFC	Hydrofluorocarbon
HFO	Hydrofluorinated olefins
HO	High oleic
IL	Ionic liquid
LAO	linear alpha olefins
NPG	Neopentylglycol
OECD	Organisation for Economic Co-operation and Development
OSP	Oil-soluble PAGs
PAO	Polyalphaolefin
PAG	Polyalkyleneglycol
PCB	Polychlorinated biphenyl
PE	Pentaerythritol
PO	Propylene oxide
POE	Polyol ester
PP	Pour point
PVE	Polyvinylether
RTIL	Room temperature ionic liquids
TMP	Trimethylolpropane
USP	The United States Pharmacopeia
VI	Viscosity index
VO	Vegetable oil

REFERENCES

Abdalla H.S., Baines W., McIntyre G. and Slade C., 2007, Development of novel sustainable neat-oil metal working fluids for stainless steel and titanium alloy machining. Formulation development, *Int. J. Adv. Manuf. Technol.*, 34, pp. 21–33.

Adhvaryu A. and Erhan S.Z., 2002, Epoxidized soybean oil as a potential source of high-temperature lubricants, *Ind. Crop. Products*, 15(3), pp. 247–254.

Adhvaryu A., Erhan S.Z. and Perez J.M., 2004, Tribological studies of thermally and chemically modified vegetable oils for use as environmentally friendly lubricants, *Wear*, 257(3–4), pp. 359–367.

Adhvaryu A., Liu Z. and Erhan S.Z., 2005, Synthesis of novel alkoxylated triacylglycerols and their lubricant base oil properties, *Ind. Crop. Prod.*, 21(1), pp. 113–119.

Al-Amin H., Brien J.O. and Lashbrook M., 2013, Synthetic ester transformer fluid: A total solution to wind park transformer technology, *Renew. Energy*, 49, pp. 33–38.

Allawzi M., Abu-Arabi M.K., Al-zoubi H.S. and Tamimi A., 1998, Physicochemical characteristics and thermal stability of Jordanian jojoba oil, *J. Am. Oil Chem. Soc.*, 75(1), pp. 57–62.

Alves S.M. and Gomes de Oliveira J.F., 2008, Vegetable based cutting fluid—An environmental alternative to grinding process, *15th CIRP International Conference on Life Cycle Engineering*, Sydney.

API Standard 1509, 2002, *Engine Classification and Guide to Crankcase Oil Selection*, Washington, DC: American Petroleum Institute.

Asadauskas S. and Erhan S.Z., 1999, Depression of pour points of vegetable oils by blending with diluents used for biodegradable lubricants, *J. Am. Oil Chem. Soc.*, 76, pp. 313–316.

Asadauskas S., Perez J.M. and Duda J.L., 1997, Lubrication properties of castor oil-potential base stock for biodegradable lubricants, *J. Lubr. Eng.*, 53(12), pp. 35–40.

Bantchev G.B., Kenar J.A., Biresaw G. and Han M.G., 2009, Free radical addition of butanethiol to vegetable oil double bonds, *J. Agric. Food Chem.*, 57(4), pp. 1282–1290.

Bartz W.J., 1998, Lubricants and the environment, *Tribol. Int.*, 31(1–3), pp. 35–47.

Bashi S.M., Abdullahi U.U., Yunu R. and Nordin A., 2006, Use of natural vegetable oils as alternative di-electric transformers coolants, *Inst. Eng.*, 67(2), pp. 4–9.

Batters N.S., 2005, Biodegradable lubricant: What does biodegradable really mean, *J. Synth. Lubr.*, 22, pp. 3–18.

Baumann H., Buhler M., Fochem H., Hirsinger F., Zoebelein H. and Falbe J., 1988, Natural fats and oils: ¼r for renewable raw materials in the chemical industry, *Angew. Chem.*, 100(1), pp. 41–62.

Biresaw G., Adhvaryu A. and Erhan S.Z., 2003, Friction properties of vegetable oils, *J. Am. Oil Chem. Soc.*, 80, pp. 697–704.

Biresaw G., Adhvaryu A., Erhan S.Z. and Carriere C.J., 2002, Friction and adsorption properties of normal and high-oleic soybean oils, *J. Am. Oil Chem. Soc.*, 79, pp. 53–58.

Biswas A., Adhvaryu A., Gordon S.H., Erhan S.Z. and Willett J.L., 2005, Synthesis of diethylamine-functionalized soybean oil, *J. Agric. Food Chem.*, 53(24), pp. 9485–9490.

Bower B., 1984, *Environmental Capabilities of Liquid Lubricants in Solid and Liquid Lubricants for Extreme Environments*, ASLE Special Publication SP-15, Denver, CO: ASLE, pp. 58–69.

Boyde S., 2002, Green lubricants—Environmental benefits and impacts of lubrication, *Green Chem.*, 4, pp. 293–307.

Bu J., Yun S. and Rhee H., 2000, Epoxidation of n-hexene and cyclohexene over titanium-containing catalysts, *Korean J. Chem. Eng.*, 17(1), pp. 76–80.

Buenemann T.F., Boyde S., Randles S. and Thompson I., 2003, Synthetic lubricants—Non aqueous, in *Fuels and Lubricants Handbook: Technology, Properties, Performance and Testing*. G.E. Totten (Ed.), West Conshohocken, PA: ASTM International.

Cermak S.C. and Isbell T.A., 2000, Biodegradable oleic estolide ester having saturated fatty acid end group useful as lubricant base stock, US09/490,360.

Cermak S.C. and Isbell T.A., 2001, Synthesis of estolides from oleic and saturated fatty acids, *J. Am. Oil Chem. Soc.*, 78(6), pp. 557–565.

Cermak S.C. and Isbell T.A., 2002, Physical properties of saturated estolides and their 2-ethylhexyl esters, *Ind. Crop. Prod.*, 16(2), pp. 119–127.

Choi U.S., Ahn B.G, Kwon O.K. and Chun Y.J., 1997, Tribological behavior of some antiwear additives in vegetable oils, *Tribol. Int.*, 30(9), pp. 677–683.

Choo H.P., Liew K.Y., Liu H.F. and Seng C.E., 2001, Hydrogenation of palm olein catalyzed by polymer stabilized Pt colloids, *J. Mol. Catal. A-Chem.*, 165(1–2), pp. 127–134.

Ciantar C., Hadfield M., Swallow A. and Smith A., 2001, The influence of POE and PVE lubricant blends within hermetic refrigerating compressors with HFC-134a refrigerant, *Wear*, 241, pp. 53–64.

Claydon D., 2014, The use of lubricity additives to maintain fuel quality in low sulphur diesel fuel, *goriva i maziva* 53(4), pp. 342–353.

Cole G., Coughlan G.S., Frey N., Hazebroek J. and Jennings C., 1998, New sunflower and soybean cultivars for novel vegetable oil types, *Fett/Lipid*, 100, pp. 177–181.

Darwin A., Perrier C. and Folliot P., 2010, The use of natural ester fluids in transformers. Mat Post 07.

Directive 2006/40/EC of the European Parliament and the Council of the European Union, 2006, Emissions from air-conditioning systems in vehicles and amending Council Directive 70/156/EEC, *Off. J. Eur. Comm.*, L161, p. 12.

Duvick D.N., 1996, Utilization of biotechnology in plant breeding in North America: A status report, in *Perspektiven nachwachsender Rohstoffe in der Chemie*. H. Eierdanz (Ed.), Weinheim: VCH Verlag, pp. 3–10.

Dwivedi M.C. and Sapre S., 2002, Total vegetable-oil based greases prepared from castor oil, *J. Synth. Lubr.*, 19, pp. 229–241.

Erhan S.Z., Sharma B.K. and Perez J.M., 2006, Oxidation and low temperature stability of vegetable oil-based lubricants, *Ind. Crop. Prod.*, 24(3), pp. 292–299.

Erhan S.Z., Sharma B.K., Liu Z. and Adhvaryu A., 2008, Lubricant base stock potential of chemically modified vegetable oils, *J. Agric. Food Chem.*, 56(19), pp. 8919–8925.

Fernandez I., Ortiz A., Delgado F., Renedo C. and Perez S., 2013, Comparative evaluation of alternative fluids for power transformers, *Electr. Pow. Syst. Res.*, 98, pp. 58–69.

Fick G.N., 1988, Novel sunflower products and methods for their production, US Patent 4743402.

Fox, N.J. and Stachowiak G.W., 2007, Vegetable oil-based lubricants—A review of oxidation, *Tribol. Int.*, 40(7), pp. 1035–1046.

Garces R., Martınez E. and Salas J.J., 2011, Vegetable oil base stocks for lubricants, *Grasas Aceites*, 62, pp. 21–28.

Gardiner J.B., Shaub H. and Tessier K.C., 1975, Mixed carboxylic acid esters as electrical insulating oils, US 3894959.

Ghazi T.I.M., Resul M.F.M.G. and Idris A., 2010, Production of an improved biobased lubricant from *Jatropha curcas* as renewable source, *Proceedings of the 3rd International Symposium on Energy from Biomass and Waste*, Venice, Italy.

Greaves M.R., 2011, Oil soluble synthetic polyalkylene glycols, a new type of Group V base oil, *Lube Mag.*, 104, pp. 21–24.

Greenspan F.P. and MacKellar D.G., 1948, Analysis of aliphatic per acids, *Analyt. Chem.* 20(11), pp. 1061–1063.

Hagemann J.W., 1988, Thermal behavior and polymorphism of acylglycerols, in *Crystallization and Polymorphism of Fats and Fatty Acids*. N. Garti and K. Sato (Eds.), New York: Marcel Dekker.

Hof M., Baehr H., Moucha M., Cooban N. and Willing A., 2008, Polyol ester for transformers, US 2008/0033201 A1.

Hwang H.S. and Erhan S.Z., 2006, Synthetic lubricant base stocks from epoxidized soybean oil and guerbet alcohols, *Ind. Crop. Prod.*, 23(3), pp. 311–317.

Hwang H.S. and Erhan S.Z., 2007, Modification of epoxidized soybean oil for lubricant formulations with improved oxidative stability and low pour point, *J. Am. Oil Chem. Soc.*, 78, pp. 1179–1184.

Isbell T.A., 2011, Chemistry and physical properties of estolides, *Igrasas y Aceites*, 62(1), pp. 8–20.

Jacob J., Bhattacharya M. and Raynor P.C., 2004, Emulsions containing vegetable oils for cutting fluid application, *Colloids Surf. A*, 237, pp. 141–150.

Jaina A.K. and Suhanea A., 2013, Capability of biolubricants as alternative lubricant in industrial and maintenance applications, *Int. J. Curr. Eng. Technol.*, 3, pp. 179–183.

Jayadas N.H. and Nair K.P., 2006, Coconut oil as base oil for industrial lubricants—Evaluation and modification of thermal, oxidative and low temperature properties, *Tribol. Int.*, 39(9), pp. 873–878.

Jimâenez A.E. and Bermâudez M.D., 2007, Ionic liquids as lubricants for steel-aluminum contacts at low and elevated temperatures, *Tribol. Lett.*, 26, pp. 53–60.

Jimenez A.E., Bermudez M.D., Iglesias P., Carrion F.J. and Martinez-Nicolas G., 2006, 1-N-alkyl-3-methylimidazolium ionic liquids as neat lubricants and lubricant additives in steel-aluminium contacts, *Wear*, 260, pp. 766–782.

Kassfeldt E. and Goran D., 1997, Environmentally adapted hydraulic oils, *Wear*, 207(1–2), pp. 41–45.

Koster R.M., Bogert M., de Leeuw B., Poels E.K. and Bliek A., 1998, Active sites in the clay catalysed dimerisation of oleic acid, *J. Mol. Catal. A: Chem.*, 134(1–3), pp. 159–169.

Lal K., 1994, Pour point depressants for industrial lubricants containing mixtures of fatty acid esters and vegetable oils, US Patent 338471.

Lansdown A., 2004, *Lubrication and Lubricant Selection: A Practical Guide*, 3rd Edition, New York: ASME Press.

Lawal S.A., Choudhury I.A. and Nukman Y., 2012, Application of vegetable oil-based metalworking fluids in machining ferrous metals—A review, *Int. J. Mach. Tools Manuf.*, 52(1), pp. 1–12.

Lawate S.S. and Lal K., 1995, High oleic polyol esters, compositions and lubricants, functional fluids and greases containing the same, EP 0712834 A1.

Lawford S., 2009, *Polyalkylene Glycols, Synthetics, Mineral Oils and Bio-Based Lubricants*, Boca Raton: CRC Press, pp. 119–138.

Liu W., Ye C., Chen Y., Ou Z. and Sun D.C., 2002, Tribological behavior of sialon ceramics sliding against steel lubricated by fluorine-containing oils, *Tribol. Int.*, 35, pp. 503–509.

Liu X., Zhou F., Liang Y. and Liu W., 2006, Tribological performance of phosphonium based ionic liquids for an aluminum-on-steel system and opinions on lubrication mechanism, *Wear*, 261, pp. 1174–1179.

Lu Q., Wang H., Ye C., Liu W. and Xue Q., 2004, Room temperature ionic liquid 1-ethyl-3-hexylimidazolium-bis(trifluoromethylsulfonyl)-imide as lubricant for steel contact, *Tribol. Int.*, 37, pp. 547–552.

McShane C.P., 1998, *Natural and Synthetic Ester Di-electric Fluids: Their Relative Environmental, Fire Safety, and Electrical Performance*, Sparks, NV: Cooper Power Systems.

McTurk W.E., 1953, Synthetic lubricants, WADC Technical Report, pp. 53–88.

Minami I., 2009, Ionic liquids in tribology, *Molecules*, 14, pp. 2286–2305.

Minami, I., Inada, T., Sasaki, R. and Nanao H., 2010, Tribo-chemistry of phosphonium-derived ionic liquids, *Tribol. Lett.*, 40, pp. 225–235.

Moser B.R. and Erhan S.Z., 2007, Preparation and evaluation of a series of α-hydroxy ethers from 9,10-epoxystreates, *Eur. Lipid Sci. Technol.*, 109(3), pp. 206–213.

Mu Z., Liu W., Zhang S. and Zhou F., 2004, Functional room-temperature ionic liquids as lubricants for an aluminum-on-steel system, *Chem. Lett.*, 33, pp. 524–525.

Mu Z., Zhou F., Zhang S., Liang Y. and Liu W., 2005, Effect of the functional groups in ionic liquid molecules on the friction and wear behavior of aluminum alloy in lubricated aluminum-on-steel contact, *Tribol. Int.*, 38, pp. 725–731.

Nagendramma P. and Kaul S., 2012, Development of ecofriendly/biodegradable lubricants: An overview, *Renew. Sustain. Energy Rev.*, 16, pp. 764–774.

Nohair B., Especel C., Lafayea G., Marécot P., Hoang L.C. and Barbier J., 2005, Palladium supported catalysts for the selective hydrogenation of sunflower oil, *J. Mol. Catal. A-Chem.*, 229(1–2), pp. 117–126.

NSF International Registration Guidelines, 2004, www.nsf.org/business/nonfood_compounds/guidelines.pdf.

Ohtani H., Ishimura S. and Kumai M., 2008, Thermal decomposition behaviors of imidazolium-type ionic liquids studied by pyrolysis-gas chromatography, *Anal. Sci. Int. J. Jpn. Soc. Anal. Chem.*, 24, pp. 1335–1340.

Omotowa B.A., Phillips B.S., Zabinski J.S. and Shreeve J.M., 2004, Phosphazene-based ionic liquids: Synthesis, temperature-dependent viscosity, and effect as additives in water lubrication of silicon nitride ceramics, *Inorg. Chem.*, 43, pp. 5466–5471.

Oomen T.V., Claiborne C.C. and Walsh E.J., 1998, Introduction of a new fully biodegradable dielectric fluid, *1998 IEEE Textile, Fiber and Film Industry Technical Conference*, Charlotte, NC.

Pal M. and Singhal S., 2000, Environmentally adapted lubricants. Part I. An overview, *J. Synth. Lubr.*, 17(2), pp. 135–143.

Panchal T., Chauhan D., Thomas M. and Patel J., 2015, Bio-based grease a value added product from renewable resources, *Ind. Crops Prod.*, 63, pp. 48–52.

Petrovic Z.S., Zlatanic A., Lava C.C. and Sinadinovic F.S., 2002, Epoxidation of soybean oil in toluene with peroxoacetic and peroxoformic acids—Kinetics and side reactions, *Eur. J. Lipid Sci. Tech.*, 104(5), pp. 293–299.

Pettersson A., 2007, High-performance base fluids for environmentally adapted lubricants, *Tribol. Int.*, 40, pp. 638–645.

Phillips B.S. and Zabinski J.S., 2004, Ionic liquid lubrication effects on ceramics in a water environment, *Tribol. Lett.*, 17, pp. 533–541.

Rangarajan B., Havey A., Grulke E.A. and Culnan P.D., 1995, Kinetic parameters of a two-phase model for in situ epoxidation of soybean oil, *J. Am. Oil Chem. Soc.*, 72(10), pp. 1161–1169.

Reich R.A., Stewart P.A., Bohaychick J. and Urbanski J.A., 2003, Base oil properties of ionic liquids, *Lubr. Eng.*, 59, pp. 16–21.

Rensselar J.V., 2011, In search of: The perfect biodegradable lubricant, *Tribol. Lubric. Tech.*, 67, p. 44.

Rhee I.S., Valez C. and Bernewitz K., 1995, Evaluation of environmentally adapted hydraulic fluids, TARDEC Tech Report 13640, pp. 1–15.

Rudnick L.R., 2006, *Polyalphaolefins, Synthetics, Mineral Oils and Bio-Based Lubricants: Chemistry and Tribology*, Boca Raton: CRC Press/Taylor & Francis, pp. 3–36.

Ruger C.W., Klinker E.J. and Hammond E.G., 2002, Abilities of some antioxidants to stabilize soybean oil in industrial use conditions, *J. Am. Oil Chem. Soc.*, 79, pp. 733–736.

Salimon J., Salih N. and Abdullah B.M., 2012, Diesters biolubricant base oil: Synthesis, optimization, characterization, and physicochemical characteristics, *Int. J. Chem. Eng.*, 2012, pp. 1–10.

Sasaki U., Ishikawa T., Hasegawa H. and Ishida N., 2003, Refrigerator oils for use with chlorine-free hydrocarbon refrigerants, US Patent No. 6,582,621.

Shah, F., Glavatskih, S. and Antzutkin O.N., 2009, Synthesis, physicochemical, and tribological characterization of S-Di-n-octoxyboron-O, O′-di-n-octyldithiophosphate, *ACS Appl. Mater. Interf.*, 1, pp. 2835–2842.

Sharma B.K., Adhvaryu A., Perez J.M. and Erhan S.Z., 2006, Biobased grease with improved oxidation performance for industrial application, *J. Agric. Food Chem.*, 54(20), pp. 7594–7599.

Short, S. and Rajewski, T., 1995, Lubricants for refrigeration and air conditioning, *Lubr. Eng.*, 53(4), pp. 269–274.

Shubkin R.L., 1993, Polyalphaolefins, in *Synthetic Lubricants and High-Performance Functional Fluids*. R.L. Shubkin (Ed.), New York: Marcel Dekker, pp. 1–40.

Sokovic M. and Mijanovic K., 2001, Ecological aspects of the cutting fluids and its influence on the quantifiable parameters of the cutting processes, *J. Mater. Process. Technol.*, 109, pp. 181–189.

Stein W. and Mead T., 1993, Refrigerant and refrigeration oil trends, *Lubrication*, 79(3), pp. 2–3.

Stoutjesdijk P.A., Hurlestone C., Singh S.P. and Green A.G., 2000, High-oleic acid Australian *Brassica napus* and *B. juncea* varieties produced by co-suppression of endogenous D12-desaturases, *Biochem. Soc. Trans.*, 28, pp. 938–940.

Swern D., 1971, Epoxidation, Chapter 5, in *Organic Peroxides*. D. Swern (Ed.), New York: Wiley-Inter-Science, pp. 355–533.

Uosukainen E., Linko Y.Y., Lämsä M., Tervakangas T. and Linko P., 1998, Transesterification of trimethylolpropane and rapeseed oil methyl ester to environmentally acceptable lubricants, *J. Am. Oil Chem. Soc.*, 75, pp. 1557–1563.

Wagner H., Luther R. and Mang T., 2001a, Lubricant base fluids based on renewable raw materials their catalytic manufacture and modification, *Appl. Catal. A: Gen.*, 221, pp. 429–442.

Wagner H., Luther R. and Mang T., 2001b, Lubricant base fluids based on renewable raw materials; their green chemistry letters and reviews 379 catalytic manufacture and modification, *Appl. Catal. A: Gen.*, 221(1–2), pp. 429–442.

Wang H., Lu Q., Ye C., Liu, W. and Cui Z., 2004, Friction and wear behaviors of ionic liquid of alkylimidazolium hexafluorophosphates as lubricants for steel/steel contact, *Wear*, 256, pp. 44–48.

Wilfried B.J., 1998, Lubricants and the environment, *Tribol. Int.*, 31, pp. 35–47.

Williamson M., 2003, Understanding food-grade lubricants, *Machinery Lubrication*, January 2003.

Wu X., Zhang X., Yang S., Chen H. and Wang D., 2000, The study of epoxidized rapeseed oil used as a potential biodegradable lubricant, *J. Am. Oil Chem. Soc.*, 77(5), pp. 561–563.

Yao M., Liang Y., Xia Y. and Zhou F., 2009, Bis-imidazolium ionic liquids as the high performance anti-wear additives in poly(ethylene glycol) for steel-steel contacts, *ACS Appl. Mater. Interf.*, 1, pp. 467–471.

Ye C., Liu W., Chen Y. and Yu L., 2001, Room-temperature ionic liquids: A novel versatile lubricant, *Chem. Commun.*, 2001, pp. 2244–2245.

Yu G., Zhou F., Liu W., Liang Y. and Yan S., 2006, Preparation of functional ionic liquids and tribological investigation of their ultra-thin films, *Wear*, 260, pp. 1076–1080.

Zoebelein H., 1992, Renewable resources. Behind in terms of basic research, *Chem. Our. Time*, 26(1), pp. 27–34.

CHAPTER 15

LASER-Based Manufacturing as a Green Manufacturing Process

Ashish K. Nath, Sagar Sarkar, Gopinath Muvvala, and Debapriya Patra Karmakar
Indian Institute of Technology Kharagpur

Shitanshu S. Chakraborty
CSIR-Central Mechanical Engineering Research Institute

Suvradip Mullick
Indian Institute of Technology Bhubaneswar

Yuvraj K. Madhukar
Indian Institute of Technology Indore

CONTENTS

15.1 INTRODUCTION

Until a few decades ago, the quest for ease, comfort, and better living standards of a large section of people in more and more economical ways has occurred at the cost of appropriation of nature. Natural resources have been treated as free gifts. Immediate profit or economic viability has been prioritized over conservation of the very environment that has given birth to and sustained life of all the living organisms including humans. Now, the resultant burning problems, such as global warming, climate change, and other pollution-related issues have shaken and awakened us to think of the environment while carrying out any activity, so as to leave the next generation with a sustainable future. The living system on earth is survived by the oxygen generated by the green plants. Many of our needs, such as food, clothing, furniture, etc., are also fulfilled with the materials from trees. Even the fossil fuel energy, that is getting exhausted to propel the current civilization within a few centuries since they had been first put to use, is actually the solar energy trapped in the plant and animal bodies, directly (in the case of plants) or indirectly (in the case of animals) with the help of photosynthesis for many years, million years back. Therefore, the activities that are significantly less polluting or having significantly high energy efficiency or linked to better sustainability through incorporation of "reduce-reuse-recycle" philosophy are marked as "green." The term "manufacturing" can be briefly defined as a process of converting raw materials into a finished material good for human utility. In modern era, manufacturing processes, materials, and tools are different as compared to ancient times. One can enlist a few broad categories of manufacturing processes. However, it is hardly possible for anyone to present an exhaustive list of manufacturing processes. New manufacturing processes are evolving day by day, and while qualifying those based on comparison to other existing processes is a new criterion—Is the new process more environment-friendly? In other words, is the process greener? In a nutshell, the term "green manufacturing" is echoing in the academic research, industrial practice, government initiatives, and newspaper reports. Green manufacturing may refer to the use of green energy for manufacturing of products, use of renewable energy systems, and clean technology equipment. However, this chapter will try to qualify laser processing as a green manufacturing technique from the sustainability and environment-friendliness point of view.

Lasers, a tool of unprecedented precision and power, are finding ever-increasing applications in a wide spectrum of material processing modalities, viz., cutting, welding, drilling, marking, scribing; surface hardening, alloying, cladding, and texturing; and shock peening, forming, micro-machining, and rapid manufacturing, with varying industrial maturity. The typical laser power density and laser-workpiece interaction time for various processing applications are presented in Figure 15.1.

Lasers, which are commonly used in various manufacturing modalities, are carbon dioxide laser, Nd:YAG (rod & disk) lasers, fiber laser, diode laser, excimer lasers, and ultra-short Ti-sapphire/fiber lasers. The main characteristics of these lasers along with the material processing applications in which they are commonly employed are summarized in Table 15.1. CO_2 laser operating at $10.6\,\mu m$ wavelength has been ruling the roost in industry for material processing applications over a long time since 1970s, but with the development of the more efficient diode pumped fiber lasers operating at $1.07\,\mu m$, in the last decade market share of CO_2 laser is gradually declining (Overton et al. 2017). Fiber laser beam owing to its lower wavelength can be focused more tightly and is absorbed better in metals and less absorbed in laser induced plasma compared to CO_2 laser (Nath 2013). Another advantage of fiber laser is that its beam can be transmitted through low-loss flexible optical fiber and delivered to the workpiece, whereas such optical fiber is not yet available for transmitting the $10.6\,\mu m$ wavelength CO_2 laser beam. Similarly, the less-efficient lamp pumped Nd:YAG lasers are getting replaced by more-efficient diode pumped Nd:YAG rod and disc lasers, except in those applications where high pulse energies in 10's J energy in 10–100 ns pulse durations are required,

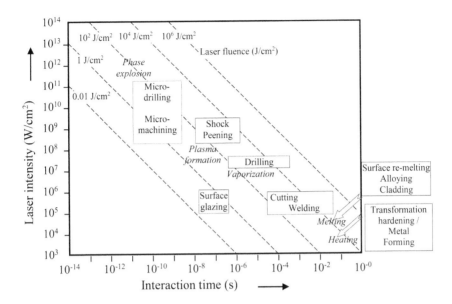

Figure 15.1 Typical laser intensities and interaction times for various laser material processing applications.

such as in laser shock peening (LSP). High-power diode lasers have the highest efficiency among all lasers, but because of their poor beam quality their applications in direct material processing are limited to those modalities which do not demand tight laser focusing and high-power densities, such as welding of plastics, brazing and surface hardening and cladding (Costa Rodrigues et al. 2014a). However, it may not be long before this scenario too gets changed.

Figure 15.2 presents a comparison of laser beam product (laser beam radius × beam divergence) parameters of different industrial lasers as a function of laser power (Costa Rodrigues et al. 2014a). The higher the beam product parameter, the less the laser beam is able to focus. This tends to increase with laser power. The desired laser power and beam product parameter range for various laser material processing applications are also indicated in Figure 15.2. This also shows how the beam quality of high-power diode lasers is being improved with new technology of power scaling. Recent advancement has yielded much improved beam quality that can produce MW/cm² range power density expanding their scope of applications in cutting and welding also (Köhler et al. 2005).

The increasing demand for fabricating miniaturized components involved in various applications, such as of MEMS, microelectronics, telecommunication, optoelectronics, and biomedical devices has created much interest in micromachining with lasers. Excimer lasers operating at short wavelengths (193–304 nm) in ultraviolet spectrum have been very popular in these applications and more recently in flat panel annealing (Overton et al. 2017). The ultrafast Ti-sapphire lasers providing intense laser pulses of sub-picoseconds and femto-second durations have revolutionized the micromachining process producing features with high spatial precision and without any detrimental thermal effects. Combining with near-field scanning optical microscope and exploiting the nonlinear properties of various photochromic materials, ultrafast lasers have been able to produce feature size smaller than even the diffraction limit (Chichkov et al. 1996; Obata et al. 2013). More recently, all fiber-based high-energy, high-average power ultrafast fiber lasers have been developed, which is expected to dominate the micromachining market due to their compactness, low maintenance, cost-effectiveness, and reduced energy consumption (Liu 2014). Thus, diode pumped fiber lasers can be rated as one of the most important developments in recent times because of their high efficiency, excellent beam quality, reliable and robust design, and very wide temporal operation range, from

Table 15.1 Common Industrial Lasers for Material Processing Applications

Parameters	CO₂ Laser	Nd:YAG (Rod/Disk) Laser	Diode Laser	Fiber Laser	Excimer Lasers	Ultrafast Lasers
Wavelength	10.06 μm	1.064 μm	0.8–1.07 μm	1.07 μm	125–351 nm	TSL: ~0.8 μm; FL: ~1.07 μm
Efficiency	10%	LP: ~3%; DP: ~20%	~50%	~30%	~4%	TSL: <1%; FL: >5%
Mode of operation	CW & pulse	CW & pulse	CW & modulated	CW & pulse	Pulse	TSL: 10 kHz–80 MHz; FL: 30–100 MHz
Output power CW/average power	Up to 20kW	Up to 16kW	Up to 25kW CW	Up to 500kW	300 W	TSL: 1–5 W (36–100fs); FL: 1–830 W (100–640fs)
Pulsed energy free running; pulse duration Q-switched; pulse energy, pulse duration	1–10 J/100 ns–10 μs; ~1 J, ~100 ns	Up to 120 J/1–20 ms; 1.2 J, ~3 ns	Single diode ~μJ/100 ns; QCW pulsed up to 50 kHz frequency	Up to 15 J/0.2–20 ms; ~1 J, 40–500 ns	Up to 1 J/20–30 ns; up to 1 kHz rep/rate	TSL: 15 nJ–290 μJ (36–100fs); FL: 30 nJ (100fs)/~10 μJ (~700fs)
Peak power	10 MW	50 MW	40 W	10 MW	10–100 MW	100 kW–10 MW
Beam quality factor: mm. mrad (typical)	3–5	0.4–20	10–100	~0.3–4	160 × 20	~0.5
Fiber delivery	No	Yes	Yes	Yes	Yes	Yes
Maintenance periods (h)	2,000	300 (lamp life)	Maintenance-free; typical life 100,000 h	Maintenance-free; pump diode life ~100,000 h	10⁸⁻⁹ pulses thyratron life	TSL: high maintenance; FL: low maintenance
Applications	C, W, D, M, SH, SC, AM	C, W, D, M, SH, SC, AM, SP	C, W, D, SH, SC, AM	C, W, D, M, SH, SC, AM, MM	SA, MM	MM

Source: Updated data from Dr. A. K. Nath, In Laser-assisted fabrications of materials, ed. J. D. Majumdar, I. Manna (2013) 69–111. Springer-Verlag: Berlin and Heidelberg.

TSL, Ti–sapphire laser; FL, fiber laser; LP, lamp pumped; DP, diode laser pumped; C, cutting; D, drilling; M, marking; W, welding; AM, additive manufacturing; SA, semiconductor annealing; SC, surface cladding; SH, surface hardening; SP, shock peening; MM, micromachining.

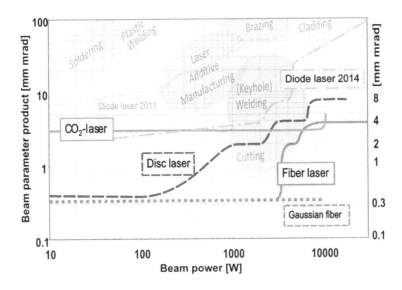

Figure 15.2 Beam parameter product vs. laser power for different laser types and typical regimes for industrial applications. (Based on data presented during the 2014 International Laser Symposium, Dresden.) (Reprinted with permission from *Physics Procedia*, Elsevier 2014 56:901–908.)

continuous wave to femto-second pulse duration. They can perform even better, most of the tasks that other lasers such as CO_2 and Nd:YAG lasers had been doing. The high brightness diode lasers are also expected to make a niche in many material processing applications in near future.

The special properties of laser beam which are exploited in material processing are its low divergence, high power and a wide range of pulse duration starting from continuous wave (CW) down to femto-seconds (1 fs = 10^{-15}s). When a high-power laser beam is focused at the surface of a material, the focal spot diameter can be as small as of the order of the laser wavelength. The laser power density (power per unit area) at the focal point can be as high as 10^{14}W/cm^2, and depending upon its magnitude the localized material at the focal point can undergo heating, melting, evaporation, plasma formation, and ablation. Controlling the laser power density and the interaction time with the material, different processing modalities mentioned above are realized as presented in Figure 15.1. Since the spatial and temporal intensity distribution of laser beam and its magnitude can be controlled very precisely over large ranges, the machining with laser beam can be realized with great precision and minimum energy. This results in no or little thermal effects like heat-affected zones, residual stresses, and distortion, thereby minimizing the requirement of post processing operation. The process is usually very fast and therefore the heat conduction loss is minimal making the process highly efficient. Moreover, this being a noncontact process, there is no tool wear. These attributes of laser material processing make laser an enabler of "green manufacturing".

As apparent from the above description laser processing of material is basically a thermal process. Depending up on the laser and processing parameters and material properties there could be heat-affected zone (HAZ), recast, spatter, oxide layer formation, and micro-cracks especially in brittle material around the processed region due to thermal stress. In other laser applications, paint stripping damage of substrates and a large amount of fume could form. In order to mitigate some of the detrimental thermal effects, laser processing is also done with water assist. Furthermore, lasers are also being used in cutting radioactive materials underwater. In recent times, several studies on water-assisted laser processing and underwater laser cutting, surface hardening, and welding have been also reported.

In this chapter, the laser material interaction at different laser power densities and pulse durations, various laser material processing mechanisms and process capability, and the present status for their readiness for industrial adaptation are presented. Since exhaustive review of all different laser material processing modalities is beyond the scope of this chapter, recent review articles have been cited for interested readers to refer. However, some interesting research work carried out by the authors and the recent research trends in material processing with high-power fiber lasers and high brightness diode lasers are also presented.

15.2 LASER MATERIAL INTERACTION

When a laser beam is incident on a workpiece surface, a part of laser power is reflected, some part is absorbed in the material and remaining is transmitted. The fractions of reflected, absorbed, and transmitted laser power depend mainly on the laser wavelength and the refractive index of the material, for laser intensity, or power density below a threshold for nonlinear absorption phenomena, which is typically of the order of $\geq$GW/cm^2. The refractive index of materials having some finite absorption is complex and for laser intensities in linear regime, i.e. <GW/cm^2, this can be expressed as follows (Nath 2014):

$$n^* = n + ik,$$ (15.1)

The reflectivity, R, is given by

$$R = \left[(n-1)^2 + k^2 \right] / \left[(n+1)^2 + k^2 \right]$$ (15.2)

The reflectivity, R, absorptivity, A, and transmitivity, T, are related as, $A + R + T = 1$

Metals being opaque, $T = 0$, and

$$A = 1 - R = 4n / \left[(n+1)^2 + k^2 \right]$$ (15.3)

When laser beam propagates inside the workpiece, it is absorbed in the material and the absorption coefficient, α (in m^{-1}), is given by

$$\alpha = 4\pi k / \lambda$$ (15.4)

where λ and k are the laser wavelength and the imaginary part of the complex refractive index, respectively.

As the laser beam propagates in the material the laser power gets attenuated following Beer–Lambert's law:

$$I(x) = I_0 (1 - R) \exp(-\alpha x)$$ (15.5)

where I_0 and $I(x)$ are the laser intensity (power per area of laser beam) incident at the surface and after propagating a distance x inside the material.

The absorption or attenuation length is given by

$$l_a = 1/\alpha = \lambda / 4\pi k$$ (15.6)

Most metals have relatively high k values than n, and therefore reflectivity, R is more than 90% and l_a is of the order of a few 10's of nm. Therefore, a major fraction of laser power is reflected from the surface and the remaining gets absorbed practically at the surface of most metallic materials. However, as the electrical conductivity of metals increases with rise in temperature, the absorptivity also increases and tends to approach nearing 100% at the melting temperature (Nath 2013).

In order to appreciate the laser material interaction with laser pulses of different time durations ranging from ms to fs, the basic mechanisms of laser energy absorption in various materials, viz., metal, semiconductor, and dielectric materials such as ceramic, quartz, and glass, etc., are reviewed here. Since ions are too heavy to follow the electric field of the laser radiation oscillating at 10^{13}–10^{15} Hz frequency, normally they do not absorb laser energy directly. Laser energy is usually absorbed by the nearly free- and bound electrons in materials. Nearly free electrons in metals absorb laser radiation of all wavelengths by inverse Bremsstrahlung process in a time scale of 10^{-16} s (Linde et al. 1997; Yao et al. 2005). The high energy electrons share their energy among other electrons very quickly through electron–electron collisions in a time scale of t_e'' 10^{-15} s and the temperature of the electron-system increases by this process. The electron energy is transferred to the ionic system through electron–ion and lattice interaction process in a time scale of t_i'' 1–10's ps ($\sim10^{-11}$–10^{-12} s), (Liu et al. 1997; Satta et al. 2000). As the electron and ionic systems come in thermal equilibrium, temperature of material rises.

In intrinsic semiconducting materials, there are few free electrons and the conduction band is empty. The laser beam having photon energy, $h\nu$, less than the band gap energy, E_g (between valance and conduction bands), is not absorbed at low intensities (<GW/cm^2), but only the photons of energy more than the band gap energy ($h\nu \geq E_g$) are absorbed by the valance band electrons through inter band (valance band to conduction band) transition. Once the conduction band is partially filled with electrons, they further absorb laser energy, as they do in metals. Dielectric materials such as glass, quartz, ceramics, and many organic polymeric materials, which too do not have free electrons are transparent to visible and near-infrared (~1 μm) laser wavelengths. However, some of these materials, such as glass, quartz, alumina, acrylic, etc., readily absorb 10.6-μm wavelength radiation of CO_2 laser beam through vibrational energy excitation.

The above description of laser absorption in materials is for laser intensities typically less than GW/cm^2. At these intensities, the laser radiation is not readily absorbed in semiconducting and dielectric materials having few free electrons and band gap energy higher than the laser photon energy ($E_g > h\nu$). At higher laser power densities ($\geq$GW/cm^2), nonlinear processes such as multiphoton ionization (MPI) and tunneling photoionization (TPI) can occur and produce a few free seed electrons, and these electrons can further multiply by avalanche ionization process in presence of the intense electric field of high-intensity laser beam and form plasma (Eliseev et al. 2005; Jiang and Tsai 2003; Liu et al. 1997). In multiphoton ionization process, when an intense laser pulse is incident on a target, several photons (N), each having $h\nu$ energy (less than the ionization potential of target material) are simultaneously absorbed by bound electrons, exciting them to the ionization level as depicted in Figure 15.3a.

The probability of simultaneous absorption of a large number of photons by a bound electron is very small; therefore, multiphoton ionization cross-section is usually very small; however, the process becomes significant at laser intensities greater than 10^9 W/cm^2. Some of the bound electrons can also tunnel across a potential barrier in the presence of high electric field of the intense laser beam (Figure 15.3b).

Whether multiphoton ionization or tunneling ionization will dominate is determined by the so-called adiabaticity parameter, as given by (Eliseev et al. 2005)

$$\gamma = \frac{\omega\sqrt{2m_e E_g}}{e\varepsilon} \tag{15.7}$$

where ε and ω are the strength and angular frequency of the electric field, m_e is the reduced mass of an electron, e is the electron charge, and E_g is the band gap energy of the target material.

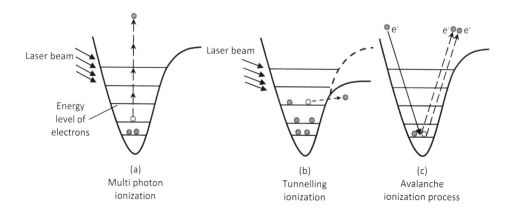

Figure 15.3   Schematic of various nonlinear absorption processes in a matter at high laser intensities; (a) multiphoton ionization, (b) tunneling ionization (c) avalanche ionization process.

If $\gamma \gg 1$, i.e., high value of ω and moderately high electric fields, the simultaneous absorption of several photons (multiphoton mechanism) is realized.

If $\gamma \ll 1$, i.e., high value of electric fields and low frequencies, the tunneling mechanism of ionization is realized (Eliseev et al. 2005).

A few free electrons generated by the above nonlinear processes can accelerate in the presence of the high electric field of laser radiation, acquire high kinetic energy, and cause further ionization in collision with neural atoms of the target material. This process is called avalanche ionization process (Figure 15.3c). Thus, at high laser intensities greater than GW/cm^2 electron-ion plasma is created through multiphoton/tunneling ionization and avalanche ionization processes in a material, irrespective of its band gap energy (metal, semiconductor, or dielectric materials). Plasma can absorb rest of the laser energy, get heated up. As the electron density in plasma reaches a critical magnitude at which the plasma frequency is equal to the laser frequency, electron plasma absorbs the laser radiation resonantly and this could cause an optical breakdown and damage in the material.

The critical electron density, n_e, is given by (Liu et al. 1997),

$$n_e = \varepsilon_0 m_e \omega^2 / e^2 \tag{15.8}$$

where e is the electric charge of electron, ε_0 is the permittivity of free space, m_e is the electron mass, and ω is the angular frequency of the laser radiation.

The absorption coefficient of plasma can be given by

$$\alpha_p \approx n_e / \omega^2 \exp\left(\Delta E_0 / k T_e\right) \tag{15.9}$$

where n_e is the electron density, k is the Boltzmann constant, T_e is the electron temperature, and ΔE_0 is the energy gap. It is apparent that at laser higher frequencies or lower laser wavelengths the absorption in plasma is less.

Heat from the plasma conducts in the surrounding material which may lead to either an irreversible damage or ablation of material depending upon the laser power density and the location of the focal point. These nonlinear processes facilitate absorption of laser energy even in transparent materials. By controlling the process, 3D artifacts are created inside transparent glass volume, and also drilling, cutting, slotting, and other processing with lasers in semiconductors and transparent materials. It may be mentioned that in the case of ns laser pulses at laser intensities of the order of 10^9W/cm^2, the absorption process through nonlinear processes could be initiated by a few stray

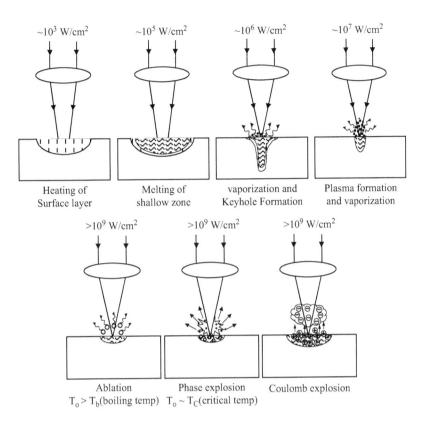

$\sim 10^3$ W/cm^2 $\sim 10^5$ W/cm^2 $\sim 10^6$ W/cm^2 $\sim 10^7$ W/cm^2

Heating of   Melting of   vaporization and   Plasma formation
Surface layer   shallow zone   Keyhole Formation   and vaporization

$>10^9$ W/cm^2 $>10^9$ W/cm^2 $>10^9$ W/cm^2

Ablation   Phase explosion   Coulomb explosion
$T_0 > T_b$(boiling temp) $T_0 \sim T_C$(critical temp)

Figure 15.4 Laser material interaction effects: heating, melting, vaporization, and keyhole formation, plasma formation, ablation, phase explosion, and Coulomb explosion.

electrons and this process is more stochastic in nature, while in the case of sub-ps and fs laser pulses at laser intensities in $>10^{10}$W/cm^2 range, the process is more deterministic as the initial electrons are produced by multiphoton and/or tunneling ionization process within a very small depth determined by the absorption length (Liu et al. 1997). The heat conduction is negligible within the ultra-short laser pulse (<1 ps) duration and temperature rise is independent of the laser pulse duration. Therefore, the threshold laser fluence for ablation with ultrafast lasers is deterministic and precise, unlike in the case of the relatively longer duration (>10 ps) laser pulses, where the depth of heating is determined by the thermal diffusivity and laser pulse duration. This effect has been exploited to machine as well as to create features of even submicron sizes in many interesting ways (Coffey 2014).

Laser material processing mechanisms:

The various material processing mechanisms can be grouped based on different phases of the material during processing, i.e., heating below melting temperature, melting, vaporization and ablation, and plasma formation (Figure 15.4).

15.3 LASER PROCESSING BY HEATING MATERIALS BELOW MELTING TEMPERATURE

Laser surface transformation hardening and laser metal forming are the two important manufacturing processes that are realized by controlled laser heating without onset of surface melting.

15.3.1 Laser Surface Hardening

This process is used to enhance the surface hardness of engineering parts made of mainly medium carbon steels and low alloy steels with carbon content less than 2 wt%, through the formation of martensitic phase by heating the surface above the phase transformation temperature up to the desired depth followed by fast self-quenching (Ashby and Easterling 1984). This has been schematically depicted in Figure 15.5.

Recently, Nath and Sarkar (2017) have presented an overview of the basic laser hardening mechanism, types of steels that have been laser hardened and some recent advances in this field. The laser power density and interaction time are so chosen that only a desired top surface layer is heated above the austenization temperature and rest of the bulk material remains near room temperature with a sharp temperature gradient between them to ensure fast cooling of the top layer by heat conduction when laser heating ceases. Typical laser power densities and interaction times are of the order of 10^3–10^4 W/cm^2 and 0.1–1 s and cooling rate of 10^3–10^5 °C/S order are usually realized.

High-power CO_2 and Nd:YAG lasers were being used traditionally; nowadays, high efficiency fiber, and diode lasers are becoming popular. High-power diode lasers with flat top power distribution are more suitable for hardening of large surface areas. The depth of laser hardening, d, depends on laser power, P_1, beam diameter, d_b, and scan speed, V, and in En8 steel it has been to vary as, $d \propto P_1 / \left(\sqrt{d_b V} \right)$ (Steen and Mazumder 2010). For practical applications, since components that get worn out beyond 100–200 µm depth are rendered unusable, hardening depth required is usually of this order. Laser beam of relatively large size compared to heating depths is normally used for hardening; therefore, the following analytical solutions of 1D heat conduction can provide the rise of temperature as a function of time and depth with reasonable accuracy (Steen and Mazumder 2010):

$$\Delta T(z,t)\big|_{\text{heating}} = [H\delta/K]ierfc(z/\delta) \qquad (15.10)$$

$$\Delta T(z,t > t_p)\big|_{\text{cooling}} = [H/K][\delta ierfc(z/\delta) - \delta * ierfc(z/\delta*)] \qquad (15.11)$$

where H = absorbed laser power density, K = thermal conductivity of steel material, k = thermal diffusivity of steel material, and t_p = laser interaction time.

For $x \ll 1$, $ierfc(x) = 1/\sqrt{\pi} - x + x^2/\sqrt{\pi}$ and $ierfc(1) \approx 0.05$

The above solutions for temperature cycle in semi-infinite workpiece has been extended for finite thickness also, with which effects of various parameters, viz., laser power, beam diameter, scan speed, sheet thickness, and absorptivity on heating and cooling cycles can be studied with relatively short computational time compared to numerical analysis (Nath et al. 2012).

The typical laser heating and cooling cycles at the surface, $z = 0$ and $z = 0.3$ mm depth ($\ll$thermal diffusion length) are shown in Figure 15.6 (Sarkar et al. 2016).

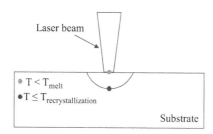

Figure 15.5 Schematic of laser transformation hardening.

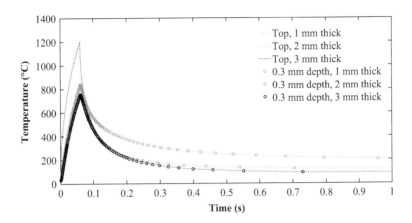

Figure 15.6   Effect of sheet thickness on (a) peak temperature at different depths from the top surface and (b) time history of temperature at the top surface and at 0.3 mm depth from it, laser power = 600 W, scan speed = 3 m/min, laser beam diameter = 3 mm, surface absorptivity = 0.65, radial heat loss factor, C_r = 0.8. (Reprinted with permission from *Surface & Coatings Technology*, Elsevier 2016 302: 344–358.)

It is apparent from the thermal history of laser heating presented in Figure 15.6, the time available above the austenization temperature for carbon to uniformly diffuse reduces along depth. Austenite having ≥0.05 wt% C transformed to martensite on self-quenching shows enhanced micro-hardness. Thus, the onset of surface melting and time available for carbon diffusion limits the hardness depth.

The depth of hardness can be increased by repetitive laser pulse irradiation (Nath et al. 2012). Since a minimum 0.05 wt% C in martensite is needed for micro-hardness to increase, laser surface hardening is seldom reported in very low carbon steels. Furthermore, self-quenching may not be effective in thin sheets having less bulk material for heat conduction. However, sheet thickness several times more than the thermal diffusion length ($2\sqrt{kt_p}$) should ensure fast quenching. Recently, Sarkar et al. (2016) have demonstrated laser surface hardening in ~1 mm thick, very low carbon (0.05–0.7 wt% C) steel sheets by proper selection of laser beam diameter, laser power, and scan speed which effected fast quenching through axial and radial heat conduction. Asnafi et al. (2016) reported tailoring the properties of 1-mm-thick boron steel sheet components by selective laser hardening process. After selective laser heat treatment which increased the micro-hardness from 200 to 700 HV the sheet was stamped to give the precise shape with high accuracy, though the maximum press load increased by ~41% due to laser hardening.

The main attributes of laser surface hardening that make the process "Green" are treatment of selected area, controlled depth, minimum power input ensuring little part distortion and minimum post operation, hardening of complex parts, and high process efficiency. Bonß (2010) has presented a comprehensive review of different online measuring techniques based on IR pyrometers and CCD camera as well as of beam shaping techniques suitable for laser hardening of different engineering parts, along with several practical examples. Presently large surface areas are covered by multiple overlapped laser tracks, and this introduces undesired tempering in the overlapped zone. Therefore, enhancing the output power capability of diode lasers will witness continuous thrust.

15.3.2 Laser Metal Forming

Laser forming is a noncontact forming technique that deforms sheet materials by inducing nonuniform strain within it through localized heating by a defocused laser beam. It is used to deform sheets mainly made of metals such as stainless steels, light alloys of aluminum, magnesium, and titanium, which have a high thermal expansion coefficient. However, brittle materials like glass, ceramics, silicon, composites and foams can also be formed by this process (Guglielmotti et al. 2009; Quadrini et al. 2010; Wu et al. 2010). Laser forming encompasses bending about or shrinkage around a straight line, forming of developable surfaces utilizing bends about multiple lines and more complicated 3D forming for generating surfaces, curved about multiple axes with the help of combined bending and shrinkage (Dahotre and Harimkar 2008; Steen and Mazumder 2010).

The laser-forming process was first successfully demonstrated by Kitamura in 1983 (Dahotre and Harimkar 2008). Three main mechanisms through which laser forming is realized are temperature gradient mechanism (TGM), upsetting mechanism (UM), and buckling mechanism (BM) (Figure 15.7) (Vollertsen 1994). In the case of TGM, the thermal diffusion length ($2\sqrt{kd/v}$, k being thermal diffusivity), corresponding to the laser interaction time, (d/v, d being the spot diameter and v the scan speed) along depth is usually maintained smaller than the sheet thickness (h). Therefore, steep temperature gradient causes relatively larger expansion of a small top layer of the laser-irradiated region. This causes initially a counter bending, i.e., away from the laser beam. As rest of the material restricts expansion a part of the thermal strain is converted into plastic strain resulting shrinkage of the top region after cooling. Therefore, a bending of the irradiated sheet, toward the laser beam, in the range of 0.3°–1° with a deviation of ±0.2° is resulted (Hennige et al. 1997; Lawrence et al. 2001). In BM, thin sheet is scanned over by a relatively large diameter (~10 times of sheet thickness) laser beam at a speed to heat uniformly through the thickness (Chakraborty et al. 2016; Liu et al. 2010). As the thermal expansion is restricted by rest of the sheet, the laser-irradiated region undergoes buckling and this bends the sheet up to 15° per scan pass either toward or away from the laser beam, usually in the later direction due to initial counter bending (Lawrence et al. 2001). Under similar conditions, a thick sheet, having sufficient moment of inertia to resist buckling, undergoes shrinkage through UM, which is manifested in thickness increment of the irradiated region due to material flow. Figure 15.7 shows schematically the three mechanisms. The value of $\frac{kd}{h^2v}$ less than unity usually favors TGM, while the other two mechanisms are favored when the value is much higher than unity (Steen and Mazumder 2010).

Research and development in the field of laser forming in recent past has been focused on experimental studies of parametric dependence of bend angle in different materials (Chakraborty et al. 2016; Edwardson et al. 2006; McBride et al. 2005; Wu et al. 2010), analytical, numerical and statistical modeling (Chakraborty et al. 2015a; Eideh et al. 2015; Maji et al. 2012, 2014; Li et al. 2000; Shen et al. 2006a,b; Shen et al. 2009), synthesis of strategies for laser forming of 3D surfaces (Chakraborty et al. 2016; Edwardson et al. 2005; Hennige 2000; Hu et al. 2012; Kim and Na 2009; Liu and Yao 2005; Yang et al. 2004), and correction of pre-deformed samples using laser forming (Chakraborty et al. 2015b; Dearden et al. 2006; Gisario et al. 2011; Magee and De Vin 2002;

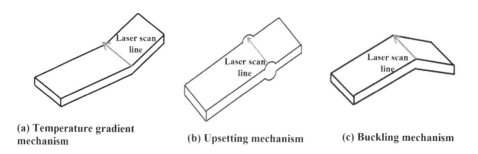

(a) Temperature gradient mechanism   **(b) Upsetting mechanism**   **(c) Buckling mechanism**

Figure 15.7   Three main mechanisms of laser-forming process: (a) temperature gradient mechanism, (b) upsetting mechanism, and (c) buckling mechanism.

Ueda et al. 2009). Though laser forming is a complex process, analytical expressions for the bending angle by different forming mechanisms provide a good idea about the parametric dependence and the bending angle. Shen and Vollertsen (2009) in their review paper presented analytical expressions for bending angle, α_b, formed by TGM (equations 15.12 and 15.13), buckling (equation 15.14), and UM (equation 15.15).

Bending angle by TGM,

$$\alpha_b = 3\alpha_{th}PA/\rho c_p vs^2 \tag{15.12}$$

where α_b is the bending angle, α_{th} is the coefficient of thermal expansion of the workpiece, P is the laser power, A is the absorption coefficient, ρ is the density, c_p is the specific heat capacity, v is the velocity, and s is the sheet thickness.

Bending angle by TGM including the effect of counter bending,

$$\alpha_b = \frac{21\alpha_{th}PA}{2\rho c_p vs^2} - \frac{36lY}{sE} \tag{15.13}$$

where l is the half length of heated zone, E is the Young's modulus, and Y is the yield stress.

Bending angle by BM,

$$\alpha_b = \left[\frac{36\alpha_{th}\sigma_s PA}{E\rho c_p vs^2}\right]^{1/3} \tag{15.14}$$

where σ_s is the flow stress in the heated region.

Bending angle by UM:

$$\alpha_b = \frac{1}{b}\left[\frac{2\alpha_{th}PAb}{\rho c_p v\left(2ds - s^2\right)} - \frac{\sigma_s d}{E}\right] \tag{15.15}$$

where b is the breadth of the bending edge and d is the laser beam diameter.

The unique advantages of laser-forming process can make its application in manufacturing, coherent with 3R philosophy: reduce, reuse, and recycle. Conventional sheet metal forming usually require a die-punch set which is produced with the help of a series of manufacturing processes including material removal processes generating a significant quantity of scraps. For forming prototypes or a small number of parts, it is hard to justify production of die-punch sets both economically as well as environmentally. Laser forming, on the other hand, is a flexible forming operation. Without dedicated fixtures being made it can form different shapes just by scanning metal sheet along proper paths and proper heating parameters. The solid-state lasers coupled with optical fiber beam delivery and laser processing head attached to computer numerical control (CNC) workstation/robot can heat along even space curves. In addition, laser forming can enable production of small bending of sheets, not possible due to spring-back in mechanical forming. Therefore, laser forming can be environmentally beneficial in producing prototypes or small number of parts made of sheet metal. Laser forming can be used for flattening the deformed sheets also for their reuse and recycle (Dearden et al. 2006; Magee and De Vin 2002). This is particularly of importance for high-valued materials like aerospace materials, etc.

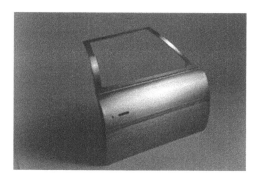

Figure 15.8 Laser formed part made to approximately 1:8 scale of real door panel. (Reprinted with permission from *Optics & Laser Technology*, Elsevier 1998 30:141–146.)

The applications of laser forming mainly center on rapid prototyping and correction of distortion induced in welding and casting (Geiger et al. 2004). Some of the important applications are listed as follows (Edwardson et al. 2001):

- Rapid prototyping of irregular shapes for turbine blade applications
- Noncontact forming for installation and adjustment of nonaccessible parts
- Automotive shapes for prototype and validation testing
- Precision shaping of tanks and pressure vessels
- Unbending techniques for repairs and alignment applications
- Tube and pipe precision forming
- Final configuration production parts for small quantities of parts

Thomson et al. (1998) performed laser forming of rear door panel of a car (Figure 15.8) shown above.

15.4 LASER MATERIAL PROCESSING BY MELTING

The material processing modalities that are realized by laser melting of materials include cutting, welding, surface modification by remelting, alloying, and cladding, and rapid manufacturing by selective laser sintering/melting and direct metal powder deposition. Laser power density and interaction time usually employed are typically of the order of 10^4–10^6 W/cm^2 and 10^{-2}–10^{-3} s. The most favored lasers for modalities requiring good quality of laser beam for tight focusing, such as cutting, welding, and sintering are high-power Nd:YAG disk laser and fiber laser; and for applications such as surface modification requiring large size laser beam of uniform power density, high-power diode lasers are most preferred.

15.4.1 Laser Cutting

Laser cutting is the most common industrial applications of the laser. In this process, a high-power focused laser beam along with a coaxial gas jet is incident on the surface of workpiece (Figure 15.9). The focused laser beam melts the material and the high-velocity gas jet ejects out the molten material. First, a hole is pierced into the workpiece and then the laser beam is moved over the surface for cutting operation. Both inert gas such as nitrogen and argon up to 20 bar pressure, and reactive gas like oxygen up to 6 bar pressure have been used. In O_2-assisted laser cutting of steel sheets, the exothermic oxidation reaction provides additional heat energy, 40%–50% of the laser power, which enhances the cutting performance.

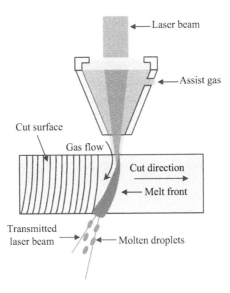

Figure 15.9 Schematic of gas-assisted laser cutting.

Lasers can cut most of the materials irrespective of their hardness and strength including various metals, alloys, glass, ceramic, and composites. This being a thermal process is governed mainly by the thermo-physical properties and absorption characteristics of the material. The cut thickness depends on the laser power and the current range with 2–5 kW is usually limited to 15–20 mm. In the case of O_2-assisted laser cutting of mild steel sheets thickness is limited to about 25 mm; beyond this uncontrolled oxidation of workpiece deteriorates the cut quality. Steel plates of higher thickness up to ~100 mm have been cut by laser-assisted oxygen cutting, known as LASOX process. In this process, a coaxial supersonic O_2 gas jet is used and the laser power is required only to sustain the exothermic reaction. The major power needed for cutting is generated by the exothermic oxidation reaction (O'Neill and Gabzdyl 2000). While the focused laser beam footprint on the workpiece is smaller than that of the gas jet in conventional gas-assisted laser cutting, this is reverse in the case of the LASOX process.

Laser cutting is usually performed at high speeds to ensure low heat conduction loss, high processing efficiency, and dross-free cut square edges. The energy loss due to heat conduction, convection, and radiation is usually very small and the process can be presented by lumped heat capacity model (Steen and Mazumder 2010):

$$AP = whv\rho\left(C_p\Delta T + L_f + m'L_v\right) \tag{15.16}$$

where A = surface absorptivity, P = laser power, w = kerf-width, h = sheet thickness, v = cutting speed, ρ = density and C_p = specific heat of sheet material, ΔT = rise in temperature, L_f = latent heat of fusion, L_v = latent heat of evaporation, and m' = fraction of vaporized molten material.

From equation 15.16, the severance energy, S, defined as laser energy required to severe unit surface area of a material can be determined as follows:

$$\text{Severance energy, } S = \frac{P}{vh} = \frac{w\rho\left(C_p\Delta T + L_f + m'L_v\right)}{A} \tag{15.17}$$

All terms of right-hand side, except for w—the kerf-width, are the material properties. For a constant focused laser beam diameter at the workpiece surface which determines the kerf-width, the severance energy, S, is constant for a given material. Steen and Mazumder (2010) have compiled experimental values for different materials with inert and O_2-assisted laser cutting, which can be used a practical guideline for determining the required laser power for cutting a sheet of a given thickness at a desired speed.

Some of the current research trend in this area is directed toward establishing process parameters for cutting various materials by high-power fiber lasers and high brightness diode laser, which were previously done by high-power CO_2 and Nd:YAG rod lasers. Fiber lasers are capable of cutting highly reflecting and thermally conductive materials, like aluminum, copper, and brass, that CO_2 lasers are not able to process without damaging the mirrors/focusing optics (Larcombe 2013). Stelzer et al. (2013) presented a comparison of inert gas fusion cutting of 1–10 mm thick AISI 304 stainless steel sheets with high-power CO_2 laser and fiber laser. While fiber laser performed equally good or even better than CO_2 laser in cutting steel sheets up to ~4–6 mm thickness, the cut quality and process efficiency of both were relatively poor in cutting sheets of higher thickness with fiber laser. Similar observations were reported by Costa Rodrigues et al. (2014b) in comparing direct diode laser fusion cutting and O_2-assisted flame cutting of aluminum alloy and steel sheets with fiber laser and CO_2 laser cutting. The Brewster angle of these metals for 10 μm wavelength being higher, closer to 90° than that for 1 μm wavelength, facilitates relatively higher absorption of CO_2 laser power at the cut front than fiber laser and diode laser power. Mullick et al. (2016a) studied effects of laser beam incident angle and location of focal point and found that the cut surface quality improved when focal point was located at the bottom surface and beam was inclined by about 4° from normal in cutting 4-mm-thick steel, sheet with O_2 assist. Laser cut surface has characteristic striations and its complete elimination has eluded the research since long. While "striation-free" cut could be realized in cutting of mild steel sheets of 1–3 mm thickness order with O_2 gas assist, but not in other materials such as stainless steel or inert gas-assisted laser cutting, this is expected remain of the active research topics (Li et al. 2007; Mullick et al. 2016b; Powell et al. 2011).

Under optimized laser processing parameters, i.e., laser power density, scan speed, standoff distance between gas-jet nozzle and workpiece surface, and gas pressure, the kerf-width, i.e., cut opening, is very narrow with very little scarp generated, cut edges are straight, clean, and smooth with no burr and little HAZ, requiring no after-processing operation and can be used directly for other operation such as welding. The process being noncontact no hard clamping of workpiece is required and there is no issue of tool wear. Using CNC program, automation and laser cutting head mounted on robotic arm, any desired 2D and 3D contour profiles are easily cut. The process is very fast and energy efficient and much economical for cutting medium size lots of varying job designs eliminating the costly die and punch otherwise required processing them. Laser can produce almost flawless cut quality without any HAZ, burr and change in microstructure, required in cutting highly precision parts like cardiovascular stents (Meng et al. 2009). Therefore, today laser cutting is established as the most important manufacturing application and many industries like automobile, aerospace, shipbuilding, ordinance, rail-transport, electronic, biomedical, and nuclear industries are using lasers in their production line as this conforms to green manufacturing as well. High-power laser beams of Nd:YAG, fiber, and diode lasers can be transmitted through low-loss optical fiber and using robots laser cutting can be done remotely in hazardous environment like in nuclear installations. Remote cutting can drastically reduce the radiation dose to the working personals (Sanyal 2014).

15.4.2 Laser Welding

Lasers have been used to weld various similar and dissimilar materials of thickness up to typically 10–15 mm, limited by laser power in a single pass without any filler material. There are three different modes of laser welding, namely conduction welding, deep penetration keyhole welding, and hybrid welding (Figure 15.10).

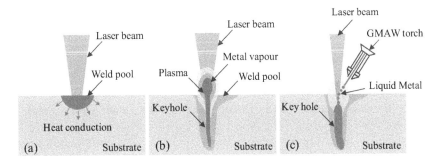

Figure 15.10 Schematic depicting (a) conduction, (b) key hole welding, and (c) hybrid laser welding.

In laser conduction welding, the melt front penetrates through heat conduction and the laser power density is of the order of ~5 × 10⁵ W/cm². This is employed to weld sheets of thickness ≤3 mm and the depth to width aspect ratio of weld is about 1–1.5. Thicker sheets are joined by laser keyhole welding, which requires laser power density ≥10⁶ W/cm². At these laser power densities, the molten material tends to evaporate and the recoil pressure of vapor pushes the molten material outward creating a keyhole. Laser beam reaches to the bottom of the keyhole and further deepens the keyhole through melting and the action of recoil pressure.

When the laser beam is moved forward the molten material fills up the keyhole, solidifies creating deep penetration weld. Laser power density and interaction time need to be controlled optimally to realize the desired penetration depth with minimum loss of material through vaporization and little porosity due to entrapment of vapor within keyhole. The laser welding capability has been further enhanced through hybridizing laser with one of the more conventional welding systems like MIG, TIG, or plasma arc. Usually MIG, TIG, or plasma arc is directed into the laser-produced keyhole from trailing side (Figure 15.10c). While laser produced plasma helps in stabilizing the arc of MIG/TIG, the energy provided by the later helps in bridging relatively larger gaps between two parts when butt welded and also strengthens the weld joint with increased fusion volume. Since per kW power cost of these welding sources are relatively less than of laser, the hybrid welding process is more cost-effective.

In conduction welding, the weld depth is determined by the thermal diffusion length given by $2\sqrt{(k\tau)}$, where k is the thermal diffusivity of material and τ is the laser interaction or dwelling time. For CW laser, the interaction time is given by $\tau = d/v$, where d and v are the laser beam diameter and welding speed, respectively. Similarly, in deep penetration keyhole welding sufficient laser dwell time needs to be provided for laser to generate keyhole up to the desired depth. During the dwelling time heat conduction loss is not negligible, unlike in laser cutting, and the following modified lumped heat capacity equation (equation 15.18) can be used as a thumb rule for determining laser power to weld a sheet at a desired speed:

$$0.483 A P_{\mathrm{l}} = w h v \rho \left(C_{\mathrm{p}} \Delta T + L_{\mathrm{f}} \right) \tag{15.18}$$

The factor 0.483 accounts for the heat conduction loss.

Some of the main characteristics of laser welding that make the process attractive include narrow and deep keyhole welding with large aspect ratio and relatively little HAZ, high processing speed, ability to weld pre-machined precision components with restricted heat input and minimal distortion, requirement of no filler material, i.e., autogenous weld, and joining of difficult to weld and dissimilar materials. By time sharing the same laser can be used to cut sheets and cut parts can be directly welded without need of any edge preparation (Kumar et al. 2006). The gap between two straight edges in butt welding needs to be maintained within typically 10% of the sheet thickness.

The process is done in open atmosphere unlike electron-beam welding; however, inert shrouding gas is used to avoid oxidation of weld in atmospheric oxygen. All types of joints including complex ones are amenable to laser weld and the process can be done remotely as well. Transmitting the high-power laser through optical fiber, welding can be done from remote distance and thus, laser is proving a boon in maintenance of nuclear installations because of much reduced human exposure to radiation (Sanyal 2014).

Most of engineering materials such as different types of steels, aluminum, titanium, and nickel alloys have been welded by lasers. With proper selection of laser power density and weld speed, laser produces good weld quality in steels, titanium, and nickel alloys; however, aluminum alloys pose difficulties in laser welding due to their high reflectivity and excessive fluidity, and the weld often has unacceptable level of porosity (Steen and Mazumder 2010). Some of the issues in laser welding being investigated are bridging the relatively large gaps in butt welding, instability of the keyhole causing nonuniform weld penetration and porosity due to entrapment of vapor, and controlling the microstructures and cooling rates to minimize the solidification cracks, especially in dissimilar materials welding having very different thermo-physical properties. Different schemes such as offsetting the laser beam from seam preferentially toward one of the materials (Casalino and Mortello 2016; Casalino et al. 2017; Zhou et al. 2017), using filler materials for joining dissimilar materials (Ezazi et al. 2015; Shanmugarajan and Padmanabham 2012; Sun et al. 2015), modulated or pulsed laser (Akman et al. 2009; Chen et al. 2016; Iwasaki et al. 2004), dual-laser beam in tandem (Xie 2002), and wobbling laser beam (Kuryntsev and Gilmutdino 2015; Samson et al. 2017) have been reported for dissimilar materials welding.

Different types of steels of thickness ranging from a mm up to 20–25 mm order have been welded with high-power fiber lasers (Bandyopadhyay et al. 2016; Sokolov et al. 2011). Titanium and other precious alloys for dental application and Ti-alloys sheets for aerospace applications have been laser welded with pulsed Nd:YAG lasers (Akman et al. 2009; Iwasaki et al. 2004). A significant improvement in weld quality was reported by Xie using dual-laser beam in tandem (Xie 2002). In steel, the improvements were in terms of reduced surface defects such as undercut, surface roughness, spatter, and underfill, centerline cracking susceptibility, and in aluminum this was in terms of smooth weld surfaces and fewer weld defects such as porosity, surface holes, and undercut. The keyhole stability was also improved in the case of dual-beam compared to single laser beam. More recently, Samson et al. (2017) reported a significant improvement in laser welding of difficult to weld materials such as aluminum and copper by wobbling the high-power high-brightness fiber laser beam across the seam. Circle, linear, eight shaped, and infinity shaped wobble modes were attempted. The wobbling laser beam welding technique could eliminate or considerably reduce other defects such as porosity, brittle intermetallic formation, and also relax the gap tolerance in butt welding (Table 15.2).

Some of the limitations of thick section welding by laser are being addressed by hybrid laser–arc welding techniques (Cao et al. 2011; Kujanpää 2014). Cao et al. successfully applied a new hybrid fiber laser–arc welding to fully penetrate 9.3-mm-thick butt joints using a single-pass process (Cao et al. 2011). As this technique combines the synergistic qualities of the laser and MAG arc welding techniques, this permits a high energy density process with fit-up gap tolerance, higher weld penetration, relatively narrow weld, and restricted HAZ, which can minimize the residual stress and distortion.

Kujanpää (2014) has reported welding of austenitic steel up to 60 mm thickness by multipass hybrid laser–arc welding, minimizing the solidification cracks by controlling the microstructure by process optimization. High-power diode lasers of improved beam quality being currently developed are also being evaluated in welding of thick metallic sections in comparison to more established CO_2, Nd:YAG and fiber lasers (Alcock and Baufeld 2017; Köhler et al. 2005). Alcock and Baufeld (2017) demonstrated autogenous welding of 304L stainless steel plates up to 12 mm thickness with 15 kW CW diode laser which could be focused to ~1.2 mm spot diameter. Their results suggest that for welding of plates in the 10 mm thickness range, the more economic diode laser systems may

Table 15.2 Wobble Shapes and Weld Characteristics

Wobble Mode	Schematic of the Mode	Weld Characteristics (Samson et al. 2017)
Circle (clockwise or counter-clockwise)		
Linear		
Eight shaped		
Infinity shaped		

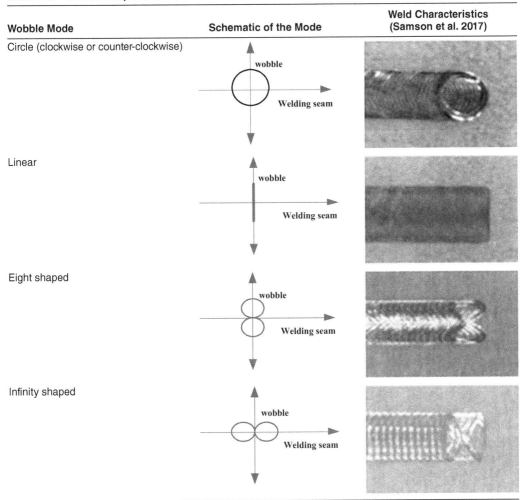

Source: Data from Bryce Samson, IPG photonics 2017.

become a competitor to other systems with high-quality laser beams which may be more expensive in both procurement and operation. Successful welding of many dissimilar materials like CP Ti and steel (Chen et al. 2016; Shanmugarajan and Padmanabham 2012), AZ31B magnesium alloy to 316 stainless steel (Casalino et al. 2017), aluminum to stainless steel with preplaced activating flux (Ezazi et al. 2015; Sun et al. 2015), tantalum to titanium (Grevey et al. 2015), niobium to Ti–6Al–4V, (Torkamany et al. 2016), steel to copper (Kuryntsev et al. 2017).

Casalino et al. (2017) reported good quality of butt welding of AZ31B magnesium alloy to 316 stainless steel with fiber laser by offsetting the laser beam from central line toward magnesium alloy, thereby controlling the formation of brittle intermetallic. Shanmugarajan and Padmanabham (2012) studied CO_2 laser welding of Ti to SS, without any filler material and also with vanadium and Ta interface layer. Chen et al. (2016) studied the effect of temporal shape of laser pulses on weld quality of CP Ti and steel and found weld quality improvement with ramped down laser pulse profiles. Grevey et al. (2015) had also suggested controlling the mixing of Ta to Ti by offsetting the laser beam away from Ta for improved weld quality. Crack-free weld could be obtained in the case with Ta interface layer only. Torkamany et al. (2016) reported successful welding of Nb to Ti alloy by first melting Ti–6Al–4V with laser and then letting the molten Ti–6Al–4V to dissolve the niobium metal. Ezazi et al. (2015) and Sun et al. (2015) reported laser welding of Al alloy and steel with filler materials for controlling the formation of Fe–Al brittle intermetallics in weld joint. Kuryntsev et al. (2017) reported welding of austenitic stainless steel to copper with fiber laser without any filler material by offsetting the laser beam toward stainless steel. Hong and Shin (2017) have presented a review on the prospects of laser welding of lightweight materials (aluminum alloys, magnesium alloys, and titanium alloys) in high volume vehicle manufacturing.

Metals have been laser welded with polymeric sheets successfully (Hussein et al. 2013; Wang et al. 2010). Wang et al. (2010) have reported laser transmission joining of biocompatible, dissimilar materials PET to Ti foil, which has the potential for application in biomedical and their encapsulation process. Hussein et al. (2013) reported welding of dissimilar materials between transparent Polymethylmethacrylate (PMMA) and stainless steel 304 sheets using a pulsed mode Nd:YAG laser in two different modes—laser transmission joining (LTJ) and conduction joining (CJ). LTJ produced joints with relatively higher strength than CJ.

Future trend in laser welding research is expected to be focused on the joining of lightweight materials for automobile and aerospace industries and dissimilar materials by spatial and temporal manipulation including wobbling of high-power fiber laser beam and welding of thick sections of structural materials with hybrid laser welding and more economical diode lasers (Kaplan et al. 2015; Köhler et al. 2005; Kujanpää 2014).

15.4.3 Laser Surface Remelting, Alloying, and Cladding

The scope of surface modification by laser has been expanded beyond transformation hardening by surface remelting, alloying, and cladding. With these methods, the mechanical, tribological, and metallurgical surface characteristics like wear, strength, friction, fatigue, oxidation, erosion, and corrosion, etc., of those materials which are not amenable to transformation hardening, can be significantly enhanced. Extensive research and development works on surface modifications of different materials by these processes carried out in last couple of decades have been presented in several review papers (Kusinski et al. 2012; Liu et al. 2017; Majumdar and Manna 2011; Quazi et al. 2015; Santo 2008; Vollertsen et al. 2005; Weng et al. 2014). A few illustrative examples and some of studies carried out by present authors are presented here.

In laser surface remelting, the top surface layer is melted up to the desired depth without the onset of surface vaporization by irradiating with either a pulsed laser or a high speed scanning laser beam (Figure 15.11). The process is done at a sufficiently fast speed to establish a steep temperature gradient across the interface between liquid and solid phases, so that after laser irradiation the molten layer is re-solidified by self-quenching at a very fast cooling rate in the range of $10^4–10^8$ K/s.

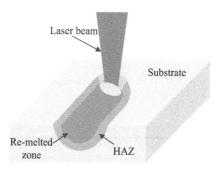

Figure 15.11 Schematic of laser remelting.

The main aim of laser surface remelting is to improve the surface roughness, wear, corrosion, friction, fatigue properties through one or more of the following processes—grain refinement, homogenization of microstructure, enhancement of solid solubility, and formation of meta-stable state/amorphous phase, i.e., surface glazing.

Typical laser power density and interaction time are in the range of 10^4–$10^7\,W/cm^2$ and 10^{-2}–$10^{-7}\,s$, respectively. Lasers which are commonly used in laser surface hardening are also employed for this process. In addition, modulated power lasers (CO_2, Nd-solid state and fiber lasers), Q-switched solid-state lasers, and excimer lasers have been used for surface remelting process. Surface properties of those materials which cannot be improved by laser hardening, such as cast iron and tool steel because of insufficient time above phase transformation temperature for dissolution of carbides and homogeneous dispersion of carbon, and several other materials such as different alloys of Al, Cu, Mg, Ti, and Ni and metal ceramic composites have been improved by laser surface remelting process (Kusinski et al. 2012; Majumdar and Manna 2011; Vollertsen et al. 2005). Laser surface remelting of cast iron is one the most popular applications to achieve a ledeburitic layer of good wear resistance and high hardness. Besides cast iron significant increase in hardness and wear resistance in alloy steels such as tool steel was obtained by surface remelting, which was stable up to 700°C (Karmakar et al. 2017; Majumdar et al. 2010a,b). Surface hardness, wear resistance, and the cavitation corrosion resistance of different kinds of bronze also were improved by laser remelting through the formation of highly refined and homogenized microstructure (Tang et al. 2004). Several novel laser surface re-solidification processes were developed to suppress sensitization in modified type 316(N) stainless steel weld metal and AISI 304 austenitic stainless steel by controlling the laser power density and interaction time, and exploiting the temporal and spatial modulations of CO_2 laser power (Kaul et al. 2008; Nath 2006; Parvathavarthini et al. 2008; Subba Rao et al. 2007). Enhanced immunity against sensitization of laser treated surface in AISI 304 austenitic SS was attributed to its duplex microstructure and higher fraction of low angle grain boundaries (Kaul et al. 2008). Better improvement in inter-granular corrosion and sensitization resistance was obtained in 316(N) SS weld metal with high frequency pulse modulated laser beam superimposed on CW laser power than with CW laser power alone. This was attributed to the growth of fine grains near the fusion boundary region under controlled thermal cycle (Parvathavarthini et al. 2008). A dramatic increase in pitting corrosion resistance, inter-granular and general corrosion near the surface of prior cold worked and aged AISI type 316L(N) SS was also realized by laser surface re-solidification (Subba Rao et al. 2007). This was attributed to homogenization of the composition and the dissolution of secondary phases in the melted and HAZs. Laser surface melting of austenitic stainless steel at high speed carries the risk of centerline solidification cracking due to the formation of tear-drop shaped melt pool. This was effectively controlled by laser beam shaping. By employing square-shape laser beam in place of circular beam the surface microstructure of laser melted region could be engineered which was less prone to crack. Figure 15.12 shows the microstructures of laser re-solidified surfaces by circular and square laser beams (Nath 2006).

In laser surface alloying a top surface layer of thickness typically within 1 mm is melted along with selected alloying elements to form a homogeneous alloyed layer (Figure 15.13). This process has all advantages of surface remelting as well. The process is attractive as this economizes use of costly alloying elements in improving the surface characteristics with no need of alloying the entire bulk material. There are typically two ways of adding external alloying elements. In single step process alloying elements in the form of gas, solid powder or filler wire is delivered continuously into the laser melted pool through a nozzle (Figure 15.14a,c,d). In the two step process alloying material is preplaced onto the surface and subsequently it is melted along with a thin surface layer by the laser beam (Figure 15.14c). The laser power density and interaction time are suitably adjusted to control the melt layer thickness and also to allow homogeneous mixing of alloying elements with the base material. Convective melt flow due to variation of surface tension with temperature across the melt pool plays an important role in creating uniform alloyed surface layer. In laser surface

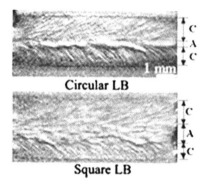

Figure 15.12 Comparison of surface microstructures produced by laser surface re-solidification with circular and square laser beams. Direction of laser scanning is from right to left. A: Axial grain region, C: Columnar grain region. (Reprinted with permission from Dr. A. K. Nath, Department of Mechanical Engineering, IIT Kharagpur.)

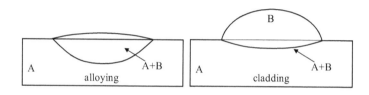

Figure 15.13 Schematic depicting difference between laser alloying and cladding.

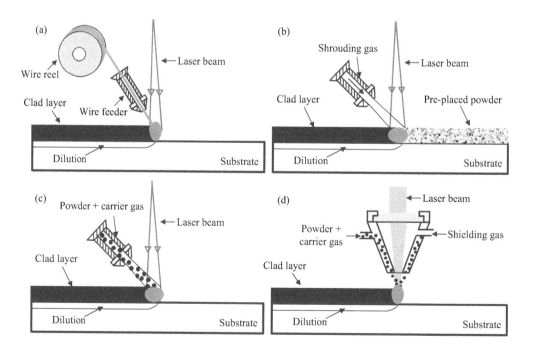

Figure 15.14 Different methods of laser cladding process (a) wire feed method, (b) preplaced powder method, (c) lateral powder feeding method, and (d) coaxial powder feeding method.

cladding, in contrast, a layer of an external material is deposited by laser melting on the surface ensuring proper metallurgical bonding with the base material and minimum dilution (Figure 15.14). Surface characteristics are mainly dictated by the clad material not the base material unlike in surface alloying.

Laser surface alloying of all types of engineering materials, viz., Al, Cu, Mg, Ti, and their different alloys, mild steel, and different grades of stainless steels have been reported to improve hardness, wear resistance, oxidation, and corrosion resistance. These were surface alloyed with different alloying materials including various elements such as Al, Cu, Co, Cr, Mn, Mo, Nb, Ni, N_2, Si, Ti, W, and ceramics like AlN, Al_2O_3, CrB_2, Cr_3C_2, Cr_2O_3, FeCr, SiC, TiC, TiB_2, TiO_2, WC, and their composites. Laser cladding is mostly done to improve surface properties of relatively cheap engineering substrate materials like iron-based substrates. Cladding materials like Co-, Ni- and Fe-based alloys and different ceramic composites have been used for mild steel and alloy steels to impart much improved surface characteristics such as high-temperature strength, wear, corrosion, and oxidation resistance. Nonferrous substrate materials such as Al-, Ti-, and Ni-based alloys substrates have been cladded with Co-based alloys and different composites to improve their surface hardness and wear resistance. Details about these studies can be found in a review paper by Vollertsen et al. (2005) and Majumdar and Manna (2011). In one of the recent studies a hard nano-structured multi-component coating of alumina, titanium carbide and titanium di-borideon mild steel (AISI 1020) substrate was deposited by combined self-propagating high-temperature synthesis and laser coating of preplaced powder of aluminum, titanium oxide and boron carbide (Masanta et al. 2009, 2010).

15.4.4 Laser Rapid/Additive Manufacturing

One of the most promising manufacturing processes that emerged to have a great potential in last few decades is Additive Manufacturing (AM) in which geometrically complex parts can be directly fabricated from computer-aided design by melting and joining metal/nonmetal powders/wires layer by layer. Different energy sources, such as laser, electron-beam, etc., can be used in this process among which laser is very popular due to its unique beam properties. Laser AM can be classified into two broad categories, Direct metal deposition and Direct Metal Laser Sintering (DMLS) process or Selective Laser Melting (SLM).

15.4.4.1 Direct Metal Deposition (DMD)

Increasing global environmental awareness, evidenced by recent worldwide calls for control of climate change and greenhouse emissions, has placed significant thrust on green technologies which are both eco-friendly as well as economical. Remanufacturing is one of the unquestionable green technologies, which reduces toxic emissions in the process of excavation of raw materials to manufacturing of final product whereas recycling could also reduce the carbon footprint contribution from mining process (Figure 15.15). Repairing and remanufacture tooling presents large reductions of 85%–90% of energy consumption, environmental emissions, and manufacturing costs of commercial and military components (Liu and Yao 2005; Morrow et al. 2007; Serres et al. 2011; Xiong et al. 2008).

In extension to remanufacturing, methods to extend the life cycle of a component is also of a great interest. In most of the mechanical components, wear is a major mechanism by which components fails. Thus, it is very vital to exploit the best technologies that can improve the life cycle of a component as well as refurbish in order to bring down the carbon foot print of the manufacturing industry.

In the recent past, researchers across the world explored several techniques to remanufacture as well as to improve the mechanical life of a component using techniques like gas tungsten arc cladding (Sharifitabar et al. 2016), plasma spraying (Xu et al. 2016), laser cladding (Nurminen et al. 2009),

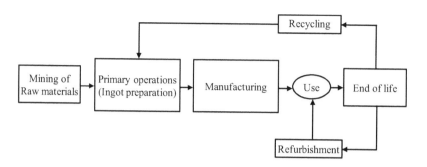

Figure 15.15 Typical manufacturing life cycle.

high-velocity oxygen fuel spraying (HVOF) (Taillon et al. 2016), etc. Among these, DMD, a process in which the component or a surface is generated by layer-by-layer welding technique using laser as a heat source, had drawn a great attention pertaining to its advantages over the conventional techniques. This includes minimum dilution, distortion and HAZ compared to other processes along with fast cooling rates experienced from self-quenching mechanism favoring fine microstructure with improved mechanical properties (Toyserkani et al. 2004). DMD, an AM process, ought to be cleaner than subtractive process like machining as this does not include generation of chips, reducing the buy-to-fly ratio (Watson et al. 2015), require no cutting fluids which is a main source of hazard in the manufacturing waste streams (Luo et al. 1999), and also toxic gas emission is very minimal compared to the conventional welding techniques (Huang et al. 2013). Furthermore, the unfused powder can be reused (Carroll et al. 2006) further reducing the buy-to-fly ratio. In view of the above advantages and high process flexibility along with advancement of high-power laser systems, DMD has drawn a significant attention of researchers and manufacturing industry.

DMD is basically a weld deposition technique, in which the feed material either in the form of powder or wire is melted using laser as heat source and deposited on to the surface of a substrate of similar or different material. The process is repeated, i.e., layer-by-layer deposition is carried out in order to achieve the required height and shape. The metal powder can be feed coaxially to the laser or laterally using a side nozzle (Figure 15.16). However, lateral powder feeding nozzle configuration has scan-direction-dependent deposition rates and thus has limited application (Li et al. 2003). The optimum particle size used in the case of laser cladding falls in the range of 44–150 μm. Too small particles cause problems with powder delivery, reducing the deposit height and efficiency, whereas coarse powder produced high surface roughness values (Kong et al. 2007).

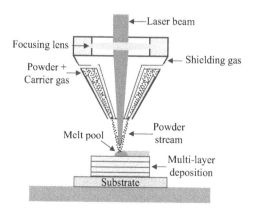

Figure 15.16 Schematic of direct metal laser deposition.

The trend toward greater energy conservation and the reduction of greenhouse gases demands for the improvement of life cycle of a component as well as extending the life by refurbishing. As mentioned earlier, wear is a major mechanism by which a component fails as well as reduces the lifetime. Apart from that, friction contributes to about 4%–15% energy loss in the engines near ring-pack/liner, piston-skirt/liner, and piston-pin/connecting-rod contacts; connecting rod and crank-shaft bearings; and the valve train subsystem (Wong et al. 2016). Thus, it is very vital to improve the wear resistant characteristic as well as lubrication properties of the surfaces in contact, improving the life cycle of a component. Repairing of rails by DMD process using hard phase materials like stellite can improve the life 15,000 cycles to more than 50,000 cycles (Lewis et al. 2015, 2016). Furthermore, it can be used as an alternative for field thermite welds. Robles Hernández et al. (2016) demonstrated that a controlled cooling in laser cladding by preheating technique can reduce the un-tempered martensite, improving the life by 1,400%. Furthermore, Liu et al. (2016) showed that the remanufacturing of cylinder heads by laser cladding could reduce the total environmental impact by 40.91%. DMD process is also used in the field of repairing turbine blade tips (Shepeleva et al. 2000; Rottwinkel et al. 2014), cladding of farm tools to improve erosion of plowing tools by soil erosion (Bartkowski et al. 2017), repairing as well as improving the wear properties of forging dies (Leunda et al. 2011), etc. It is also used to clad varieties of ceramic as well as hard phase materials and super alloys to improve the surface properties like wear and corrosion resistance, solid lubrication proper-ties, etc. (Quazi et al. 2016; Weng et al. 2014).

Despite decades of research, some of the problems associated with the laser cladding still persist, where reproducibility is one of the major concerns. DMD process involves more than 19 process parameters (Kahlen et al. 2001), viz., laser beam power, scan speed, beam spot diameter, powder layer thickness, powder mass flow rate, carrier and shielding gas pressure, powder particle size and shape, standoff distance, thermal conductivity of powder and substrate, hatch spacing, percentage of overlapping, and interlayer ideal time, etc., and a small change in any of them could significantly change the quality of alloying/cladding layer (Gopinath et al. 2017a). Furthermore, many process parameters are system-dependent like design of the nozzle, wavelength and intensity distribution of the laser system used, thermo-physical properties of the materials, etc., making the process more complex and difficult to reproduce the results elsewhere. With the available laser powers, the width of a single laser clad track is typically of the order of a few mm. Therefore, large surface areas are covered by a number of overlapped tracks, and to build up the height of the layer multi-overlaid lay-ers are deposited. Each layer undergoes several thermal cycles when other adjacent overlapped layers are deposited. This combined with small fluctuations in process parameters make the repeatability and reproducibility of deposited layers' characteristics like geometry which includes clad height, width and contact angle, percentage of dilution, surface roughness, microstructure, and mechanical properties, etc., a significant challenge. Thus, it is vital to have online monitoring system to have reproducibility and improved product quality. The laser cladding process involves wide range of temperatures, fast thermo-cycles, i.e., rapid heating and cooling rates (up to $10^{6\circ}$C/s) demanding for monitoring systems with fast sampling rates. Further, the monitoring system should be isolated from the laser radiation. Researchers across the world have used various types of noncontact type detectors like acoustic sensors, CCD cameras, photodiodes, IR pyrometers, etc., a detailed list of which was presented in their work by Purtonen et al. (2014). In laser cladding process, it is very essential to have reproducibility and sound quality in terms of geometry as well as mechanical properties of the component generated. As discussed, the DMD process involves large number of process parameters, including system dependent parameters, and thus it is necessary to select such a parameter that is an outcome of all the process parameters. Molten pool temperature is one such monitoring parameter using which one can assess the quality of the deposit. Several researchers have studied the dependency of molten pool temperature and shape as a function of process param-eters in order to assess the quality of the clad track (Bi et al. 2006a,b, 2007; Doubenskaia et al. 2004, 2006; Ignatiev et al. 1996, 1997; Pavlov et al. 2011; Smurov et al. 1996, 2013). In multilayer thin wall deposition, the heat conduction mechanism transforms from 3D to 2D as the number of layers increases, resulting in heat accumulation and inhomogeneous wall thickness. Furthermore, absence of material at the edges to conduct the heat results in edge build up as well as shrinkage of the wall. The authors further extended their work in which laser beam power was varied during the process with change in IR-temperature signal from a pre-set value using a PID controller-based closed-loop control system (Robles Hernández et al. 2016). Apart from the geometrical aspects, monitoring of metallurgical aspects and detecting the phase changes online is also very vital which minimizes the post processing analysis required. The formation different phases depend upon the thermal history especially the cooling rate that a molten pool under goes. Depending upon the material and application, one needs to select the process parameters that would yield an optimum cooling rate. Recently, Gopinath et al. (2017b,c) reported online monitoring of thermo-cycles including heating and cooling rates, melt pool lifetime in preplaced Inconel 718 with and without ceramic (WC and TiC) additives on AISI 304 stainless steel substrate, their dependence on various process parameters and its effect on different quality aspects of the clad layer, like microstructure, elemental segregations, ceramic particle wetting characteristics and decomposition, and mechanical properties (Figure 15.17). The authors also demonstrated the possibility of online identification of phase change (Figure 15.18 and 15.19). In order to make laser AM by DMD an industrial process to produce functional components online process monitoring and feedback control is vital.

15.4.4.2 DMLS Process or SLM

Among various other AM processes, DMLS process or SLM is very popular which uses metal powder as raw materials to build the final part. The schematic of DMLS process is shown in Figure 15.20.

Using precise movements of the scanning mirrors, laser unit selectively melts few regions (cross-section of the part to be built) of the powder bed. After that, the build plate moves down by a distance equal to powder layer thickness for the next layer to be built. Re-coater puts a uniform thin layer of powder (few microns) above the melted parts and excess powders are collected in powder collection container. All the movements of powder container, collection container, and build plates are controlled using CNC. The above steps get repeated till the entire parts get built, final part being

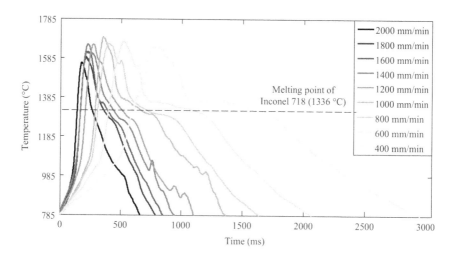

Figure 15.17   Variation in thermo-cycles with change in laser scan speed. (P = 1200 W, preplaced powder layer thickness = 2 mm.) (Reprinted with permission from *Optics and Lasers in Engineering*, Elsevier 2017 88:139–152.)

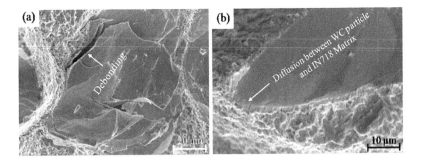

Figure 15.18 Fracture surfaces of clad layer showing de-bonding between the ceramic particles and the metal matrix deposited at scan speed of (a) 1200 mm/min and (b) 400 mm/min. (1200 W laser power.) (Reprinted with permission from *Journal of Alloys and Compounds*, Elsevier 2017 714:514–521.)

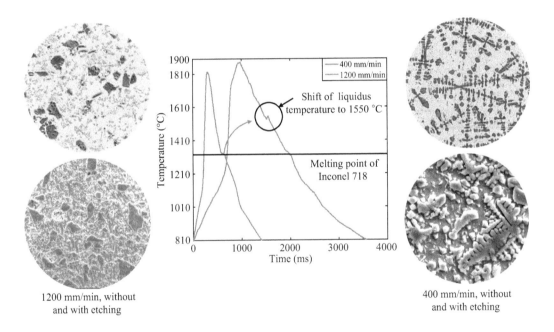

1200 mm/min, without and with etching

400 mm/min, without and with etching

Figure 15.19 Online identification of ceramic particles (TiC) decomposition from shift in liquidus temperature. (Reprinted with permission from *Materials and Design*, Elsevier 2017 121:310–320.)

submerged in the powder. Due to layer-by-layer addition of metals, parts built using DMLS process tends to be anisotropic in nature (Yadollahi et al. 2017). Therefore, in order to obtain desired mechanical properties, different post processing techniques such as heat treatments, remelting, and shock peening are usually carried out (Hooreweder et al. 2017).

DMLS/SLM process can be considered as a green manufacturing process form both process and supply chain perspective as summarized below:

From process perspective

- In DMLS/SLM, LASER is used as energy source so the energy source does not emit any fumes itself. Also the process is done in a closed chamber, so there is no chance of pollution or spread of toxic materials to the environment while building the parts.

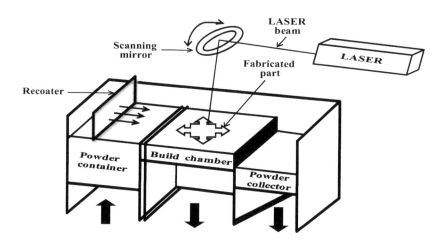

Figure 15.20 Schematic of Direct Metal Laser Sintering process.

- In DMLS/SLM, directly the final product can be obtained without any need to remove the excess materials and unmelted powders can be reused after sieving; therefore, there is a little waste of material, hence the process is economic.
- Die/tool less manufacturing with wide range of customization which reduces secondary manufacturing processes which may emit fumes.
- Using DMLS/SLM, geometrically complex parts can be built in a single step which may need several steps of conventional manufacturing processes which in turn reduce cost associated with the process and make the process economic.
- Also DMLS/SLM is a process in which parts can be built without supervision; therefore, less man power is required and hence economic.

From supply chain perspective
- Using DMLS/SLM, onsite part production (for both new and repair case) is possible. Therefore, no/ less intermediate inventory storage and transportation of parts is required which in turn reduces fixed and variable carbon emission associated with the supply chain.
- On demand part production (leading to high cost due to customization) with reduced lead time is possible in DMLS/SLM and it can tackle the cases where sudden change in design (generally associated cost of sudden change in design is very high) is required to meet customer expectation. Hence, the process is economic.

Some of the popular materials which can be processed in DMLS/SLM are Maraging steel, Stainless Steel (15-5 PH, 17-4 PH, 316L, tooling grade steel), Nickel-based super alloy (Inconel 625, Inconel 718), Cobalt chrome alloy, Aluminum alloy, etc. (EOS GmbH Electro Optical Systems 2017). Apart from the above specified materials, different custom-made metal powder combinations, ceramic materials, cermet, nano composites, etc., can be processed with judicious selection of laser process parameters. Parts built using DMLS process can be used in a variety of applications such as jet engine parts, impeller (Figure 15.21a), biomedical implants (Figure 15.21b), functional prototypes, different engineering components including high-temperature turbine components, etc. Associated flexibilities in designing the component using DMLS/SLM process allow users to design custom made unit cell structure having specific mechanical properties depending upon the applications (Huynh et al. 2016). These unit cells could be used to produce porous structures having major applications in medical, aerospace, and automobile industries. Process engineers can also play with various process parameters to create functionally graded parts which typically have gradual variations in microstructures and pores along a particular direction of the part built.

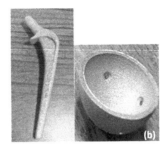

Figure 15.21 (a) Impeller built using Al alloy and (b) hip stem and acetabular cup made of Stainless Steel PH1. (Copyright and Courtesy: Direct Digital Manufacturing Laboratory, Department of Mechanical Engineering, IIT Kharagpur, India.)

Finally, the best possible outcome from DMLS/SLM can be achieved following multidisciplinary approach such as combining inputs from design optimization of parts for specific mechanical properties using finite element-based simulation (design aspect), thermal simulation and microstructure modeling using different heat transfer equations to predict final strength and quality (warping/distortion, etc.) of the built parts (thermal aspect) and design for manufacturing (build orientations, type of support structure selections, different post processing techniques, etc.) which is a manufacturing aspect.

IIT Kharagpur has state-of-the art facility 200 W Yb-fiber-integrated EOSINT M270 DMLS machine (Figure 15.22a). Although a lot of research is being carried out in the area of DMLS process, it is not yet industry ready. One of the reasons for the same is insufficient databases on fatigue properties of DMLS parts (Shamsaei et al. 2015). In recent years, the investigation on the fatigue behavior of TiAl6V4 components produced by SLM process has received much attention. However, associated manufacturing costs become a huge constraint for developing countries where economic viability is of major concern (Mudali et al. 2003). Therefore, research on biocompatible stainless steel such as stainless steel 316L, precipitation hardened grade stainless steel, is in focus to manufacture bio-implants (Manivasagam et al. 2010; Stevenson et al. 2002). Keeping the above point in mind, study on fatigue properties of DMLS parts (Figure 15.22b) made of 15-5 PH Stainless Steel are being carried out thoroughly.

Leuders et al. (2013) studied fatigue properties of TiAl6V4 alloy to investigate microstructure–defect–property relationships under cyclic loading and found that fatigue strength is mainly affected by micron-sized pores, whereas residual stresses have a major impact on fatigue crack growth. Edwards et al. (2014) evaluated fatigue properties of SLM processed Ti–6Al–4V specimens built in different orientations and reported relatively poor fatigue life as compared to wrought Ti–6Al–4V

Figure 15.22 (a) EOSINT M270 DMLS machine at IIT Kharagpur and (b) specimens built perpendicular to laser scan path. (Copyright and Courtesy: Direct Digital Manufacturing Laboratory, Department of Mechanical Engineering, IIT Kharagpur, India.)

specimens. Reasons for this reduction were attributed to the microstructure formed, surface roughness, porosity, and residual stress associated with it. Stress relieving after the sample building and post-processing such as HIP process has been suggested to improve the fatigue life of SLM processed Ti–6Al–4V further. Li et al. (2016) presented a critical assessment of the fatigue performance of additively manufactured Ti–6Al–4V parts in their study. Based on the literatures available, they found out that although fatigue performance can be qualitatively correlated with microstructure and defect population, quantitative correlations for the same are not available. Riemer et al. (2014) investigated crack initiation and crack growth behavior of Stainless Steel 316L specimens manufactured by SLM and found out that its fatigue life is similar to the conventionally processed material in its as-built condition. It was reported that columnar grains are present in the as-built condition whereas equiaxed grains are formed by HIP processing leading to isotropic properties in terms of crack propagation. A comparison on tensile, fatigue, and fracture behavior of Ti–6Al–4V and 15-5 PH stainless steel parts manufactured by SLM has been given by Rafi et al. (2013). They found that tensile strength of horizontally built specimens for both the cases is slightly better than the vertically built specimens.

Most of the available studies report fatigue life of DMLS parts under constant amplitude loading under zero mean stress (Shamsaei et al. 2015, Spierings et al. 2013). However, in actual service conditions, there may be presence of under different mean stresses as well as variable amplitude loadings (Shamsaei et al. 2015). Results (Figure 15.23) show that presence of compressive mean stress increases fatigue life, whereas tensile mean stress is detrimental to fatigue life of DMLS parts (Sarkar et al. 2017a).

Also, load sequence has a strong impact on fatigue life under variable amplitude loading with nonzero mean stress, and leads to completely different fracture surface morphologies (Figure 15.24) (Sarkar et al. 2018a).

A recent report by Sarkar et al. (2017b) shows that a wide range of desired mechanical properties such as tensile strength, wear rate, impact toughness (Sarkar et al. 2018b), and electro-chemical properties (Sarkar et al. 2018c) can be obtained by carrying out different heat treatment processes of DMLS specimens. Surface properties of the specimen play a very important role in the fatigue life (Spierings et al. 2013). Therefore, effect of different surface properties on fatigue life of DMLS parts under rotation bending fatigue conditions is under investigation. Since parts built through DMLS process tends to be anisotropic in nature (Edwards et al. 2014; Yadollahi et al. 2017),

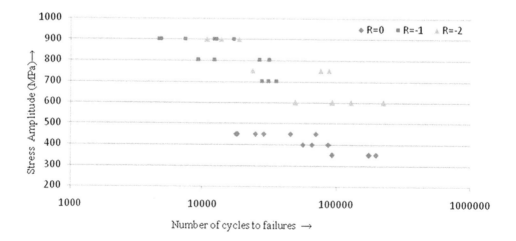

Figure 15.23 Plot for stress amplitude vs. number of cycles to failure. (Reprinted with permission from *Materials Science and Engineering A*, Elsevier 2017 700:92–106.)

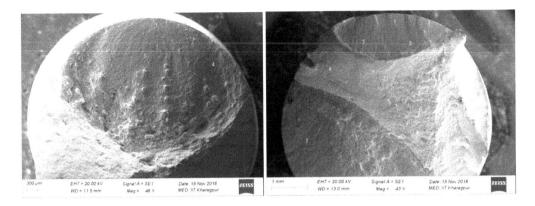

Figure 15.24 (a) Fracture surface for high stress cycle followed by low stress cycle and (b) fracture surface for low stress cycle followed by high stress cycle. (Copyright and Courtesy: Direct Digital Manufacturing Laboratory, Department of Mechanical Engineering, IIT Kharagpur, India.)

the effects of build orientation on fatigue life of DMLS are also being carried out. It is reported that one of the reasons which affects the fatigue life of DMLS parts is micron-level pores in DMLS parts (Hooreweder et al. 2017; Leuders et al. 2013; Tang and Pistorius 2017) and therefore micro-CT-based analysis of DMLS parts (Figure 15.25) and its correlation with the fatigue life is another interesting area that are also being investigated in detail.

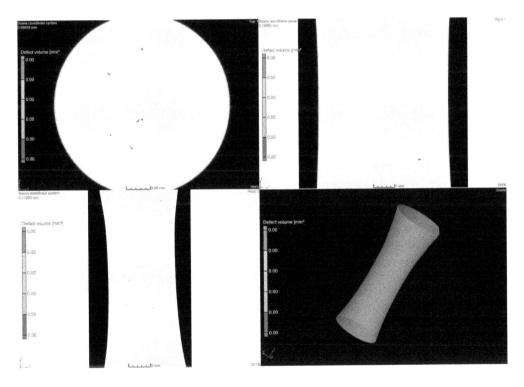

Figure 15.25 Micro-CT images of different cross-sections of a DMLS specimen showing micron-level pores. (Copyright and Courtesy: Direct Digital Manufacturing Laboratory, Department of Mechanical Engineering, IIT Kharagpur, India.)

In developed countries, much research is being carried out on different aspects of DMLS/SLM process. However, in developing countries very less attention has been given to this process. In a nut shell, in spite of having several advantages of DMLS/SLM process, this process does not seem to be yet industry-ready. The following could be the possible reasons for that:

- high initial cost of the equipment
- high cost of metal powder
- proprietary nature of the process and lack of sharing of knowledge in this area
- lack of mindset to adopt new technology
- lack of industry participation specially in developing countries
- lack of rigorous analysis of different mechanical properties of DMLS/SLM processed parts and standardization of database, etc.

In order to tackle the high initial investment and high cost of metal powders, some researchers are working on the hybrid manufacturing processes that involve combination of traditional manufacturing processes combined with laser-based AM processes. With increase in more cost-effective metal powder production and increased number of DMLS/SLM machines in the market along with sufficient amount of industry participation in research activities related to DMLS/SLM process, true potential of this process can be realized in years to come. Although DMLS/SLM may not completely replace the traditional manufacturing processes, this green manufacturing process has a great potential to emerge as "Future of Manufacturing" process.

15.5 LASER MATERIAL PROCESSING BY VAPORIZATION AND ABLATION

Manufacturing operations such as drilling and micromachining that require material removal are usually carried out using high-power pulsed lasers by vaporization and ablation. Cutting of nonmetallic materials such as polymeric materials, wood, textiles, and leather, etc., are also done by the process of vaporization with CW and pulsed lasers. Laser power densities and interaction time involved are typically in 10^7–10^{12} W/cm^2 and ms to fs ranges, respectively, and commonly used lasers include Nd:YAG, Fiber, CO_2, Excimer and ultrafast lasers.

15.5.1 Laser Drilling

Historically, laser drilling was the first industrial application introduced by Western Electric in 1985 for drilling of diamond wire-drawing dies using ruby laser (Anon 1966). Since then laser drilling is finding ever-increasing applications in drilling precision holes in a wide variety of engineering and exotic materials in many industries such as aerospace, automobile, electronic, and medical industries (Dahotre and Harimkar 2008; Steen and Mazumder 2010; Yeo et al. 1994). In this process, a focused laser beam of high intensity is incident on the job along with a coaxial gas jet to drill a hole. The main attributes of laser drilling which make it attractive include (Dahotre and Harimkar 2008; Steen and Mazumder 2010),

- Any material, irrespective of its hardness can be drilled without tool wear
- High quality of holes with high precision and minimum burr and spatter
- Holes of any size and shape at any angle
- Very high drilling speed
- Cost-effective process with all the above qualities

There are four methods of laser drilling, namely single pulse drilling, percussion drilling, trepanning, and helical trepanning (Dausinger et al. 2003). These are schematically shown in Figure 15.26.

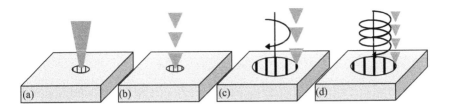

Figure 15.26   (a) Single pulse, (b) percussion, (c) trepanning, and (d) helical drilling.

In single-pulse drilling, a sufficiently high energy laser pulse is used to drill a hole of diameter typically less than 1 mm in a thin sheet. Drilling a deep hole with a single high energy laser pulse usually results in poor tolerances due to thick recast, spatter, and taper. Therefore, a series of short-duration laser pulses are used to drill a through hole in a relatively thick plate in percussion drilling. Laser trepanning is used to produce large size holes, by drilling a series of overlapping holes around their perimeter with either a high repetition rate pulsed laser, CW or quasi CW laser. In helical trepanning the focused laser beam is rotated around the perimeter and gradually deepened the hole lowering down the focal position with each rotation spirally. Molten material is ejected upward in the process by the recoil pressure of vapor and the gas jet. Focus of the laser beam is adjusted so that it is always at the bottom of the hole. Once the through hole is created, laser is moved around the hole a few times to smooth out the edges. This process can produce very large, deep, high-quality holes.

Lasers that are commonly used for drilling are pulsed TEA (Transversely Excited Atmospheric pressure) CO_2 lasers; free-running and Q-switched pulsed Nd:YAG lasers; quasi CW Q-switched fiber lasers; and excimer lasers (Meijer 2004; Westphäling 2010; Yeo et al. 1994). The choice of a laser is decided by the absorption characteristic of material to be drilled and the quality and precision of hole required. Pulsed Nd:YAG lasers are most commonly used for percussion drilling in thick metallic as well as in semiconducting and ceramic materials because of their higher pulse energy capability (Low et al. 2001; McNally et al. 2004). The 9.6- and 10.6-μm wavelengths of CO_2 laser are readily absorbed in nonmetals, viz., glass, quartz, ceramics, glass-epoxy, organic polymers such as polyimide, polypropylene, polyvinyl chloride (PVC) and poly-methyl methacrylate (PMMA), etc. Therefore, pulsed TEA CO_2 lasers are used in drilling holes in these materials (Dyer et al. 1997; Takeno et al. 2009). Excimer lasers operating in the 159–351 nm UV wavelength range are used to machine semiconducting materials, glass, quartz, ceramics, and organic materials. Latest introduced ultrafast femto-second Ti–sapphire laser and fiber laser operating in 0.7–1 and 1.03–1.08 μm wavelength range, respectively, are also gaining popularity in micro-drilling of high-quality holes and micromachining of different materials. As mentioned earlier, at high laser intensities typical of ultrafast lasers, all types of materials including transparent ones, absorb laser energy by nonlinear absorption mechanisms. Thus, any material can be machined by ultrafast lasers.

Various phenomena happening during laser drilling are depicted in Figure 15.27a. During the heating process by lasers of pulse durations typically in ms–ns range, melting and vaporization take place maintaining electrons and ions in thermal equilibrium.

The basic material removal mechanism is vaporization and melt-ejection by the recoil pressure of vapor, and their contribution depends upon the laser power density and thermo-physical properties of material (Figure 15.27b). At high-power densities, larger fraction of material is removed by vaporization with less recast producing holes of improved quality. Furthermore, the thermal diffusion length ($2\sqrt{k\tau}$) which depends on laser pulse duration is smaller for ns duration pulses than ms duration pulses; therefore, shorter duration pulses are preferred in laser drilling, as they produce less deleterious thermal effects like micro-cracks and HAZ with less recast. In addition to the thermal process, excimer lasers can ablate organic materials whose chemical bond energy is less than the laser photon energy by photochemical processes, i.e., direct bond breaking.

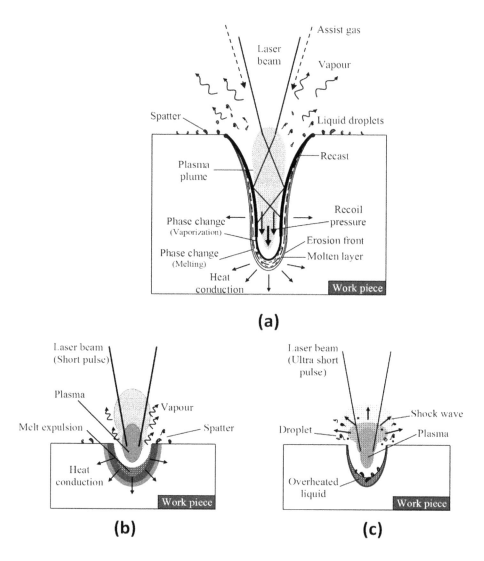

Figure 15.27 (a) Different processes involved during laser drilling process, (b) classical beam matter interaction with laser pulses of sub-μs and ns ranges, and (c) ultrafast beam matter interaction.

At laser power densities >10^9 W/cm^2 typical of ns and shorter duration laser pulses, the rate of energy deposition is so fast that some of the electrons and ions get enough kinetic energy before complete thermalization between electron and ion sub-systems establishes, and some of them escape out of the surface even before melting takes place. As the laser continues, heating, melting, and evaporation set in. Vapor exerts intense recoil pressure on the molten pool which increases the boiling temperature. As heat conducts inside the melt layer its temperature increases even beyond the normal boiling temperature. At the end of the laser pulse, as the recoil pressure ceases, superheated liquid and vapor comes out violently. This process is usually called as the ablation process.

The recoil pressure, p_s, can be approximately given by (Semak and Matsunawa 1997)

$$p_s = p_0 \exp\left[\left(\frac{L_p}{k_B T_v}\right)\left(1 - \frac{T_v}{T_0}\right)\right] \tag{15.19}$$

where T_v, L_p, and p_0 are the evaporation temperature, the heat of evaporation per particle, and the ambient pressure, respectively, and k_B is the Boltzmann constant.

The thermal diffusion length is usually negligible compared to the absorption length in the case of ultrafast laser pulses and at laser power densities of 10^{10}–10^{12} W/cm^2 range, typical of these lasers, the laser energy is mainly absorbed by multiphoton ionization and avalanche processes. During ps and sub-ps laser pulse duration, electron and ion sub-systems do not thermalize completely, and, therefore, phenomena like heating, melting, and vaporization cannot take place during the laser pulse. Material is ablated by several nonthermal and thermal processes continuing for several 10's of ns after the end of laser pulse (Dausinger et al. 2003; Rethfeld et al. 2004). Very high intensity fs (10^{-14}–10^{-13} s) laser pulse sometimes causes disturbing nonlinear effects which could adversely affect the machining quality, therefore, laser pulse duration 1–10 ps order is considered as optimal to minimize thermal damage and also to avoid nonlinear effects (Dausinger et al. 2003).

Since the electron subsystem gets heated to very high temperatures in 10^4–10^5 K range before thermalization with ion subsystems, a chain of events occurs during the laser pulse and thereafter as the electron-ion energy transfer takes place, and ultimately leads to ablation of the material. It depends strongly on material properties and laser parameters. A number of thermal and nonthermal processes occur, which contribute significantly in material removal (Lewis et al. 2009; Stoian et al. 2000). The phase explosion is of the important material removal process at laser intensities above threshold (Miotello and Kelly 1999). During ultrafast laser irradiation as the high-temperature electron subsystem thermalizes with the ion subsystem, temperature of small irradiated volume increases at a very fast rate. The material is transferred into a state of superheated liquid and homogeneous nucleation of gas bubble forms inside the liquid. With increased temperature, the pressure in the volume also increases. Assisted by the tensile stress, a high-pressure mixture of liquid droplets and vapor ablates out (Figure 15.27c) (Leitz et al. 2011). The normal vaporization can also take place in a time scale greater than ns and continue long after (~100 ns) the end of laser pulse (Dausinger et al. 2003; Stoian et al. 2000).

Laser-drilled holes are inherently associated with some geometrical defects related with hole size, taper, circularity, and the metallurgical defects such as spatter, recast layer, micro-cracks, and microstructural changes in HAZ. Although the basic mechanism of laser drilling and the influences of various processing parameters on the quality of drilled holes are well researched, the process has been adopted to drill several different engineering components successfully in many industries. R&D is still continuing directed toward furthering the drilling quality by eliminating recast and spatter, and minimizing taper; drilling holes in fragile and brittle materials; and enhancing the repeatability, drilling speed, and aspect ratio of holes, etc. Recently, several novel techniques such as hybrid drilling (EDM), micro gas jet and twin gas-jet-assisted laser drilling, and ultrafast laser drilling have been developed to tackle some of these issues (Li et al. 2006; Okasha et al. 2010; Rekow and Murison 2011). Devoid of any detrimental thermal effects such as recast, HAZ, and micro-cracks in holes drilled by fs laser pulses, their quality is much better than that drilled by ns duration laser pulses (Figure 15.28) (Chichkov et al. 1996).

Some of the popular industrial applications are drilling large numbers of closely spaced cooling holes in jet engine components such as turbine blades, nozzle guide vanes, combustion chambers, and afterburners; fine holes in automobile injection nozzles, inkjet nozzles, biomedical devices, sensors, irrigation pipes, and interconnecting micro-vias in electronic packages and solar cells (Meijer 2004; Yeo et al. 1994). The domain of laser drilling ranges from producing micro-holes in thin films for solar-cell applications (Rekow and Murison 2011) to drilling large holes in rocks for natural gas and oil exploration (Xu et al. 2003). Figure 15.29 shows some typical laser-drilled components. More details about the laser drilling of different types of metallic and nonmetallic substrates are presented in a recent review article by Nath (2014).

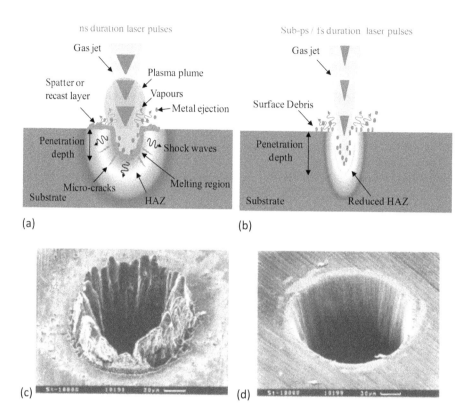

Figure 15.28   Schematic of laser drilling with (a) ns and (b) fs pulses and holes drilled in a 100 μm thick steel foil with (c) 3.3 ns, 1 mJ, and F = 4.2 J/cm² laser pulses (d) 200 fs, 120 μJ and F = 0.5 J/cm². (Reprinted with permission from *Applied Physics A*, Elsevier 1996 63:109–115.)

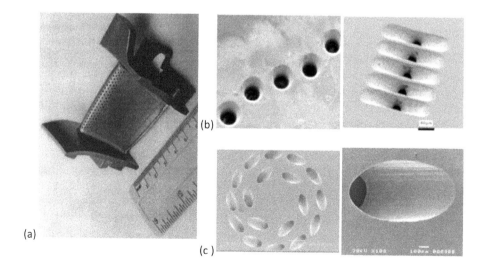

Figure 15.29   (a) Laser-drilled holes in an aero-engine component. (Reprinted with permission from Mohammed Naeem, Laser Institute of America, Vol. 95, Publication No. 595, 2003. ISBN: 0-912035-75-7.) (French et al. 2003). (b) Some geometries of laser-drilled holes in inkjet printer head. (Reprinted with permission from Rebecca Robinson, The Optical Society, USA.) (Gower 2000). (c) Example of a laser-drilled injector with 60° orifice angle. (Reprinted with permission from D. Karnakis, Oxford Lasers, UK.) (Rutterford et al. 2005)

15.5.2 Laser Marking

According to Annual Laser Market Review & Forecast, 2017, laser marking has second largest market share, next to laser cutting, and is a well-established process (Overton et al. 2017). This has been very popular for marking product information and logo on various products ranging from engineering components, medical products including capsules, implants, and surgical instruments catheters, electronic components, labels, beverage cans, plastic, and wood furniture, etc.

Laser marking is done by two methods. First is the direct writing with a tightly focused laser beam scanned over the surface with the help of two orthogonally rotating galvanic mirrors (Figure 15.30a). This is a serial process, however, much flexible. Second is the writing through masks with a broad laser beam of uniform intensity (Figure 15.30b). This is a parallel process, but relatively, less flexible.

Lasers which are very common are the pulsed CO_2 laser for marking on mostly the nonmetallic surfaces and Q-switched solid state and fiber lasers for metallic and other materials. For organic materials excimer lasers are also often used.

Depending upon the type of material laser marking is created by different mechanisms. Materials such as plastic, glass, ceramic, rubber, and metals are laser marked by controlled vaporization/ablation of a very thin surface layer. A thin layer of plastic film, paper, ink, or paint is vaporized exposing the under-layer of a different color. Metals, epoxies, and glass are often marked by melting also. In metals, marking contrast is produced due to oxidation or incorporation of impurities into the melt. In plastics, material melts, and forms ridges. Marking can be done through chemical change by either photo- or thermal-induced color change, e.g., excimer laser induced photochemical process for aircraft cable (white to black), and CO_2 induced thermal-chemical color change on PVC (gray to red-brown). The color change is due to changes in chemical composition or in molecular structures. Popularity of laser marking is because of its fine quality, flexibility, high production rate and above all not easy to duplicate. Thus, laser marking ensures genuineness of the product.

15.5.3 Laser Micro- and Nano-Machining

The ability to focus a laser beam to micron range spot and development of high-intensity short-duration (ns–ps) laser pulses which eliminates the deleterious thermal effects have made the laser an ideal tool for micromachining applications. As mentioned earlier, at high laser intensities

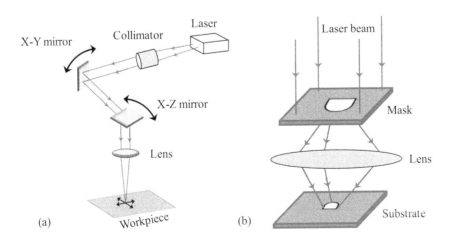

Figure 15.30 Schematic configuration of (a) beam deflecting marking system and (b) mask marking.

($>10^9$ W/cm^2) the multiphoton ionization, tunneling ionization, and avalanche processes are the dominant absorption mechanisms. These nonlinear absorption processes have well defined threshold laser fluence; therefore, when an ultrafast laser pulse is focused on a material, the radial distance from the focal point up to which laser fluence is above the ablation threshold, enclosed area of that central region is ablated and this region can be smaller than the focal spot size (Figure 15.31) (Liu et al. 1997). Thus, controlling the laser fluence feature size smaller than focal spot diameter can be produced with high precision. Furthermore, the thermal diffusion length corresponding to sub-ps duration laser pulse is negligible, and thus thermal effects such as HAZ and micro-cracks are also absent. Therefore, the ultrafast lasers are becoming increasingly popular in precision micromachining.

Laser micromachining is usually done for drilling micro-holes, grooving, and formation of different patterns including micro-channels, and also generating micro- and nano-scale features through polymerization. Over the past decade, many types of laser, viz., CO_2 laser, solid-state lasers, fiber laser, excimer lasers, and most important ultrafast lasers have been exploited in micromachining of miniaturized components involved in various applications, such as MEMS, microelectronics, telecommunication, optoelectronics, and biomedical devices. A few illustrative examples are presented here. Both CW and pulsed CO_2 lasers have been used to micro-machine polymeric materials, glass, wood, etc. Cheng et al. demonstrated the fabrication of microfluidic device by direct-write CO_2 laser micromachining of poly-methyl methacrylate (PMMA) (Figure 15.32) (Cheng et al. 2004). As-laser-machined trench walls were very ragged which was improved by thermal heat treatment at 170°C for 30 min.

Figure 15.33 (Cheng et al. 2004) shows a comparison of as-laser machined and thermal annealed trench walls. Recently, Kant et al. (2015) presented a detailed review of CO_2 laser machining of PMMA for micro-fluidic applications.

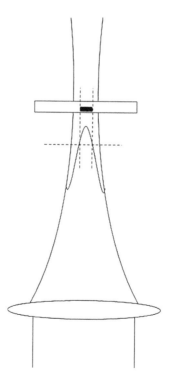

Figure 15.31 Schematic of spatial extent near the center of focused laser beam within which laser intensity is above the absorption threshold by nonlinear processes.

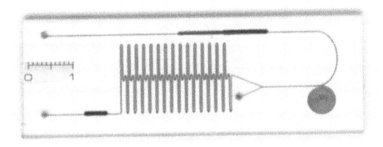

Figure 15.32   A sample pattern showing the various pattern fabricated simultaneously in the same PMMA substrate. The scale bar is 1 cm. (Reprinted with permission from *Sensors and Actuators B*, Elsevier 2004 99:186–196.)

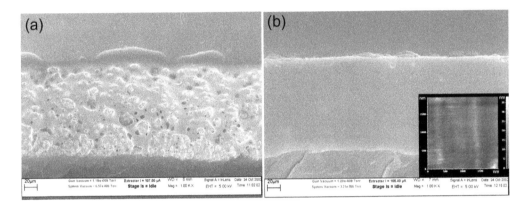

Figure 15.33   The SEM pictures showing the rugged interior surface of the trench after laser machining (a) and smooth surface after thermal annealing (b). The AFM topography of the annealed surface shown in the inset with full scale of 38 nm in the Z-axes. The viewing angle is perpendicular to the plane of the side wall. (Reprinted with permission from *Sensors and Actuators B*, Elsevier 2004 99:186–196.)

Naeem et al. (2012) demonstrated good quality micro-holes drilled by percussion drilling with pulsed Nd:YAG laser and modulated power single mode fiber laser in metals and nonmetals, Figure 15.34a–c. In this work, 1.5-mm-thick nickel-based alloy was percussion-drilled using a pulsed Nd:YAG laser at 150-µm, 0.6-mm-thick alumina ceramic was percussion-drilled using a pulsed Nd:YAG laser at 50-µm, and 0.5-mm-thick holes in 304 stainless steel were percussion-drilled using a JK100FL laser with modulated output at 50 µm. In a recent technological breakthrough, Rekow and Murison (2011) reported micro—via drilling at a rate up to 12,500 holes per second, each of 25-µm diameter in 182-µm-thick Si wafer, with a 1,064-nm wavelength laser of 20–600 ns pulse duration at 10–700 kHz repetition frequency and 22 W average power for the fabrication of Emitter-Wrap through (EWT) solar cells. Klotzbach et al. (2011) reported better quality micromachining of silicon by 355 nm of pulsed Nd:YAG laser compared to its fundamental 1,064 nm wavelength.

Figure 15.35 (Schoonderbeek and Ostendorf 2008) shows various laser processing modalities in solar-cell production. Laser processing modalities in solar-cell production including hole drilling of back-contact solar cells with pulsed fiber laser, groove making by SiO_2 removal, laser welding/soldering of contacts, edge isolation, wafer cutting, contact performance improvement by removal

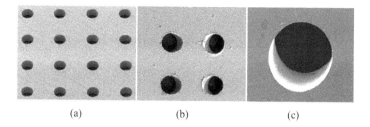

(a)   (b)   (c)

Figure 15.34 (a) Pulsed lasers are good at percussion drilling metals and nonmetals. This 1.5-mm-thick nickel-based alloy was percussion-drilled using a pulsed Nd:YAG laser at 150 μm; (b) this 0.6-mm-thick alumina ceramic was percussion-drilled using a pulsed Nd:YAG laser at 50 μm; (c) these 0.5-mm-thick holes in 304 stainless steel were percussion-drilled using a JK100FL laser with modulated output at 50 μm. (Reprinted with permission from Mohammed Naeem, JK Lasers, Photonic spectra, Oct 2012.)

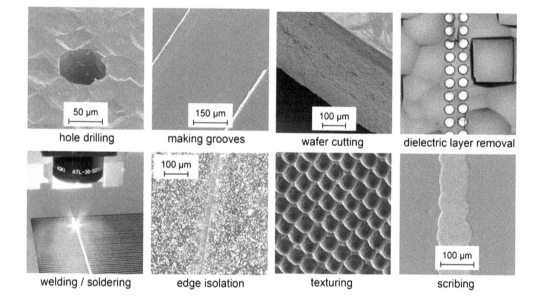

Figure 15.35 Laser applications for solar-cell production: hole drilling of back-contact solar cells with pulsed fiber laser, SiO_2 removal for groove production, laser welding/soldering of contacts, edge isolation, wafer cutting, removal of dielectric layers for improved contact performance, texturing to increase the absorption of sunlight and therefore to enhance the efficiency of the cell, and scribing of thin-film cells. (Reprinted with permission from Nicole Harris, SPIE 2008. doi: 10.1117/2.1200804.1132.)

of dielectric layers, enhancing the efficiency of cell by texturing, and scribing of thin-film cells were performed.

The high precision of laser micromachining in different materials by ultrafast laser can be illustrated in a few examples presented in Figure 15.36. Figure 15.36a–d (Muhammad et al. 2012a) shows precision dross-free cutting of metallic (nitinol, platinum–iridium alloy, cobalt chromium stent) as well as nonmetallic (low melting point polymer stent) by picosecond laser. Figure 15.36e shows micro-hole drilled in a 100 μm thick steel foil by femtosecond-pulse laser ablation with a small HAZ (Chichkov et al. 1996).

Advances in the ultrafast laser technology and development of a better understanding of laser–material interaction have led to several interesting innovations in micro- and nano-manufacturing

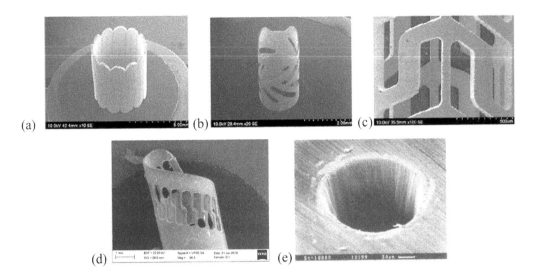

Figure 15.36 Examples of precision machining by ultrafast lasers; precision cutting of (a) nitinol, (b) platinum–iridium alloy, (c) cobalt chromium stent, and (d) low melting point polymer stent material by pico-second laser. (Reprinted with permission from *Applied Physics A*, Elsevier 2012 106:607–617.) (e) Micro-hole drilled in a 100 μm thick steel foil by femtosecond-pulse laser ablation. (Reprinted with permission from *Applied Physics A*, Elsevier 1996 63:109–115.)

beyond the diffraction limit (Hwang et al. 2009; Hong et al. 2008; Li et al. 2011; Liao et al. 2013; Ostendorf and Chichkov 2006). Li et al. presented an overview of advances in laser-based nano-manufacturing technologies including surface nano-structure manufacturing, production of nano materials (nano-particles, nano-tubes and nano-wires) and 3D nanostructures manufacture through multiple layer additive techniques and nano-joining/forming (Li et al. 2011). Exploiting various nonlinear effects induced by ultrafast laser pulses, several new techniques have been developed to produce sub-diffraction limited structures. In one of such techniques nano-level etching has been demonstrated by coupling a high-intensity, short-duration (ns–fs) laser beam to aperture-based and aperture-less near-field scanning optical microscopic (NSOM) probes of diameter typically in the range of 100–200 nm (Figure 15.37) (Hwang et al. 2009). The evanescent wave emerging out of the probe-tip produced nano-patterns on a surface placed in near-field, with ~10 nm separation between the tip and the target.

More recently, combining NSOM and polymer pen lithography a new technique, named mul-tiplexed beam pen lithography has been developed for generating a large number of diffraction-unlimited structures in parallel (Liao et al. 2013). Figure 15.38 illustrates the basic principle of this technique presented by Liao et al. (2013).

Hong et al. (2008) reported development of parallel laser nano-patterning to achieve large area and high-speed nano-fabrication with laser irradiation through a micro-lens array and transparent micro-particles exploiting the optical resonance and near-field effects in the interaction of micron and submicron size transparent particles on a substrate with laser light. They reported fabrication of 100-nm functional periodic nanostructures on substrate surfaces by laser interference lithography also. Also Yang et al. (2010) reported that nano-wires with a pitch of 430 nm (Figure 15.39a: top view, Figure 15.39b: side view) and nano-pillars with a period of 450 nm (Figure 15.39c: top view, Figure 15.39d: side view) can be produced by this technique.

Ostendorf and Chichkov (2006) presented an overview of 3D writing and structuring of nano-level features by two-photon polymerization (TPP) of photosensitive materials using ultrafast laser. Figure 15.40a shows the basic setup for TPP technique. A 3D scaffold fabricated by TPP of SR610

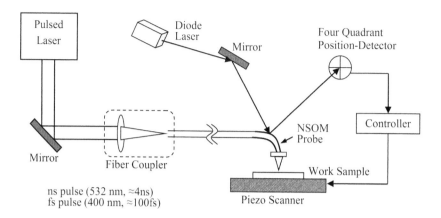

ns pulse (532 nm, ≈4ns)
fs pulse (400 nm, ≈100fs)

Figure 15.37   Schematics of experimental setup for ns and fs laser micro- and nano-machining with aperture-based near-field optical scanning microscope (NSOM).

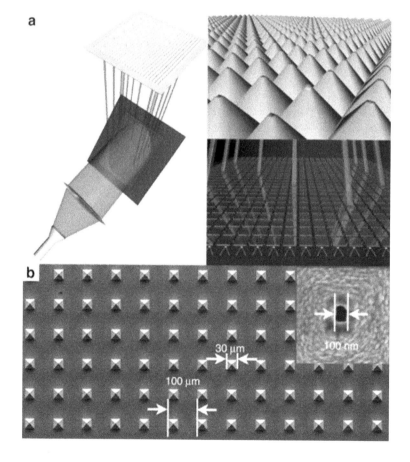

Figure 15.38   Schematic of actuated Beam Pen Lithography (BPL). (a) Schematic of the principle of operation of actuated BPL. A digital micro-mirror device (DMD) is illuminated with ultraviolet light, which is selectively directed onto the back of a near-field aperture array. (b) Scanning electron microscope (SEM) image of section of the BPL tip array. Each pyramidal pen in the array has a sub-wavelength aperture that has been opened in a parallel fashion. (Reprinted with permission from *Nature communications* 2013 4:2103. doi: 10.1038/ncomms3103.)

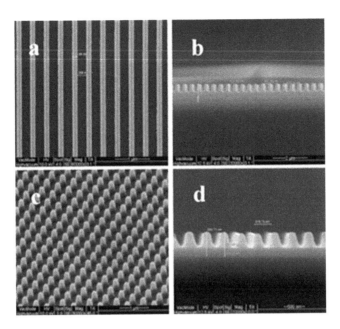

Figure 15.39 SEM images of (a: top view, b: side view) nano-wires and (c: top view, d: side view) nano-pillars made by laser interference lithography. (Reprinted with permission from *Applied Surface Science*, Elsevier 2010 256:3683–3687.)

material is shown in Figure 15.40b,c (Ovsianikov et al. 2011). Furthermore, Kawata et al. (2001) showed that diffraction limit can be exceeded by nonlinear effects using this process and "micro-bull" sculptures (Figure 15.40d), smallest model animals ever made artificially and are about the size of a red blood cell, were fabricated using this process.

Menon et al. have developed a technique for confining light-to-deep subwavelength dimensions by exploiting the absorbance modulation characteristics of photo-chromatic molecules which could be alternately switched from opaque to transparent state by irradiating at two different wavelengths and demonstrated creating features of 36 nm with simultaneous application of 325- and 633-nm wavelengths laser beams (Andrew et al. 2009).

Figure 15.41 shows the principle of producing a nano-scale optical beam by this technique to write nano-features on a photo-resist film (Andrew et al. 2009).

Fourkas et al. developed another novel technique named as resolution augmentation through photo-induced deactivation (RAPID) photolithography to fabricate features with resolution down to 40 nm along laser beam axis (Fourkas 2010). The basic principle of RAPID is illustrated in Figure 15.42.

These developments have demonstrated that the visible lasers hold great potential for nano-scale lithography, though much development needs to be done for enhancing feature density and through-put and compete with more established lithographic techniques based on electron- and ion beams and synchrotron radiation.

15.5.4 Laser Shock Peening

LSP is a process in which high-intensity laser pulses are used to generate high residual compressive stresses. This is similar to mechanical shot peening for improving the wear, fatigue, and corrosion properties including stress corrosion crack (SCC) resistance of the surface. For LSP, surface of the component is coated with a laser-absorbent coating and this is covered by a thin transparent

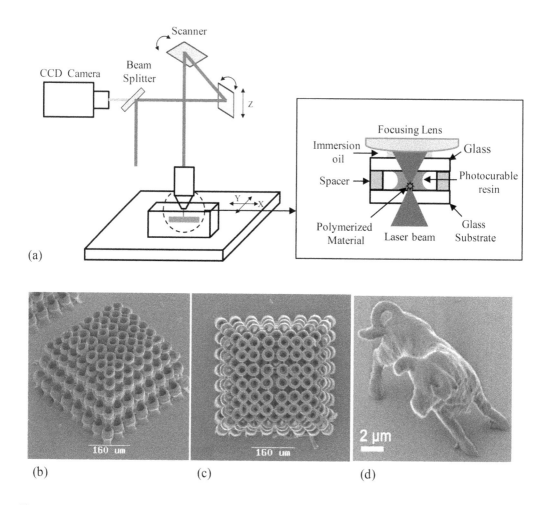

(a)

(b)   (c)   (d)

Figure 15.40 (a) Principal setup for laser 3D writing; SEM images of a 3D fabricated by two-photon polymerization of SR610 material, (b) perspective view, and (c) top view. (Reprinted with permission from *Acta Biomaterialia*, Elsevier 2011 7:967–974.) (d) "micro-bull" sculpture-smallest model animal ever made artificially and is about the size of a red blood cell. (Reprinted with permission from *Nature* 2001 412:697–698.)

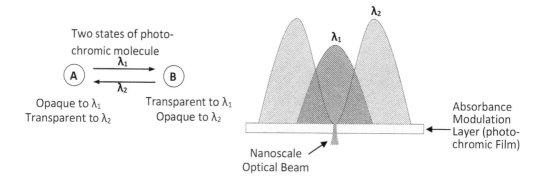

Figure 15.41 Schematics of (a) two reversible states of photochromic material used for absorption modulation photolithography and (b) illumination configuration for creating sub-wavelength aperture at λ_1 by modulating absorption with λ_2.

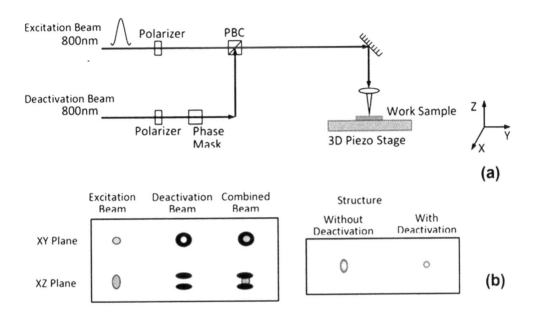

Figure 15.42 Schematics of (a) experimental setup for resolution augmentation through photo-induced deactivation (RAPID) lithography, CW: continuous wave. PBC: polarizing beam cube. (b) Intensity distribution of the excitation beam and deactivation beam at the focal region.

layer of either glass, quartz or water layer, and then exposed to high-intensity (typical $>10^9$ W/cm^2), short-duration (10–50 ns) laser pulses. The laser intensity is maintained below the optical breakdown threshold of transparent layer. Usually a thin layer (~1 mm) of water is flown over the surface and the maximum laser intensity is of the order of 5 GW/cm^2. The high laser intensity laser pulse is absorbed in the absorbent coating which gets ablated and forms plasma. The high-temperature plasma tends to expand; however, this is constrained by the water layer. In this process, a pulse of very high pressure is developed at the surface. The pulse of pressure propagates as a shock wave deep into the material and generates compressive residual stresses. The basic arrangement for the LSP process is schematically shown in Figure 15.43. The use of thin transparent water layer increases the shock wave intensity propagating into the metal by up to two orders of magnitude, as compared to plasmas generated in a vacuum where it can expand freely (Montross et al. 2002).

The use of laser absorbent sacrificial coatings is also found to increase the shock wave intensity in addition to protecting the metal surface from laser ablation and melting. Among the absorbent coatings, commercially available black paint or tape has been found to be the most practical and effective, as compared to other coating materials.

The lasers that have been used for LSP are mainly the Q-switched Nd:Glass and Nd:YAG solid-state lasers operating at fundamental frequency or higher (2nd) harmonics; however, high-power pulsed CO$_2$ laser and excimer lasers can be also used. The typical laser pulse energy of solid-state lasers is in 0.1–100 J energy in 10–50 ns pulse duration operating at a few Hz repetition frequencies. Since the laser energy is deposited onto the absorbing coating in a very short time the diffusion of thermal energy away from the interaction zone is limited to a couple of microns and is preferred to be less than the thickness of the absorbent coating to maintain protection of the workpiece surface. The hydrodynamic expansion of the hot plasma in the confined region between the metal target and the transparent overlay creates a high-amplitude short-duration pressure pulse. This critical pressure P_{sw} is related to the density of the material (ρ) and its elastic wave sound velocity (v), where $P_{sw} = \rho v^2$, and the maximum pressure generated is of magnitude of the order of 10^3–10^4 atmosphere.

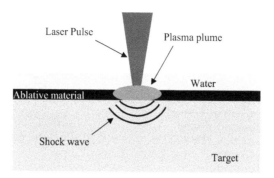

Figure 15.43   Schematic of laser shock peening process.

A portion of this energy propagates as a shock wave into the metal. When the pressure of the shock wave exceeds the dynamic yield strength of the material, plastic deformation occurs which consequently modifies the near-surface microstructure and properties. Compared to the conventional shot peening, LSP can generate much higher residual compressive stresses and up to higher depths (1–2 mm) from the irradiated surface. The surface roughness in LSP is also significantly less than that in conventional shot peening. Additionally, there is no geometrical restriction of the area for LSP to be implemented as long as it is at the line of sight.

LSP has been successfully employed to enhance the fatigue life and strength of turbine blades, weld joints in various metals and alloys, include different types of steel, and alloys of Mg, Al, Ti, and Ni, etc. (Brar et al. 2000). Sundar et al. (2014) reported significant enhancement of surface compressive stress (−260 to −390 MPa) which can enhance fatigue resistance of turbine blades with an in-house developed laser peening with Nd:YAG laser pulses of 2.5 J energy per 7 ns pulse duration. Figure 15.44 presents a photograph of laser peening of the root of the turbine blade in progress, and surface profiles of the residual stress on laser-peened and un-peened regions of the root region of the turbine blade.

Significant reduction of fatigue crack propagation in aluminum alloy specimens have been reported (Kashaev et al. 2017; Peyre et al. 1996). Figure 15.45 (Brar et al. 2000) shows the fatigue crack growth behavior of as received, shot peened, and laser-peened specimens. Laser peening

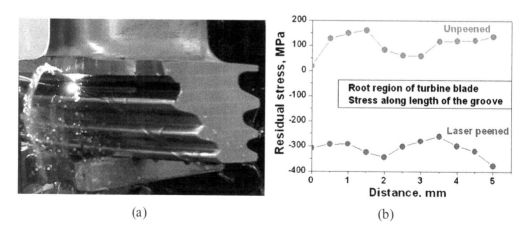

(a)   (b)

Figure 15.44   (a) Laser shock peening of the root of the turbine blade in progress and (b) surface profiles of the residual stress on laser-peened and un-peened regions of the root region of the turbine blade. (Reprinted with permission from *Pramana Journal of Physics*, Springer 2014 82:347–351.)

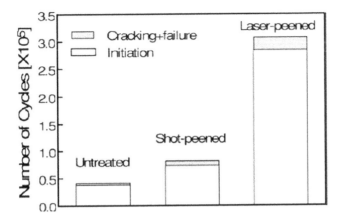

Figure 15.45 Comparison of initiation (crack development to a specified size, and usually the detectable limit) and later cracking stages at σ_{max} = 260 MPa for crack detection tests on 7075-T7351 aluminum alloy. (Reprinted with permission from *Materials Science and Engineering A*, Elsevier 1996 210:102–113.)

dramatically improved the fatigue life with clear differences in the early and later stages of crack growth. The difference in behavior between laser peening and shot peening is attributed to surface embrittlement and surface roughening due to the shot peening process which creates sites at which cracks develop rapidly and tends to reduce the beneficial effects of the compressive residual stresses.

Sano et al. (2006) have reported laser peening of SUS304 (Type 304) and SUS316L (Type 316L) austenitic stainless steels without protective coating (LPwPC) with a Q-switched and frequency-doubled Nd:YAG laser. In the case of no protective coating on the surface, the residual stress is introduced by the thermo-mechanical effect. The compressive stress introduced by the laser peening without protective coating in steel completely prohibited the initiation of SCC and the propagation of small pre-cracks on SUS304 in an environment that accelerates SCC and enhanced the fatigue strength of SUS316L. More recently, Lu reported improvement in radiation withstanding capacity along with improvement of SCC resistance of austenitic stainless steels by LSP (Lu et al. 2016). The significant reduction in SCC susceptibility by laser peening has been demonstrated through actual applications as preventive maintenance against SCC in operating nuclear power reactors since long (Zhu et al. 2012). In nuclear industry, one of the other important applications of LSP is to enhance the fatigue resistance of weld joints of storage canister's lids of radioactive nuclear burn fuel waste for its safe storage for 100–1,000 s of years till radioactive decays down to safe level.

More recently, Sathyajith et al. (2013) and Karthik and Swaroop (2016) reported LSP of Ni-based super alloy (Inconel 600) without an absorbent coating at the surface using a Q-switched Nd:YAG laser operating at 1,064 nm wavelength. They studied the influence of Laser Peening without Protective Coating (LPwPC) on residual stresses, hardness, microstructure, surface roughness, and topography of Inconel 600 specimens. With optimized LPwPC condition maximum compressive residual stresses of nearly 672 MPa and 1.4 mm depth of work-hardening of about 1,400 mm were reported. The average roughness was within 1 μm and no grain refinement was observed. There was a 10^6 fold decrease in corrosion rate of LPwC specimen compared to that of untreated condition. Recently, Tani et al. (2011) have reported laser peening of a pre-warmed workpiece surface, combining the thermal effect of the pre-heated surface and the mechanical phenomenon of the recoil shock pressure and the dynamic aging of the surface microstructure has been obtained. Precipitates surrounded by dense dislocation together with residual stress considerable increase the mechanical properties of the workpiece. Kamkarrad et al. (2014) and more recently Caralapatti and

Narayanswamy (2017) reported a significant improvement in biocompatibility and corrosion resistance in commercially pure Mg by high repetition LSP.

The main advantages of LSP over conventional shot peening which make LSP attractive are less surface roughness, no embedded particles on the treated surface, less environment pollution as there is no material to collect, grade, clean and recycle, and no contamination created while treating radioactive material.

15.6 WATER-ASSISTED AND UNDERWATER LASER PROCESSING

Lasers have been utilized along with water assist also and in underwater environment. Water-assisted laser processing methods depend on the physical phenomena like transmission and absorption of light in water, vaporization of water, formation of bubble, its growth and collapse, shock wave formation and propagation, scattering of laser by water vapor formed at the processing zone, convective cooling of workpiece, etc. For most of the applications of water-assisted material processing like cutting, surface hardening, welding, cleaning, shock peening, minimum absorption of laser radiation in water is desirable. However, for biomedical applications, high absorption is desired for the ablation of water-containing materials, such as biological tissues.

Absorption and transmission characteristics of laser beam in water depend on the wavelength of laser radiation. Water is most transparent in green light region, at copper vapor laser (511 nm) and frequency-doubled Nd:YAG laser (532 nm) wavelengths. For high-absorption, Raman-shifted Nd:YAG lasers (2,900 nm), Er:YAG lasers (2,940 nm) and Er:YSGG lasers (2,790 nm) are often the best choice because of the high absorption of their beam in water and the possibility to deliver the beam through low-loss optical fiber. However, for underwater laser cutting of thick materials the commonly used laser sources are high-power CO_2 laser ($\lambda = 10.6$ µm), Nd:YAG laser ($\lambda = 1.06$ µm) and Yb-Fiber ($\lambda = 1.07$ µm). The absorption coefficient of water at 10.6 µm and 1.06/1.07 µm wavelengths are ~1,000 and ~0.135 cm^{-1}, respectively; therefore, CO_2 laser beam is readily absorbed within ~10 µm of water column length, while Nd:YAG/fiber laser beam travels up to ~75 mm before it gets attenuated by 67% (Curcio and Petty 1951; Kruusing 2004; Mullick et al. 2011).

Different configurations of underwater and water-assisted laser cutting along with the transmission of beam through water have been reported and these are schematically shown in Figures 15.46–15.49 (Chida et al. 2003; Matsumoto et al. 1992; Muller et al. 1992; Mullick et al. 2013; Richerzhagen et al. 2004; Tsai and Li 2009).

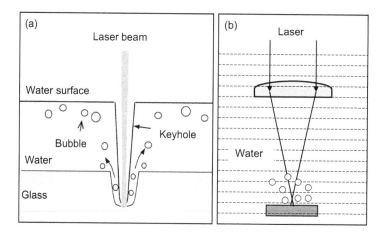

Figure 15.46 Transmission of different wavelengths, (a) 10.6 micron and (b) 1.06 micron, through water, in case of underwater laser cutting without assist fluid (Matsumoto et al., 1992; Muller et al., 1992).

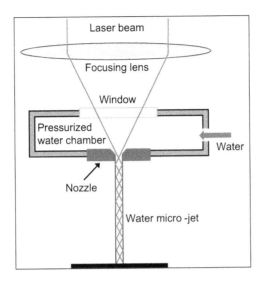

Figure 15.47 Schematic of water micro-jet guided laser cutting (Richerzhagen et al. 2004).

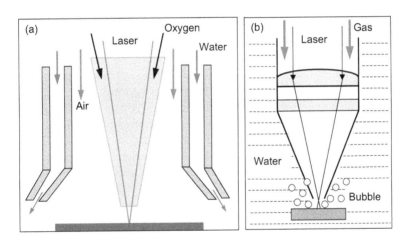

Figure 15.48 Schematic of gas-assisted underwater laser cutting (Chida et al. 2003; Matsumoto et al. 1992).

In one case, laser is directly transmitted through water (Figure 15.46), where the propagation of laser beam depends on its absorption in water. In the case of CO_2 laser, which is readily absorbed in water, forms a conical keyhole, as well as vapor bubbles (Figure 15.46a). Laser beam reaches the workpiece through the keyhole. However, in the case of Nd^+-based solid-state lasers (about 1 μm wavelength), or Yb–Fiber laser, the laser beam may be transmitted through water for a longer distance with small absorption (Figure 15.46b). Another type of water-assisted laser cutting is water micro-jet guided laser cutting (Figure 15.47), where laser beam is focused into a very fine water-jet of diameter in the range of 25–150 μm which guides and transmits the beam through multiple internal reflections (Richerzhagen et al. 2004). This technique enhances the working distance to about 50 mm; however, this is limited to cutting thin sheets up to a few mm thickness. In gas-assisted underwater laser processing, a high-pressure gas jet applied along with the laser beam creates a dry condition around the processing zone (Figure 15.48). Recently, water jet-assisted underwater laser cutting technique is developed (Figure 15.49).

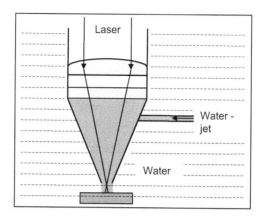

Figure 15.49   Water-jet-assisted underwater laser cutting (Mullick et al. 2013).

Since the underwater cutting provides effective cooling of the workpiece during laser processing, this has been exploited to cut and drill holes in brittle materials such as Si wafer, glass, ceramic
tiles, etc. (Tsai and Li 2009; Yan et al. 2011). Tsai and Li (2009) reported drilling and cutting of
LCD glass and alumina substrates kept 1 mm below water surface using a CO_2 laser and found
much better quality than that in air. Yan et al. (2011) reported underwater laser machining of deep
cavities in alumina with CO_2 laser. Lu et al. (2004) and Nath et al. (2010) studied underwater laser
drilling in metal sheets with pulsed Nd:YAG laser and reported significant increase in recoil pressure of plasma due to water confinement, high impact force generated by cavitation bubble collapse,
elimination of re-solidification around the hole surface and the effect of Marangoni force induced
by surface tension gradient in melt pool. Muhammad et al. (2010) and Muhammad et al. (2012b)
reported improvement in surface quality of steel and nitinol tubes by fiber laser and ultrafast laser
in presence of water. In water-assisted laser cutting the surface roughness, HAZ, dross and debris
on back surface of the tube were completely eliminated, therefore, the process could be exploited
for fabricating cardiovascular stents (Figure 15.50) (Muhammad et al. 2010)

Madhukar et al. 2016b compared the cut surface quality of conventional gas-jet-assisted and
water-jet-assisted laser cutting of mild steel and titanium sheets and reported that while in the case of
mild steel O_2-assisted and water-jet-assisted laser cutting produced similar quality, in the case of Ti
water-jet-assisted laser cutting produced better quality of cut surface. Several authors (Das and Saha
2010; Jiao et al. 2011; Madhukar et al. 2016a; Kaakkunen et al. 2011; Tangwarodomnukun et al. 2012;

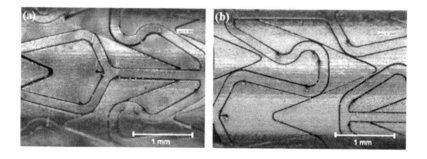

Figure 15.50   Thermal discoloration in stent cutting before removing the excess material: (a) dry cutting and
(b) wet cutting. (Reprinted with permission from *Journal of Materials Processing Technology*,
Elsevier 2010 210:2261–2267.)

Figure 15.51   Comparison of laser drill quality in ceramic material for (a) drilling in air, and (b) underwater drilling. (Reprinted with permission from *Journal of Materials Processing Technology*, Elsevier 2009 209:2838–2846.)

Tsai and Li 2009) reported water-assisted/underwater laser machining of brittle substrates like glass, Si wafer, and ceramics and observed less HAZ, reduced tendency of micro-crack formation, less debris, reduced taper and better surface topology. Superiority of water-assisted laser cutting/drilling of ceramic material over conventional laser processing could be seen in a comparison presented in Figure 15.51 (Tsai and Li 2009).

In the case of water-assisted machining of Si, the HAZ and micro-cracks were significantly reduced compared to conventional laser machining. Figure 15.52 presents a comparison of Si grooving by both these methods (Madhukar et al. 2016a). Severe cracks in the machined bottom groove surface in the case of gas-assisted laser grooving are apparent, while in water-assisted grooving such crack is completely eliminated.

In the case of laser paint striping also, water-assisted processing produced much cleaner surface without any trace of residual ashes, while the ashes could not be completely removed in gas-assisted laser paint removal process and some post cleaning operation was needed (Madhukar et al. 2013a, Madhukar et al. 2013b).

Another application of underwater laser cutting is in nuclear power plants for the cutting of burned fuel rods. Nuclear burnt fuel rods are highly radioactive and are generally stored under water to allow their radioactivity to decay down and minimize environmental contamination. They need to cut into small pieces for their easy storage. High-power CO_2, Nd:YAG, and fiber lasers have been pressed for this task. Usually an assist gas is supplied to displace water from the processing zone to create a dry zone at the workpiece surface, a condition similar to that of processing in air

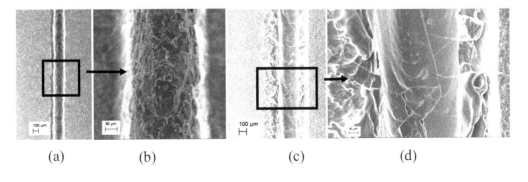

(a)   (b)   (c)   (d)

Figure 15.52   SEM images of a typical Si groove machined by water-assisted fiber laser (a) top surface (b) bottom surface of groove and by gas-assisted fiber laser (c) top surface (d) bottom surface of groove. (Reprinted with permission from *Journal of Materials Processing Technology*, Elsevier 2016 227:200–215.)

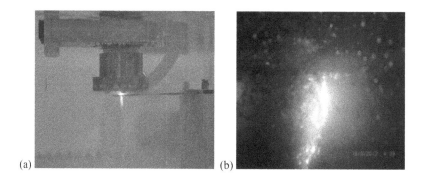

(a)   (b)

Figure 15.53 Photograph of (a) water-jet-assisted underwater laser cutting and (b) gas-assisted underwater laser cutting of 1 mm thick stainless steel sheet. (Reprinted with permission from *International Journal of Machine Tools & Manufacture*, Elsevier 2013 68:48–55.)

(Figure 15.48). The assist gas also provides the shear force for melt-ejection from the cutting zone. Chida et al. (2003) reported underwater cutting technology of thick stainless steel with YAG laser in order to cut used control rod. Preliminary tests were performed on flat plate test-pieces with 4 kW laser and creating a local dry area between nozzle and test-piece by blowing high-pressure air through the nozzle. Stainless steel plates up to 14 mm were cut with gas assist. Okado et al. (1999) reported underwater cutting of SS plate by chemical oxygen–iodine laser at 1–7 kW power with oxygen assist. Jain et al. (2010) reported gas-assisted underwater laser cutting of 4.2 mm thick zircaloy pressure tubes and up to 6-mm-thick steel sheets using pulsed Nd:YAG laser of 250 W average power with nitrogen and oxygen as assist gas. Same concept of producing a dry space for laser welding and cladding for repairing underwater nuclear reactor components has been developed (Morita et al. 2006; Miura et al. 2009; Yamashita et al. 2001).

A major drawback of gas-assisted underwater cutting of radioactive materials is the generation of gas bubbles and aerosols causing environmental radioactive contamination. In order to mitigate this drawback, Mullick et al., (2013, 2016a, 2016c) developed water-jet-assisted underwater laser cutting process replacing gas-jet by water-jet for providing shear force to remove molten material from the kerf (Figure 15.49). A comparison of gas-assisted and water-jet-assisted underwater laser cutting is presented in Figure 15.53 (Mullick et al. 2013). Use of water minimizes the formation of gas bubble and aerosols, and thereby the environmental pollution. However, this technique has comparatively lower efficiency than the previous one due to the absorption and scattering losses of laser power in water. More recently, underwater laser welding and surface hardening by creating laser-induced direct channel also have attracted attention of researchers (Jin et al. 2015; Guo et al. 2017).

The advantages of underwater laser processing include improved quality, reduction, or elimination of HAZ, spatter, and debris formation, elimination of cracks in brittle materials, thereby reduction in rejection rate, minimization of fume, and radioactive contamination in the case of water-jet-assisted underwater cutting of nuclear components.

15.7 CONCLUDING REMARKS

As apparent from the above presentation, in the last couple of decades, many interesting and novel developments have been witnessed in the area of laser material processing, especially in welding of dissimilar materials, surface engineering, AM, underwater processing, and diffraction-unlimited micro- and nano-fabrications. The trend in the development of laser systems has been to improve the beam quality of high-power fiber and diode lasers at multi-kilowatt level and it is expected that this trend will continue. In regard to ultrafast lasers more users friendly all—optical fiber-based systems are becoming more popular over Ti–sapphire-based systems. More efficient and relatively long-life high energy Q-switched fiber laser in ns pulse duration level could edge the relatively less-efficient flash lamp pumped Nd:YAG lasers out in applications like shock peening. Industrial application of lasers in surface modification and welding has witnessed development of laser beam-shaping system, online process monitoring, and feedback control to facilitate reproducible processing of any complex shape engineering components. For the industrial readiness of laser AM of complex functional parts with different materials including composites online geometry and thermal history monitoring and feedback control for ensuring repeatability and reproducibility in multilayer deposition is much needed. Therefore, research focus is expected to be also on the online temperature monitoring and correlating with microstructure and mechanical and metallurgical properties, and developing feedback control. In addition, research on laser material processing in controlled environment, hybridization with other conventional and nonconventional machining process and by manipulating the spatial and temporal beam characteristics to furthering the processing quality, efficiency, productivity, cost-effectiveness; minimizing the use of costly materials by selective area processing, post-processing operations and scarps, and extending the present processing boundaries will continue unabated. Thus, lasers will be increasingly exploited as a powerful tool in manufacturing industry in many interesting ways and the attributes highlighted for various laser material processing modalities make the manufacturing "green".

REFERENCES

Akman, E., Demir, A., Canel, T., Sınmazcelik, T. 2009. Laser welding of Ti6Al4V titanium alloys. *Journal of Materials Processing Technology* 209: 3705–3713.

Alcock, J.A., Baufeld, B. 2017. Diode laser welding of stainless steel 304L. *Journal of Materials Processing Technology* 240: 138–144.

Andrew, T.L., Tsai, H.Y., Menon, R. 2009. Confining light to deep sub-wavelength dimensions to enable optical nanopatterning. *Science* 324: 917–921. doi:10.1126/science.1167704.

Anon, 1966. Laser punches holes in diamond wire-drawing dies. *Laser Focus* 2: 4–7.

Ashby, M.F., Easterling, K.E. 1984. The transformation hardening of steel surfaces by laser beams-I, hypo-eutectoid steels. *Acta Metallurgica* 32: 1935–1948.

Asnafi, N., Andersson, R., Persson, M., Liljengren, M. 2016. Tailored boron steel sheet component properties by selective laser heat treatment. *IOP Conference Series: Materials Science and Engineering* 159: 012023. doi:10.1088/1757-899X/159/1/012023.

Bandyopadhyay, K., Panda, S.K., Saha, P. 2016. Optimization of fiber laser welding of DP980 steels using RSM to improve weld properties for formability. *Journal of Materials Engineering and Performance* 25: 2462–2477.

Bartkowski, D., Bartkowska, A. 2017. Wear resistance in the soil of Stellite-6/WC coatings produced using laser cladding method. *International Journal of Refractory Metals and Hard Materials* 64: 20–26.

Bi, G., Gasser, A., Wissenbach, K., Drenker, A., Poprawe, P. 2006a. Identification and qualification of temperature signal for monitoring and control in laser cladding. *Optics and Lasers in Engineering* 44: 1348–1359.

Bi, G., Gasser, A., Wissenbach, K., Drenker, A., Poprawe, R. 2006b. Characterization of the process control for the direct laser metallic powder deposition. *Surface and Coatings Technology* 201: 2676–2683.

Bi, G., Schurmann, B., Gasser, A., Wissenbach, K., Poprawe, R. 2007. Development and qualification of a novel laser-cladding head with integrated sensors. *International Journal of Machine Tools and Manufacture* 47: 555–561.

Bonß, S. 2010. Laser transformation hardening of steel. In *Advances in Laser Materials Processing*, 1st Ed., eds. J.R. Lawrence, J. Pou, D.K.Y. Low, E. Toyserkani, 291–326. Elsevier Woodhead Publishing, Cambridge, UK, ISBN 9780081012529.

Brar, N.S., Hopkins, A., Laber, M.W. 2000. Laser shock peening of titanium 6-4 alloy. *AIP Conference Proceedings* 505: 435–438. doi:10.1063/1.1303510.

Cao, X., Wanjara, P., Huang, J., Munro, C., Nolting, A. 2011. Hybrid fiber laser—Arc welding of thick section high strength low alloy steel. *Materials & Design* 32: 3399–3413.

Caralapatti, V.K., Narayanswamy, S. 2017. Effect of high repetition laser shock peening on biocompatibility and corrosion resistance of magnesium. *Optics and Laser Technology* 88: 75–84.

Carroll, P.A., Brown, P., Ng, G., Scudamore, R., Pinkerton, A.J., Syed, W., Sezer, H., Li, L., Allen, J. 2006. The effect of powder recycling in direct metal laser deposition on powder and manufactured part characteristics. *Proceedings of the AVT-139 Specialists Meeting on Cost Effective Manufacture via Net Shape Processing*, pp. 1–10, Amsterdam.

Casalino, G., Mortello, M. 2016. Modeling and experimental analysis of fiber laser offset welding of Al-Ti butt joints. *The International Journal of Advanced Manufacturing Technology* 83: 89–98.

Casalino, G., Guglielmi, P., Lorusso, V.D., Mortello, M., Peyre, P., Sorgente, D. 2017. Laser offset welding of AZ31B magnesium alloy to 316 stainless steel. *Journal of Materials Processing Technology* 242: 49–59.

Chakraborty, S.S., Maji, K., Racherla, V., Nath, A.K. 2015a. Investigation on laser forming of stainless steel sheets under coupling mechanism. *Optics & Laser Technology* 71: 29–44.

Chakraborty, S.S., More, H., Racherla, V., Nath, A.K., 2015b. Modification of bent angle of mechanically formed stainless steel sheets by laser forming. *Journal of Materials Processing Technology* 222: 128–141.

Chakraborty, S.S., More, H., Nath, A.K., 2016. Laser forming of a bowl shaped surface with a stationary laser beam. *Optics and Lasers in Engineering* 77: 126–136.

Chen, H.C., Bi, G., Lee, B.Y., Cheng, C.K. 2016. Laser welding of CP Ti to stainless steel with different temporal pulse shapes. *Journal of Materials Processing Technology* 231: 58–65.

Cheng, J.Y., Wei, C.W., Hsu, K.H., Young, T.H. 2004. Direct-write laser micromachining and universal surface modification of PMMA for device development. *Sensors and Actuators B* 99: 186–196.

Chichkov, B.N., Momma, C., Nolte, S., von Alvensleben, F., Tünnermann, A. 1996. Femtosecond, picosecond and nanosecond laser ablation of solids. *Applied Physics A* 63: 109–115.

Chida, I., Okazaki, K., Shima, S., Kurihara, K., Yuguchi, Y., Sato, L. 2003. Underwater cutting technology of thick stainless steel with YAG laser. *Proceedings of SPIE 4831*, Washington, USA.

Coffey, V.C. 2014. Ultrafast & ultrashort: Some recent advances in pulsed lasers. *Optics & Photonics News* May 30–35. www.osa-opn.org/home/articles/volume_25/may_2014/.

Costa Rodrigues, G., Pencinovsky, J., Cuypers, M., Duflou, J.R. 2014a. Theoretical and experimental aspects of laser cutting with a direct diode laser. *Optics and Lasers in Engineering* 61: 31–38.

Costa Rodrigues, G., Vanhove, H., Duflou, J.R. 2014b. Direct diode lasers for industrial laser cutting: A performance comparison with conventional fiber and CO_2 technologies. *Physics Procedia* 56: 901–908.

Curcio, J.A., Petty, C.C. 1951. The near infrared absorption spectrum of liquid water. *Journal of the Optical Society of America* 41: 302–304.

Dahotre, N.B., Harimkar, S. 2008. *Laser Fabrication and Machining of Materials.* Springer Science & Business Media, Boston, MA.

Das, A.K., Saha, P. 2010. Excimer laser micromachining of silicon in air and water medium. *International Journal of Manufacturing Technology and Management* 21: 42–53.

Dausinger, F., Hugel, H., Konov, V. 2003. Micro-machining with ultrashort laser pulses: From basic understanding to technical applications. *Proceedings of SPIE* 5147: 106. doi:10.1117/12.537496.

Dearden, G., Edwardson, S.P., Abed, E., Watkins, K.G., 2006, September. Laser forming for the correction of distortion and design shape in aluminium structures. In *AILU Industrial Technology Programme, Photon'06 International Conference on Optics and Photonics*, Manchester.

Doubenskaia, M., Bertrand, P., Smurov, I. 2004. Optical monitoring of Nd:YAG laser cladding. *Thin Solid Films* 453–454: 477–485.

Doubenskaia, M., Bertrand, P., Smurov, I. 2006. Pyrometry in laser surface treatment. *Surface and Coatings Technology* 201: 1955–1961.

Dyer, P.E., Waldeck, I., Roberts, G.C. 1997. Fine-hole drilling in Upilex polyimide and glass by TEA laser ablation. *Journal of Physics D: Applied Physics* 30:L19–L21.

Edwards, P., Ramulu, M. 2014. Fatigue performance evaluation of selective laser melted Ti–6Al–4V. *Material Science and Engineering A* 598: 327–337.

Edwardson, S.P., Watkins, K.G., Dearden, G., Magee, J., 2001. 3D laser forming of saddle shapes. *Proceedings of Laser Assisted Net Shaping* 559–568.

Edwardson, S.P., Abed, E., French, P., Dearden, G., Watkins, K.G., McBride, R., Hand, D.P., Jones, J.D.C., Moore, A.J. 2005. Developments towards controlled three-dimensional laser forming of continuous surfaces. *Journal of Laser Applications* 17: 247–255.

Edwardson, S.P., Abed, E., Bartkowiak, K., Dearden, G., Watkins, K.G., 2006. Geometrical influences on multi-pass laser forming. *Journal of Physics D: Applied Physics* 39: 382.

Eideh, A., Dixit, U.S., Echempati, R. 2015. A simple analytical model of laser bending process. In *Lasers Based Manufacturing, Topics in Mining, Metallurgy and Materials Engineering*, ed. S.N. Joshi, U.S. Dixit, 1–15. Springer, New Delhi. doi:10.1007/978-81-322-2352-8_1.

Eliseev, P.G., Krokhin, O.N., Zavestovskaya, I.N. 2005. Nonlinear absorption mechanism in ablation of transparent materials by high power and ultrashort laser pulses. *Applied Surface Science* 248: 313–315.

Ezazi, M.A., Yusof, F., Sarhan, A.A.D., Shukor, M.H.A., Fadzil, M. 2015. Employment of fiber laser technology to weld austenitic stainless steel 304 l with aluminum alloy 5083 using pre-placed activating flux. *Materials & Design* 87: 105–123.

Fourkas, J.T. 2010. Nanoscale photolithography with visible light. *Journal of Physical Chemistry Letters* 1: 1221–1227.

French, P.W., Naeem, M., Watkins, K.G. 2003. Laser percussion drilling of aerospace material using a 10 kW peak power laser using a 400 mm optical fibre delivery system (Paper no. 503). *22nd International Congress on Applications of Lasers & Electro-Optics (ICALEO 2003)*, Jacksonville, FL, October 13–16, Laser Institute of America, Vol. 95, Publication No. 595, 2003, ISBN 0-912035-75-7.

Geiger, M., Merklein, M., Pitz, M., 2004. Laser and forming technology—An idea and the way of implementation. *Journal of Materials Processing Technology* 151: 3–11.

Gisario, A., Barletta, M., Conti, C., Guarino, S., 2011. Springback control in sheet metal bending by laser-assisted bending: Experimental analysis, empirical and neural network modelling. *Optics and Lasers in Engineering* 49: 1372–1383.

Gopinath, M., Karmakar, D.P., Nath, A.K. 2017a. Online monitoring of thermo-cycles and its correlation with microstructure in laser cladding of nickel based super alloy. *Optics and Lasers in Engineering* 88: 139–152.

Gopinath, M., Karmakar, D.P., Nath, A.K. 2017b. Monitoring and assessment of tungsten carbide wettability in laser cladded metal matrix composite coating using an IR pyrometer. *Journal of Alloys and Compounds* 714: 514–521.

Gopinath, M., Karmakar, D.P., Nath, A.K. 2017c. Online assessment of TiC decomposition in laser cladding of metal matrix composite coating. *Materials & Design* 121: 310–320.

Gower, M.C. 2000. Industrial applications of laser micromachining. *Optics Express* 7: 56–67.

Grevey, D., Vignal, V., Bendaoud, I., Erazmus-Vignal, P., Tomashchuk, I., Daloz, D., Sallamand, P. 2015. Microstructural and micro-electrochemical study of a tantalum–titanium weld interface. *Materials & Design* 87: 974–985.

Guglielmotti, A., Quadrini, F., Squeo, E.A., Tagliaferri, V., 2009. Laser bending of aluminum foam sandwich panels. *Advanced Engineering Materials* 11: 902–906.

Guo, N., Xing, X., Zhao, H., Tan, C., Feng, J., Deng, Z., 2017. Effect of water depth on weld quality and welding process in underwater fiber laser welding. *Materials & Design* 115: 112–120.

Hennige, T., 2000. Development of irradiation strategies for 3D-laser forming. *Journal of Materials Processing Technology* 103: 102–108.

Hennige, T., Holzer, S., Vollertsen, F., Geiger, M., 1997. On the working accuracy of laser bending. *Journal of Materials Processing Technology* 71: 422–432.

Hong, K.M., Shin, Y.C. 2017. Prospects of laser welding technology in the automotive industry: A review. *Journal of Materials Processing Technology* 245: 46–69.

Hong, M., Lim, C.S., Zhou, Y., Tan, L.S., Shi, L., Chong, T.C. 2008. Surface nano-fabrication by laser precision engineering. *The Review of Laser Engineering* Supplemental Volume 36: 1184–1187.

Hooreweder, B.V., Apers, Y., Lietaert, K., Kruth, J. 2017. Improving the fatigue performance of porous metallic biomaterials produced by selective laser melting. *Acta Biomaterialia* 47: 193–202.

Hu, J., Dang, D., Shen, H., Zhang, Z. 2012. A finite element model using multi-layered shell element in laser forming. *Optics & Laser Technology* 44: 1148–1155.

Huang, S.H., Liu, P., Mokasdar, A., Hou, L. 2013. Additive manufacturing and its societal impact: A literature review. *The International Journal of Advanced Manufacturing Technology* 67: 1191–1203.

Hussein, F.I., Akman, E., Oztoprak, B.G., Gunes, M., Gundogdu, O., Kacar, E., Hajim, K.I., Demir, A. 2013. Evaluation of PMMA joining to stainless steel 304 using pulsed Nd:YAGlaser. *Optics & Laser Technology* 49: 143–152.

Huynh, L., Rotella, J., Sangid, M.D. 2016. Fatigue behavior of IN718 microtrusses produced via additive manufacturing. *Materials & Design* 105: 278–289.

Hwang, D., Ryu, S.G., Misra, N., Jeon, H., Grigoropoulos, C.P. 2009. Nanoscale laser processing and diagnostics. *Applied Physics A* 96: 289–306.

Ignatiev, M., Smurov, I., Flamant, G., Senchenko, V. 1996. Surface temperature measurements during pulsed metallic and ceramic materials laser action on metallic and ceramic materials. *Applied Surface Science* 96–98: 505–512.

Ignatiev, M., Smurov, I., Flamant, G., Senchenko, V. 1997. Two-dimensional resolution pyrometer for real-time monitoring of temperature image in laser materials processing. *Applied Surface Science* 109/110: 498–508.

Iwasaki, K., Ohkawa, S., Uo, M., Akasaka, T., Watari, F. 2004. Laser welding of titanium and dental precious alloys. *Materials Transactions* 45: 140–1146.

Jain, R.K., Agrawal, D.K., Vishwakarma, S.C., Choubey, A.K., Upadhyaya, B.N., Oak, S.M. 2010. Development of underwater laser cutting technique for steel and zircaloy for nuclear applications. *Pramana* 75: 1253–1258.

Jiang, L., Tsai, H.L. 2003. Femtosecond laser ablation: Challenges and opportunities, 1–15. http://galilei.chem.psu.edu/wtatiana/PUBLIS/My_papers/prlpaper/femtosecond_rev_NSFWorkshop.pdf.

Jiao, L.S., Ng, E.Y.K., Wee, L.M., Zheng, H.Y. 2011. The effect of assist liquid on the hole taper improvement in femtosecond laser percussion drilling. *Physics Procedia* 24: 1021–1036.

Jin, B., Li, M., Hwang, T. W., Moon, Y.H. 2015. Feasibility studies on underwater laser surface hardening process. *Advances in Materials Science and Engineering*, Article ID 845273, 6 pages. doi:10.1155/2015/845273.

Kaakkunen, J.J.J., Silvennoinen, M., Paivasaari, K., Vahimaa, P. 2011. Water-assisted femtosecond laser pulse ablation of high aspect ratio holes. *Physics Procedia* 12: 89–93.

Kahlen, F.J., Kar, A. 2001. Tensile strengths for laser-fabricated parts and similarity parameters for rapid manufacturing. *Journal of Manufacturing Science and Engineering* 123: 38–44.

Kamkarrad, H., Narayanswamy, S., Tao, X.S. 2014. Feasibility study of high-repetition rate laser shock peening of biodegradable magnesium alloys. *The International Journal of Advanced Manufacturing Technology* 74: 1237–1245.

Kant, R., Gupta, A., Bhattacharya, S. 2015. Studies on CO_2 laser micromachining on PMMA to fabricate micro channel for microfluidic applications. In *Lasers Based Manufacturing, Topics in Mining, Metallurgy and Materials Engineering*, ed. S.N. Joshi, U.S. Dixit. Springer, New Delhi. doi:10.1007/978-81-322-2352-8_13.

Kaplan, A.F.H., Frostevarg, J., Ilar, T., Bang, H.S., Bang, H.S. 2015. Evolution of a laser hybrid welding map. *Physics Procedia* 78: 2–13.

Karmakar, D.P., Gopinath, M., Nath, A.K. 2017. Effect of tempering temperature on hardness and microstructure of laser surface remelted AISI H13 tool steel. *Proceedings of the ASME 2017 12th International Manufacturing Science and Engineering Conference*, Paper ID MSEC2017-3014, Los Angeles, CA.

Karthik, D., Swaroop, S. 2016. Influence of laser peening on phase transformation and corrosion resistance of AISI 321 steel. *Journal of Materials Engineering and Performance* 25: 2642–2650.

Kashaev, N., Ventzke, V., Horstmann, M., Chupakhin, S., Riekehr, S., Falck, R., Maawad, E., Staron, P., Schell, N., Huber, N. 2017. Effects of laser shock peening on the microstructure and fatigue crack propagation behaviour of thin AA2024 specimens. *International Journal of Fatigue* 98: 223–233.

Kaul, R., Mahajan, S., Kain, V., Ganesh, P., Chandra, K., Nath, A.K., Prasad, R.C. 2008. Laser surface treatment for enhancing intergranular corrosion resistance of AISI 304 stainless steel. *Corrosion* 64: 755–763.

Kawata, S., Sun, H.B., Tanaka, T., Takada, K. 2001. Finer features for functional microdevices. *Nature* 412: 697–698.

Kim, J., Na, S.J., 2009. 3D laser-forming strategies for sheet metal by geometrical information. *Optics & Laser Technology* 41: 843–852.

Klotzbach, U., Lasagni, A.F., Panzner, M., Franke, V. 2011. Laser micromachining in fabrication and characterization in the micro-nano range. In *Advanced Structured Materials*, eds. F.A. Lasagni, A.F. Lasagni, Vol. 10, 29–26. Springer-Verlag, Berlin and Heidelberg. doi:10.1007/978-3-642-17782-8_2.

Köhler, B., Biesenbach, J., Brand, T., Haag, M., Huke, S., Noeske, A., Seibold, G., Behringer, M., Luft, J. 2005. High-brightness high-power kW-system with tapered diode laser bars. *Proceedings of SPIE 5711, High-Power Diode Laser Technology and Applications III*. doi:10.1117/12.589044, Washington USA.

Kong, C.Y., Carroll, P.A., Brown, P., Scudamore, R.J. 2007. The effect of average powder particle size on deposition efficiency, deposit height and surface roughness in the direct metal laser deposition process, *14th International Conference on Joining of Materials 2007*, Helsingør, Denmark.

Kruusing, A. 2004. Underwater and water-assisted laser processing: Part 1—General features, steam cleaning and shock processing. *Optics and Lasers in Engineering* 41: 307–327.

Kujanpää, V. 2014. Thick-section laser and hybrid welding of austenitic stainless steels. *Physics Procedia* 56: 630–636.

Kumar, H., Ganesh, P., Kaul, R., Rao, B.T., Tiwari, P., Brajpuriya, R. 2006. Laser welding of 3 mm thick laser-cut AISI 304 stainless steel sheet. *Journal of Materials Engineering and Performance* 15: 23–31.

Kuryntsev, S.V., Gilmutdino, A.K. 2015. The effect of laser beam wobbling mode in welding process for structural steels. *The International Journal of Advanced Manufacturing Technology* 81: 1683–1691.

Kuryntsev, S.V., Morushkin, A.E., Gilmutdinov, A.K. 2017. Fiber laser welding of austenitic steel and commercially pure copper butt joint. *Optics and Lasers in Engineering* 90: 101–109.

Kusinski, J., Kac, S., Kopia, A., Radziszewska, A., Rozmus-górnikowska, M., Major, B., Major, L., Marczak, J., Lisiecki, A. 2012. Laser modification of the materials surface layer—A review paper. *Bulletin of the Polish Academy of Sciences, Technical Sciences* 60: 711–728. doi:10.2478/v10175-012-0083-9.

Larcombe, D. 2013. Fiber versus CO_2 laser cutting. Industrial laser solutions for manufacturing. www.industrial-lasers.com/articles/2013/11/fiber-versus-co2-laser-cutting.html.

Lawrence, J., Schmidt, M.J., Li, L., 2001. The forming of mild steel plates with a 2.5 kW high power diode laser. *International Journal of Machine Tools and Manufacture* 41: 967–977.

Leitz, K.H., Redlingshöfer, B., Reg, Y., Otto, A., Schmidt, M. 2011. Metal ablation with short and ultrashort laser pulses. *Physics Procedia* 12: 230–238.

Leuders, S., Thöne, M., Riemer, A., Niendorf, T., Tröster, T., Richard, H.A., Maier, H.J. 2013. On the mechanical behaviour of titanium alloy TiAl6V4 manufactured by selective laser melting: Fatigue resistance and crack growth performance. *International Journal of Fatigue* 48: 300–307.

Leunda, J., Soriano, C., Sanz, C., García Navas, V. 2011. Laser cladding of vanadium-carbide tool steels for die repair. *Physics Procedia* 12: 345–352.

Lewis, L.J., Perez, D. 2009. Laser ablation with short and ultrashort laser pulses: Basic mechanisms from molecular-dynamics simulations. *Applied Surface Science* 255: 5101–5106.

Lewis, S.R., Lewis, R., Fletcher, D.I. 2015. Assessment of laser cladding as an option for repairing/enhancing rails. *Wear* 330–331: 581–591.

Lewis, S.R., Fretwell-Smith, S., Goodwin, P.S., Smith, L., Lewis, R., Aslam, M., Fletcher, D.I., Murray, K., Lambert, R. 2016. Improving rail wear and RCF performance using laser cladding. *Wear* 366–367: 268–278.

Li, W., Yao, Y.L., 2000. Numerical and experimental study of strain rate effects in laser forming. *Journal of Manufacturing Science and Engineering* 122: 445–451.

Li, Y., Yang, H., Lin, X., Huang, W., Li, J., Zhou, Y. 2003. The influences of processing parameters on forming characterizations during laser rapid forming. *Materials Science and Engineering A* 360: 18–25.

Li, L., Diver, C., Atkinson, J., Giedl-Wagner, R., Helml, H.J. 2006. Sequential laser and EDM micro-drilling for next generation fuel injection nozzle manufacture. *Annals of the CIRP* 55: 1–4.

Li, L., Sobih, M., Crouse, P.L. 2007. Striation-free laser cutting of mild steel sheets. *CIRP Annals: Manufacturing Technology* 56: 193–196.

Li, L., Hong, M., Schmidt, M., Zhong, M., Malshe, A., Huis in'tVeld, B., Kovalenko, V. 2011. Laser nanomanufacturing-state of the art and challenges. *CIRP Annals: Manufacturing Technology* 60: 735–755.

Li, P., Warner, D.H., Fatemi, A., Phan, N. 2016. Critical assessment of the fatigue performance of additively manufactured Ti–6Al–4V and perspective for future research. *International Journal of Fatigue* 85: 130–143.

Liao, X., Brown, K.A., Schmucker, A.L., Liu, G., He, S., Shim, W., Mirkin, C.A. 2013. Desktop nanofabrication with massively multiplexed beam pen lithography. *Nature Communications* 4: 2103. doi:10.1038/ncomms3103.

Linde, D.v.d., Tinten, K.S., Bialkowski, J. 1997. Laser solid-interaction in the femtosecond time regime. *Applied Surface Science* 109/100: 1–10.

Liu, J. 2014. Fiber lasers: Ultrafast fiber lasers reach millijoule energies with PC and LMA fibers. www.laserfocusworld.com/articles/print/volume-50/issue-04/features/fiber-lasers-ultrafast-fiber-lasers-reach-millijoule-energies-with-pc-and-lma-fibers.html.

Liu, C., Yao, Y.L. 2005. FEM-based process design for laser forming of doubly curved shapes. *Journal of Manufacturing Processes* 7: 109–121.

Liu, X., Du, D., Mourou, G. 1997. Laser ablation and micromachining with ultrashort laser pulses. *IEEE Journal of Quantum Electronics* 33: 1706–1716.

Liu, J., Sun, S., Guan, Y., Ji, Z., 2010. Experimental study on negative laser bending process of steel foils. *Optics and Lasers in Engineering* 48: 83–88.

Liu, Z., Jiang, Q., Li, T., Dong, S., Yan, S., Zhang, H. 2016. Environmental benefits of remanufacturing: A case study of cylinder heads remanufactured through laser cladding. *Journal of Cleaner Production* 133: 1027–1033.

Liu, J., Yu, H., Chen, C., Weng, F., Dai, J. 2017. Research and development status of laser cladding on magnesium alloys: A review. *Optics and Lasers in Engineering* 93: 195–210.

Low, D.K.Y., Li, L., Byrd, P.J. 2001. The influence of temporal pulse train modulation during laser percussion drilling. *Optics and Lasers in Engineering* 35: 149–164.

Lu, Q. 2016. Improving radiation and stress corrosion cracking resistance of austenitic stainless steels by laser shock peening. Mechanical (and Materials) Engineering—Dissertations, Theses, and Student Research. Paper 97. http://digitalcommons.unl.edu/mechengdiss/97.

Lu, J., Xu, R.Q., Chen, X., Shen, Z.H., Ni, X.W. 2004. Mechanism of laser drilling of metal plates underwater. *Journal of Applied Physics* 95: 3890–3894.

Luo, Y., Ji, Z., Leu, M.C., Caudill, R. 1999. Environmental performance analysis of solid freeform fabrication processes. *IEEE International Symposium on Electronics and the Environment*, 1–6, IEEE, New York.

Madhukar, Y.K., Mullick, S., Shukla, D.K., Kumar, S., Nath, A.K. 2013a. Effect of laser operating mode in paint removal with a fiber laser. *Applied Surface Science* 264: 892–901.

Madhukar, Y.K., Mullick, S., Nath, A.K. 2013b. Development of a water-jet assisted laser paint removal process. *Applied Surface Science* 286: 192–205.

Madhukar, Y.K., Mullick, S., Nath, A.K. 2016a. A study on co-axial water-jet assisted fiber laser grooving of silicon. *Journal of Materials Processing Technology* 227: 200–215.

Madhukar, Y.K., Mullick, S., Nath, A.K. 2016b. An investigation on co-axial water-jet assisted fiber laser cutting of metal sheets. *Optics and Lasers in Engineering* 77: 203–218.

Magee, J., De Vin, L.J., 2002. Process planning for laser-assisted forming. *Journal of Materials Processing Technology* 120: 322–326.

Maji, K., Pratihar, D.K., Nath, A.K. 2012. Analysis and synthesis of laser forming process using neural networks and neuro-fuzzy inference system. *Soft Computing* 17: 849–865.

Maji, K., Pratihar, D.K., Nath, A.K. 2014. Laser forming of a dome shaped surface: Experimental investigations, statistical analysis and neural network modeling. *Optics and Lasers in Engineering* 53: 31–42.

Majumdar, J.D., Manna, I. 2011. Laser material processing. *International Materials Reviews* 56: 341–388.

Majumdar, J.D., Nath, A.K., Manna, I. 2010a. Studies on laser surface melting of tool steel-Part I: Surface characterization and it's electrochemical behavior. *Surface and Coatings Technology* 204: 1321–1325.

Majumdar, J.D., Nath, A.K., Manna, I. 2010b. Studies on laser surface melting of tool steel-Part II: Mechanical properties of the surface. *Surface and Coatings Technology* 204: 1326–1329.

Manivasagam, G., Dhinasekaran, D., Rajamanickam, A. 2010. Biomedical implants: Corrosion and its prevention—A review. *Recent Patents on Corrosion Science* 2: 40–54.

Masanta, M., Ganesh, P., Kaul, R., Nath, A.K., Choudhury, A.R. 2009. Development of a hard nano-structured multi-component ceramic coating by laser cladding. *Materials Science and Engineering: A* 508: 134–140.

Masanta, M., Shariff, S.M., Choudhury, A.R. 2010. Tribological behavior of TiB_2-TiC-Al_2O_3 composite coating synthesized by combined SHS and laser technology. *Surface & Coatings Technology* 204: 2527–2538.

Matsumoto, O., Sugihara, M., Miya, K. 1992. Underwater cutting of reactor core internals by CO_2 laser using local-dry-zone creating nozzle. *Journal of Nuclear Science and Technology* 29: 1074–1079.

McBride, R., Bardin, F., Gross, M., Hand, D.P., Jones, J.D.C., Moore, A.J., 2005. Modelling and calibration of bending strains for iterative laser forming. *Journal of Physics D: Applied Physics* 38: 4027.

McNally, C.A., Folkes, J., Pashby, I.R. 2004. Laser drilling of cooling holes in aeroengines: State of the art and future challenges. *Materials Science and Technology* 20: 805–813.

Meijer, J. 2004. Laser beam machining, state of the art and new opportunities. *Journal of Materials Processing Technology* 149: 2–17.

Meng, H., Liao, J., Zhou, Y., Zhang, Q. 2009. Laser micro-processing of cardiovascular stent with fiber laser cutting system. *Optics & Laser Technology* 41: 300–302.

Miotello, A., Kelly, R. 1999. Laser-induced phase explosion: New physical problems when a condensed phase approaches the thermodynamic critical temperature. *Applied Physics A* 69: 67–73.

Miura, T., Kono, W., Chida, I., Hino, T., Yamamoto, S., Yamamoto, S., Yoda, M., Ochiai, M. 2009. Development of multifunction laser welding head as maintenance technologies against stress corrosion cracking for nuclear power reactors. www.ndt.net/article/jrc-nde2009/papers/54.pdf.

Montross, C.S., Wei, T., Ye, L., Clark, G., Mai, Y.W. 2002. Laser shock processing and its effects on microstructure and properties of metal alloys: A review. *International Journal of Fatigue* 24: 1021–1036.

Morita, I., Owaki, K., Yamaoka, H., Kim. C.C. 2006. Study of underwater laser welding repair technology. *Welding in the World* 50: 37–43.

Morrow, W.R., Qi, H., Kim, I., Mazumder, J., Skerlos, S.J. 2007. Environmental aspects of laser-based and conventional tool and die manufacturing. *Journal of Cleaner Production* 15: 932–943.

Mudali, U.K., Sridhar, T.M., Raj, B. 2003. Corrosion of bio implants. *Sadhana* 28: 601–637.

Muhammad, N., Whitehead, D., Boor, A., Li, L. 2010. Comparison of dry and wet fibre laser profile cutting of thin 316L stainless steel tubes for medical device applications. *Journal of Materials Processing Technology* 210: 2261–2267.

Muhammad, N., Whitehead, D., Boor, A., Oppenlander, W., Liu, Z., Li, L. 2012a. Picosecond laser micromachining of nitinol and platinum–iridium alloy for coronary stent applications. *Applied Physics A* 106: 607–617.

Muhammad, N., Li, L. 2012b. Underwater femtosecond laser micro machining of thin nitinol tubes for medical coronary stent manufacture. *Applied Physics A* 107: 849–861.

Muller, R.E, Bird, J., Duley, W.W. 1992. Laser drilling into an absorbing liquid. *Journal of Applied Physics* 71: 551–556.

Mullick, S., Madhukar, Y.K., Kumar, S., Shukla, D.K., Nath, A.K. 2011. Temperature and intensity dependence of Yb-fiber laser light absorption in water. *Applied Optics* 50: 6319–6326.

Mullick, S., Madhukar, Y.K., Roy, S., Kumar, S., Shukla, D.K., Nath, A.K. 2013. Development and parametric study of a water-jet assisted underwater laser cutting process. *International Journal of Machine Tools & Manufacture* 68: 48–55.

Mullick, S., Agrawal, A.K., Nath, A.K. 2016a. Effect of laser incidence angle on cut quality of 4 mm thick stainless steel sheet using fiber laser. *Optics & Laser Technology* 81: 168–179.

Mullick, S., Priyadarshini, A., Gopinath, M., Nath, A.K. 2016b. Striation-free cutting of mild steel and stainless steel by Yb-fiber laser. *Proceedings of 25th DAE-BRNS National Laser Symposium*, December 20–23, 2016, Bhubaneswar, India.

Mullick, S.,Madhukar, Y.K., Roy, S., Nath, A.K. 2016c. Performance optimization of water-jet assisted underwater laser cutting of AISI 304 stainless steel sheet. *Optics and Lasers in Engineering* 83: 32–47.

Naeem, M. 2012. Different lasers offer different features for micromachining, Photonics Spectra, JK LASERS. www.photonics.com/Article.aspx?PID=5&VID=100&IID=640&AID=52004.

Nath, A.K. 2006. Micro-structural engineering during laser surface remelting through laser beam shaping, 10. www.cat.ernet.in/newsletter/NL/nl2006/issue2/pdf/L13.pdf.

Nath, A.K. 2013. High power lasers in material processing applications: An overview of recent developments. In *Laser-Assisted Fabrications of Materials*, eds. J.D. Majumdar, I. Manna, 69–111. Springer-Verlag, Berlin and Heidelberg.

Nath, A.K. 2014. Laser drilling of metallic and nonmetallic substrates, comprehensive materials processing. In *Laser Machining and Surface Treatment*, ed. B. Yilbas, 115–175. Elsevier, Amsterdam.

Nath, A.K., Sarkar, S. 2017. Laser transformation hardening of steel. In *Advances in Laser Materials Processing*, 2nd Ed., ed. J. Lawrence. Elsevier Woodhead Publishing, Cambridge, UK, ISBN 9780081012529.

Nath, A.K., Hansdah, D., Roy, S., Choudhury, A.R. 2010. A study on laser drilling of thin steel sheet in air and under-water. *Journal of Applied Physics* 107: 1–9.

Nath, A.K., Gupta, A., Benny, F. 2012. Theoretical and experimental study on laser surface hardening by repetitive laser pulses. *Surface and Coatings Technology* 206: 2602–2615.

Nurminen, J., Näkki, J., Vuoristo, P. 2009. Microstructure and properties of hard and wear resistant MMC coatings deposited by laser cladding. *International Journal of Refractory Metals and Hard Materials* 27: 472–478.

Obata, K., El-Tamer, A., Koch, L., Hinze, U., Chichkov, B.N. 2013. High-aspect 3D two-photon polymerization structuring with widened objective working range. *Light: Science & Applications* 2: e116. doi:10.1038/lsa.2013.72.

Okado, H., Sakurai, T., Adachi, J., Miyao, H., Hara, K. 1999. Underwater cutting of stainless steel with the laser transmitted through optical fiber. *High-Power Lasers in Civil Engineering and Architecture. Conference*, Osaka, Japan 3887: 152–161.

Okasha, M.M., Mativenga, P.T., Driver, N., Li, L. 2010. Sequential laser and mechanical micro-drilling of Ni superalloy for aerospace application. *CIRP Annals: Manufacturing Technology* 59: 199–202.

O'Neill, W., Gabzdyl, J.T. 2000. New developments in laser-assisted oxygen cutting. *Optics and Lasers in Engineering* 34: 355–367.

Ostendorf, A., Chichkov, B.N. 2006. Two-photon polymerization: A new approach to micromachining, *Photonics Spectra*. www.photonics.com/Article.aspx?AID=26907.

Overton, G., Nogee, A., Belforte, D., Holton, C. 2017. Annual laser market review & forecast: Where have all the lasers gone? Laser Focus World, no. 53(1). www.laserfocusworld.com/articles/print/volume-53/issue-01/features/ annual-laser-market-review-forecast-where-have-all-the-lasers-gone.html.

Ovsianikov, A., Malinauskas, M., Schlie, S., Chichkov, B., Gittard, S., Narayan, R., Löbler, M., Sternberg, K., Schmitz, K.P., Haverich, A. 2011. Three-dimensional laser micro- and nano-structuring of acrylatedpoly(ethylene glycol) materials and evaluation of their cytoxicity for tissue engineering applications. *Acta Biomaterialia* 7: 967–974.

Parvathavarthini, N., Dayal, R.K., Kaul, R., Ganesh, P., Khare, J., Nath, A.K., Mishra, S.K., Samajdar, I. 2008. A novel laser surface treatment approach to suppress sensitization in modified type 316(N) stainless steel weld metal. *Science & Technology of Welding & Joining* 13: 335–343.

Pavlov, M., Novichenko, D., Doubenskaia, M. 2011. Optical diagnostics of deposition of metal matrix composites by laser cladding. *Physics Procedia* 12: 674–82.

Peyre, P., Fabbro, R., Merrien, P., Lieurade, H.P. 1996. Laser shock processing of aluminum alloys. Application to high cycle fatigue behavior. *Materials Science and Engineering A* 210: 102–113.

Powell, J., Al-Mashikhi, S.O., Kaplan, A.F.H., Voisey, K.T. 2011. Fibre laser cutting of thin section mild steel: An explanation of the 'striation free' effect. *Optics and Lasers in Engineering* 49: 1069–1075.

Purtonen, T., Kalliosaari, A., Salminen, A. 2014. Monitoring and adaptive control of laser processes. *8th International Conference on Photonic Technologies LANE 2014, Physics Procedia* 56: 1218–1231.

Quadrini, F., Guglielmotti, A., Squeo, E.A., Tagliaferri, V., 2010. Laser forming of open-cell aluminium foams. *Journal of Materials Processing Technology* 210: 1517–1522.

Quazi, M.M., Fazal, M.A., Haseeb, A.S.M.A., Yusof, F., Masjuki, H.H., Arslan, A. 2015. Laser-based surface modifications of aluminum and its alloys. *Critical Reviews in Solid State and Materials Sciences* 41: 106–131. doi:10.1080/10408436.2015.1076716.

Quazi, M.M., Fazal, M.A., Haseeb, A.S.M.A., Yusof, F., Masjuki, H.H., Arslan, A. 2016. Effect of rare earth elements and their oxides on tribo-mechanical performance of laser claddings: A review. *Journal of Rare Earths* 34: 549.

Rafi, H.K., Starr, T.L., Stucker, B.E. 2013. A comparison of the tensile, fatigue, and fracture behavior of Ti–6Al–4V and 15-5 PH stainless steel parts made by selective laser melting. *The International Journal of Advanced Manufacturing Technology* 69: 1299–1309.

Rekow, M., Murison, R. 2011. High speed laser drilling of solar cells, vol. 26. www.industrial-lasers.com/content/ils/en/articles/print /volume-26/issue-6/features/high-speed-laser-drilling-of-solar-cells.html.

Rethfeld, B., Sokolowski-Tinten, K., Von Der Linde, D., Anisimov, S.I. 2004. Time scales in the response of materials to femtosecond laser excitation. *Applied Physics A* 79: 767–769.

Richerzhagen, B., Housh, R., Wagner, F., Manley, J. 2004. Waterjet guided laser cutting: A powerful hybrid technology for fine cutting and grooving, eds. D. Roessler, N. Uddin, 175–181. *Proceedings of the 2004 Advanced Laser Applications Conference and Exposition (ALAC,2004),USA.*

Riemer, A., Leuders, S., Thöne, M., Richard, H.A., Tröster, T., Niendorf, T. 2014. On the fatigue crack growth behavior in 316L stainless steel manufactured by selective laser melting. *Engineering Fracture Mechanics* 120: 15–25.

Robles Hernández, F.C., Okonkwo, A.O., Kadekar, V., Metz, T., Badi, N. 2016. Laser cladding: The alternative for field thermite welds life extension. *Materials & Design* 111: 165–173.

Rottwinkel, B., Nölke, C., Kaierle, S., Wesling, V. 2014. Crack repair of single crystal turbine blades using laser cladding technology. *3rd International Conference on Through-life Engineering Services, Procedia CIRP* 22: 263–267.

Rutterford, G., Karnakis, D., Webb, A., Knowles, M. 2005. Optimization of the laser drilling process for fuel injection components. www.oxfordlasers.com_files_laser_drilling_fuel_injection.2005.

Samson, B., Hoult, T., Coskun, M. 2017. Fiber laser welding technique joins challenging metals, vol. 32, issue 2. www.industrial-lasers.com/articles/print/volume-32/issue-2/features/fiber-laser-welding-technique-joins-challenging-metals.html.

Sano, Y., Obata, M., Kubo, T., Mukai, N., Yoda, M., Masaki, K., Ochi, Y. 2006. Retardation of crack initiation and growth in austenitic stainless steels by laser peening without protective coating. *Materials Science and Engineering A* 417: 334–340.

Santo, L. 2008. Laser cladding of metals: A review. *International Journal of Surface Science and Engineering* 2. doi:10.1504/IJSURFSE.2008.021345.

Sanyal, D.N. 2014. Laser applications in nuclear power plants. *Pramana* 82: 135–141.

Sarkar, S., Gopinath, M., Chakraborty, S.S., Syed, B., Nath, A.K. 2016. Analysis of temperature and surface hardening of low carbon thin steel sheets using Yb-fiber laser. *Surface & Coatings Technology* 302: 344–358. doi:10.1016/j.surfcoat.2016.06.045.

Sarkar, S., Kumar, C.S., Nath, A.K. 2017a. Effect of mean stresses on mode of failures and fatigue life of selective laser melted stainless steel. *Materials Science and Engineering A* 700: 92–106.

Sarkar, S., Kumar, C.S., Nath, A.K. 2017b. Effect of different heat treatments on mechanical properties of laser sintered additive manufactured parts. *ASME Journal of Manufacturing Science and Engineering.* doi:10.1115/1.4037437.

Sarkar, S., Kumar, C.S., Nath, A.K. 2018a. Investigation on the mode of failures and fatigue life of laser-based powder bed fusion produced stainless steel parts under variable amplitude loading conditions. Additive Manufacturing. doi:10.1016/j.addma.2018.10.044.

Sarkar, S., Dubey, S., Nath, A.K. 2018b. Effect of heat treatment on impact toughness of selective laser melted stainless steel parts. ASME. *International Manufacturing Science and Engineering Conference,* Volume 1: Additive Manufacturing; Bio and Sustainable Manufacturing ():V001T01A005. doi:10.1115/MSEC2018-6418.

Sarkar, S., Jha, S.R., Nath, A.K. 2018c. Effect of heat treatment on corrosion properties of selective laser melted stainless steel parts. ASME. *International Manufacturing Science and Engineering Conference,* Volume 1: Additive Manufacturing; Bio and Sustainable Manufacturing ():V001T01A007. doi:10.1115/MSEC2018-6429.

Sathyajith, S., Kalainathan, S., Swaroop, S. 2013. Laser peening without coating on aluminum alloy Al-6061-T6 using low energy Nd:YAG laser. *Optics & Laser Technology* 45: 389–394.

Satta, M., Ermer, D.R., Papantonakis, M.R., Flamini, C., Haglund, R.F., Mele, A. 2000. Time-resolved studies of electron–phonon relaxation in metals usinga free-electron laser. *Applied Surface Science* 154–155: 172–178.

Schoonderbeek, A., Ostendorf, A. 2008, Novel industrial approaches in solar-cell production, *SPIE Newsroom,* June 5, 2008. doi:10.1117/2.1200804.1132.

Semak, V., Matsunawa, A. 1997. The role of recoil pressure in energy balance during laser materials processing. *Journal of Physics D: Applied Physics* 30: 2541–2552.

Serres, N., Tidu, D., Sankare, S., Hlawka, F. 2011. Environmental comparison of MESO-CLAD process and conventional machining implementing life cycle assessment. *Journal of Cleaner Production* 19: 1117–1124.

Shamsaei, N., Yadollahi, A., Bian, L., Thompson, S.M. 2015. An overview of direct laser deposition for additive manufacturing; Part II: Mechanical behavior, process parameter optimization and control. *Additive Manufacturing* 8: 12–3.

Shanmugarajan, B., Padmanabham, G. 2012. Fusion welding studies using laser on Ti–SS dissimilar combination. *Optics and Lasers in Engineering* 50: 1621–1627.

Sharifitabar, M., Khaki, J.V., Sabzevar, M.H. 2016. Microstructure and wear resistance of in-situ TiC–Al$_2$O$_3$ particles reinforced Fe-based coatings produced by gas tungsten arc cladding. *Surface & Coatings Technology* 285: 47–56.

Shen, H., Vollertsen, F. 2009. Modelling of laser forming—An review. *Computational Materials Science* 46: 834–840.

Shen, H., Yao, Z., Shi, Y., Hu, J., 2006a. An analytical formula for estimating the bending angle by laser forming. *Proceedings of the Institution of Mechanical Engineers, Part C: Journal of Mechanical Engineering Science* 220: 243–247.

Shen, H., Shi, Y., Yao, Z., Hu, J., 2006b. An analytical model for estimating deformation in laser forming. *Computational Materials Science* 37: 593–598.

Shepeleva, L., Medres, B., Kaplan, W.D., Bamberger, M., Weisheit, A. 2000. Laser cladding of turbine blades. *Surface and Coatings Technology* 125: 45–48.

Smurov, I., Ignatiev, M. 1996. Real time pyrometry in laser surface treatment. *Proceedings of the Conference of Laser Processing: Surface Treatment and Film Deposition*, pp. 529–564, The Netherlands.

Smurov, I., Doubenskaia, M., Zaitsev, A. 2013. Comprehensive analysis of laser cladding by means of optical diagnostics and numerical simulation. *Surface and Coatings Technology* 220: 112–121.

Sokolov, M., Salminen, A., Kuznetsov, M., Tsibulskiy, I. 2011. Laser welding and weld hardness analysis of thick section S355 structural steel. *Materials & Design* 32: 5127–5131.

Spierings, A.B., Starr, T.L., Wegener, K. 2013. Fatigue performance of additive manufactured metallic parts. *Rapid Prototyping Journal* 19: 88–94.

Steen, W.M., Mazumder, J. 2010. *Laser Material Processing*, 4th ed. Springer-Verlag, London.

Stelzer, S., Mahrle, A., Wetzig, A., Beyer, E. 2013. Experimental investigations on fusion cutting stainless steel with fiber and CO$_2$ laser beams, *Physics Procedia* 41: 399–404.

Stevenson, M.E., Barkey, M.E., Bradt, R.C. 2002. Fatigue failures of austenitic stainless steel orthopedic fixation devices. *Journal of Failure Analysis and Prevention* 2: 57–64.

Stoian, R., Ashkenasi, D., Rosenfeld, A., Campbell, E.E.B. 2000. Coulomb explosion in ultrashort pulsed laser ablation of Al$_2$O$_3$. *Physical Review B* 62: 13167.

Subba Rao, R.V., Parvathavarthini, N., Pujar, M.G., Dayal, R.K., Khatak, H.S., Kaul, R., Ganesh, P., Nath, A.K. 2007. Improved pitting corrosion resistance of cold worked and thermally aged AISI type 316L(N) SS by laser surface modification. *Surface Engineering* 23: 83–92.

Sun, J., Yan, Q., Gao, W., Huang, J. 2015. Investigation of laser welding on butt joints of Al/steel dissimilar materials. *Materials & Design* 83: 120–128.

Sundar, R., Pant, B.K., Kuamr, H., Ganesh, P., Nagpure, D.C., Haedoo, P., Kaul, R., Ranganathan, K., Bindra, K.S., Oak, S.M., Kukreja, L.M. 2014. Laser shock peening of steam turbine blade for enhanced service life. *Pramana Journal of Physics* 82: 347–351.

Taillon, G., Pougoum, F., Lavigne, S., Ton-That, L., Schulz, R., Bousser, E., Savoie, S., Martinu, L., Klemberg-Sapieha, J.E. 2016. Cavitation erosion mechanisms in stainless steels and in composite metal–ceramic HVOF coatings. *Wear* 364–365: 201–210.

Takeno, S., Moriyasu, M., Kuzumoto, M., Kurosawa, M., Hirata, Y. 2009. Laser drilling of epoxy-glass printed circuit boards. *Journal of Laser Micro / Nanoengineering* 4: 118–123.

Tang, M., Pistorius, P.C. 2017. Oxides, porosity and fatigue performance of AlSi10Mg parts produced by selective laser melting. *International Journal of Fatigue* 94: 192–201.

Tang, C.H., Cheng, F.T., Man, H.C. 2004. Improvement in cavitation erosion resistance of a copper-based propeller alloy by laser surface melting. *Surface and Coatings Technology* 182: 300–307.

Tangwarodomnukun, V., Wang, J., Huang, C.Z., Zhu, H.T. 2012. An investigation of hybrid laser-waterjet ablation of silicon substrate. *International Journal of Machine Tools and Manufacture* 56: 39–49.

Tani, G., Orazi, L., Fortunato, A., Ascari, A., Campana, G. 2011. Warm laser shock peening: New developments and process optimization. *CIRP Annals: Manufacturing Technology* 60: 219–22.

Thomson, G., Pridham, M., 1998. Improvements to laser forming through process control refinements. *Optics & Laser Technology* 30: 141–146.

Torkamany, M.J., Ghaini, F.M., Poursalehi, R. 2016. An insight to the mechanism of weld penetration in dissimilar pulsed laser welding of niobium and Ti–6Al–4V. *Optics & Laser Technology* 79: 100–107.

Toyserkani, E., Khajepour, A., Corbin, S. 2004. *Laser Cladding*. CRC Press, New York.

Tsai, C.H., Li, C.C. 2009. Investigation of underwater laser drilling for brittle substrates. *Journal of Materials Processing Technology* 209: 2838–2846.

Ueda, T., Sentoku, E., Wakimura, Y., Hosokawa, A., 2009. Flattening of sheet metal by laser forming. *Optics and Lasers in Engineering* 47: 1097–1102.

Vollertsen, F., 1994. An analytical model for laser bending. *Lasers in Engineering* 2: 261–276.

Vollertsen, F., Partes, K., Meijer, J. 2005. State of the art of laser hardening and cladding. *Proceedings of the Third International WLT-Conference on Lasers in Manufacturing*, pp. 1–25, Munich.

Wang, X., Li, P., Xu, Z., Song, X., Liu, H. 2010. Laser transmission joint between PET and titanium for biomedical application. *Journal of Materials Processing Technology* 210: 1767–1771.

Watson, J.K., Taminger, K.M.B. 2015. A decision-support model for selecting additive manufacturing versus subtractive manufacturing based on energy consumption. *Journal of Cleaner Production*. doi:10.1016/j. jclepro.2015.12.009.

Weng, F., Chen, C., Yu, H. 2014. Research status of laser cladding on titanium and its alloys: A review. *Materials & Design* 58: 412–425.

Westphäling, T. 2010. Pulsed fiber lasers from ns to ms range and their applications. *Physics Procedia* 5: 125–136.

Wong, V.W., Tung, S.C. 2016. Overview of automotive engine friction and reduction trends-effects of surface, material, and lubricant-additive technologies. *Friction* 4: 1–28.

Wu, D., Zhang, Q., Ma, G., Guo, Y., Guo, D., 2010. Laser bending of brittle materials. *Optics and Lasers in Engineering* 48: 405–410. www.oxfordlasers.com_files_laser_drilling_fuel_injection.2005.

www.eos.info/material-m, accessed on 27th May, 2017.

www.titanovalaser.com/remanufacturing.html.

Xie, J. 2002. Dual beam laser welding. *Welding Journal* 81: 223–230.

Xiong, Y., Lau, K., Zho, X., Schoenung, J.M. 2008. A streamlined life cycle assessment on the fabrication of WC-Co cermets. *Journal of Cleaner Production* 16: 1118–1126.

Xu, Z., Reed, C.B., Konercki, G., Parker, R.A., Gahan, B.C., Batarseh, S. 2003. Specific energy for pulsed laser rock drilling. *Journal of Laser Applications* 15: 25–30.

Xu, J., Zou, B., Tao, S., Zhang, M., Cao, X. 2016. Fabrication and properties of Al_2O_3-TiB_2-TiC/Al metal matrix composite coatings by atmospheric plasma spraying of SHS powders. *Journal of Alloys and Compounds* 672: 251–259.

Yadollahi, A., Shamsaei, N., Thompson, S.M., Elwany, A., Bian, L. 2017. Effects of building orientation and heat treatment on fatigue behavior of selective laser melted 17-4 PH stainless steel. *International Journal of Fatigue* 94: 218–235.

Yamashita, Y., Kawano, T., Mann, K. 2001. Underwater laser welding by 4 kW CW YAG laser. *Journal of Nuclear Science and Technology* 38: 891–895.

Yan, Y., Li, L., Sezer, K., Wang, W., Whitehead, D., Ji, L., Bao, Y., Jiang, Y. 2011. CO_2 laser underwater machining of deep cavities in alumina. *Journal of the European Ceramic Society* 31: 2793–2807.

Yang, L.J., Wang, Y., Djendel, M., Qi, L.T., 2004. Experimental investigation on 3D laser forming of metal sheet. *Materials Science Forum* 471: 568–572.

Yang, Y.L., Hsu, C.C., Chang, T.L., Kuo, L.S., Chen, P.H. 2010. Study on wetting properties of periodical nanopatterns by a combinative technique of photolithography and laserinterference lithography. *Applied Surface Science* 256: 3683–3687.

Yao, Y.L., Chen, H., Zhang, W. 2005. Time scale effects in laser material removal: A review. *The International Journal of Advanced Manufacturing Technology* 26: 598–608.

Yeo, C.Y., Tam, S.C., Jana, S., Lau, M.W.S. 1994. A technical review of the laser drilling of aerospace materials. *Journal of Materials Processing Technology* 42: 15–49.

Zhou, L., Li, Z.Y., Song, X.G., Tan, C.W., He, Z.Z., Huang, Y.X., Feng, J.C. 2017. Influence of laser offset on laser welding-brazing of Al/brass dissimilar alloys. *Journal of Alloys and Compounds* 717: 78–92.

Zhu, J., Jiao, X., Zhou, C., Gao, H. 2012. Applications of underwater laser peening in nuclear power plant maintenance. *Energy Procedia* 16: 153–158.

CHAPTER 16

Models and Approaches for Sustainable Performance Measurement

S. Vinodh and K. J. Manjunatheshwara
National Institute of Technology

CONTENTS

16.1 INTRODUCTION

Sustainability concepts are being applied in the context of manufacturing to deal with changing industrial conditions and preferences of customers. Sustainability is based on Triple Bottom Line Approach – Environment, Economy and Society. Tools/techniques from the viewpoint of sustainability, namely Quality Function Deployment for Environment (QFDE), Design for Environment (DfE), and Life Cycle Assessment (LCA), are being developed. In order to make organizations sustainable, the decision-makers seek tools that enable performance assessment. Sustainability performance measurement ensures effectiveness of the planned sustainability improvement activities. This chapter presents two prominent approaches of sustainability measurement. Modeling approaches of sustainability to tackle the inefficient use of resources are being developed by researchers globally. Two widely used approaches such as Interpretive Structural Modeling (ISM) and system dynamics modeling are discussed in the chapter. Additionally, a case study on sustainability assessment of a manufacturing enterprise situated in Tamil Nadu is presented. This performance measurement involves computation of sustainability index, identification of weaker areas, and proposals for improvement.

16.2 SUSTAINABLE DEVELOPMENT

Industrialization has been damaging environment. Ecology and economy are interlaced both locally and globally. Sustainable development enables decision-making by merging environmental and economy factors. Sustainable development is a much-debated topic over the past two decades since its conception during Bruntland commission in 1987 (Giddings et al., 2002). There has been discrepancy in its interpretation since then. Bruntland's report on global environment and development stated – Sustainable development as the competence to satisfy the demands of the present generation, without compromising the anticipated future demands. Sustainable development enables long-term usage of natural resources. It addresses environmental issues and socioeconomic problems and benefits human health. The concept has been adapted by governments, businesses, environmental and social activists.

A scheme for sustainable development was approved during the 1992 Earth summit at Rio de Janerio. It was rejuvenated in 2012 Rio Summit and the UN members decided to accomplish sustainable development goals by 2030. Sustainable development by the member nations is extensively targeted at improving living conditions of the under privileged.

The policy concept of sustainable development is enabled through several standards (Rachuri et al., 2009). ISO 14000, ISO 14020, ISO 14040, ISO 14064, and ISO 19011 ensure reduction in environmental impact, eco labeling, LCA, reduction in greenhouse gases (GhG) and implementation of environmental management and quality systems, respectively.

Waste Electric and Electronic Equipment (WEEE) directive ensures manufacturing facilities for Reuse, Reprocess, or Recycling of waste electronic equipment at the end-of-life of products. Restriction for Hazardous Substances (RoHS) directive imposes restriction on usage of harmful materials in electronic equipments. Registration, Evaluation, Authorization and Restriction of Chemicals (REACH) restricts usage of harmful chemicals. End-of-Life of Vehicles (ELV) norms ensures Reuse, Remanufacture, and Recycling of automotive products.

With industrial practitioners across the globe exploring methods of sustainability assessment, sustainable development forms the prime area of research. Logical and robust measurement approaches based on persistent standards would support environmental directives from the legal authorities.

16.3 SUSTAINABILITY

16.3.1 Principles

Sustainability principles enable complementing different zonal and development policies with the sustainable development horizontal policy. They offer an extensive guidance in resolving the distinct objectives and activities, models and modes of execution, in a synchronized and balanced way. The basic principles have been prepared, elucidated, and incorporated at the top positions by bodies of the UN.

Price of products must include the actual costs of production, utilization, and impacts. Any act of affecting human health or damaging natural systems must be compensated appropriately. Sufficient stocks of natural, human, and social capital must be maintained considering long-term benefits. Impartial sharing of capital forms across groups of the present-day and country and covering generations must be guaranteed.

During enhancement, assessment, and execution of regulations and schemes, three aspects, such as financial, environmental, and social considerations, and their interrelationships must be ensured for mutual strengthening. Coordination of local and countrywide activities must be ensured. Sustainability is a system of interconnected elements/systems. Even simple intrusion is disseminated to distant systems. The global conception will consequently enable addressing the disputes in the local setting.

The objectives of sustainable development must be driven toward human welfare. Long-term sustainability goals should be formulated keeping in mind the interests of the future generations. Basic human rights and better living conditions must be ensured to all. Right to acquire education and technical competence must be ensured to improve the nobility of the human society. People must be educated about decision-making processes. Knowledge pertaining to the factors influencing society, economy, and environment must be disseminated.

Production and utilization patterns must be sustainable. Organizations must exhibit better social responsibility along with strong public–private sectors cooperation. Measures should be taken to meet the necessities of the local population using the resources available in the locality. Local biodiversity should be conserved along with its characteristic features. Safeguarding the cultural heritage and careful exploitation of human constructions are crucial for sustainable development.

Natural resources should be handled judiciously in order to avoid causing irreversible damages to the environment. Sustainable management of natural resources will enable conservation of resources and their utilization for a long term. It is important to protect rich biodiversity from extinction.

Preventing/reducing activities that cause harm to the environment and affect human health should be considered. Extreme efforts should be taken to restore the natural settings in case of any harm. If severe damage is perceived during the onset of any activity, precautionary measures must be taken. Absence of absolute scientific evidence does not justify delaying the damage prevention activities.

16.3.2 Concepts

Sustainability concepts are strategies for sustainability implementation. Various concepts of sustainability have been developed over the past few years. The concepts are briefly discussed as follows:

Biosphere stability and toughness—It is the degree of turbulence that can be absorbed before the system alters constitution. This characteristic will allow the ecosystem to favor human society with regard to economic and social gains. Various indicators are quantified in order to assess the ecosystem stability.

Integrated system—Ecological, social, and economic constituents form interrelated, interacting, or interdependent group of a complex whole sustainable system. This immensely secured idea of coalescence sets apart sustainability from other policies.

Investment—The investment for sustainable development can be fit into a five capitals framework including natural resources, manufactured goods, monetary stocks, human assets, and social assets. These capitals must be maintained and increased through sustainable development.

Justness—Sustainable development should ensure intra-generational and inter-generational equity, i.e., equity within generation and over generations. This concept reiterates the classic definition of sustainability, framed by the Bruntland commission in 1987.

Economic capability—Profitability and management solvency are essential for financial stability of organizations. Insolvent organizations would first require reordering of the organization to attain sustainability. Sustainability is driven by several financial factors; only careful inspection of every business aspect would lead to financial achievement of the organization.

16.3.3 TBL Approach

Initially, a two-pillar model which echoed environmental and developmental issues was demonstrated at the Bruntland commission. Later the concept of Triple Bottom Line (TBL) approach was introduced by John Elkington in 1994. This three-pillar TBL model divided the development issues into social and economic factors. It highlighted that the material gains do not sufficiently indicate the welfare of human society.

TBL demonstrated the indicators for evaluating and articulating the sustainability of an organization. It is a measurement model with economic, environmental, and societal dimensions. These three components of sustainability often refer to alliterated terms, People, Planet, and Profit, making them the 3Ps of sustainability. The scheme of TBL identifies the environmental and social values that the organizations put in along with the gain of economic value. The three aspects are obviously connected and are good fit theoretically for evaluating sustainability. The dimensions are examined differently and integrated at the end. The typical accounting framework for TBL concept

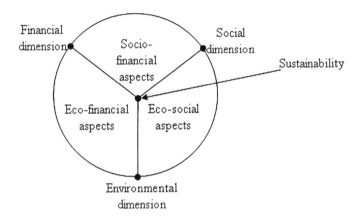

Figure 16.1   TBL concept.

is depicted in Figure 16.1. Effecting TBL approach in any organization is challenging. Three dimensions need to be appropriately measured and their contribution to sustainability should be reported.

16.4 SUSTAINABILITY TOOLS/TECHNIQUES

The section presents the following three sustainability tools, QFD for environment, Life Cycle Assessment and Design for Environment.

16.4.1 QFD for Environment (QFDE)

QFDE enables environmentally conscious design in the early stages of product design and development. It handles both environmental and conventional product quality requirements concurrently. QFDE phases I and II deal with parts determination which are vital for improving environmental consciousness of their products. QFDE phases III and IV are focused on evaluating the design improvement effects on environmental parameters. QFDE starts with gathering customer voice and engineering metrics (EMs). In phase I, customer voices are weighted. Voices of customers are mapped with EMs and the relative weight of EMs to be computed to identify potential EMs from environment viewpoint. In phase II, EMs are mapped with components and relative weight of components are computed to identify components from environment viewpoint. With the identified potential EMs and components, options need to be formed from the viewpoint of material substitution, product design modification, and process substitution. In order to evaluate the design improvements, in phase III, the effect on EMs due to a set of design changes is estimated. In phase IV, the effects of design changes on EM into environmental requirements are computed. The equations used in phases III and IV to determine the effect of improvement of EMs and improvement rate of customer requirements could be referred to Masui et al. (2003) and Vinodh and Rathod (2010).

16.4.2 Life Cycle Assessment (LCA)

An assessment of environmental impact needs to be done to ascertain the product design from environment viewpoint. Society of Toxicology and Chemistry (SETAC) developed a four-step process that enables understanding the environmental impact of a product from manufacture through use through disposal. The four phases include goal definition, inventory analysis, impact assessment, and interpretation as shown in Figure 16.2.

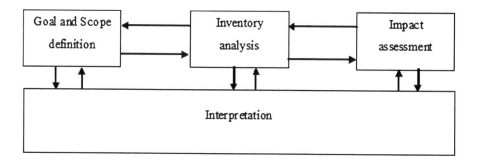

Figure 16.2 Phases of LCA.

16.4.2.1 Goal and Scope Definition

This phase is concerned with setting the purpose of environmental analysis and scope of LCA. With proper definition of objectives, EMs must be defined appropriately. A clear definition of overall product function and functional unit is needed.

16.4.2.2 Life Cycle Inventory Analysis

This phase creates stock flow, into and out of the system boundary for a product. Input stocks include water, energy, and raw materials. The outputs are releases to environmental media, such as air, land, and water. Based on these inputs and outputs, a technical system flow model is configured.

16.4.2.3 Life Cycle Impact Assessment

Based on the inventory flow results, the critical environmental effects are assessed in this phase of LCA. It includes the identification of impact categories, category indicators, characterization models, and impact measurement.

16.4.2.4 Interpretation

It is a technique to systematically identify, measure, and assess the key information, based on the results of inventory analysis and impact assessment. Sensitivity and consistency checks will be done including the recommendations.

Benefits of LCA include the recognition of significant environmental impacts of products, identify improvement opportunities, compliance with regulations, and facilitate eco-design practices.

16.4.3 Design for Environment

DfE includes phases such as assessment of present state, improvement redesign for recognition, and implementation of refined design. DfE includes design guidelines for recovery, disassembly, waste minimization, energy conservation, optimal resource consumption, and risk reduction. DfE guidelines could be applied during product development phases such as concept development and system level design. Sample DfE guidelines include materials incompatible in recycling are not to be combined, no surface treatments are to be done, reduce weight from shipping, trim down time for disassembly, reduce number of parts or components to reduce usage of raw materials, ensure that design facilitates ease of cleaning, inspection, disassembly, and reassembly.

16.5 SUSTAINABILITY MEASURES

Performance measurement enables practitioners to determine target areas and measures to be taken for improvement. Sustainability is measured along its three dimensions—Economy, Society, and Environment.

16.5.1 Economy

Several frameworks of sustainability from past studies have emphasized economic sustainability in different context. The following criteria have been identified in conjunction with economic sustainability.

16.5.1.1 Economic Stability

This indicator involves the facets evaluating the financial health of the organizations and includes conventional secondary indicators of economy, viz., solvency, liquidity, and profitability.

16.5.1.2 Effective Economy

This indicator evaluates value of the organization as identified by stakeholders, administrators, and regulatory bodies and includes secondary indicators such as market share performance, share profitability, and contribution to GDP.

16.5.1.3 Prospective Financial Gains

This economic indicator of sustainability measures economic benefits apart from profits.

16.5.1.4 Business Potential

The indicator evaluates susceptibility of networks of the business and also the possible risk associated with the system it is rooted in. The local and global business links serve as secondary indicators.

16.5.2 Society

Contemporary organizations are keen on improving societal aspects of sustainability. Fewer studies from the literature emphasized on social sustainability. Social issues can be considered from micro and macro perspectives. Micro perspective is related to the organizational people, whereas macro perspective is related to the society within which the organization operates. The level of social sustainability of an organization is often difficult to measure. Social sustainability factors ensure overall organizational sustainability.

16.5.2.1 Organizational People

This includes people along the supply chain including suppliers/vendors, financiers, and employees. Health and safety of employees is the major concern. Training and growth needs to be ensured as the organizational people are directly responsible for the outcome. It is also critical in connection with human rights.

16.5.2.2 Customers

Customer satisfaction is the benchmark for any business. Customer-related factors ensure safety during product usage. The factors account for customer privileges. Customer opinions are sought and are transformed into managerial decisions in order to empower the stakeholder.

16.5.2.3 External Community

The concerned factors pertain to consequences of organizational actions on the community in which the organization operates. The organizations should ensure better health and education of the human capital. The quality of life should be improved by ensuring proper housing, infrastructure, and transport facilities. The community's cultural priorities should be valued and a social cohesion should be maintained.

16.5.3 Environment

Environmental sustainability ensures well-being of mankind by safeguarding the natural resources exploited for the needs of the society and seeking to decrease environmental wastages to avert damage of the ecosystem. Exhaustion of finite resources and ecological effects due to gaseous emissions, release of liquid wastes, and huge amounts of solid contaminants are the major environmental issues. Carbon footprint and energy efficiency are also significant.

Environmental factors indicate the critical issues of environment and their severity. Life cycle thinking enables development of indicators for environmental sustainability which is considered to be a comprehensive approach for tackling sustainable development. LCA is instrumental method in measuring the effects of products and processes with respect to environment. The factors of environmental sustainability quantify the usage of natural resources.

16.5.3.1 Land Resources

The emphasis is on usage of land and transformation and the pollutants released to soil. Contamination of land can create health hazards to humans or effect pollution of surface water.

16.5.3.2 Air Resources

The emphasis is on quality of local air. Different organizational activities emit air pollutants such as sulphur and nitrous oxides, which in turn result in acid rain. Global effects such as ozone layer depletion and greenhouse effect are other major concerns.

16.5.3.3 Water Resources

The emphasis is on quality of water and water eutrophication occurs when nutrients are released in abundance to aquatic system causing damage to aquatic lives.

16.5.3.4 Energy Resources

The emphasis is on depletion of nonrenewable energy resources. The energy consumed during product's lifecycle and upstream energy required to produce these resources are taken into account.

16.6 SUSTAINABILITY MEASUREMENT METHODS

This section presents multigrade fuzzy and fuzzy logic methods for sustainability performance measurement.

16.6.1 Multigrade Fuzzy Method

The principle of multigrade fuzzy approach is briefed as follows: the sustainability index is denoted by I which is the product of overall assessment factor (R) and overall weight (W) (Vinodh, 2011; Yang and Li, 2002). The sustainability index is given by

$$I = W \times R \tag{16.1}$$

The evaluation is divided into five grades since every factor of sustainability involves fuzzy evaluation, $I = \{10, 8, 6, 4, 2\}$

8–10 represent extremely sustainable
6–8 represent sustainable
4–6 represent generally sustainable
2–4 indicate not sustainable
<2 indicate extremely unsustainable

Inputs from experts need to be obtained in terms of ratings and weights for the developed model with indicators. Ratings need to be assigned using 1–10 scale (1 represents least important, 10 represents most important). Weights need to be assigned using 0–1 scale. For attributes, both ratings and weights need to be assigned; and for criteria and enablers, weights to be assigned.

16.6.1.1 Primary Assessment

The index pertaining to criterion is given by

$$I_{11} = W_{11} \times R_{11} \tag{16.2}$$

where W_{11} and R_{11} represent weights and ratings pertaining to criterion 1 of enabler 1.

16.6.1.2 Secondary Assessment

Equation for calculating the index pertaining to an enabler is given by,

$$I_1 = W_1 \times R_1 \tag{16.3}$$

W_1, R_1—represent weight and rating pertaining to enabler 1.

16.6.1.3 Tertiary Assessment

The overall index is given by

$$I = W \times R \tag{16.4}$$

where W and R are overall weight and assessment vector
Also weaker areas could be identified as attributes having performance rating lesser than 50%.

16.6.2 Fuzzy Logic Method

Linguistic terms are used in this approach to evaluate the performance ratings and importance of weights of sustainability attributes to assign the score for vague attributes. Linguistic terms namely excellent, very good, good, fair, poor, very poor, and worst are used to evaluate the performance ratings of sustainability capabilities. Linguistic variables such as very high, high, fairly high, medium, fairly low, low, and very low are used to assign the importance weights of sustainability attributes (Lin et al., 2006). Utilizing the concept of fuzzy theory, fuzzy numbers are used to approximate the linguistic variables. By applying the relation between linguistic terms and fuzzy numbers, the linguistic terms are transformed into fuzzy numbers. Fuzzy Sustainability Index (FSI) represents the overall enterprise level sustainability. Fuzzy index needs to be determined at certain level and then extended to enabler level. The fuzzy index of level 2 sustainability criteria is computed as follows:

$$\text{SC}_{ij} = \frac{\sum_{k=1}^{n} \left(W_{ijk} \otimes \text{SC}_{ijk} \right)}{\sum_{k=1}^{n} W_{ijk}} \tag{16.5}$$

where SC_{ijk} and W_{ijk} represent performance rating and importance weight of sustainability attributes.

By applying equation 16.5, the fuzzy index of level 2 sustainability criteria is obtained. For validating the results obtained using fuzzy method, Euclidean distance is computed. FSI is to be matched with relevant sustainability level. FSI is matched with linguistic label. Euclidean distance from FSI to each member is set. Sustainability level is computed as follows:

$$D(\text{FSI}, \text{SL}_i) = \left\{ \sum_{x \in P} \left[f_{\text{FSI}}(x) - f_{\text{SL}_i}(x) \right]^2 \right\}^{1/2} \tag{16.6}$$

By relating the linguistic label with minimum D, the sustainability level is to be identified. Weaker areas to be identified using Fuzzy Performance Importance Index (FPII). FPII integrates performance ratings and importance weights of sustainability attributes. Higher the FPII, higher is the contribution of the attribute:

$$\text{FPII}_{ijk} = W_{ijk} \otimes \text{SC}_{ijk}$$

$$W'_{ijk} = \left[(1,1,1) - W_{ijk} \right] \tag{16.7}$$

where W_{ijk} denotes the fuzzy importance weight of the sustainability attribute. This is followed by fixing the management threshold. Those attributes having FPII lower than management threshold are considered to be weaker attributes.

16.7 ADVANTAGES AND DRAWBACKS

Multigrade fuzzy method overcomes the drawbacks with conventional method as it works with combined ratings and weights thereby vagueness and uncertainty associated with measurement could be minimized to certain extent. But identifying the weaker attributes will be relatively difficult. Since the computation is matrix based, industry practitioners also feel comfortable to deploy this approach.

Fuzzy logic approach deploys triangular or trapezoidal fuzzy numbers for evaluation. Inputs will be obtained through linguistic variables and linguistic variables are transformed into fuzzy numbers. This approach deals with uncertainty and vagueness associated with inputs. The computations behind fuzzy logic need to be clearly understood. Identification of weaker areas could be systematically done by the computation of FPII. Fuzzy logic is most preferred performance measurement approach to deal with vagueness and impreciseness associated with measurement.

16.8 MODELING FOR SUSTAINABLE MANUFACTURING

The structural modeling and system dynamics modeling approaches are discussed in the following sections.

16.8.1 Structural Modeling

For developing structural model, ISM is a widely used approach. The basic idea behind ISM is to incorporate practical expertise of experts to decompose a complex system into several subsystems and develop a structural model with multiple levels to depict the dominant factors associated with implementation of sustainable manufacturing. Some of the key attributes of ISM include the following: the methodology is imperative; relationships can be depicted in a digraph model; enhances both group and individual learning.

The key steps involved in ISM technique are as follows: analysis of the problem by identification of variables (factors/barriers), development of contextual relationship among factors, development of Structural Self-Interaction Matrix (SSIM), and reachability matrix and checking for transitivity, derivation of digraph, categorization of factors using MICMAC and model review to check for consistency (Kannan et al., 2009). ISM has several applications in sustainable manufacturing such as analysis of factors for sustainable product development, sustainability implementation in small and big size organizations, sustainable supply chains, environmentally conscious manufacture, and so on. To statistically validate the constructs, Structural Equation Modeling (SEM) could be applied wherein relationship among constructs and latent variables could be analyzed based on statistical modeling. SEM includes two classes of models, namely measurement and structural model. The measurement model shows the fact that indicates how measured latent variables represent the stated theory and the structural model implies the relationships between various constructs of the system. SEM includes two types of analysis, namely Partial Least Square and Covariance-Based Analysis. SEM also has potential application in the context of sustainable design, manufacturing, and determination of relation between sustainable manufacturing indicators and organizational performance.

The application studies of structural and statistical modeling in the context of sustainable manufacturing are depicted in Table 16.1.

Table 16.1 Important Contributions Pertaining to Structural Modeling of Sustainability

Research Study	Model	Applications
Vinodh and Joy (2012)	SEM	Study of critical sustainable manufacturing factors influencing organizational performance
Zailani et al. (2012)	SEM	Study of influence of eco-design on environmental performance
Mehregan et al. (2014)	ISM	Study pertaining to interrelations among sustainable supplier selection criteria
Vinodh et al. (2016)	ISM	Lean sustainable system implementation in automotive component manufacturing organizations
Girubha et al. (2016)	ISM	Sustainable supplier selection
Thirupathi and Vinodh (2016)	ISM and SEM	Sustainable automotive component manufacturing
Vasantha kumar et al. (2016)	ISM	Lean remanufacturing of automotive components

16.8.2 System Dynamics Modeling

System Dynamics is a modeling approach focused on modeling system with complex characteristics and evaluates its performance over time period. It is quite different from other approaches as it utilizes feedback concept, and evaluate dynamic behavior in loops. System Dynamics modeling is vital for sustainable manufacturing as it focuses on analysis of the system behavior and reaction to trends (Sterman, 2001). The casual feedback loop diagram forms the fundamental for system dynamics modeling and allows the users to simulate the system variables. The casual feedback loop diagram allows the users to define the system pattern, key variables, causal relationship, and the direction of system variables. After the causal feedback loop diagram is configured, the system is modeled using simulation software.

16.9 SUSTAINABILITY APPLICATIONS

Applications of sustainable development are widespread across areas such as manufacturing, healthcare, construction, agriculture, etc. The manufacturing applications include benefits such as reduction in consumption of resources required for production, improvement in efficiency of manufacturing operations, reduction in impact of the product during usage, and end-of-life phases.

Construction applications ensure lower environmental impact of constructions throughout their lifetime. Developing sustainable practices for construction would enable reduction in material usage, reduction in operational cost, improvement employee health, and conserve energy.

Sustainable practices can enable significant savings in the healthcare sector while reducing the operational cost. Sustainable ways of waste disposal at hospitals ensures more savings. Reuse of medical appliances and energy savings contribute to the reduction in impact on environment.

Agricultural sustainability is the task of farming through adaptation of ecological principles, the study of relationships between organisms and their environment. It is a holistic system of plant and animal cultivation strategies possessing site-specific application with long-term focus. The benefits include cost reduction, pollution prevention, etc.

16.9.1 Application Domain

Environmentally conscious production and consumption is possible through sustainable product design (Charter, 1998). Environmentally conscious design tools are being developed for this purpose (Vinodh et al., 2017). The concepts of sustainability are being integrated with manufacturing aspects. Profit maximization, environmental impact minimization, and social well-being by effective management of operations, funds, resources and information of supply chains is called sustainable supply chain management (Hassini et al., 2012). Sustainable supply chain metrics have been developed in various studies for industry specific applications. Some major eco indicators include water effluents, air emissions, energy usage, land emissions, and human toxicity. Efforts to take back the end-of-life products are being made in order to restrain environmental effects. Models are being developed for end-of-life management of products (Sodhi and Reimer, 2001).

Table 16.2 briefs the research studies in the areas such as design and development, manufacturing, supply chain, end-of-life analyses, with reference to sustainability of manufacturing organizations.

16.9.2 Case Study on Sustainability Measurement

The performance measures applicable to sustainability of manufacturing enterprises are derived from the benchmark studies. A framework of sustainability measurement is developed. The framework consists of three sustainability enablers, ten sustainability criteria, and 36 sustainability subcriteria

Table 16.2 Research Studies Pertaining to Application of Sustainable Manufacturing

Application Area	Research Study
Design and development	Charter (1998); McDonough and Braungart (2002); Maxwell and Van der Vorst (2003); Bereketli et al. (2009); Shi and Chew (2012); Tian and Chen (2014); Pal (2017); Saride et al. (2017); Vinodh and Manjunatheshwara (2017); Vinodh et al. (2017); Schöggl et al. (2017)
Manufacturing	Jovane et al. (2008); Wu et al. (2008); Huang et al. (2009); Mohammed and Sadique (2011); Al-Oqla and Sapuan (2014); Hallstedt et al. (2015)
Supply chain	Linton et al. (2007); Carter and Rogers (2008); Ninlawan et al. (2010); Gold et al. (2010); Carter and Liane Easton (2011); Hassini et al. (2012); Ahi and Searcy (2013); Seuring (2013); Mota et al. (2015); Papadopoulos et al. (2017); Ansari and Kant (2017)
End-of-life	Sodhi and Reimer (2001); Daniels et al. (2004); Fisher et al. (2005); Gehin et al. (2008); Genovese et al. (2017)

(Table 16.3). The three enablers of the developed framework reiterate the three pillars of sustainability. The broad categories form the criteria and specific measures form the subcriteria of the developed framework. The sustainability evaluation case study is conducted for a manufacturing enterprise situated in Tamil Nadu, India. Assessment based on the developed framework is carried out using Fuzzy logic method and sustainability index is computed. Interpretations are derived from the results.

Table 16.3 Framework for Sustainability Performance Assessment

Enabler	Criteria	Subcriteria	References
Environment (SE_1)	Product weight (SC_{11})	Material type (SSC_{111})	Boks and Stevels (2003)
		Components (SSC_{112})	Despeisse et al. (2012)
		Assembly (SSC_{113})	Despeisse et al. (2012)
		Functional parts and accessories (SSC_{114})	Boks and Stevels (2003)
		Connections (SSC_{115})	Boks and Stevels (2003)
	Energy (SC_{12})	Available energy sources (SSC_{121})	Vinodh (2011); Vinodh et al. (2017)
		Consumer usage (SSC_{122})	Vinodh et al. (2017)
	Toxic substances (SC_{13})	Carbon footprint (SSC_{131})	Despeisse et al. (2012)
		Air pollutants (SSC_{132})	Vinodh (2011); Despeisse et al. (2012); Rosen and Kishawy (2012); Vinodh et al. (2017)
		Water pollutants (SSC_{133})	Vinodh (2011); Despeisse et al. (2012); Rosen and Kishawy (2012); Vinodh et al. (2017)
		Land contamination (SSC_{134})	Rosen and Kishawy (2012)
		Noise emission (SSC_{135})	Choi et al. (1997)
		Human health hazards (SSC_{136})	Vinodh et al. (2017)
	Waste (SC_{14})	Material type (SSC_{141})	Despeisse et al. (2012); Joung et al. (2012)
		Recyclability (SSC_{142})	Despeisse et al. (2012)
		Reusability (SSC_{143})	Rosen and Kishawy (2012)
		Remanufacturability (SSC_{144})	Joung et al. (2012)
		Packaging materials (SSC_{145})	Vinodh et al. (2017)

(continued)

Table 16.3 (*Continued*)   Framework for Sustainability Performance Assessment

Enabler	Criteria	Subcriteria	References
Society (SE_2)	Organizational people (SC_{21})	Management commitment (SSC_{211})	Law and Gunasekaran (2012)
		Employee safety and health (SSC_{212})	Vinodh (2011); Rosen and Kishawy (2012); Labuschagne et al. (2005)
		Job satisfaction (SSC_{213})	Joung et al. (2012)
		Training and empowerment (SSC_{214})	Labuschagne et al. (2005); Joung et al. (2012)
	Involvement of stakeholders (SC_{22})	Empowerment of stakeholders (SSC_{221})	Vinodh (2011); Labuschagne et al. (2005)
		Selective and collective audiences (SSC_{222})	Vinodh (2011); Labuschagne et al. (2005)
		Decision influence potential (SSC_{223})	Vinodh (2011); Labuschagne et al. (2005)
	Larger community (SC_{23})	Government regulations (SSC_{231})	Vinodh (2011); Labuschagne et al. (2005)
		Ethical responsibility (SSC_{232})	Labuschagne et al. (2005); Vinodh (2011); Vinodh et al. (2017)
Economy (SE_3)	Financial well-being (SC_{31})	Profit maximization (SSC_{311})	Labuschagne et al. (2005); Vinodh (2011); Joung et al. (2012)
		Liquidity (SSC_{312})	Labuschagne et al. (2005); Vinodh (2011)
		Solvency (SSC_{313})	Labuschagne et al. (2005); Vinodh (2011)
	Financial benefits (SC_{32})	Technological advancements (SSC_{321})	Labuschagne et al. (2005); Vinodh (2011)
		Leaner organizational structure (SSC_{322})	Larson and Greenwood (2004); Mohammed and Sadique (2011); Vinodh et al. (2017)
		National/global grants (SSC_{323})	Labuschagne et al. (2005); Vinodh (2011)
	Economic performance (SC_{33})	Effect on Gross Domestic Product (GDP) (SSC_{331})	Labuschagne et al. (2005); Vinodh (2011)
		Performance of market share (SSC_{332})	Labuschagne et al. (2005); Vinodh (2011)
		Profitability of shares (SSC_{333})	Labuschagne et al. (2005); Vinodh (2011)

16.9.2.1 *Sustainability Measurement Using Fuzzy Logic*

The sustainability index is a measurement value of sustainability and is calculated as follows:

$$SI = \frac{\sum_{x=1}^{n} R_x \times W_x}{\sum_{x=1}^{n} W_x} \qquad (16.8)$$

where R_x and W_x are sustainability performance rating and sustainability importance weight.

The importance weights and performance ratings are represented in linguistic terms and expressed as fuzzy numbers following the scale of linguistic variable shown in Table 16.4.

Table 16.4   Linguistic Variable Scale

Linguistic Variable[a]	Fuzzy Number
W	(0, 0.5, 1.5)
VP	(1, 2, 3)
P	(2, 3.5, 5)
F	(3, 5, 7)
G	(5, 6.5, 8)
VG	(7, 8, 9)
E	(8.5, 9.5, 10)
VL	(0, 0.05, 0.15)
L	(0.1, 0.2, 0.3)
FL	(0.2, 0.35, 0.5)
M	(0.3, 0.5, 0.7)
FH	(0.5, 0.65, 0.8)
H	(0.7, 0.8, 0.9)
VH	(0.85, 0.95, 1.0)

[a] W, worst; VP, very poor; P, poor; F, fair; G, good; VG, very good; E, excellent; VL, very low; L, low; FL, fairly low; M, medium; FH, fairly high; H, high; VH, very high.

A cross-functional team of three experts provided inputs for sustainability assessment. Through series of discussions, linguistic ratings were conducted for importance weights and performance ratings were gathered as shown in Table 16.5.

The linguistic terms are replaced with relevant fuzzy numbers as shown in Table 16.6.

16.9.2.2 First Level Computation

The initial level finding R_{ij} is calculated as follows:

$$R_{ij} = \sum_{k=1}^{n} \left(W_{ijk} \times R_{ijk} \right) \bigg/ \sum_{k=1}^{n} W_{ijk} \tag{16.9}$$

where

R_{ij} is the sustainability score of jth criterion with respect to ith enabler,
W_{ijk} is the weight of kth attribute with respect to jth criterion and ith enabler, and
R_{ijk} is the performance rating of kth attribute with respect to jth criterion and ith enabler.

The sample calculation for "Product weight" sustainability criterion is as follows:
Product weight (SC_{11}) includes five subcriteria:

$$R_{11} = \begin{bmatrix} (0.5, 0.65, 0.8) & \otimes & (5, 6.5, 8) & \otimes \\ (0.5, 0.65, 0.8) & \otimes & (4.33, 6, 7.67) & \otimes \\ (0.5, 0.65, 0.8) & \otimes & (4.33, 6, 7.67) & \otimes \\ (0.3, 0.5, 0.7) & \otimes & (3.67, 5.5, 7.33) & \otimes \\ (0.3, 0.5, 0.7) & \otimes & (3.67, 5.5, 7.33) & \end{bmatrix} \bigg/ \begin{bmatrix} (0.5, 0.65, 0.8) \otimes \\ (0.5, 0.65, 0.8) \otimes \\ (0.5, 0.65, 0.8) \otimes \\ (0.3, 0.5, 0.7) \otimes \\ (0.3, 0.5, 0.7) \end{bmatrix}$$

$$R_{11} = (4.30, 5.94, 7.61)$$

The same method is applied to calculate indices pertaining to other criteria.

Table 16.5 Assigned Weights and Performance Ratings

Enabler	W_i	Criteria	W_{ij}	Subcriteria	W_{ijk}	R_{ijk} (1)	R_{ijk} (2)	R_{ijk} (3)
SE$_1$	VH	SC$_{11}$	FH	SSC$_{111}$	FH	G	G	G
				SSC$_{112}$	FH	G	F	G
				SSC$_{113}$	FH	G	G	F
				SSC$_{114}$	M	F	G	F
				SSC$_{115}$	M	F	F	G
		SC$_{12}$	FH	SSC$_{121}$	FH	G	F	G
				SSC$_{122}$	M	F	F	G
		SC$_{13}$	FL	SSC$_{131}$	FL	F	F	F
				SSC$_{132}$	FL	F	G	F
				SSC$_{133}$	FL	F	F	G
				SSC$_{134}$	FL	F	P	F
				SSC$_{135}$	M	F	G	F
				SSC$_{136}$	M	F	F	G
		SC$_{14}$	M	SSC$_{141}$	FH	F	F	G
				SSC$_{142}$	H	G	G	F
				SSC$_{143}$	H	G	F	G
				SSC$_{144}$	H	F	G	G
				SSC$_{145}$	M	F	F	G
SE$_2$	FH	SC$_{21}$	H	SSC$_{211}$	H	VG	G	G
				SSC$_{212}$	H	G	G	F
				SSC$_{213}$	VH	G	G	VG
				SSC$_{214}$	H	G	G	F
		SC$_{22}$	H	SSC$_{221}$	H	G	F	G
				SSC$_{222}$	FH	G	F	G
				SSC$_{223}$	FH	F	G	F
		SC$_{23}$	FH	SSC$_{231}$	FH	G	F	G
				SSC$_{232}$	H	G	VG	G
SE$_3$	H	SC$_{31}$	FH	SSC$_{311}$	H	G	VG	G
				SSC$_{312}$	H	G	F	G
				SSC$_{313}$	FH	G	F	G
		SC$_{32}$	H	SSC$_{321}$	H	VG	G	G
				SSC$_{322}$	FH	G	F	G
				SSC$_{323}$	H	G	F	G
		SC$_{33}$	FH	SSC$_{331}$	VH	VG	G	G
				SSC$_{332}$	H	G	F	G
				SSC$_{333}$	H	G	F	G

16.9.2.3 Second Level Computation

The performance rating of ith enabler is calculated as follows:

$$R_i = \sum_{j=1}^{n} \left(W_{ij} \times R_{ij} \right) \Big/ \sum_{j=1}^{n} W_{ij} \qquad (16.10)$$

where

R_i is the sustainability score of ith enabler,
W_{ij} is the weight of jth criterion with respect to ith enabler, and
R_{ij} is the performance rating of jth criterion with respect to ith enabler.

Table 16.6 Fuzzy Numbers Representing Linguistic Variables

Enabler	W_i	Criteria	W_{ij}	Subcriteria	W_{ijk}	R_{ijk}
SE$_1$	(0.85, 0.95, 1.0)	SC$_{11}$	(0.5, 0.65, 0.8)	SSC$_{111}$	(0.5, 0.65, 0.8)	(5, 6.5, 8)
				SSC$_{112}$	(0.5, 0.65, 0.8)	(4.33, 6, 7.67)
				SSC$_{113}$	(0.5, 0.65, 0.8)	(4.33, 6, 7.67)
				SSC$_{114}$	(0.3, 0.5, 0.7)	(3.67, 5.5, 7.33)
				SSC$_{115}$	(0.3, 0.5, 0.7)	(3.67, 5.5, 7.33)
		SC$_{12}$	(0.5, 0.65, 0.8)	SSC$_{121}$	(0.5, 0.65, 0.8)	(4.33, 6, 7.67)
				SSC$_{122}$	(0.3, 0.5, 0.7)	(3.67, 5.5, 7.33)
		SC$_{13}$	(0.2, 0.35, 0.5)	SSC$_{131}$	(0.2, 0.35, 0.5)	(3, 5, 7)
				SSC$_{132}$	(0.2, 0.35, 0.5)	(3.67, 5.5, 7.33)
				SSC$_{133}$	(0.2, 0.35, 0.5)	(3.67, 5.5, 7.33)
				SSC$_{134}$	(0.2, 0.35, 0.5)	(2.67, 4.5, 6.33)
				SSC$_{135}$	(0.3, 0.5, 0.7)	(3.67, 5.5, 7.33)
				SSC$_{136}$	(0.3, 0.5, 0.7)	(3.67, 5.5, 7.33)
		SC$_{14}$	(0.3, 0.5, 0.7)	SSC$_{141}$	(0.5, 0.65, 0.8)	(3.67, 5.5, 7.33)
				SSC$_{142}$	(0.7, 0.8, 0.9)	(4.33, 6, 7.67)
				SSC$_{143}$	(0.7, 0.8, 0.9)	(4.33, 6, 7.67)
				SSC$_{144}$	(0.7, 0.8, 0.9)	(4.33, 6, 7.67)
				SSC$_{145}$	(0.3, 0.5, 0.7)	(3.67, 5.5, 7.33)
SE$_2$	(0.5, 0.65, 0.8)	SC$_{21}$	(0.7, 0.8, 0.9)	SSC$_{211}$	(0.7, 0.8, 0.9)	(5.67, 7, 8.33)
				SSC$_{212}$	(0.7, 0.8, 0.9)	(4.33, 6, 7.67)
				SSC$_{213}$	(0.85, 0.95, 1.0)	(5.67, 7, 8.33)
				SSC$_{214}$	(0.7, 0.8, 0.9)	(4.33, 6, 7.67)
		SC$_{22}$	(0.7, 0.8, 0.9)	SSC$_{221}$	(0.7, 0.8, 0.9)	(4.33, 6, 7.67)
				SSC$_{222}$	(0.5, 0.65, 0.8)	(4.33, 6, 7.67)
				SSC$_{223}$	(0.5, 0.65, 0.8)	(3.67, 5.5, 7.33)
		SC$_{23}$	(0.5, 0.65, 0.8)	SSC$_{231}$	(0.5, 0.65, 0.8)	(4.33, 6, 7.67)
				SSC$_{232}$	(0.7, 0.8, 0.9)	(5.67, 7, 8.33)
SE$_3$	(0.7, 0.8, 0.9)	SC$_{31}$	(0.5, 0.65, 0.8)	SSC$_{311}$	(0.7, 0.8, 0.9)	(5.67, 7, 8.33)
				SSC$_{312}$	(0.7, 0.8, 0.9)	(4.33, 6, 7.67)
				SSC$_{313}$	(0.5, 0.65, 0.8)	(4.33, 6, 7.67)
		SC$_{32}$	(0.7, 0.8, 0.9)	SSC$_{321}$	(0.7, 0.8, 0.9)	(5.67, 7, 8.33)
				SSC$_{322}$	(0.5, 0.65, 0.8)	(4.33, 6, 7.67)
				SSC$_{323}$	(0.7, 0.8, 0.9)	(4.33, 6, 7.67)
		SC$_{33}$	(0.5, 0.65, 0.8)	SSC$_{331}$	(0.85, 0.95, 1.0)	(5.67, 7, 8.33)
				SSC$_{332}$	(0.7, 0.8, 0.9)	(4.33, 6, 7.67)
				SSC$_{333}$	(0.7, 0.8, 0.9)	(4.33, 6, 7.67)

The sample calculation for environmental enabler is shown below.

Environment (SE$_1$) includes four criteria:

$$R_1 = \begin{bmatrix} (0.5, 0.65, 0.8) & \otimes & (4.30, 5.94, 7.61) \otimes \\ (0.5, 0.65, 0.8) & \otimes & (4.08, 5.78, 7.51) \otimes \\ (0.2, 0.35, 0.5) & \otimes & (3.43, 5.28, 7.13) \otimes \\ (0.3, 0.5, 0.7) & \otimes & (4.15, 5.83, 7.54) \end{bmatrix} / \begin{bmatrix} (0.5, 0.65, 0.8) \otimes \\ (0.5, 0.65, 0.8) \otimes \\ (0.2, 0.35, 0.5) \otimes \\ (0.3, 0.5, 0.7) \end{bmatrix}$$

$$R_1 = (4.08, 5.76, 7.48)$$

The performance ratings of different criteria and enabler are listed in Table 16.7.

16.9.2.4 Third Level Computation

FSI from the criteria performance ratings is computed as follows:

$$\text{FSI} = \sum_{i=1}^{v} (W_i \times R_i) \bigg/ \sum_{i=1}^{v} W_i \tag{16.11}$$

where W_i = Weight of ith enabler and R_i = performance rating of ith enabler
Sustainability is measured through three enablers:

$$\text{FSI} = \begin{bmatrix} (0.85, 0.95, 1) & \otimes & (4.08, 5.76, 7.48) \otimes \\ (0.5, 0.65, 0.8) & \otimes & (4.72, 6.29, 7.85) \otimes \\ (0.7, 0.8, 0.9) & \otimes & (4.82, 6.36, 7.90) \end{bmatrix} \bigg/ \begin{bmatrix} (0.85, 0.95, 1) \otimes \\ (0.5, 0.65, 0.8) \otimes \\ (0.7, 0.8, 0.9) \end{bmatrix}$$

$$\text{FSI} = (4.49, 6.10, 7.73)$$

Appropriate sustainability level corresponding to FSI should be identified. Euclidean distance method is used as superior method compared to other existing methods.

The linguistic expressions for different sustainability levels are expressed as follows:

$$\text{SL} = \big\{ \text{Slightly Sustainable (SS), Fairly Sustainable (FS), Sustainable (S),}$$

$$\text{Very Sustainable (VS) and Extremely Sustainable (ES)} \big\}.$$

The sustainability level using Euclidean distance method is calculated as follows:

$$D(\text{FSI}, \text{SL}_i) = \left\{ \sum_{xep} \left(f_{\text{FSI}}(x) - f_{\text{SL}_i}(x) \right)^2 \right\}^{\frac{1}{2}} \tag{16.12}$$

where

Table 16.7 Fuzzy Index for Different Levels of Sustainability

SE$_i$	SC$_{ij}$	R$_i$	R$_{ij}$
SE$_1$	SC$_{11}$	(4.08, 5.76, 7.48)	(4.30, 5.94, 7.61)
	SC$_{12}$		(4.08, 5.78, 7.51)
	SC$_{13}$		(3.43, 5.28, 7.14)
	SC$_{14}$		(4.14, 5.83, 7.54)
SE$_2$	SC$_{21}$	(4.72, 6.29, 7.85)	(5.03, 6.52, 8.00)
	SC$_{22}$		(4.13, 5.84, 7.56)
	SC$_{23}$		(5.11, 6.55, 8.01)
SE$_3$	SC$_{31}$	(4.82, 6.36, 7.89)	(4.82, 6.35, 7.89)
	SC$_{32}$		(4.82, 6.35, 7.89)
	SC$_{33}$		(4.83, 6.37, 7.90)

$D(\text{FSI}, \text{SL}_i)$ is the Euclidean distance between FSI and SL_i,

FSI is Fuzzy sustainability index,

SL_i is corresponding fuzzy number for natural language expression,

$f_{\text{FSI}}(x)$ FSI fuzzy triangular number,

$f_{\text{SL}_i}(x)$ SL_i fuzzy triangular number

The sample calculation for $D(\text{FSI}, \text{SS})$ is depicted as follows:

$$D(\text{FSI},\text{SS}) = \left\{ (4.49-0)^2 + (6.10-1.5)^2 + (7.73-3)^2 \right\}^{1/2}$$

Thus, $D(\text{FSI}, \text{SS}) = 7.98$

Euclidean distances for other conditions are as follows:

$$D(\text{FSI},\text{FS}) = 5.38$$

$$D(\text{FSI},\text{S}) = 1.93$$

$$D(\text{FSI},\text{VS}) = 1.55$$

$$D(\text{FSI},\text{ES}) = 4.14$$

Therefore, the minimum Euclidean distance $D(\text{FSI}, \text{VS})$ indicates that the manufacturing enterprise is "very sustainable." The approach is further extended to identify the critical issues in practicing sustainability. FPII is computed using performance ratings and weights representing the capability of an attribute to contribute for sustainability. Factors with lower FPII contribute less toward organization progress toward sustainability. FPII is computed as follows:

$$\text{FPII}_{ijk} = W'_{ijk} \otimes R_{ijk} \tag{16.13}$$

FPII_{ijk} is the FPII of kth attribute with respect to ith enabler and jth criteria:

$$W'_{ijk} = \left[(1,1,1) - W_{ijk} \right] \tag{16.14}$$

where W_{ijk} is the weight of kth attribute with respect to ith enabler and jth criteria.

The FPII value for each sustainability factor is computed and presented in Table 16.8. Ranking of fuzzy numbers is carried out using centroid method as follows:

$$\text{Ranking score} = \frac{l + 4m + n}{6} \tag{16.15}$$

where l, m, and n are lower, middle, and upper values of triangular fuzzy membership functions.

A threshold of 2.02 is fixed by the industry management. Sixteen attributes are identified as weaker sustainability factors and actions are planned for improving the sustainability performance of the potential attributes. Employee safety and health (SC_{212}), Job satisfaction (SC_{213}), Training and empowerment (SC_{214}), Empowerment of stakeholders (SC_{221}), Selective and collective audiences (SC_{222}), Decision influence potential (SC_{223}), Government Regulations (SC_{231}), Ethical responsibility (SC_{232}), Profit maximization (SC_{311}), Liquidity (SC_{312}), Solvency (SC_{313}), Technological advancements (SC_{321}), Leaner organizational structure (SC_{322}), National/Global grants (SC_{323}), Effect on

Table 16.8 FPII Corresponding to Different Subcriteria

Attribute	R_{ijk}	$W'_{ijk} = \left[(1,1,1) - W_{ijk} \right]$	FPII	Ranking Score
SSC_{111}	(5, 6.5, 8)	(0.2, 0.35, 0.5)	(1, 2.275, 4)	2.35
SSC_{112}	(4.33, 6, 7.67)	(0.2, 0.35, 0.5)	(0.86, 2.1, 3.83)	2.18
SSC_{113}	(4.33, 6, 7.67)	(0.2, 0.35, 0.5)	(0.86, 2.1, 3.83)	2.18
SSC_{114}	(3.67, 5.5, 7.33)	(0.3, 0.5, 0.7)	(1.1, 2.75, 5.13)	2.87
SSC_{115}	(3.67, 5.5, 7.33)	(0.3, 0.5, 0.7)	(1.1, 2.75, 5.13)	2.87
SSC_{121}	(4.33, 6, 7.67)	(0.2, 0.35, 0.5)	(0.86, 2.1, 3.83)	2.18
SSC_{122}	(3.67, 5.5, 7.33)	(0.3, 0.5, 0.7)	(1.1, 2.75, 5.13)	2.87
SSC_{131}	(3, 5, 7)	(0.5, 0.65, 0.8)	(1.5, 3.25, 5.6)	3.35
SSC_{132}	(3.67, 5.5, 7.33)	(0.5, 0.65, 0.8)	(1.83, 3.575, 5.867)	3.67
SSC_{133}	(3.67, 5.5, 7.33)	(0.5, 0.65, 0.8)	(1.83, 3.575, 5.87)	3.67
SSC_{134}	(2.67, 4.5, 6.33)	(0.5, 0.65, 0.8)	(1.33, 2.925, 5.067)	3.02
SSC_{135}	(3.67, 5.5, 7.33)	(0.3, 0.5, 0.7)	(1.1, 2.75, 5.13)	2.87
SSC_{136}	(3.67, 5.5, 7.33)	(0.3, 0.5, 0.7)	(1.1, 2.75, 5.13)	2.87
SSC_{141}	(3.67, 5.5, 7.33)	(0.2, 0.35, 0.5)	(0.73, 1.925, 3.67)	2.02
*SSC_{142}	(4.33, 6, 7.67)	(0.1, 0.2, 0.3)	(0.43, 1.2, 2.3)	1.26
*SSC_{143}	(4.33, 6, 7.67)	(0.1, 0.2, 0.3)	(0.43, 1.2, 2.3)	1.26
*SSC_{144}	(4.33, 6, 7.67)	(0.1, 0.2, 0.3)	(0.43, 1.2, 2.3)	1.26
SSC_{145}	(3.67, 5.5, 7.33)	(0.3, 0.5, 0.7)	(1.1, 2.75, 5.13)	2.87
*SSC_{211}	(5.67, 7, 8.33)	(0.1, 0.2, 0.3)	(0.567, 1.4, 2.5)	1.44
*SSC_{212}	(4.33, 6, 7.67)	(0.1, 0.2, 0.3)	(0.43, 1.2, 2.3)	1.26
*SSC_{213}	(5.67, 7, 8.33)	(0, 0.05, 0.15)	(0, 0.35, 1.25)	0.44
*SSC_{214}	(4.33, 6, 7.67)	(0.1, 0.2, 0.3)	(0.43, 1.2, 2.3)	1.26
*SSC_{221}	(4.33, 6, 7.67)	(0.1, 0.2, 0.3)	(0.43, 1.2, 2.3)	1.26
SSC_{222}	(4.33, 6, 7.67)	(0.2, 0.35, 0.5)	(0.86, 2.1, 3.83)	2.18
*SSC_{223}	(3.67, 5.5, 7.33)	(0.2, 0.35, 0.5)	(0.73, 1.925, 3.67)	2.02
SSC_{231}	(4.33, 6, 7.67)	(0.2, 0.35, 0.5)	(0.86, 2.1, 3.83)	2.18
*SSC_{232}	(5.67, 7, 8.33)	(0.1, 0.2, 0.3)	(0.567, 1.4, 2.5)	1.44
*SSC_{311}	(5.67, 7, 8.33)	(0.1, 0.2, 0.3)	(0.567, 1.4, 2.5)	1.44
*SSC_{312}	(4.33, 6, 7.67)	(0.1, 0.2, 0.3)	(0.43, 1.2, 2.3)	1.26
SSC_{313}	(4.33, 6, 7.67)	(0.2, 0.35, 0.5)	(0.86, 2.1, 3.83)	2.18
*SSC_{321}	(5.67, 7, 8.33)	(0.1, 0.2, 0.3)	(0.567, 1.4, 2.5)	1.44
SSC_{322}	(4.33, 6, 7.67)	(0.2, 0.35, 0.5)	(0.86, 2.1, 3.833)	2.18
*SSC_{323}	(4.33, 6, 7.67)	(0.1, 0.2, 0.3)	(0.43, 1.2, 2.3)	1.26
*SSC_{331}	(5.67, 7, 8.33)	(0, 0.05, 0.15)	(0, 0.35, 1.25)	0.44
*SSC_{332}	(4.33, 6, 7.67)	(0.1, 0.2, 0.3)	(0.43, 1.2, 2.3)	1.26
*SSC_{333}	(4.33, 6, 7.67)	(0.1, 0.2, 0.3)	(0.43, 1.2, 2.3)	1.26

Gross Domestic Product (GDP) (SC_{331}), Performance of market shares (SC_{332}), and Profitability of shares (SC_{333}) are weaker attributes.

16.10 SUMMARY

The present chapter describes the importance, principles, and concepts of sustainable development in modern manufacturing scenario. Widely used tools/techniques of sustainability are presented with sustainability measurement approaches. Modeling approaches for sustainable manufacturing, namely structural and system dynamics modeling and a case study on sustainability performance measurement, are discussed in detail. The importance of sustainability performance measurement in the modern manufacturing scenario is also highlighted.

REFERENCES

Ahi, P., & Searcy, C. (2013). A comparative literature analysis of definitions for green and sustainable supply chain management. *Journal of Cleaner Production*, 52, 329–341.

Al-Oqla, F. M., & Sapuan, S. M. (2014). Natural fiber reinforced polymer composites in industrial applications: Feasibility of date palm fibers for sustainable automotive industry. *Journal of Cleaner Production*, 66, 347–354.

Ansari, Z. N., & Kant, R. (2017). A state-of-art literature review reflecting 15 years of focus on sustainable supply chain management. *Journal of Cleaner Production*, 142, 2524–2543.

Bereketli, İ., Genevois, M. E., & Ulukan, H. Z. (2009). Green product design for mobile phones. *World Academy of Science, Engineering and Technology*, 58, 213–217.

Boks, C., & Stevels, A. (2003). Theory and practice of environmental benchmarking in a major consumer electronics company. *Benchmarking: An International Journal*, 10(2), 120–135.

Carter, C. R., & Rogers, D. S. (2008). A framework of sustainable supply chain management: Moving toward new theory. *International Journal of Physical Distribution & Logistics management*, 38(5), 360–387.

Charter, M. (1998). Sustainable product design. In Kostecki M. (ed.) *The Durable Use of Consumer Products* (pp. 57–68). Boston, MA: Springer

Choi, A. C. K., Kaebernick, K., & Lai, W. H. (1997). Manufacturing processes modelling for environmental impact assessment. *Journal of Materials Processing Technology*, 70, 231–238.

Carter, C. R., & Liane Easton, P. (2011). Sustainable supply chain management: Evolution and future directions. *International Journal of Physical Distribution & Logistics Management*, 41(1), 46–62.

Daniels, E. J., Carpenter, J. A., Duranceau, C., Fisher, M., Wheeler, C., & Winslow, G. (2004). Sustainable end-of-life vehicle recycling: R&D collaboration between industry and the US DOE. *JOM*, 56(8), 28–32.

Despeisse, M., Mbaye, F., Ball, P. D., & Levers, A. (2012). The emergence of sustainable manufacturing practices. *Production Planning & Control*, 23(5), 354–376.

Fisher, M. M., Mark, F. E., Kingsbury, T., Vehlow, J., & Yamawaki, T. (2005). Energy recovery in the sustainable recycling of plastics from end-of-life electrical and electronic products. In *Proceedings of the 2005 IEEE International Symposium on Electronics and the Environment* (pp. 83–92). IEEE. New Orleans, LA, USA.

Gehin, A., Zwolinski, P., & Brissaud, D. (2008). A tool to implement sustainable end-of-life strategies in the product development phase. *Journal of Cleaner Production*, 16(5), 566–576.

Genovese, A., Acquaye, A. A., Figueroa, A., & Koh, S. L. (2017). Sustainable supply chain management and the transition towards a circular economy: Evidence and some applications. *Omega*, 66, 344–357.

Giddings, B., Hopwood, B., & O'brien, G. (2002). Environment, economy and society: Fitting them together into sustainable development. *Sustainable Development*, 10(4), 187–196.

Girubha, J., Vinodh, S., & KEK, V. (2016). Application of interpretative structural modelling integrated multi criteria decision making methods for sustainable supplier selection. *Journal of Modelling in Management*, 11(2), 358–388.

Gold, S., Seuring, S., & Beske, P. (2010). Sustainable supply chain management and inter-organizational resources: A literature review. *Corporate Social Responsibility and Environmental Management*, 17(4), 230–245.

Hallstedt, S. I., Bertoni, M., & Isaksson, O. (2015). Assessing sustainability and value of manufacturing processes: A case in the aerospace industry. *Journal of Cleaner Production*, 108, 169–182.

Hassini, E., Surti, C., & Searcy, C. (2012). A literature review and a case study of sustainable supply chains with a focus on metrics. *International Journal of Production Economics*, 140(1), 69–82.

Huang, Y. A., Weber, C. L., & Matthews, H. S. (2009). Carbon footprinting upstream supply chain for electronics manufacturing and computer services. In *IEEE International Symposium on Sustainable Systems and Technology, 2009. ISSST'09* (pp. 1–6). IEEE.

Joung, C. B., Carrell, J., Sarkar, P., & Feng, S. C. (2012). Categorization of indicators for sustainable manufacturing. *Ecological Indicators*, 24, 148–157.

Jovane, F., Yoshikawa, H., Alting, L., Boër, C. R., Westkamper, E., Williams, D., & Paci, A. M. (2008). The incoming global technological and industrial revolution towards competitive sustainable manufacturing. *CIRP Annals-Manufacturing Technology*, 57(2), 641–659.

Kannan, G., Pokharel, S., & Kumar, P. S. (2009). A hybrid approach using ISM and fuzzy TOPSIS for the selection of reverse logistics provider. *Resources, Conservation and Recycling*, 54(1), 28–36.

Labuschagne, C., Brent, A. C., & Van Erck, R. P. (2005). Assessing the sustainability performances of industries. *Journal of Cleaner Production*, 13(4), 373–385.

Larson, T., & Greenwood, R. (2004). Perfect complements: Synergies between lean production and eco-sustainability initiatives. *Environmental Quality Management*, 13(4), 27–36.

Law, K. M., & Gunasekaran, A. (2012). Sustainability development in high-tech manufacturing firms in Hong Kong: Motivators and readiness. *International Journal of Production Economics*, 137(1), 116–125.

Lin, C.-T., Chiu, H., & Tseng, Y.-H. (2006). Agility evaluation using fuzzy logic. *International Journal of Production Economics*, 101(2), 353–368.

Linton, J. D., Klassen, R., & Jayaraman, V. (2007). Sustainable supply chains: An introduction. *Journal of Operations Management*, 25(6), 1075–1082.

Masui, K., Sakao, T., Kobayashi, M., & Inaba, A. (2003). Applying quality function deployment to environmentally conscious design. *International Journal of Quality & Reliability Management*, 20(1), 90–106.

Maxwell, D., & Van der Vorst, R. (2003). Developing sustainable products and services. *Journal of Cleaner Production*, 11(8), 883–895.

McDonough, W., & Braungart, M. (2002). Design for the triple top line: New tools for sustainable commerce. *Corporate Environmental Strategy*, 9(3), 251–258.

Mehregan, M. R., Hashemi, S. H., Karimi, A., & Merikhi, B. (2014). Analysis of interactions among sustainability supplier selection criteria using ISM and fuzzy DEMATEL. *International Journal of Applied Decision Sciences*, 7(3), 270–294.

Mohammed, J., & Sadique, A. (2011). Lean value stream manufacturing for sustainability. In Seliger G., Khraisheh M., Jawahir I. (eds.) *Advances in Sustainable Manufacturing* (pp. 365–370). Berlin and Heidelberg: Springer.

Mota, B., Gomes, M. I., Carvalho, A., & Barbosa-Povoa, A. P. (2015). Towards supply chain sustainability: Economic, environmental and social design and planning. *Journal of Cleaner Production*, 105, 14–27.

Ninlawan, C., Seksan, P., Tossapol, K., & Pilada, W. (2010). The implementation of green supply chain management practices in electronics industry. In *Proceedings of the International Multi Conference of Engineers and Computer Scientists* (Vol. 3, pp. 17–19). Hong Kong.

Pal, R. (2017). Sustainable design and business models in textile and fashion industry. In Muthu S. (ed.) *Sustainability in the Textile Industry* (pp. 109–138). Singapore: Springer.

Papadopoulos, T., Gunasekaran, A., Dubey, R., Fosso Wamba, S., & Childe, S. J. (2017). World Class Sustainable Supply Chain Management: Critical Review and Further Research Directions.

Rachuri, S., Sriram, R. D., & Sarkar, P. (2009). Metrics, standards and industry best practices for sustainable manufacturing systems. In *IEEE International Conference on Automation Science and Engineering, CASE 2009* (pp. 472–477). IEEE.

Rosen, M. A., & Kishawy, H. A. (2012). Sustainable manufacturing and design: Concepts, practices and needs. *Sustainability*, 4(2), 154–174.

Saride, S., George, A. M., Avirneni, D., & Basha, B. M. (2017). Sustainable design of indian rural roads with reclaimed asphalt materials. In Sivakumar Babu G., Saride S., Basha B. (eds.) *Sustainability Issues in Civil Engineering* (pp. 73–90). Singapore: Springer.

Schöggl, J. P., Baumgartner, R. J., & Hofer, D. (2017). Improving sustainability performance in early phases of product design: A checklist for sustainable product development tested in the automotive industry. *Journal of Cleaner Production*, 140, 1602–1617.

Seuring, S. (2013). A review of modeling approaches for sustainable supply chain management. *Decision Support Systems*, 54(4), 1513–1520.

Shi, L., & Chew, M. Y. L. (2012). A review on sustainable design of renewable energy systems. *Renewable and Sustainable Energy Reviews*, 16(1), 192–207.

Sodhi, M. S., & Reimer, B. (2001). Models for recycling electronics end-of-life products. *OR Spectrum*, 23(1), 97–115.

Sterman, J. D. (2001). System dynamics modeling: Tools for learning in a complex world. *California Management Review*, 43(4), 8–25.

Thirupathi, R. M., & Vinodh, S. (2016). Application of interpretive structural modelling and structural equation modelling for analysis of sustainable manufacturing factors in Indian automotive component sector. *International Journal of Production Research*, 54(22), 6661–6682.

Tian, J., & Chen, M. (2014). Sustainable design for automotive products: Dismantling and recycling of end-of-life vehicles. *Waste Management*, 34(2), 458–467.

Vasanthakumar, C., Vinodh, S., & Ramesh, K. (2016). Application of interpretive structural modelling for analysis of factors influencing lean remanufacturing practices. *International Journal of Production Research*, 54(24), 7439–7452.

Vinodh, S. (2011). Assessment of sustainability using multi-grade fuzzy approach. *Clean Technologies and Environmental Policy*, 13(3), 509–515.

Vinodh, S., & Joy, D. (2012). Structural equation modeling of sustainable manufacturing practices. *Clean Technologies and Environmental Policy*, 14(1), 79–84.

Vinodh, S., & Manjunatheshwara, K. J. (2017). Application of fuzzy QFD for environmentally conscious design of mobile phones. In Machado C., Davim J. (eds) *Green and Lean Management* (pp. 149–160). Cham: Springer International Publishing.

Vinodh, S., & Rathod, G. (2010). Application of QFD for enabling environmentally conscious design in an Indian rotary switch manufacturing organisation. *International Journal of Sustainable Engineering*, 3(2), 95–105.

Vinodh, S., Manjunatheshwara, K. J., Sundaram, S. K., & Kirthivasan, V. (2017). Application of fuzzy quality function deployment for sustainable design of consumer electronics products: A case study. *Clean Technologies and Environmental Policy*, 19(4), 1021–1030. doi:10.1007/s10098-016-1296-7.

Vinodh, S., Ramesh, K., & Arun, C. S. (2016). Application of interpretive structural modelling for analysing the factors influencing integrated lean sustainable system. *Clean Technologies and Environmental Policy*, 18(2), 413–428.

Wu, B. Y., Chan, Y. C., Middendorf, A., Gu, X., & Zhong, H. W. (2008). Assessment of toxicity potential of metallic elements in discarded electronics: A case study of mobile phones in China. *Journal of Environmental Sciences*, 20(11), 1403–1408.

Yang, S. L., & Li, T. F. (2002). Agility evaluation of mass customization product manufacturing. *Journal of Materials Processing Technology*, 129(1), 640–644.

Zailani, H.M.S., Eltayeb, T. K., Hsu, C. C., & Choon Tan, K. (2012). The impact of external institutional drivers and internal strategy on environmental performance. *International Journal of Operations & Production Management*, 32(6), 721–745.

CHAPTER 17

Sustainable Manufacturing—Getting Future Ready

Prasad Modak
Ekonnect Knowledge Foundation

CONTENTS

17.1 THE EVOLUTION

Sustainability of this planet depends heavily on the availability of resources. Today, resources are under threat due to severe depletion and degradation. Depletion has been on a steep rise due to increasing population, urbanization, and rising consumption. Degradation of the resources has been a result of reckless disposal of the residues.

Oddly enough, the national governments, particularly the Ministries of Environment, have focused more on the management of residues rather than management of the resources. Legislation was evolved to set limits on the residues that were required to be met prior to disposal. Unfortunately, not much attention was given on the limits of extraction of resources and resource pricing.

Limits on residues became stricter over the years as our understanding of the adverse impacts and risks to the environment improved. We realized that residues when not properly disposed could lead to considerable damage to the humans and the ecosystems. There were severe economic implications both on damage and restoration. Many of the impacts were found to be long term and irreversible and further compounded with risks that were not easy to anticipate.

Most national governments followed a precautionary approach, following a "do no harm principle" in setting the limits on residues. Advances in monitoring of pollutants in the residues and the technologies for their removal made tightening of the limits on residues possible.

Having framed the legislation and limits or standards on the residues, the national governments established institutions for monitoring and enforcement. Procedures and practices of documentation were laid down. Most legislations began with addressing management of wastewater but soon air emissions, solid, and hazardous wastes were included. Consequently, the investments on the end of pipe management of residues increased. The cost structure of the manufacturing industries got skewed with significant share of costs now taken by pollution control, essentially in the form of end of pipe treatment.

The manufacturing industries realized that to reduce cost of the end of pipe treatment and remain competitive, efforts were required to reduce generation of residues at the source. The concepts such as waste minimization and pollution prevention therefore emerged and the industries did every effort to reduce residue generation by deploying better housekeeping and practicing reuse, recycling, and recovery to the extent possible. This required a behavioral change, application of management systems, use of productivity improvement tools, and adoption to modern technologies. The investments for management of residues essentially moved upstream leading to "ecological modernization."

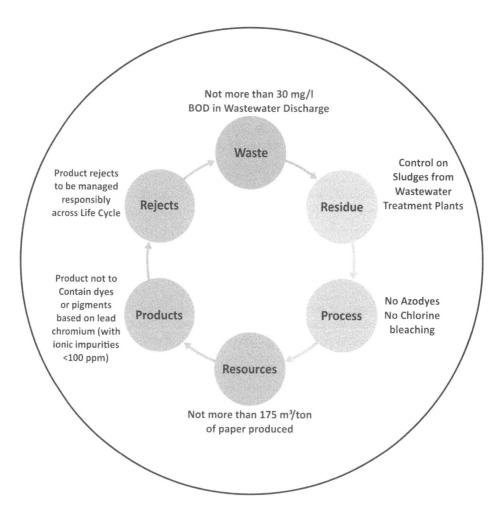

Figure 17.1   Evolution of environmental standards across life cycle—illustration of paper and pulp industry. (From Modak, P., *Environmental Management toward Sustainability*, 2017, CRC Press, Taylor & Francis Group.)

Unlike end of pipe investments, the "upstream" investments had a payback or economic returns. Strategies such as Cleaner Production, Green Productivity (GP), and Eco-efficiency were promising in this perspective. These strategies showed a link between resources (in specific the resource use efficiency) and the residues that could be converted as a resource.

Over a period of time, the legislation on residues expanded and became more comprehensive. Figure 17.1 shows an illustration of evolution of limits, expectations, and requirements for the pulp and paper sector across the "life cycle" [1,2].

Clearly, enforcement of such limits could not be done solely by the government. It required a partnership approach where the markets (consumers, retailers) and investors were also involved. The new paradigm of governance emphasized the G–B–FI–C (Government, Business, Financing Institutions and Communities) partnership (see Figure 17.2).

To guide and motivate industries toward sustainable manufacturing, many leading institutions and national governments, across the world, launched programs where interests of profitability could be integrated with the protection of the environment. The importance of product design was also realized that connected resources and residues across various stages of the life cycle, i.e., extraction, transportation, processing, packaging, distribution, use, disposal. The two R's (viz. Resources and Residues) were thus integrated with the opportunities of 3R (Reduce, Reuse, and Recycle). Reduce was an upstream strategy that influenced product design as well as consumption while reuse and recycle were downstream

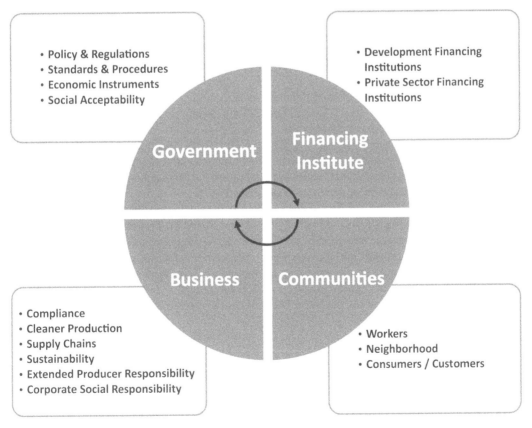

Figure 17.2 New paradigm of G–FI–B–C. (From Modak, P., *Environmental Management toward Sustainability*, 2017, CRC Press, Taylor & Francis Group.)

strategies that were built upon the involvement of informal sector. The requirement of Extended Producer Responsibility (EPR)[1] emphasized the role of sustainability considerations across the life cycle.

Launched in 1990, United Nations Environment Programme's (UNEP) Cleaner Production was one such "smart" Program. The concept of Cleaner Production was established by the UNEP's Division of Technology, Industry and Economics (DTIE) in Paris.

The term "Cleaner Production" was defined as follows: "The continuous application of an integrated environmental strategy to processes, products and services to increase efficiency and reduce risks to humans and the environment." This definition was rather "deep" yet "expansive": as it covered both processes and products covering segments of manufacturing and services. Previously, in the United States, the term "Pollution Prevention" prevailed that was introduced in the ISO 14001 Environmental Management System.

In 1992, the World Business Council for Sustainable Development (WBSCD) came with the concept of Eco-Efficiency. The concept was based on creating more goods and services while using fewer resources and creating less waste and pollution. The term "Eco-efficiency" was measured as the ratio between the (added) values of what has been produced (e.g., GDP) and the (added) environment impacts of the product or service (e.g., SO_2 emissions). WBSCD in its 1992 publication "Changing Course," introduced this term and at the 1992 Earth Summit, eco-efficiency was endorsed as a new business concept and means for companies to implement Agenda 21 in the private sector. The term "Factor 4" emerged that was later replaced by Factor 10.

In 1994, Asian Productivity Organization (APO) came up with a definition of GP. GP was defined as a strategy for enhancing productivity and environmental performance for overall socioeconomic development. GP was considered as the application of appropriate productivity and environmental management policies, tools, techniques, and technologies in order to reduce the environmental impact of an organization's activities. GP integrated several productivity-related tools along with methodologies followed in Cleaner Production Opportunity Assessments.

More recently, UNIDO along with UNEP came up with the concept of Resource Efficient Cleaner Production (RECP). RECP was defined as continuous application of preventive environmental strategies to processes, products, and services to increase efficiency and reduce risks to humans and the environment. RECP works specifically to advance production efficiency, management of environment, and human development.

The recently promulgated concept of Circular Economy added additional 3Rs—Repair, Refurbish, and Remanufacture. These additional 3Rs introduced three significant components, viz., social (employment), investment, and innovation. Once material flows become circular, compliance becomes a necessity and of interest to every stakeholder (see Figure 17.3).

China legislated Circular Economy Law as early as in 2007 focusing on industrial estates. Japan promoted this concept at Eco-Towns and introduced the term Sound Material Society. European Union came up with country specific targets, indicators, and reporting requirements on Circular Economy.

Circular Economy brought management of resources and residues together in the interest of economy, livelihoods, and the environment. Furthermore, if implemented well, then Circular Economy was expected spur innovation and stimulate investments.

Figure 17.4 presents the above evolution. We have to view the concept of sustainable manufacturing in this context.

---

[1] Extended Producer Responsibility is a concept where manufacturers and importers of products should bear a significant degree of responsibility for the environmental impacts of their products throughout the product life-cycle, including upstream impacts inherent in the selection of materials for the products, impacts from manufacturers' production process itself, and downstream impacts from the use and disposal of the products. Producers accept their responsibility when designing their products to minimise life-cycle environmental impacts, and when accepting legal, physical or socio-economic responsibility for environmental impacts that cannot be eliminated by design. (*Source*: Fact Sheet on Extended Producer Responsibility—Online source www.oecd.org/env/waste/factsheetextendedproducerresponsibility.htm.)

Figure 17.3 The 6Rs of circular economy.

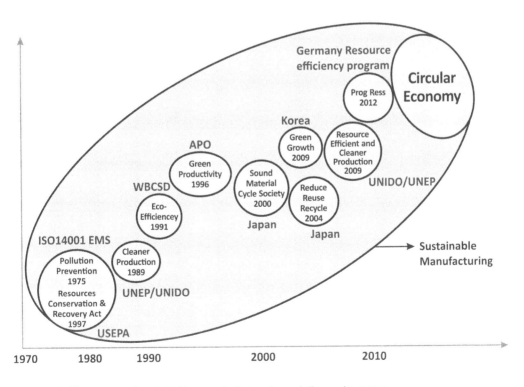

Figure 17.4 The canvas of sustainable manufacturing: Its evolution and expanse.

17.2 SUSTAINABLE MANUFACTURING

According to US Environmental Protection Agency (US EPA), the term "Sustainable manufacturing" is the creation of manufactured products through economically sound processes that minimize negative environmental impacts while conserving energy and natural resources. Sustainable manufacturing also enhances employee, community, and product safety.

To put simply, sustainable manufacturing is all about minimizing the diverse business risks inherent in any manufacturing operation while maximizing the new opportunities that arise from improving your processes and products.

"Sustainable manufacturing" is no longer just nice-to-have an "option." Today, a business imperative to be "future ready." Companies across the world are facing increased costs in sourcing materials, energy, and meeting compliance. In addition, there are higher expectations of customers, investors, and local communities. There are also caps or impositions on greenhouse gas (GHG) emissions due to the threat of climate change that demand right choice of energy sources, technologies, and materials.

Integration of economic, social, and environmental perspectives is the foundation of sustainability. Figure 17.5 shows the concept of sustainable manufacturing in these three dimensions.

Many businesses have already started to take steps toward sustainable manufacturing. Their experiences show that environmental improvements go hand in hand with profit-making and improved competitiveness. However, many small and medium-sized businesses (SMEs) that account for more than 90% of all enterprises have not yet embraced these great opportunities. Their enterprises often struggle with their short-term survival, or cost pressure from clients, or lack of knowledge and resources to invest in manufacturing sustainably, or simply not know where to start.

In the early phase of the sustainability revolution, the business was asking for the "evidence" that would prove that it was profitable to integrate business with environmental and social considerations. UNEP's International Cleaner Production Information Clearinghouse (ICPIC) came up with 400+ case studies across more than 20 industrial sectors covering not just large but medium and small-scale industries. Similar case studies emerged at the WBCSD, APO, and UNIDO. Today, we do not need any more convincing. We want to know more about "how to go about sustainable manufacturing."

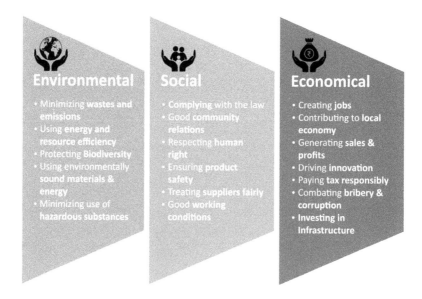

Figure 17.5 Perspectives of sustainable manufacturing. (From www.oecd.org/innovation/green/toolkit/about-sustainablemanufacturingandthetoolkit.htm.)

17.3 IMPORTANCE OF SUSTAINABLE PRODUCT DESIGN

Businesses started exploring means to make their processes and products more sustainable. The first step was to innovate and alter the product design. Concepts such as Design for Environment (DfE) and Design for Sustainability (D4S) emerged. D4S involves designing a product, service, or process while considering its social and environmental impact across the product's lifecycle. For instance, the lifecycle of a product spans across stages like raw material procurement, manufacturing, packaging, transportation, sales, usage, reuse, and disposal. Use of renewable, biodegradable, or recycled material, reducing the quantity of material required by altering design, designing for easy disassembly, reuse and recycling, minimizing quantity or increasing reuse of packaging, and enhancing resource efficiency of processes are some D4S strategies.

Use of materials other than virgin material for manufacturing or altering their products has been a prime focus of DfE and D4S strategies. In this approach, there were dual-objectives such as to improve the bottom line by procuring locally produced and easily available material and to lower the ecological impact by using sustainable or recycled material. Some of the material sources that are rapidly gaining popularity are renewable and natural sources (like bamboo, hemp, cork, agri-fibers) and use of pre-consumer or post-consumer recycled material (like steel, plastic from used products or ocean waste). In designing the sustainable products, the following principles are proposed:

- Use of low-impact or low material intensity materials
- Minimum use of nonbiodegradable or recalcitrant substances
- Use of local materials to the extent possible without adversely affecting local sustainability
- Design such that there is low energy demand/consumption and a higher energy efficiency
- Maximum possible use of renewable energy both in the making and use
- Design of the components for easy dismantling, reuse, and recycling
- Low carbon footprint through the products lifecycle

There are several examples of sustainable products [3,4]. The Bedol Water Clock is one case where clock operates on water without batteries or electricity. The electrodes within the water reservoir of the clock convert ions into a current strong enough to power the clock for at least 3 months. The second example is the EcoKettle, which was invented in the UK. It contains a special compartment in which the water is stored. You can then transfer desired amount of water in the second compartment, which will be the only one to actually boil the water. This prevents energy that is wasted by boiling more water than you actually need.

Many designers are now looking to prolong the product's shelf life and make product and packaging refillable. An example is that of Bobble that is free of bisphenol A and complete with a carbon filter to remove contaminants from ordinary tap water. Bobble filters tap water and is 100% recyclable once it needs replacing that is typically over 2 months. It is a smart solution to the current problem of disposable bottled water filling up our landfills.

Products are also designed in a modular style with flexibility to allow different configurations depending on the interest and use. AMAC's Rhombin Desktop Storage units are bins in the shape of equilateral triangles, and are designed to stack as well as cluster together in limitless configurations. These units are made in California using Cereplast, a plant-based bioplastic.

There is a lot of interest to make products using existing waste streams. In 2016, Adidas collaborated with Parley for the Oceans and innovated the world's first shoe using ocean plastic waste. Other remarkable examples in this category come from startups Terracycle and Loopworks. One concerns customers have is that there are no environmental standards prescribed for products made from waste to ensure health and safety of the consumer.

We present in Box 1 some interesting examples of DfE/D4S.

BOX 1 EXAMPLES OF DFE FOR RECYCLING, REUSE, OR DISASSEMBLY

- Xerox established a Recycling Design Guideline way back in 1995 that focused on strengthening design, easing product disassembly, standardizing parts to allow reuse in subsequent models, using sustainable materials and improving product longevity and reusability. One example is redesigning of the caster (the wheel's part used at the bottom of printers or copiers) for disassembly. The wheel of the caster was made separable from the wheel to allow replacement and reuse of the caster instead of tossing the caster away in case of wheel rupture [5]
- PUMA's Clever Little Bag was a packaging innovation that used 65% less cardboard required in shoeboxes. The shoebox structure was redesigned to use a recycled plastic bag that held the cardboard container that required no top. The new design also eliminated the need for plastic bags and tissue paper that is typically stuffed in with shoes. The innovative design drastically reduced paper consumption and use of resource like energy, diesel and water use by 60% (as estimated in 2010) [6]
- In 2008, Philips committed to designing sustainable products. Senseo Up, its first one-cup coffee machine, is an outcome of this commitment. Senseo Up designers successfully incorporated recycled plastics into the design despite the aesthetic challenges associated with use of recycled plastic in a high-end product. The product uses 13% recycled plastic in its internal frame and baseplate. These are parts that do not come in contact with food or in direct line of vision of users [7]

ReTuna Återbruksgalleria in Sweden is a shopping mall that sells only repaired or upcycled products and has gone beyond the local drop-off centers. Here, the dropped off goods are sorted into various workshops where they are refurbished or repaired. There are 14 workshops that include furniture, computers, audio equipment, clothes, toys, bikes, gardening, and building materials; all garnered from second-hand products. Several of these shops function as "do-it-yourself" showrooms, where customers can learn how to repair or refurbish. The products then reach the mall. Visitors can enroll in a one-year Design–Recycle–Reuse program.

Sustainable products are also made by following crowdsource-based innovation processes. One prominent example is Unilever's Foundry IDEAS platform, which acts as a digital hub for consumers and entrepreneurs to work together on tackling global sustainability challenges.

Products are now designed to take advantage of densely populated areas. Pavegen, a flooring tile company, converts kinetic energy from people's footsteps into renewable energy. Pavegen has set up streets with such tiles next to places of convention and sports where the commuters walk and generate electricity for street lighting.

Sustainable product design is sometimes more than the "design." It can even include a strategy. On an average, 30% of garments in our closets are many times not worn for almost a year. Lease A Jeans is a guilt-free solution from the US-based company, MUD Jeans, for people who have a desire for newness and are conscious about the environment. This company leases jeans instead of selling them, encouraging customers to return them after use. The old jeans are recycled to make new wonderful items. This leads to a considerable saving of water and energy, and leads to reduced generation of waste.

17.4 SUSTAINABLE PACKAGING

When we look at greenness of a product, type, extent, and material of packaging should also a part of the perspective. While a product could be green, its packaging may not. Packaging accounts for almost 10% of the environmental impact of anything bought. In estimating the "environmental impact" of the package, we assess the material (i.e., embodied intensity), its biodegradability and recyclability, and the label and method of printing (e.g., ink used). Today, increased consumer demand has compelled many companies to make packaging as sustainable as possible.

The global packaging industry is estimated to be $429 billion. According to a study from Pike Research, sustainable packaging is a fast-growing segment of the global packaging industry, and will grow to 32% of the total market by 2014, up from just 21% in 2009.

It is not surprising therefore that you see commitments and innovations in eco-friendly packaging. EnviroPAK in North America, for example, uses complex recycled paper pulp for packing electronic goods. By opting for paper pulp in the place of expanded polystyrene, the company has claimed to save 70% in packaging and shipping costs.

Take example of the Clever Little Bag from Puma Shoes, designed by Yves Béhar. Yvers's design completely ignored the traditional route of a box and lid combination and instead developed a new design significantly reducing the amount of waste and CO_2 emissions normal shoe packaging creates. It also cut down the packaging waste. It was estimated that the new design reduced 8,500 tons of paper, saved 20 million megajoules of electricity and 1 million liters less fuel oil and water over a year. In addition, 500,000 L of diesel and 275 tons of plastic were saved during transportation. Therefore, clever designs save resources, make profits, and build a brand.

Sustainable packaging often leads to innovations. While use of recycled paper in packaging is one of the common options, Dell has pioneered the use of bamboo to protect certain devices. Two-thirds of Dell's portable devices will ship in bamboo by the end of 2011. Bamboo is local (hence has less GHG emissions), grows quickly, and is strong and durable. In addition, bamboo packaging is biodegradable and can be composted after use.

However, "underpackaging" is also not desirable. A recent report by the Global Packaging Project states that the environmental risks of underpackaging can be greater than excessive packaging. By reducing packaging excessively, products get damaged in transit, requiring re-manufacturing and re-distribution in order to replace the original products. Furthermore, there are costs and liabilities for disposal of off-spec, discarded, or foul products. Thus, by trying to reduce the environmental impact of packaging, companies may simply be shifting, and potentially increasing, the adverse impact to the environment. Therefore, a careful balance is needed and let us not "overdo" sustainability. For coming up with a sustainable packaging design therefore, every manufacture has to think "out of the box" and yet be "practical."

17.5 PRODUCT ECO-LABELS

According to the Global Ecolabeling Network, "Eco-labeling" is a voluntary method of environmental performance certification and labeling that is practiced around the world. An eco-label identifies products or services proven environmentally preferable based on Life Cycle, within a specific product or service category.

The manufacturing industry, especially with exports to some of the sustainability conscious markets in the world, need to reflect requirements of eco-labels in the product making.

Germany's Blue Angel was one of the first world's first eco-label some four decades ago. Organization for Economic Cooperation and Development (OECD) analysis from 2013 found that the number of labels increased roughly fivefold between 1988 and 2009. As of 2015, the Ecolabel Index lists 377 schemes in 211 countries and 25 industry sectors [8]. Increasing number of eco-labels is a response to the pressure for greater environmental sustainability of production and consumption systems.

The eco-labels have evolved through four "waves" (Figure 17.6). The first wave focused on greenness of the product addressing resources (inputs) and wastes/emissions (outputs) showing

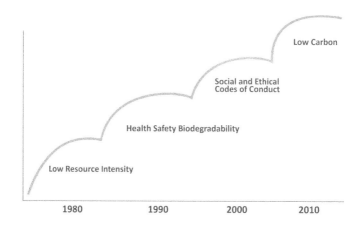

Figure 17.6 Waves in the evolution of eco-labels influencing manufacturing.

Figure 17.7 Benefits of eco-label.

preference to products that had low "resource intensity[2]" or low "ecological rucksack.[3]" The next wave addressed the health, safety, and biodegradability-related considerations result of which led to phasing and substitution of harmful and nonbiodegradable substances. The third wave looked at social issues such as management of labor (working hours, fair wages, and child labor) and came up with requirements to meet the codes of conduct[4] and ethical practices across the supply chains.

---

[2] Resource intensity is a measure of the resources (e.g. water, energy, materials) needed for the production, processing and disposal of a unit of good or service, or for the completion of a process or activity; it is therefore a measure of the efficiency of resource use.

[3] An Ecological Rucksack is the total quantity (in kg) of materials moved from nature to create a product or service, minus the actual weight of the product. That is, ecological rucksacks look at hidden material flows. Ecological rucksacks take a life cycle approach and signify the environmental strain or resource efficiency of the product or service.

[4] Code of conduct. A code of conduct is a set of rules outlining the social norms and rules and responsibilities of, or proper practices for, an individual, party or organization. Related concepts include ethical, honor, moral codes and religious laws.

The fourth wave came up with a need to reduce carbon footprints in product making, packaging, and transportation showing preference to low-carbon goods and services. Today, the consumer demand for lower-carbon products and services is growing, despite the tough economic climate.

In terms of "downstream" of the supply chains, especially the SMEs, adherence to criteria set for the eco-labels became part of the requirements in "vendor registrations." Eco-labels became like a benchmark that companies must meet if they want to continue to be suppliers to multinational brands.

Eco-labels offer many benefits. Fulfilling the eco-label criteria can be beneficial for businesses by encouraging them to adopt more environmentally sound management practices and business models, and by helping them to improve efficiencies. Eco-labels can also help in better branding and widening of the market—especially sectors like organic food. Figure 17.7 illustrates the benefits of eco-labeling. Among all, product innovation is considered to be the principal opportunity.

We illustrate in Box 2 a case study of economic and environmental and economic benefits of eco-labeling at Misr Mahalla in Egypt.

BOX 2 ECONOMIC AND ENVIRONMENTAL BENEFITS OF ECO-LABELING AT MISR MAHALLA IN EGYPT [9,10]

Misr Mahalla is the largest textile company in Egypt. The company had to meet the requirement to comply with Oeko-Tex 100 standard as imposed by the German market. Misr Mahalla under the support of the SEAM Program achieved this requirement with significant economic and environmental benefits. These benefits are elucidated below.

CHEMICAL AND DYE SUBSTITUTION

Meeting of Oeko-Tex standard required a chemical audit followed by substitution of hazardous chemicals and certain objectionable dyes. The chemical substitution and process modifications resulted in annual savings of LE30,456. Savings resulted from modifications to the bleaching process were LE89,820 on annual basis. Against these gains, purchasing eco-friendly's chemicals and dyes resulted in a yearly increase of LE59,364.

REDUCED SHIPMENT ANALYTICAL COSTS

Previously, a number of German clients required that all shipments without an eco-label had to be tested, resulting in analytical costs of LE1,000 per shipment. This requirement was not required for articles with an eco-label. The annual savings to the company by reducing the analytical costs was around LE20,000.

POTENTIAL EXPORT MARKET GAINS

The annual value of the factory export market is around LE383 million. Of this, almost 15%, or LE57.5 million in value was export to Germany. Because of the grant of eco-label, the market was expected to increase by 5%, equivalent to gain of LE2.9 million per annum.

IMPROVED PRODUCTION EFFICIENCY

Financial benefits also resulted from the following:

- 5% improvement in the Right First Time in the dyeing process;
- 20% reduction in the processing time;
- 14% reduction in steam consumption.

(Continued)

**BOX 2 (*Continued*) ECONOMIC AND ENVIRONMENTAL BENEFITS
OF ECO-LABELING AT MISR MAHALLA IN EGYPT [9,10]**

IMPROVED PRODUCT QUALITY

There was a noticeable improvement in the product quality. When sodium hypochlorite was used for bleaching, there were occasional incidences of low tensile strength, which at times was 20% lower than the required standard. Since the elimination of hypochlorite with hydrogen peroxide such incidences have not occurred.

ENVIRONMENTAL IMPROVEMENTS

- Complete phase-out of sodium. This resulted in the elimination of AOX (Adsorbable Organic Halogens) and a reduction of Total Dissolved Solids (TDS) in the effluent.
- Improvement in working conditions through the elimination of hazardous chlorine compounds.
- Reduction in water consumption leading to reduction in volume of generated effluent.
- Reduction in energy requirements.

17.6 CONCEPT OF MINIMUM ENVIRONMENTAL CARE SIZE

Industrial production is important to meet our needs and generate employment. To ensure that the industrial production is economically feasible, the industry must operate above a Minimum Efficient Scale (MES).

MES can be computed by equating average cost with marginal cost. The rationale behind this is that if an industry were to produce a small number of units, its average cost per output would be high because the bulk of the costs would come from fixed costs. But if the industry produces more units, then the average cost incurred per unit will be lower as the fixed costs are spread over a larger number of units. In such a case, the marginal cost is below the average cost, pulling the latter down. An efficient scale of production is reached when the average cost is at its minimum and therefore the same as the marginal cost. If we exceed the MES, then the marginal costs may increase due to pressures on product distribution (logistics), additional labor oversight and need for tapping more resources that are not locally available.

Many industries do not perform well because they arrive at MES without considering or sometimes not adequately internalizing the costs that they must incur on environmental pollution control. When they approach the regulator for a permit or consent, they are stipulated several conditions on permissible pollution discharge. Compliance to these conditions often upsets the overall profitability of their operations. Consequently, many industries receive closure orders from the regulator and judiciary directives due to noncompliance.

Here the concept of Minimum Environmental Care Size (MECS) becomes relevant (see Figure 17.8). MECS is essentially the minimum level of production that is needed to be compliant to the environmental norms or standards while remaining economically viable. Both scale and manufacturing technology play a significant role in arriving at the MECS. Of course, there are other equally important variables such as the location (where resources are extracted and processed) and the demand on the products (especially the green products) from the market.

Figure 17.8 leads us to the following observations

- Industries that operate on the MES are often unable to do an environmentally sound or responsible business. Perhaps scales higher than MES allow use of more resource minimal and efficient technologies. An MECS may therefore be higher than a conventional MES.

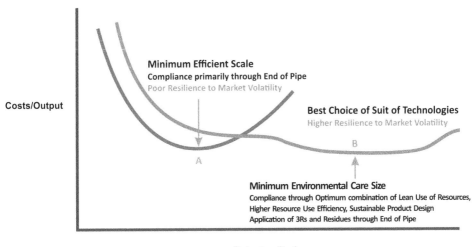

Costs/Output

Minimum Efficient Scale
Compliance primarily through End of Pipe
Poor Resilience to Market Volatility

Best Choice of Suit of Technologies
Higher Resilience to Market Volatility

B

A

Minimum Environmental Care Size
Compliance through Optimum combination of Lean Use of Resources,
Higher Resource Use Efficiency, Sustainable Product Design
Application of 3Rs and Residues through End of Pipe

Output or Scale

Figure 17.8 Minimum economic and environmental care size.

- At this scale, the costs/output would be lower and hence even if the costs of investments may be higher, the overall economic returns will be impressive.
- In addition, the MECS will exhibit higher resilience to the volatility of the markets.

An analysis of the cost of production break down between a large forest-based pulp mill in India with chemical recovery and an agro-based small mill without chemical recovery has shown that the chemical cost alone is 30% of the total cost of production against a figure of 21% for forest-based mills. A decade ago, Indian machinery manufacturing companies have shown that, when the mills reach a level of 100 TPD (Tons Per Day) black liquor solids, it is viable to set up a chemical recovery plant. Today, this threshold could be lower.

Pulp mills of small sizes (20–30 TPD capacity) cannot afford a chemical recovery unit and they would continue to discharge harmful chemicals into the environment. As the society and the regulator cannot allow continuation of discharge of polluted effluent, either the industry will have to close down or find out alternative methods production to stop pollution or take production to higher scale. The latter is often not possible due to shortage of finance.

In such cases, one may conceive a central or common chemical recovery for a number of pulp mills, where black liquor of individual mills can be collected and processed in a Central Recovery Plant. The white cooking liquor produced in the Central Chemical recovery plant can be transported to the individual mills for their use. Again, the Central chemical recovery unit shall be of a capacity which is technically desirable and is viable financially. The Centralized Recovery unit can either be an independent unit or an integrated unit with one of large mills in the cluster.

In Figure 17.9, note the points B and C carefully. The MECS with support of a Common Resource Recovery Center (CRRC) and a common end of pipe solution could be competitive compared to MECS for a larger industry. The small and medium enterprises (SMEs) under a cooperative agreement can thus still do business sustainably on a smaller production scale. What you need is a proper industrial planning, right institutional set up and an interested technology provider/investor for a joint venture on CRRC and a common end of pipe solution.

You can see potential of this concept for chrome recovery in tannery clusters, metal recovery in the cluster of electroplating industries and spent acid recovery in chemical industries. We want to see more manufacturing (in the interest of economic development and employment) but on a scale that will ensure environmentally and socially sound production.

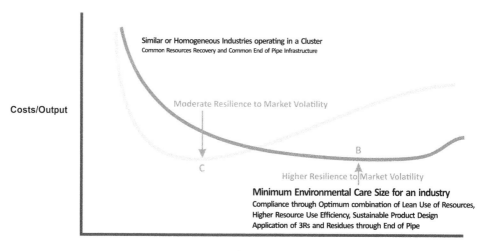

Costs/Output

Similar or Homogeneous Industries operating in a Cluster
Common Resources Recovery and Common End of Pipe Infrastructure

Moderate Resilience to Market Volatility

C

B

Higher Resilience to Market Volatility

Minimum Environmental Care Size for an industry
Compliance through Optimum combination of Lean Use of Resources,
Higher Resource Use Efficiency, Sustainable Product Design
Application of 3Rs and Residues through End of Pipe

Output or Scale

Figure 17.9 Common infrastructure and minimum environmental care size.

17.7 SUSTAINABILITY CONSIDERATIONS WHILE MAKING TECHNOLOGY CHOICES

Technologies are always evolving and present several options while making decisions. Generally, costs are given prime consideration as a selection criterion while meeting the required target of efficiency and yields. Costs include the capital and operating costs on a life time basis. We calculate the present worth of the two options that amortizes the operating costs and help us to make a comparison to select the technology of least cost.

It has been now realized that costs alone may not be the sole criteria. While cost could be one of the critical and deciding factors, there could be other considerations as well, especially on the environmental and social frontiers.

One of the important environmental factors is the generation of wastes and emissions. We consider this perspective in adding the **costs of pollution control and monitoring** to the base costs of the technology. These costs include installation as well as operating costs. A technology may not be preferred when costs of pollution control are added to the base costs.

Another example is the emissions of greenhouse gases (GHG). While there are no emission limits legally prescribed, most would like to choose technology option that has minimum emissions of GHG. Such technologies are often called as **low carbon technology** options. Technologies that use more of renewable energy get preferred as compared with those which may use of fossil fuels.

Exclusion is another environmental criterion that could decide the choice of technology. For example, we may not like to use technologies that use mercury or chlorine in any form. On the other hand, there could be preferences. We may prefer technologies that can make **maximum use of the locally sourced materials**.

We also need to factor the **costs of disposal of the end of life of the equipment**. In some cases, because of the use of nonbiodegradable or potentially hazardous materials—costs of disposal of the used or abandoned equipment could be substantial. This aspect could influence the final decision between technology options.

Environmental risk is another perspective. Some technologies may be compliant to the waste/emission standards that are legally prescribed but may pose risks of process upsets even leading to disaster during operations. These risks may arise due to process abnormalities that may occasionally arise—albeit rarely, posing certain restrictions on continuous operations or requiring online monitoring with process control or investments on the back-up systems. These considerations could add to the total costs.

Generation of odor and noise and safety become additional considerations in the interest of workers and the neighborhood. Technologies that lead to nuisance during manufacturing, shut down, and startup operations are not preferred.

Social considerations could also play an important role in the selection of technologies. These considerations include **local employment**. When one of the objectives of investment is to improve livelihoods of people, then technology options that generate local jobs may be preferred. Here low-to-moderate scale of technology and associated investments, especially in offering decentralized solutions, get preferred. These technologies should be easy to operate.

When investments are coming from public funds, a transparent process of technology assessment needs to be followed. The International Environment Technology Center (IETC) at the UNEP came up with a methodology called Sustainability Assessment of Technologies (SAT). Figure 17.10 shows the steps followed.

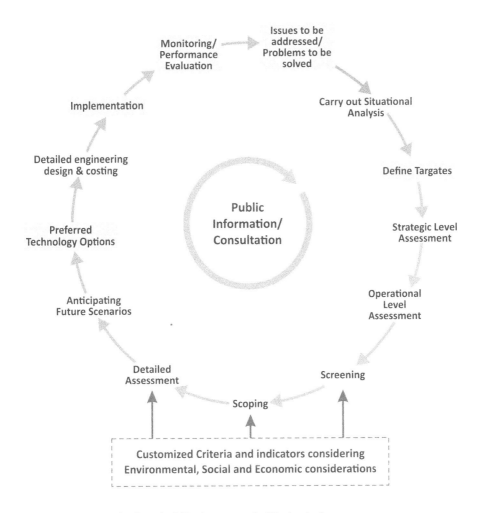

Figure 17.10 Methodology for Sustainability Assessment of Technologies.

It may be observed that the SAT methodology is in tiers such as screening, scoping, and detailed assessment. Furthermore, it follows a PDCA (Plan–Do–Check–Act) cycle as implemented in the most management systems. Importantly it involves expert and if appropriate a public consultation. Lastly, the methodology could be used to select options at both strategic and operational levels. More details on SAT can be accessed from the manual available online [11].

Applications of SAT methodology have been carried out for making choice of technologies across sectors such as water treatment, wastewater recycling, waste recovery, etc. The SAT manual cited above includes a solved example that could be perused for better understanding.

17.8 CONCLUDING REMARKS

Businesses always look for Compliance, Competitiveness, Continuity, and Communication, i.e., the 4Cs. These 4Cs pave the path toward sustainability. Sustainable manufacturing is now seen as an opportunity—to cut costs, to increase revenue, to innovate, to look good, and to stand apart.

Innovation and sustainability are closely linked. Leading organizations are teaming up with academia and investing in research, startups, and such efforts, to develop new technologies. A sizable amount of investment is spent on R&D for such innovation which is expected to save money as well as lessen the adverse impact on society and environment. Some organizations are highlighting such innovations on the canvas of their corporate philosophy on sustainability. However, resources for upscaling the innovation are not always available. There is also a low appetite from stakeholders—less encouragement from the government, not enough of a push from the consumer and low interest from employees.

Unfortunately, sustainability considerations in manufacturing have not yet penetrated in the graduate-level education programs, especially in the developing world. The ocean of resources created and the practice experience documented have not yet reached the student and community of young professionals. We need to run continuing education programs on these topics especially for the mid-level industry professionals. Those on the top layer are generally aware of the benefits of integration. However, we have definitely a long way to go for mainstreaming sustainability in the business.

We still need more evidence to prove that sustainability is not just ethos or a responsibility, but that it is material and can be monetized. This realization will send ripples across the supply chains influencing SMEs. Very few organizations have begun looking at sustainability across their supply chains. There are few leaders who have put in place policies and requirements to bring up the standards of their vendors and suppliers. Organizations however recognize that it is a risk to ignore the supply chain since any slip up on this front will mar the company image and hurt the brand. The requirement of sustainable manufacturing is thus creeping across the supply chains influencing the SMEs.

Certainly, the doctrine on sustainable manufacturing is not going to be limited to large corporates and multinationals. We will also see many organizations investing in sustainability to spur innovation—there will be an increase in innovative technologies and materials which will improve products and processes. Sustainable manufacturing will be the business strategy to be future-ready.

17.9 USEFUL TOOLKITS

17.9.1 The OECD Sustainable Manufacturing Toolkit

The OECD Sustainable Manufacturing Toolkit [12] provides ideas to improve the efficiency of industrial manufacturing processes and products such that sustainable and green manufacturing is taken seriously. This also includes set of indicators that help industries rank their performance in sustainable manufacturing.

The toolkit includes a guidance report and a web portal explaining the indicators and details on performance management. The facility will help beginners and experts, or those in between. The toolkit is actually meant for small and moderate manufacturing initiatives, though it will be helpful for other size and type of companies in the world.

17.9.2 RECPnet and PRE-SME Resource Kit

RECPnet is a global network for Resource Efficient and Cleaner Production. RECP was formulated in November 2010. The RECP Project developed a web-based knowledge management system and undertook several capacity building initiatives at the regional chapters.

The main objective of this network is to mainstream RECP concepts, methods, policies, practices, and technologies in developing and transition economies. The network encourages cross-sharing of RECP's relevant knowledge, experiences, and technologies globally through multiple collaborations.

The PRE-SME Resources Kit at RECPNet [13] is a generic electronic resource kit to assist SMEs to implement RECP. The kit is organized by resource category (water, energy, chemicals, wastes, and materials) with a clear methodological guidance, based on an in-depth survey and review of existing tools and techniques.

The PRE-SME resource kit is targeted to (1) **CEOs**, (2) SME **operations' Managers**, and (3) RECP consultants. This kit is available on the web and can be downloaded [14].

REFERENCES

1. The European Ecolabel for Copying and Graphic Paper - europa.eu, http://ec.europa.eu/environment/ecolabel/documents/copying_paper.pdf.
2. CPCB Pulp, Paper Regulatory Overview, www.gpcb-kp.in/live/hrdpmp/hrdpmaster/igep/content/e48745/e49028/e51797/e52844/02_PulpPaperSector-RegulatoryOverview_BRNaidu_CPCB.pdf.
3. 5 Types of Sustainable Products to Follow in 2015 and Beyond, www.sustainablebrands.com/news_and_views/blog/dimitar_vlahov/5_types_sustainable_products_follow_2015_beyond.
4. 25 Eco-friendly Products for Your Daily Life, www.conserve-energy-future.com/25-green-eco-friendly-products.php.
5. Fuji Xerox Reuse/Recycling Design, www.fujixerox.com/eng/company/ecology/cycle/newstyle/design/.
6. Puma's 'Clever Little Bag' Slashes Sneaker Packaging, www.greenbiz.com/news/2010/04/13/puma-clever-little-bag-slashes-packaging.
7. The Small Coffee Maker that's Big on Sustainability, www.90yearsofdesign.philips.com/article/6.
8. Global Ecolabelling Network (GEN), www.globalecolabelling.net.
9. The SEAM Project - An Introduction, www.eeaa.gov.eg/seam/Manuals/ecolabell/CONTENT-1.html.
10. Modak, P., 2017, *Environmental Management Towards Sustainability*, CRC Press, Taylor & Francis Group.
11. Application of the Sustainability Assessment of Technologies Methodology: Guidance Manual, https://wedocs.unep.org/rest/bitstreams/17340/retrieve.
12. OECD Sustainable Manufacturing Toolkit, www.oecd.org/innovation/green/toolkit/.
13. The Global Network for Resource Efficient and Cleaner Production, www.recpnet.org.
14. UNEP Resource Efficiency Toolkit for SMEs, http://wedocs.unep.org/bitstream/handle/20.500.11822/7961/-Promoting%20resource%20efficiency%20in%20small%20and%20medium%20sized%20enterprises_%20industrial%20training%20handbook-2010PRE-SME_handbook_2010.pdf?sequence=3&isAllowed=y.

CHAPTER 18

Dissemination of Sustainability and Education

R. Ganesh Narayanan
IIT Guwahati

Jay S. Gunasekera
Ohio University

CONTENTS

18.1 RELEVANCE OF MANUFACTURING TO OTHER SUBJECTS

Although the objective of the chapter is to highlight the importance of education in disseminating knowledge on sustainability in manufacturing, there are allied subjects in mechanical engineering such as machine and product design, structural and fluid mechanics, thermal engineering, and aerodynamics, which are related to manufacturing in many ways. From the industry's perspective, the subjects cannot be seen separately. Any modification in the life cycle of a product toward sustainability would require modifications at least in a few of the allied aspects, along with manufacturing and materials. Some of the typical examples are presented here.

Hallstedt et al. (2015) summarized the impact of decision-making system developed on the assessment of production of an aerospace engine component. The decision-making system involves an environmental impact assessment system which has suggested a nondestructive testing using hazardous substances and electrochemical machining (ECM) process generating hazardous substances (when applied for nickel-based alloys) as "hotspots" and more analyses should be included to minimize such problems and help in sustainable production. The strategic sustainability assessment finally proposed mechanical milling as a replacement for ECM, although it is costly. The mechanical milling involves only one hazardous substance, nickel, and produces less toxic material when compared to the ECM.

Hermatic reciprocating compressors generally seen in refrigerators use oil to provide adequate lubrication and should be compatible with the refrigerant. In order to have energy efficiency, reliability, and durability, the friction and thermodynamics losses should be minimized with appropriate oil usage. Oil acts as a heat transfer medium in the refrigerator circuits. Prata and Barbosa (2009) highlighted the importance of oil in the thermodynamic, heat and mass transfer, and fluid flow processes in such compressors. The thermodynamic aspects are related to mixing of oil with refrigerant, heat transfer aspects to temperature distribution on the compressor shell, and heat removal rate, and the fluid mechanics aspects to mechanism of oil pumping and circulation to the constituents of the bearings system.

The design and manufacturing of wind turbine blades involves aerodynamics analyses and design. Such analyses are inevitable to decide the suitability of material, dimensional design, and in-situ performance. The research investigation done by Jureczko et al. (2005), Kong et al. (2005), Maalawi and Negm (2002), and Habali and Saleh (2000) discusses such aspects using computations (e.g., optimization, FEM software, etc.) and experiments. Material and blade geometry design, fatigue life investigation, and frequency design are aimed in these works for sustainable production of turbine blades.

Another thermal component which is of importance from design and manufacturing point of view is heat exchanger. The design and manufacturing of micro-components are difficult in terms of mechanics, manufacturing technology, accuracy, tolerance, handling equipment, etc. The fabrication of micro heat exchangers is essential for thermal management in the areas such as processing, telecommunication, robotics, automotive, electronics, and medical industries. Foli et al. (2006) and Friedrich and Kang (1994) discussed about the design and fabrication aspects of micro heat exchangers that include fluid dynamic analyses, optimization of channel geometry, micromachining, and test performance of micro-channels. The fabricated micro heat exchanger could function with a minimum volumetric heat transfer coefficient of 45 MW/m^3 K under conservative design conditions. This is 20 times that of what is seen in conventional compact heat exchangers. Mehendale et al. (2000) through their critical literature review suggested the future prospects in designing ultra-compact heat exchangers including the need for flow visualization, inconsistency in friction and heat transfer data, and proper design of headers.

The structural performance and materials development cannot be separated. Jamshidi et al. (2015) showed that the blast furnace slags can be used in cement pavements for airports. The compressive strength of the cement mixture increased with increase in slag content. It depends on the curing periods as well. A decision-making system proposed by Fuente et al. (2016) suggests the appropriate sustainable material and pipe diameter for sanitary piping used to remove wastes in cities. The aerostatic bearings should be designed optimally for precise manufacturing, specifically if it is installed in machines involving nanometer precise movements. The parts should be made with high level of tolerances required for sustainable manufacturing. Through their investigation, Stout and Barrans (2000) presented the prominence of bearing clearance, variations in tolerance and geometry, orifice dimension tolerance, selection of supply pressure on the design, and manufacture of aerostatic bearings through some design criteria.

18.2 CURRICULA AND SYLLABI MODIFICATION

Typical undergraduate courses in the fields of thermal engineering, machine design, and materials and manufacturing sciences that constitute the mechanical engineering discipline are given below. Importance given to sustainable manufacturing in the course content is restricted. It is imperative to convey the significance of sustainability concepts to present generation of students as they form the future society. It should be made clear that the sustainability concepts are for their safe, hygienic, and secured future. It is mandatory to include these concepts at the undergraduate level, as the course content for postgraduate students involves advanced engineering/technical concepts. Some modifications in such courses are also suggested.

18.2.1 Materials and Manufacturing Processes

18.2.1.1 Existing Course Content

Crystal systems and lattices; dislocations; bonds in solids; deformation mechanisms; strengthening mechanisms; solidification; phase diagrams: iron carbon equilibrium phase diagrams, TTT and CCT diagrams, phase transformations; heat treatment processes and hardenability; hot working and cold working of metals; properties and applications of steel, cast irons, copper base alloys, aluminum base alloys, nickel base alloys, composites, ceramics, and polymers.

Casting: molding materials; patterns design; processes: sand casting, investment casting, pressure die-casting, centrifugal casting, continuous casting, thin roll casting; casting defects; joining processes: brazing, soldering, fusion and solid state welding, welding defects; metal forming processes: forging, rolling, extrusion, wire drawing, sheet metal working, spinning, swaging, thread rolling; super plastic deformation; metal forming defects; powder metallurgy and its applications; metal cutting: mechanics, tool design, surface finish and machinability; machine tool: operations on machines: lathe, milling, shaping, slotting, planning, drilling, boring, broaching, grinding, thread rolling and gear cutting machines; tooling: jigs and fixtures; CNC machines; finishing: microfinishing; unconventional machining; rapid prototyping and rapid tooling.

18.2.1.2 Modifications Required

Meaning of green manufacturing and sustainability, history of environment degradation, efficient process and product design, product life cycle analyses, efficient material selection, relating materials and manufacturing processes, lightweight concepts, anti-corrosion properties, energy and waste minimization, recycling of wastes, waste disposal mechanisms, pollution control, resource economics, eco-human relations, eco-related policies, eco-protection, models for eco-maintenance and eco-protection, advantages of computations in materials and manufacturing sciences, agile and sustainable manufacturing.

The topics mentioned here should be described with respect to each manufacturing processes and material systems, and not as separate units. A separate subject/course on sustainability and green issues can also be created for the benefit of staff and students.

18.2.2 Machine Design

18.2.2.1 Existing Course Content

Mechanisms; analysis of planar mechanisms; dimensional synthesis for motion; function and path generation; cam profile synthesis; gears; gear trains; static and dynamic force analysis;

flywheel; gyroscope and gyroscopic effects; governers: types and applications; analysis of cam and follower; free and force vibrations; vibrations of rotor systems; geared system; multi-degree of free-dom system: forced harmonic vibration, vibration absorber and damper; properties of vibrating system, Rayleigh damping.

Principles of mechanical design; factor of safety and material considerations; stress concentra-tions; design for fatigue; limits and fits; design of joints and couplings; belt and chain drives; power screws; shafts; keys; clutches; brakes; axles; springs; design of gears; bearings; wear and lubrica-tion; approaches to design: simulation of mechanical systems, robust design, design planning, value analysis, optimization.

18.2.2.2 Modifications Required

Principles of sustainable design, component design for energy minimization, material selection, strategies for fatigue life enhancement and vibration reduction, selection of lubrication, relevance of sustainable design and lubrication, strategies for wear reduction, optimized design procedures, design for sustainable manufacturing, computational design, and its advantages toward robustness and sustainability.

The topics mentioned here should be described with respect to design of each machine element, and not as a separate unit.

18.2.3 Thermal Engineering

18.2.3.1 Existing Course Content

Vapor power cycles: Carnot cycle, Rankine cycle, reheat cycle, regenerative cycle, steam cycles for nuclear power plant; steam generator: boilers; condenser; cooling tower; steam turbine: impulse and reaction stage, degree of reaction, velocity triangle, velocity and pressure compounding, effi-ciencies, reheat factor, governing, nozzles; heat pump and refrigeration cycles: reversed Carnot cycle, vapor compression and vapor absorption refrigerators, gas cycles, refrigerants and environ-mental issues; Air-conditioning; reciprocating air compressors; I. C. engines: classification, operat-ing characteristics, air standard cycles—Otto, diesel and dual, real air-fuel engine cycles, fuels and combustion—fuel types, alternate fuels, etc., combustion in S.I. and C.I. engines, conservation of mass in a combustion process, fuel injector and carburetor, ignition, lubrication, heat transfer and cooling; gas power cycles: simple gas turbine cycle, intercooling, reheating, regeneration, closed cycles, combined gas and steam cycles; axial-flow gas turbine; centrifugal and axial-flow compres-sors; combustion chambers; jet propulsion; rocket propulsion; direct energy conversion: thermionic and thermoelectric converters, photovoltaic generators, fuel cells.

Conduction: steady conduction; unsteady conduction; fins. Convection: momentum and energy equations; hydrodynamic and thermal boundary layers; free and forced convection; external and internal flows; phase change. Radiation: Stefan–Boltzmann law; Planck's law; emissivity and absorptivity; heat exchangers: analyses methods, heat transfer enhancement techniques. Mass trans-fer: molecular diffusion; Fick's law; analogy between heat and mass transfer.

18.2.3.2 Modifications Required

Design of thermal equipments (refrigerator, air conditioner, boiler, engines, heat exchanger) for sustainability, energy efficient fuels and lubricants, fuels and pollution, nano and bio fuels, relevance of green lubrication in thermal engineering, air quality analyses, life cycle assessment, strategies for efficient heat transfer, nonconventional energy sources, comparison of conventional and nonconventional energy sources, sustainable batteries, exergy calculation.

18.3 EXAMPLES OF MODIFICATIONS IN COURSE CONTENTS

In this section, some existing course content modification and strategies undertaken to educate sustainability issues by various institutes and groups across the globe are highlighted. Cai (2010) has discussed about green computing—the study of utilizing computing resources in an energy efficient and eco-friendly fashion. He has quoted that lack of interest, lack of staff training and expertise, lack of training and priority as barriers to integrate sustainability to undergraduate computing education. Later an undergraduate course on "green computing" was developed. In this course, the first module deals with the introduction and motivation for green computing. In the second module, emphasis on the server virtualization technology and green data center is given. In the third module, energy efficient computing and smart power management techniques are discussed. In the last module, electronic waste recycling and recovery, policy issues are discussed. During study practice, students were asked to develop solar powered laptop, and wireless wattmeters for power consumption measurement from computing systems. At the end of the course, the students were asked to complete a questionnaire that includes many questions to know their understanding about the subject. Some of the questions are as follows:

In Part 1: Understanding concepts

What is green computing?
What is sustainability?
What are the best ways to build energy efficient data centers?
How to recycle e-waste?

In Part 2: Items to carry forward in life

Understanding the concepts of sustainability and green computing
Ease of using computational thinking
I want to help improving the world around me with green techniques
I want to learn more green technologies

Out of all the questions in part 1, "How to recycle e-wastes?" has received the minimum average of 3.1 in a 5-point scale. So students should be made to have a closer look at this particular issue seriously and be given more importance while teaching the course. In part 2, the last two items—learning green technologies and improving the surroundings with green techniques—received only 3.6 points. The students should be motivated to learn more on sustainable and green aspects and follow that to maintain their nearby surroundings.

Handy et al. (2005) had presented a similar concern about introducing sustainability concepts in a manufacturing technology course. They have introduced "environmental sustainability" content to one of the manufacturing technology courses. As part of the content, they discussed about environmental impact of different casting processes and also compared that with other manufacturing processes for finding a suitable process for making a product. Assignments were given to identify a suitable method to reuse the spent green sand during sand casting process. In future, the course will be used to address the concerns like use of dry spraying for manufacturing, use of alternative energy sources like burning manufacturing by-products to produce power, net shaped casting to reduce power/cost of melting metal that will be machined, etc.

Murphy et al. (2009) through their review and critical survey on sustainability in education has concluded that lots of activities are going on at a "grassroots" level on education and research related to sustainable engineering. Though this is the case, only a little overall organization is observed among different institutes. This has made the requirement for a consortium on "sustainable and green engineering" that can bring together all the institutes and industries involved in such issues to develop curricula and policies, deliver expert lectures, and fund research projects related to sustainable engineering. It is also recorded by Murphy et al. that out of all the project themes, energy and power generation, life cycle assessment, business and economics belong to first three ranks, respectively, in terms of number of projects, while industrial processes and materials, relevant to manufacturing processes, hold 7th and 15th ranks, respectively (Table 18.1).

Kumar et al. (2005) surveyed the importance and status of infusing sustainability concepts into mechanical/manufacturing education in some of the US universities. As part of the discussion, they have quoted three criteria that are suggested by accreditation committee to consider including sustainability into the course content. The criteria set the requirement for a detailed course objectives including sustainability, improving the capability of students to design and make products considering the economic, environmental, social, political, ethical, health and safety, manufacturability, and sustainability needs, and finally the quality and competence of a faculty. They have also found from their survey that though students understand the importance of sustainability in mechanical engineering education, along with basic sciences, they are neutral in accepting the requirement of arts, social sciences and humanities in engineering education. Table 18.2 shows US universities supported by National Science Foundation for providing innovative

Table 18.1   Project Themes and Their Ranks

Project Themes	Rank of Theme in Terms of Project Importance	Project Themes	Rank of Theme in Terms of Project Importance
Energy and power generation	*1*	Pollution prevention and transport	11
Life cycle assessment	*2*	Transportation	12
Business and economics	*3*	Biochemical systems	14
Industrial ecology	4	*Materials*	*15*
Systems, metrics, and information management	5	Building and construction	16
Water	6	Urbanism and urban systems	17
Industrial processes	*7*	Climate change	18
Humanities	8	Agriculture and land use	19
Waste management	9	Material flow examination	21
Design	10	Human health	22

Source: With permission from Murphy et al. (2009), ACS publication.

Table 18.2   Mechanical Engineering Curricula Modifications Supported by NSF

Institution	Curricula Modification
Johns Hopkins University	Incorporated nontechnical topics, floated new courses and modules, created a committee to assess changes
University of Illinois at Chicago	Introduced computer science courses and methods to specialize in selected topics
University of Notre Dame	Created industry oriented courses with microprocessor-based mechanical systems as part of subjects
Rice University	Implementation of virtual laboratory to model dynamic systems
Kettering University	Incorporated problem-based learning by using computer simulations and animations to teach thermodynamics courses

Source: With permission from Kumar et al. (2005), Elsevier.

Table 18.3   Initiatives Undertaken at US and Other Universities in Sustainable Engineering (Kumar et al., 2005)

Institute	Department/Center Involved	Initiatives Undertaken
Georgia Institute of Technology, USA	Institute for Sustainable Technology and Development	• Modified existing courses to have sustainability content • Did not develop new courses on sustainability • Provides research and guidance support to help bringing about curricula change
University of Washington, USA	Design for Environment Lab	• Developed interdisciplinary courses on energy and environment • Created discussion forum on environment issues
University of Michigan, USA	Centre for Sustainable Systems	• Developed hydrogen technology curriculum for undergraduates • Developed curriculum through a program in environment, a collaborative effort between School of Natural Resources and Environment, and College of Literature Arts, and Science
Michigan Tech, USA	Sustainable Futures Institute	• Focused on environment decision-making by integrating eco-engineering, environment assessment and modeling, economic and human behavior elements of social sciences • Established undergraduate/graduate courses on Engineering for Environment, Sustainable Futures, Life Cycle Engineering and Environmentally Responsible Design and Manufacturing. Seminars on Sustainable Futures are also designed
University of Technology, Sydney	Institute of Sustainable Futures	• Conducts research in sustainability and green issues • Introduced sustainability-related courses at graduate level, but not at undergraduate level

educational approaches to students and curricula improvement. It should be noted that none of these are relevant to including sustainability in their education system, rather they concentrate on including computing and industry oriented education. Table 18.3 shows the course modification initiatives done by some of the US and other universities in sustainable engineering, as surveyed by Kumar et al.

A typical summary of number of green and sustainability-related courses in "Environmental science and engineering" in small European countries, large European countries, and in US is given in Glavic (2006). The small European countries include Austria, Denmark, Finland, Ireland, Netherlands, Norway, Slovenia, Sweden, and Switzerland. The large European countries include France, Great Britain, Italy, and Spain. The undergraduate and graduate level courses in environmental science and engineering are divided into groups of subjects like pollution control, pollution prevention, resource minimization, green manufacturing, eco-design, environmental management and economics, and special topics like environment, health and safety, law and policies, public relations, etc. Out of all the subject groups, "green manufacturing" related courses form almost 50% of the total number of subjects floated to the students in all the countries considered. While this is encouraging to manufacturing and materials field, the "eco-design" related courses formed the least, which ideally should be given more importance than manufacturing as we speak for "design for environment and eco-friendly," rather than conventional design practices these days. The article also suggests that out of all the sub-subjects in "green manufacturing," "green science" has been taught predominantly in European universities, while it is "green engineering" in USA (Glavic, 2006).

In the same line of thought, Holmberg et al. (2008) summarized the curricula development of sustainable education in three European universities—Chalmers University of Technology at Sweden, Delft University of Technology at The Netherlands, and Technical University of Catalonia at Spain. Table 18.4 shows the courses that are introduced in the universities.

Table 18.4 Course Modifications in the Three European Universities

Degree	Chalmers University of Technology		Delft University of Technology		Technical University of Catalonia	
	Course	Since When	Course	Since when	Course	Since When
Undergraduate	—	—	Program in Sustainable Molecular Science and Technology	2005	—	—
M.Sc.	Environmental Measurements and Assessments	1995	Sustainable Development branch within each M.Sc. program	2000	**Four developments**: Building, Regional and Urban Planning, Cooperation for Development, Policies and Evaluation	2007
	Innovative and Sustainable Chemical Engineering	1997	Industrial Ecology	2003	Environmental Engineering	2007
	Industrial Ecology	2004	Sustainable Energy Technologies	2005	Architecture and Environment	2007
	Sustainable Energy Systems	2007				
	Design for Sustainable Development	2007				
Minor (both at undergraduate and graduate levels)	Environment and Sustainable Development (graduate level)	2008	Sustainable Development for engineers (undergraduate level)	2007	Many courses were developed from 2002 and undergraduate and postgraduate levels	
Optional	Many		Many		Many	

Source: With permission from Holmberg et al. (2008), Taylor & Francis.

Environment, energy, construction, and materials are some of the topics on which courses were developed. Through their joint experiences, four crucial strategies were pointed out for successful implementation of sustainability into undergraduate and graduate courses. They are as follows: (1) interaction-based learning between individuals, (2) setting good examples and representatives to motivate younger talents, (3) separate and permanent coordinate platforms within university for organized activities, and (4) prolonged effort in motivating and coaching through seminars and lectures.

Jones et al. (2008) explored and presented the perceptions of academicians and students on implementing "sustainable engineering education" in the School for Earth, Ocean and Environmental Science at the University of Plymouth. They have also briefly highlighted the courses such as earth materials, introductory fieldwork and skills, research methods, earth surface systems, earth sciences fieldwork and skills, applied geology, geoscience trends and applications, global change, applied geoscience, geo-resources, geo-hazards, in which modifications were introduced in the university to take care of sustainability and green concepts. There are subjects like introductory fieldwork and skills, research methods, geological fieldwork and skills, earth sciences fieldwork and skills, geoscience research project and professional skills, in which additional education for sustainable development topics were identified.

Chau (2007) highlighted the incorporation of sustainability concepts in the civil engineering curriculum in Hong Kong. In the initial stage of curriculum development, emphasis was given on environmental, economic and social sustainability. They identified two important barriers for this—(1) lack of broad knowledge on the sustainability issues, and (2) difficulty involved in replacing the existing course contents with sustainability topics as the existing syllabus is already overloaded. Finally, as per suggestions from ASCE, it was decided that the students should focus on the problem solving skills, improving the decision-making ability with incomplete information, skill to work in multidisciplinary and multiactivity teams, fine tuning the presentation skills, and developing wider knowledge on different engineering topics throughout the program. It was aimed to modify the existing syllabus of the courses to include sustainability concepts, instead of introducing new courses. In the content modification, new topics of interest are introduced in the subjects like environmental science, construction management, construction materials, economics and law. Incorporating urban water recycling principles and sustainable urban drainage into "drainage engineering," and sustainable sludge management into "water and waste management" are typical examples. It has been concluded that problem-based learning will improve understanding about sustainability concepts to some extent in students and staffs.

In the University of Valencia, Spain, a questionnaire-based analysis was performed to know about the views of teachers in incorporating sustainability-related modifications to course contents. Some important responses are summarized in Table 18.5 (Minguet et al., 2011). In general, it is observed that most of the teachers are ready to introduce sustainability and green concepts into the university syllabus. They are ready to reorient their thinking and teaching, in a self-taught manner, to accommodate sustainability in the courses they teach. Lack of knowledge and training about such programs act as barrier to such initiatives. This should be considered for motivating teachers by the university committee through seminars, lectures, and training, focusing on a particular degree.

Once a course is developed on "sustainability" in a major discipline, it should be assessed for its relevance and competence. In this aspect, Malkki et al. (2015) proposed two indexes, "cumulative competence" and "relevance ratio," to identify the competence and relevance, and demonstrated the same in the study path of four courses, namely, Energy and Environmental Technology, Heat and Ventilation Technology, Urban Energy Systems and Energy Economics, and Combustion Engine Technology, in the department of energy technology at Aalto University. The indexes are based on recognizing certain terms, verbs, and words from the surveyed data based on learning

Table 18.5 Responses to Questions Posed to Teachers at the University of Valencia, Spain (Minguet et al. 2011)

Responses	Percentage of Support (%)
View of environmental problem statements that teachers identify themselves with most	
It is a fashionable statement, but will clear	2.4
We should focus on developing sustainability methods that minimize our negative impacts on the ecological, social and cultural environments around us	44.1
Implementing sustainability is professional and is eco-friendly	48.9
Other opinions	3.3
No response	1.2
Were you trained at your university prepared you to work with environmental and sustainability criteria to your profession?	
Received training	68.7
No training received	30.2
No response	0.9
In the degrees you teach, is there any training given to students to face environmental problems in their profession?	
Yes to acceptable level	10.3
Yes but significant work to be done in the field	46.5
Not much	16.3
Unaware of such training	25.1
No response	1.8
Do you deal with sustainability and environmental aspects in the courses you teach?	
Yes, dealing in all the courses that I am involved	16.6
Yes, but occasionally	40.2
Not much. It is irrelevant to the course I am teaching	17.2
Not at the moment	23.9
No response	2.1
Would you be willing to introduce environment-related content in the course you teach?	
Yes	29.9
Yes, only if I receive material or information to do so	54.7
Not willing or possible	10.9
No response	4.5
What is your opinion about devising a curricular environmental plan for the courses you teach?	
It is needed and should be done	43.8
It is a good idea, but it should not cause too much of upheaval	25.4
It is not a good idea	0.9
No opinion	21.5
Other responses	3
No response	5.4

outcomes. The terms, verbs, and words are related to sustainability and renewable energy as given in Table 18.6. The final analyses reveal that the "Energy and Environmental Technology" major had a 49% share of renewable energy and a 16% sustainability share, out of the total cumulative competence for the major. It is clear from this that including renewable energy in the course content does not mean that sustainability is studied. The other three courses had relevance ratio much less than that of "Energy and Environmental Technology" indicating their irrelevance in terms of sustainability and renewable energy.

Table 18.6   Selected Keywords, Relevant Terms, and Verbs for Recognizing the Sustainability and Renewable Energy Content of the Learning Outcomes

Keywords Of		Relevant Terms	Relevant Verbs
Sustainability	Renewable Energy		
Sustainability	Renewable energy	Energy resources	Understand
Emission control	Biofuel	Energy process	Know
Environment	Biomass	Energy systems	Recognize
Climate change	Fuel cells	Energy technologies	Identify
Environmental impacts	Geothermal energy	Eco-efficiency	Search
Ecological impacts	Hydropower	Energy efficiency	Compare
Economical impacts	Solar power	Waste treatment	Classify
Societal impacts	Wave power		Evaluate
Global impacts	Wind power		Estimate
Health	Wood energy		Analyze
Life cycle assessment			Apply

Source: With permission from Malkki et al. (2015), Elsevier.

18.4 BARRIERS FOR IMPLEMENTING SUSTAINABLE ENGINEERING IN UNIVERSITY SYLLABI

The following are some of the barriers faced while incorporating sustainable engineering in university curricula as recognized by those pursuing research in the topic.

- Difficulty in replacing the existing course content by sustainability issues as the existing syllabus itself is substantial
- Lack of broad knowledge and training among staffs in the sustainability and green concepts
- Avoiding interdisciplinary teaching and research
- Mind set evolution in the university system
- Lack of sufficient time in modifying the syllabus
- Waiting to get reaction from "others" to modify course contents and initiation from university management
- No commitment from university management
- Unable to decide whether to include a new course on sustainability or to modify the existing course content—choice between holistic approach and separated approach
- Unable to decide whether the course needs modification to include sustainability or not; in other words, the compatibility between sustainability concepts and the course nature is doubtful
- Resistance from teachers to accept modification in the course contents related to sustainability
- Unable to reach people and spread responsibility about including sustainability in the university syllabus
- Unstructured education system and stubborn culture unable to take care of such course content modification
- Inefficient models that quantify sustainability inclusion in the existing courses
- Considering sustainability and green education as fashionable statement
- Lack of motivation among students to know the seriousness of sustainability engineering in the courses
- Overlooking the relationship between course content modification and employability as part of marketing the degree program
- Unavailability of forum or consortium to implement modifications in courses
- Lack of resources to assist and support academics and encourage students
- Absence of demonstration of getting benefits because of such inclusion
- Lack of industry participation in such implementation activities

18.5 STRATEGIES FOR DISSEMINATING SUSTAINABILITY AND GREEN CONCEPTS

The following are some of the strategies followed to motivate students and staffs to disseminate the idea of sustainability among them.

For students
- They should be taught about the importance of green and sustainability concepts at the start of the course. They should feel that the topic is meant for their safe and hygienic environment in future. Some industrial product development case studies highlighting the importance of sustainability can be presented. The following are some examples.
  i.   The coatings unit of PPG industries (www.ppgtruefinish.com) developed a citric-acid-based gel cleaner to remove rust and mill scale like metallic oxides, heat scale, heat-treated stains, etc., from the hot rolled or cold rolled steel sheets. The cleaner is safer for customers (as it does not contain phosphates and strong acids), and environment as compared to grinding, sanding, and existing chemical-based products (Sustainable manufacturer magazine, November 2015 issue).
  ii. Development of a thermal camera (www.flir.com) that locates energy waste sources, locations of structural defects, and plumbing problems in buildings. This is easy to use, and uses heat patterns to locate such details, which otherwise is difficult to find (Sustainable manufacturer magazine, January 2015 issue).
  iii. To show how minimum quantity lubrication (MQL) is used during machining operation. MQL is nothing but a nearly dry machining operation that uses right volume of lubricity at the tool-specimen interface. By installing MQL, the Ford Van Dyke transmission plant has attained green savings of seven figures in initial cost, produced machines parts at a low variable cost, reduced the cost involved in cooling tanks and high pressure supply systems, reduced operating costs, improved the plant air quality as compared to wet conventional machining.
- Students should be asked to deliver classroom presentations on topics related to green and sustainability concepts, in the chosen subject. Some of the topics are sustainable disposal of printer cartridges, energy efficient batteries, application of bio-lubricants in automotive industries, and efficient casting methods for turbine blades. Many such topics can be suggested.
- Course projects that run for few months to a year can be given to students. If done successfully by the student during the course, the same can be extended as summer or winter internship. Some example topics are (1) comparing the process efficiency and material behavior during hydroforming and traditional forming method to fabricate bicycle components, (2) material wastage comparison during two different casting processes with respect to sand casting, and (3) eco-friendly vehicle washes and washing methods, etc.
- Application of computations to teach sustainability including evaluating load requirement, energy and power savings, recycling, efficient material utilization, etc., during any manufacturing operation.
- Barriers for adopting sustainability in industries like life understanding, cultural difference, history of entertainment followed, attitude evolved, sporting activities, climatic and weather conditions, raw material and natural resources available with that community, etc.

For staffs
- Firstly, the staff members should understand the significance of sustainability and its dissemination among students. They should perform research on sustainability issues in mechanical engineering, most of those are recently developed, so that the course content is not new for them.
- Proposing elective courses specifically on green and sustainability issues in mechanical engineering, so that the course contents proposed earlier can be taught authoritatively.
- Performing collaborative research and teaching with social science researchers, so that interdisciplinary research and subject prospers. This will be helpful for a staff with mechanical engineering background to understand the societal issues related to sustainability in a better fashion.

18.6 SUGGESTED TOPICS RELATED TO SUSTAINABILITY FOR SEMINARS AND PROJECTS

A comprehensive list of topics for seminars and projects related to sustainability and green engineering and sciences is provided here. Many such topics can be thought about for students, probably a better way to convey the importance of sustainability to engineers.

- Mechanical joining vs. traditional joining processes toward sustainability
- Replacement for phosphate metal pre-treatment system in metal fabrication
- Solar power: a clean energy approach to power houses and workplace
- Approaches to reduce paper usage in workplace and implementation strategies
- Efficient dust collectors and controls for appropriate maintenance
- New packaging technology in flight and train services
- High performance laundry cleaning with cold water using clean chemicals
- Bio-grease made from vegetables oils
- Safe and environmental friendly rust cleaners
- Improving the working life and reducing energy consumption of LED products
- Design of spray nozzle for energy efficient and clean coating
- Dawn of new car era, the electric car and electric hybrid technology
- Biomass for sustainable fuels and chemicals
- Standard metrics to analyze sustainability initiatives in industries
- Recycling CO_2 into fuels using renewable or nuclear energy
- Alternative fuels for aviation industries
- Hydrogen as a future fuel for sustainable energy system
- Efficient methods to recycle/reuse e-wastes
- Recycling and reusing thermosetting and thermoplastics
- Sustainable methods of removal of chrome from plated plastics
- Carbon nanotubes for controlling thermal conductivity of engineering materials
- Applications of carbon nanotubes in biomaterials and efficient heat transfer
- MQL-based machining vs. cryogenic machining vs. dry machining
- Sustainability of heat pumps
- Heat transfer enhancement in energy storage systems
- Environment waste management in resource industries
- Cryogenic machining of engineering materials
- Economics of cryogenic machining including sustainability costs
- Chip formation improvement during cryogenic machining
- Developing sustainability model for manufacturing processes
- Computer aided process planning toward sustainable manufacturing
- Improving product sustainability through surface integrity during machining
- Developing sustainability and green indicators/measures for manufacturing processes
- Reducing material wastage during manufacturing processes
- Coatings to improve the lifetime of die-casting dies
- Life cycle assessments for achieving sustainability in the raw material making industries
- Using agricultural wastes (such as fly ash, coconut shell ash, rice husk ash, etc.) in metal matrix composites for sustainable materials
- Enhancing the sustainability of building materials
- Sustainable and green food packaging
- Bauxite residue management and reuse in alumina industries

18.7 SUSTAINABILITY IN UNIVERSITY CAMPUS

This topic is slightly different from what has been discussed till now. Generally, sustainability and green policies implemented in industries are important and discussed in literature. This is directly related to the quality of products made and their impact on the environmental pollution, and power consumption. But sustainability within university campus has not been discussed much. This is very much crucial if sustainability concepts have to be introduced into the university curricula. If not followed, the same can be shown as a reason for neglecting course modifications suggested before. Moreover, universities can be seen as a city or industry because of large population and size, multicultural interactions that can ruin the environment. In this section, some example cases describing the environment policy management strategies developed in the universities will be discussed. It is expected that the universities not executing such strategies will get motivation after looking at such successful milestones.

In this context, Balsas (2003) through his meticulous survey about the transportation system of eight different US university campuses namely Cornell University, University of Wisconsin at Madison, University of Colorado at Boulder, University of California at Santa Barbara, Stanford University, University of California at Davis, University of Oregon at Eugene, and University of Washington at Seattle, and presented the importance of bicycling and pedestrian walking in order to maintain sustainability in the university campus. In all the campuses, strong measures are taken to avoid or minimize usage of cars for transportation. Though percentage (weighted) of car trips are considerable in all the campuses, 48% bicycle trips and 49% walking trips are seen in UC Davis and UW Madison, respectively. The car parking spaces range from 3,300 to 21,000 in all the campuses showing the dependence of transportation on car availability. The weighted percentage of bicycle commute in the university campuses of four localities, Boulder, Eugene, Madison, Seatle, are 12, 12, 15, and 5, respectively, which are encouraging figures. It is projected that about 14,000 people travel to UC Santa Barbara by bicycle on a given day during working hours. A small city, Davis, located closer to Sacramento, has another university campus of the University of California where assessments indicate that minimum 15,000 bicycles can be found on its bicycle grounds every day. There are cycle and pedestrian committee, full time coordinators, committee that conducts regular user survey, and committee that spreads cycle/pedestrian planning knowledge in all the university campuses in one or other form. There are facilities like cycle lanes, cycling roads, cycle routes, cycle parking space in all the university campuses.

Alshuwaikhat and Abubakar (2008) suggest that the university campus sustainability can be achieved by three modes: namely (1) university environment management system, (2) public participation and social responsibility, and (3) sustainability in teaching and research. In the first mode, the main aim is to recycle and minimize wastes, reduce the negative impacts of operations, resource conservation and pollution prevention, and providing green solutions in the form of green buildings, and green transportation. The green building initiative can be successful by using efficient HVAC systems with improved building control systems, generating power for illuminating class rooms, and heat during winter through wind, solar and geothermal energy systems, providing energy efficient lighting systems in class rooms, hostels and other campus locations, along with centralized control systems, etc. In the second mode, involvement of campus community and alumni in sustainable development, creating awareness through lectures and community projects, maintaining social justice through equity, helping elderly people, handicap and weaker community, etc. The third mode is discussed previously of how to implement sustainability concepts in course curricula.

Curtin Environment Awareness Team (CEAT), a volunteering group at Curtin University of Technology in Perth, Western Australia, decided to sort out declining bird and animal habitat at a lake in the university campus through an UG design project in the School of Architecture. Many student groups participated in the project. Finally, six different proposals were assessed by a committee based on the factors like proposal practicality, compatibility between ecosystem and recreation facilities including human facilities, signage locations, materials usage and maintenance predicted, etc. Two of the proposals are described by Karol (2006), including (1) Nodal intervention method that emphasizes the appropriate locations of access and discharge of water from the lake and creates observation and resting platforms at these locations, and (2) Interlude scheme that creates more importance on people using elevated footpaths above lake and through a tree shade thus leaving the ground level undisturbed, are noteworthy.

At the Massey University, New Zealand, a survey was conducted to understand and improve the recycling scheme within the university campus (Kelly et al., 2006). This was achieved through a questionnaire focused on chances of improving the recycling participation and source separation performance, and on attitudes within the university community toward recycling. It is observed from the results that about 60%–70% of the respondents supported that the recycling performance is independent of size of recycle bins, while 50% supported that it depends on usage of brighter colors for bins. About 90% disagreed to have bin colors blended in more with the environment. One important factor that contribute majorly is providing bins with signage at appropriate locations in the campus, providing some sort of incentives to promote recycling, and avoid discouraging events like transferring contents of one or more bins to single bin making the whole recycling initiative absurd.

As depicted by Mason et al. (2003), a zero waste program was initiated at the Massey University campus in order to take care of environmental decay and management issues. Support from university management, paid academic researchers, volunteers, committee were obtained for successful implementation of policies.

The University of British Columbia Food System Project initiative in University of British Columbia, Canada, identified a series of vulnerabilities that the food system faces, its disconnection with ecosystem, promotion of awareness of within the UBC community about the operating farm with its economic unsustainability, harmful environmental, social, and economic impacts associated with high food miles from the source station to the university campus, and many more (Rojas et al., 2007). The remedial measures were sorted out through projects, seminars, pedagogical activities, and research activities. The measures are still continued in the same direction to take care of sustainability issues in the campus.

Smyth et al. (2010) reported waste characterization studies that were conducted at the Prince George campus of the University of Northern British Columbia, Canada, as part of waste management system. They suggested waste minimization strategies and recycling ideas to improve the overall sustainability of the campus. Table 18.7 gives the mean composition (percentage by weight) of the recyclable material recovered from waste stream in the campus. From the university campus, totally, 1,359 kg of waste was dealt with, in which, 640 kg was recyclable, 338 kg was compostable and 370 kg was nonrecyclable material. The electronics and hazardous by-products formed the

Table 18.7   Mean Composition of the Recyclable Material Recovered from Waste Stream (Smyth et al., 2010)

Recyclable Products	Weight%
Ferrous metals	1.83
Nonferrous metals	0.11
Printed paper	3.54
Mixed paper	17.55
Old corrugated cardboard	9.63
Newspaper	4.3
Paper towel	13.19
Disposable hot beverage cups	15.21
Plastics	11.03
Beverage containers	10.68
Durable plastics	1.77
Milk containers	8.36
Expanded polystyrene	1.78
Glass	1.02

remaining 14 kg of waste. From the final analyses, paper and paper products, disposable cups and compostable organic material constituted the most significant material types to reduce waste and for recycling.

A similar study was made by Vega et al. (2008) at Campus Mexicali of the Autonomous University of Baja California, Mexico, toward implementing recovery, reduction and recycling waste management programs at the university campus. Out of the total 4,800 kg of waste segregated, 2,567 kg originated from buildings, 1,360 kg from gardens, 238 kg from the community center, and 673 kg from unknown locations. Out of this, 68% belongs to recyclable or potentially recyclable wastes, and 34% belongs to nonrecyclable wastes. It was found that the local recyclable market is good enough to absorb most of the recyclable wastes.

18.8 SUMMARY

The overall aim of the chapter is to highlight the importance of including sustainability and green manufacturing concepts in the existing courses. Initially the relationship between manufacturing subject and other subjects like design and thermal engineering is discussed with the help of case studies from available literature. Later course modifications in three major mechanical engineering subjects are suggested. Many such modifications can be identified by discussions with experts and industrialists in the areas. Similar changes are possible in other engineering subjects too. Some existing course modifications across different universities and the barriers faced while implementation are discussed elaborately. Further some ideas for disseminating the sustainability concepts among students and staffs are listed. Numerous seminar and project topics relevant to sustainability and green manufacturing are listed. Finally, the strategies adopted by different universities to have a sustainable campus in the day-to-day life have been portrayed with a few case studies.

REFERENCES

Alshuwaikhat H.M., Abubakar I., 2008, An integrated approach to achieving campus sustainability: Assessment of the current campus environmental management practices, *Journal of Cleaner Production*, 16, pp. 1777–1785.

Balsas C.J.L., 2003, Sustainable transportation planning on college campuses, *Transport Policy*, 10, pp. 35–49.

Cai Y., 2010, Integrating sustainability into undergraduate computing education, *Proceedings of the 41st ACM Technical Symposium on Computer Science Education*. Milwaukee, WI, pp. 524–528.

Chau K.W., 2007, Incorporation of sustainability concepts into a civil engineering curriculum, *Journal of Professional Issues in Engineering Education and Practice*, 133, pp. 188–191.

Foli K., Okabe T., Olhofer M., Jin Y., Sendhoff B., 2006, Optimization of micro heat exchanger: CFD, analytical approach and multi-objective evolutionary algorithms, *International Journal of Heat and Mass Transfer*, 49, pp. 1090–1099.

Friedrich C.R., Kang S.D., 1994, Micro heat exchangers fabricated by diamond machining, *Precision Engineering*, 16, pp. 56–59.

Fuente A.D.L, Pons O., Josa A., Aguado A., 2016, Multi-criteria decision making in the sustainability assessment of sewerage pipe systems, *Journal of Cleaner Production*, 112, pp. 4762–4770.

Glavic P., 2006, Sustainability engineering education, *Clean Technologies and Environmental Policy*, 8, pp. 24–30.

Habali S.M., Saleh I.A., 2000, Local design, testing and manufacturing of small mixed airfoil wind turbine blades of glass fiber reinforced plastics part I: Design of the blade and root, *Energy Conversion & Management*, 41, pp. 249–280.

Hallstedt S.I., Bertoni M., Isaksson O., 2015, Assessing sustainability and value of manufacturing processes: A case in the aerospace industry, *Journal of Cleaner Production*, 108, pp. 169–182.

Handy R.G., French R.M., Jackson M.J., 2005, Introducing environmental sustainability in a manufacturing processes course, *35th ASEE/IEEE Frontiers in Education Conference*, USA, SID, pp. 7–9.

Holmberg J., Svanstrom M., Peet D.J., Mulder K., Ferrer-Balas D., Segalas J., 2008, Embedding sustainability in higher education through interaction with lecturers: Case studies from three European technical universities, *European Journal of Engineering Education*, 33, pp. 271–282.

Jamshidi A., Kurumisawa K., Nawa T., Hamzah M.O., 2015, Analysis of structural performance and sustainability of airport concrete pavements incorporating blast furnace slag, *Journal of Cleaner Production*, 90, pp. 195–210.

Jones P., Trier C.J., Richards J.P., 2008, Embedding education for sustainable development in higher education: A case study examining common challenges and opportunities for undergraduate programmes, *International Journal of Educational Research*, 47, pp. 341–350.

Jureczko M., Pawlak M., Mezyk A., 2005, Optimisation of wind turbine blades, *Journal of Materials Processing Technology*, 167, pp. 463–471.

Karol E., 2006, Using campus concerns about sustainability as an educational opportunity: A case study in architectural design, *Journal of Cleaner Production*, 14, pp. 780–786.

Kelly T.C., Mason I.G., Leiss M.W., Ganesh S., 2006, University community responses to on-campus resource recycling, *Resources, Conservation and Recycling*, 47, pp. 42–55.

Kong C., Bang J., Sugiyama Y., 2005, Structural investigation of composite wind turbine blade considering various load cases and fatigue life, *Energy*, 30, pp. 2101–2114.

Kumar V., Haapala K.R., Rivera J.L., Hutchins M.J., Endres W.J., Gershenson J.K., Michalek D.J., Sutherland J.W., 2005, Infusing sustainability principles into manufacturing/Mechanical Engineering curricula, *Journal of Manufacturing Systems*, 24, pp. 215–225.

Maalawi K.Y., Negm H.M., 2002, Optimal frequency design of wind turbine blades, *Journal of Wind Engineering and Industrial Aerodynamics*, 90, pp. 961–986.

Malkki H., Alanne K., Hirsto L., 2015, A method to quantify the integration of renewable energy and sustainability in energy degree programmes: A Finnish case study, *Journal of Cleaner Production*, 106, pp. 239–246.

Mason I.G., Brooking A.K., Oberender A., Harford J.M., Horsley P.G., 2003, Implementation of a zero waste program at a university campus, *Resources, Conservation and Recycling*, 38, pp. 257–269.

Mehendale S.S., Jacobi A.M., Shah R.K., 2000, Fluid flow and heat transfer at micro- and meso-scales with application to heat exchanger design, *Applied Mechanics Review*, 53, pp. 175–193.

Minguet P.A., Martinez-Agut M.P., Palacios B., Pinero A., Ull M.A., 2011, Introducing sustainability into university curricula: An indicator and baseline survey of the views of the views of university teachers at the University of Valencia, *Environmental Education Research*, 17, pp. 145–166.

Murphy C.F., Allen D., Allenby B., Crittenden J., Davidson C.I., Hendrickson C., Matthews H.S., 2009, Sustainability in engineering education and research at U.S. Universities, *Environmental Science and Technology*, 43, pp. 5558–5564.

Prata A.T., Barbosa Jr. J.R., 2009, Role of the thermodynamics, heat transfer, and fluid mechanics of lubricant oil in hermetic reciprocating compressors, *Heat Transfer Engineering*, 30, pp. 533–548.

Rojas A., Richer L., Wagner J., 2007, University of British Columbia Food System Project: Towards sustainable and secure campus food systems, *EcoHealth*, 4, pp. 86–94.

Smyth D.P., Fredeen A.L., Booth A.L., 2010, Reducing solid waste in higher education: The first step towards 'greening' a university campus, *Resources, Conservation and Recycling*, 54, pp. 1007–1016.

Stout K.J., Barrans S.M., 2000, The design of aerostatic bearings for application to nanometre resolution manufacturing machine systems, *Tribology International*, 33, pp. 803–809.

Vega C.A.D., Benitez S.O., Barreto M.E.R., 2008, Solid waste characterization and recycling potential for a university campus, *Waste Management*, 28, pp. S21–S26.

CHAPTER 19

Sustainability, Health, and Environment
A Case Study of Waste Management Sector

Radha Goyal and Ashish Jain
Indian Pollution Control Association

Shyamli Singh
Indian Institute of Public Administration

CONTENTS

19.1 INTRODUCTION

All living organisms produce waste as part of their processes throughout their life. Waste is the second nature of the existence of life form on this Earth. Increased consumerism and rapid sprawl in urbanization has contributed to this menace manifolds. It is also observed that the inherent nature of waste, both the quality and quantity, has witnessed a drastic change in the recent years. These enormous quantum of waste generated, especially in the urban locales, has reached peak condition. This increased waste generation looks somehow out of bound and is beyond the control of the urban local bodies (ULBs). There are many causes of this failure: a few indicative reasons are lack of infrastructure, dearth of fund, privatization of technology, want of human resources, etc. The huge quantity of waste thus generated creates unsustainability in environment leading to threat for all organisms. This chapter discusses the case study of waste management in urban cities of India and shows the effect of good practices of waste management on health and environment.

19.2 WASTE MANAGEMENT SCENARIO IN URBAN INDIA

India is generating 68.8 million tons of solid waste annually: out of which 5.6 million tons is plastic waste, 0.17 million tons biomedical waste, 7.90 million tons hazardous waste, and 15 lakh tons e-waste (Javadekar, 2016). The waste generation from 366 Indian cities, representing 70% of

India's urban population, is 130,000 tons per day (TPD) or 47.2 million tons per year (TPY) at per capita waste generation rate of 500 g/day (Kaushal et al., 2012).

It was found that a major fraction of urban municipal solid waste (MSW) generated in India consists of organic matter (50%), recyclables (35%) (paper, plastic, metal, and glass), and inert material (15%). The moisture content of urban MSW is 47% and the average calorific value is 7.3 MJ/kg (1745 kcal/kg). However, this composition of waste is at the source of waste generation. The composition of waste at a dump yard in the cities of India is different, which includes 51% organics, 17.5% recyclables, and 31% inert matter. This suggests that almost 50% of the recyclables are separated from the waste reaching to the dump yards by informal sector (IPCA Newsletter, 2014).

The current waste management practices in urban India depict a complete lack of source segregation in Indian cities, as shown in Figure 19.1. According to Central Pollution Control Board (CPCB), overall, only 70% of the total generated wastes are collected, and further only 12.45% of the total wastes are processed and treated (CPCB, 2013). The waste processing and treatment technologies practiced in India have 80 units of composting plants, 8 units of Sanitary Landfills (SLF), 7 units of Refused Derived Fuel (RDF) and Waste to Energy (WTE) plants as per the latest available data (Annepu, 2012).

The existing system of waste management has a number of limitations. For instance, the compost yield from mixed waste composting facilities (MBTs) is only 6%–7% of the total feed material. Almost 60% of the input waste is discarded as composting "reject" and sent to the landfills. Further remains of composting processes are in form of carbon dioxide and water vapor which are released in the environment. In addition, the quality of the compost from such facilities is very poor with

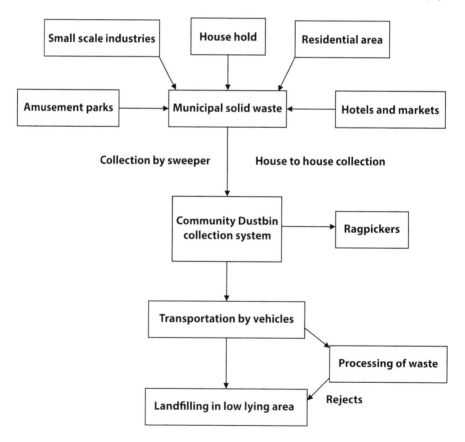

Figure 19.1 Current waste management practices in urban India.

heavy-metal contamination. If such poor-quality compost is introduced to Indian agricultural lands, it would add to 73,000 tons of heavy metals to agricultural soil (Kaushal et al., 2012). Furthermore, a review of the present status of solid waste management (SWM) in India from a material and energy recovery perspective shows that in the absence of source segregation, proper reuse, recycling policies practices, and pollution control regulations in India. Landfills are receiving 6.7 million TPY of recyclable material which could have been used as secondary raw materials in manufacturing industries; 9.6 million tons of compost, which could have been a potential fertilizer supplement; and 58 million barrels of oil energy equivalent in residues of composting operations that could have been a source to generate electricity (DEA report, 2009).

In contrast to the present situation of waste management in India, according to the future projections made by Niti Aayog of India, the waste generation would reach up to ~165 million tons/year by 2031 and ~up to 436 million tons/year by 2050. Although these projections are on conservative side, keeping 1.33% annual per capita waste generation growth rate, to accommodate this huge amount of waste generated, about 23.5×10^7 cubic meter of landfill space or 1,175 hectare of land/year will be required by 2,031 and 43,000 hectares for landfills piled in 20 meter height by 2050 if ULBs in India continue to rely on landfilling to dispose the solid waste (Joshi and Ahmad, 2016). If the annual growth in per capita generation of waste will become 5%, landfill area required for disposal of waste could be manifold as per the records of CPCB (2013). In a *business as usual* scenario, India does not have the capacity to dispose of these enormous quanta of wastes properly.

Apart from the operational limitations, policy level hindrances also exist in India. There is a policy paralysis and the results are thus fractured, fragmented, and lack coherence. There is no holistic policy document, which emphasizes upon the whole production–consumption and recovery pattern of waste.

19.3 INDIAN WASTE MANAGEMENT: ENVIRONMENTAL AND PUBLIC HEALTH IMPACTS

Although the waste management rules exist in India (MSW (Management and Handling) Rule, 2000 and revised SWM Management Rule, 2016), their poor enforcement and implementation, the ill-fitted pieces of policies, poor infrastructural, operational, and performance related issues of technologies, the urban realms of environment comprising of air, water, and soil in India are at greater risk of pollution and so as the health of public. The well-established delirious environmental impacts of various waste management activities, such as transportation, landfilling, incineration, composting, and recycling, have a great number of *direct* and *indirect* health effects. The *direct* public health impacts are in the form of occupational health hazards and injuries to the people working in SWM sector, viz., waste pickers/collectors/recyclers, etc. The activities of the waste collectors/rag-pickers/scavengers often pose a pronounced health risk. Wastes are often contaminated with degraded food waste, dust, and sometimes medical discards too, so the scavengers collecting such wastes are constantly exposed to the danger of accidents, injuries, cuts, burns, allergies, poisonous substances, wounds through sharp materials as they scrounge with bare hands and sometimes even bare feet. The risk is equally high for the children, who are involved with their parents in waste collection and segregation. Exposure to the hazardous materials might be even more severe for them. For instance, children have a faster rate of breathing than adults which may make them more vulnerable to airborne hazards (such as gases given off during burning of waste materials). They have thinner dermal layers than adults, making them more susceptible to chemical absorption and burns. Furthermore, the softness of children's bones due to lack of calcium may lead to skeletal deformities resulting from carrying heavy loads (IPCA Newsletter, 2015). The *indirect* impact is in the form of exposure via inhalation (especially due to emissions from incinerators and landfills), consumption of water and food in cases of contamination due to landfill leachate, bacteria, and viruses from land spreading of sewage and manure.

Waste management practices like landfilling and incineration are one of the main factors responsible for the release of greenhouse gases, i.e., CO_2 and CH_4 in the atmosphere (Goyal, 2017). Understanding the current impacts and presuming the future disastrous impacts of MSW management in India, it has become imperative to manage the huge quantity of waste generated in urban India in a more sustainable way in order to avoid any further deterioration of public health, air, water, and land resources, which are ultimately going to affect the quality of life in Indian cities. A dire need is to opt for the appropriate strategy and approach to manage the MSW in such a way to reduce its impact on environment and public health in India.

19.4 SUSTAINABLE WASTE MANAGEMENT: KEY ISSUES AND WAY FORWARD

More than two decades ago, Chapter 21 of Agenda 21 of the Rio Earth Summit 1992 had identified four key issues of sustainable waste management:

1. Reduction in the amount of waste
2. Maximization of segregation at source
3. Promotion of small-scale waste recycling industries
4. Integration of recycling with formal waste management

However, we are still trailing behind to achieve the goals set so many years ago. The reasons could be the lack of awareness, poor public participation, faulty schemes and policies of government, poor implementation of policies, weak law and order in the country etc. Therefore, the task that lies ahead is to simulate the situation and set up a participatory framework of action, so that all the stakeholders can voice their opinions and can articulate their specific issues and concerns. Consequently, the focus must build upon the development of correct policy and on the design and implementation of a robust and resilient system for the recycling and reuse of solid waste to reduce its quantity.

The identified key issues of existing waste management, which needs immediate attention for finding sustainable solutions, are as follows:

- *Paradigm shift.* In the Indian scenario, SWM is a public necessity, which is provisioned by the respective municipalities. The sole objective of contemporary urban waste management policy is that it is "disposal-centric," i.e., the waste is sent largely to unsanitary landfills or open dumps where it is disposed off and covered over or concealed under the rubble. As a result, most of such waste collection sites represent a very unsightly scene besides adding to other environmental worries and troubles, *viz.,* odor, breeding ground for vectors leading to various vector-borne diseases, issues of leachate, dioxins, and furans in case the waste is incinerated. This calls for an urgent paradigm shift from a "disposal-centric" approach to a "recovery-centric" approach. Such a transition in the approach will divert 93.5% of MSW from landfilling, and increase the life span of a landfill from 20 to 300 years. It will also decrease disease, improve the quality of life of urban Indians, and avoid environmental pollution (Singh, 2017).
- *Change in perception.* At present, waste is being perceived as a disposal problem. The outlook has to be changed and it has to be seen as a material flow issue and a resource conservation issue. A resource is required to reprocess materials rather than to extract and process virgin materials. Today, waste management is considered as cost-recurring project for any ULB or municipalities and no one considers this as a cost recovery project. If we rework the entire supply chain of waste management from its collection to disposal and explore the possibility of maximum recovery of waste material through different processes and channels, then it becomes a revenue model and each stakeholder brings his interest to make it more efficient and viable (Jain, 2017).
- *Lacunae in different tiers of waste collection.* There exists an unsatisfactory collection efficiency (only 50%–70%), uncontrolled street collection points and improper disposal practice in open dumps, which allows refuse to be readily available for informal waste recycling. The scavenging/ waste picking can be unreliable, inadequate and may conflict with the other urban services and often makes the waste

management systems very poor and of low grade. Therefore it is critical to integrate and mainstream this informal sector of waste management with the formal system. These waste pickers can be trained for efficient door to door collection which would strengthen their knowledge about the real value of each recyclable commodity they have collected and to further sell them to the authorized recyclers. As per the estimates, every ton per day of recyclables collected informally by the waste pickers saves 500 USD/YEAR (INR 24,500) for the ULBs and avoids the emission of 721 Kg of CO_2/year into the atmosphere (Annepu, 2012).

- *Formalization of the informal.* Unemployment is a vast problem in countries like India and there are more than 1 million people engaged in waste management system in informal way. They are generating their livelihood from waste by segregating it at different levels from road sides to dump yards. This large number of human resources is available to the municipalities or ULB at zero cost. If the concerned authority gives some recognition to this informal sector and engage them in door-to-door collection of waste and provide training and education on segregation practices. This way, we can solve many issues of waste management, i.e., littering, segregation, recycling, and disposal at the dump yard.
- *Decentralize waste management system.* Residents, restaurant owners, hotel industries, and food processing industries should be encouraged to install small-scale biomethanation/composting/biogas units for the treatment of their biodegradable organic waste. This practice may reduce pressure on landfill site and may also reduce cost of collection and transportation.
- *Inclusion of closed-loop economy.* It is presumed that improved and increased efficiency leads to sustainability. However, it has been realized that it just leads to short-term gains, which are bound to get extinct eventually under the current linear model of development. This model can never be sustainable as it leads to continuous consumption, which eventually leads to continuous generation of waste. Therefore, an alternative system of sustainability such as the Circular Economy Model, as shown in Figure 19.2, should be employed as best practice. It is based on the elimination of

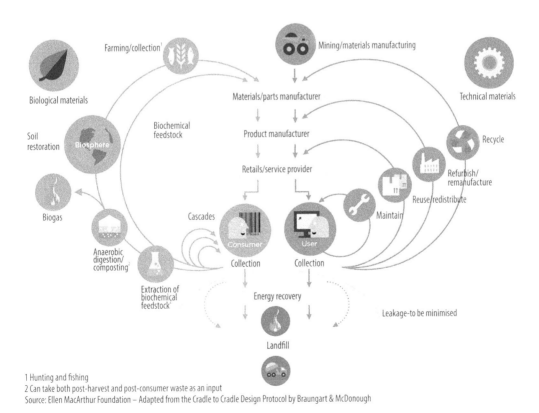

1 Hunting and fishing
2 Can take both post-harvest and post-consumer waste as an input
Source: Ellen MacArthur Foundation – Adapted from the Cradle to Cradle Design Protocol by Braungart & McDonough

Figure 19.2   Circular economy model.

waste generation, thus reducing the need to recycle or seek another means of disposal (The Ellen MacArthur Foundation, 2013). It will be based on closing the loop of current open system of waste management so that no waste will go out. The product should be designed in such a manner that it makes clear provision for reprocessing and reclaiming. The product design approach includes incorporation of consumability and durability of the product. The product design enforces a stringent distinction between replaceable and resilient components of the product. The replaceable in a circular model generally consists of organic elements that are unswervingly returned to the natural environment, whereas the resilient substances, comprising durable substances such as plastics and metals, which do not find a direct entry into the natural biosphere, are designed for reuse. Renewable sources of energy are preferred for reclaiming, and these give further benefits in reducing the energy quotient required for the reclamation. A closed loop recycling system should be adopted to solve the problem of MSW disposal and to function the waste management system properly, i.e., consumers, recyclers, and manufacturers must work together to reclaim valuable materials from the waste stream and use them to make new products (Figure 19.3). Waste management should not be limited to housekeeping (in collection) and discarding (in incineration or landfilling). The proper resource allocation must be encouraged for social and ecological gains to be achieved.

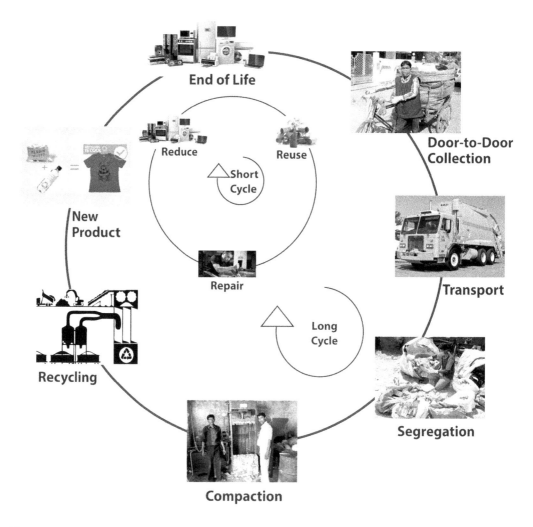

Figure 19.3   A short- and long-term closed-loop recycling system.

- *Fixation of accountability.* Ultimately, the polluters should owe the responsibility for their actions. The defaulters should be listed and should be embarrassed in public for their committed action. This pubic embarrassment would ward off others to follow the same negligence and would be exemplary. Thus, public participation can play an important role in sustainable waste management in India.
- *Policy take aways.* The government interventions of transition from present linear model to a circular model should highlight: rewarding best infrastructure practices; set up success models; devise punitive measures; develop a transparent monitoring system and implement Extended Producer Responsibility (EPR).

19.5 SUMMARY

There is a lack of strategic planning and poor implementation of government regulatory framework in India coupled with lack of awareness and motivation amongst the people to understand the problem of waste and their attitude is the major hindrance in improving the status of waste management in India. The MSW management is also facing the challenges with allocation of funds, economically viable technologies, and skilled manpower in waste management sector. As a result, impacts abide to the environment and public health with increased incidences of skin-related diseases, allergies, respiratory illnesses, reduced immunity, and other microbial infections. Hence, to obtain the sustainability in waste management sector in India, a strong regulation enforcement is needed to improve the current system with involvement of all ULBs, private sector, NGOs, and public. The role and responsibility of each individual for waste management, needs to be defined to understand the waste as a resource, not as a problem. The capacity building is needed at all levels to make the waste management a sustainable option.

REFERENCES

Annepu, R. K. (2012). Sustainable solid waste management in India, Waste-to-Energy Research and Technology Council (WTERT). New York: Columbia University. Available at www.seas.columbia.edu/earth/wtert/sofos/Sustainable%20Solid %20Waste%20Management%20in%20India_Final.pdf.

CPCB. (2013). Status report on municipal solid waste management. Available at www.cpcb.nic.in/divisionsofheadoffice/pcp/MSW_Report.pdf; http://pratham.org/images/paper_on_ragpickers.pdf.

DEA. (2009). Position paper on the solid waste management sector in India, Department of Economic Affairs, Ministry of Finance, Government of India. Retrieved from www.indiaenvironmentportal.org.in/files/ppp_position_paper_solid_waste_mgmt.pdf.

The Ellen MacArthur Foundation. (2013). Towards the circular economy: Economic and business rationale for an accelerated transition. Available at www.ellenmacarthurfoundation.org/assets/downloads/publications/Ellen-MacArthur-Foundation-Towards-the-Circular-Economy-vol.1.pdf.

Goyal, R. (2017). Air pollution from solid waste management practices: Evolving an environmental and public health crisis in urban India, Chapter 13, *"The Urban Environmental Crisis in India: New Initiatives in Safe Water and Waste Management,"* eds. Singh, S., Goyal, R., and Jain, A., Cambridge Scholar Publishing, Cambridge, pp. 208–230; ISBN (10): 1-4438-7960-6.

IPCA Newsletter. (January–March, 2014). Rag pickers: Informal vs formal sector, Volume 1, Issue 1. Available at http://ipcaworld.co.in/uploads/News%20Letter1.pdf.

IPCA Newsletter. (October–December, 2015). Rag pickers health: A need more than a right, Volume 1, Issue 4. Available at http://ipcaworld.co.in/uploads/News%20Letter%20Edition4%20volume%201,%20 Issue%204,%20October-December%202015.pdf.

Jain, A. (2017). Waste recycling: A sustainable solution to urban solid waste management, Chapter 7, *"The Urban Environmental Crisis in India: New Initiatives in Safe Water and Waste Management,"* eds. Singh, S., Goyal, R., and Jain, A., Cambridge Scholar Publishing, Cambridge, pp. 91–105; ISBN (10): 1-4438-7960-6.

Javadekar, P. (2016). Office of the Principal Scientific Advisor to the GOI. "Report on opportunity for green chemistry initiatives: Pulp and paper industry". Available at http://psa.gov.in/sites/default/files/pulp_paper_final.pdf. Accessed March 12, 2016.

Joshi, R., & Ahmad, S. (2016). Status and challenges of municipal solid waste management in India: A review. *Cogent Environmental Science*, 2, 1139434.

Kaushal, R. K., Varghese, G. K., & Chabukdhara, M. (2012). Municipal solid waste management in India-current state and future challenges: A review. *International Journal of Engineering Science and Technology*, 4, 1473–1489.

MSW (Management and Handling) Rule. (2000). Available at www.cpcb.nic.in/MSW_AnnualReport_2001-02.pdf.

Singh, S. (2017). Urban waste management in India: A revisit of policies, Chapter 1, "*The Urban Environmental Crisis in India: New Initiatives in Safe Water and Waste Management*," eds. Singh, S., Goyal, R., and Jain, A., Cambridge Scholar Publishing, Cambridge, pp. 1–11; ISBN (10): 1-4438-7960-6.

SWM Rule. (2016). Available at www.moef.nic.in/content/so-1357e-08-04-2016-solid-waste-management-rules-2016?theme=moef_blue.

Author Index

Subject Index